WAVES AND FIELDS IN OPTOELECTRONICS

PRENTICE-HALL SERIES IN SOLID STATE PHYSICAL ELECTRONICS

Nick Holonyak, Jr., *Editor*

GREEN Solar Cells: Operating Principles, Technology, and System Applications
HAUS Waves and Fields in Optoelectronics
NUSSBAUM Contemporary Optics for Scientists and Engineers
STREETMAN Solid State Electronic Devices, 2nd ed.
UMAN Introduction to the Physics of Electronics
VAN DER ZIEL Solid State Physical Electronics, 3rd ed.
VERDEYEN Laser Electronics

WAVES AND FIELDS IN OPTOELECTRONICS

HERMANN A. HAUS
Massachusetts Institute of Technology

Prentice-Hall, Inc., Englewood Cliffs, New Jersey 07632

Library of Congress Cataloging in Publication Data

Haus, Hermann A.
Waves and fields in optoelectronics.

(Prentice-Hall series in solid state physical electronics)
Includes bibliographical references and index.
1. Optoelectronics. 2. Quantum electronics.
3. Electromagnetic waves. 4. Electromagnetic fields. I. Title. II. Series.
TA1750.H38 1984 621.36′6 83-8645
ISBN 0-13-946053-5

Editorial/production supervision: Ellen Denning
Interior design: Anne Simpson
Manufacturing buyer: Anthony Caruso

To my family.

Printed in the United States of America

10 9 8 7 6 5 4 3 2 1

ISBN 0-13-946053-5

Prentice-Hall International, Inc., *London*
Prentice-Hall of Australia Pty. Limited, *Sydney*
Editora Prentice-Hall do Brasil, Ltda., *Rio de Janeiro*
Prentice-Hall Canada Inc., *Toronto*
Prentice-Hall of India Private Limited, *New Delhi*
Prentice-Hall of Japan, Inc., *Tokyo*
Prentice-Hall of Southeast Asia Pte. Ltd., *Singapore*
Whitehall Books Limited, *Wellington, New Zealand*

CONTENTS

PREFACE

There is no absolute measure of success for an introductory text to a field that is unfamiliar, or only partially familiar to the reader. There are subjective measures, however. Any particular exposition or explanation may strike some as having clarified the material, whereas others may not find it particularly helpful. The subjectivity of these measures is a reflection of the subjectivity of the learning and thinking processes.

The present text is, in part, an exposition of the work of many individuals as assimilated by the author and taught to his students. It is also an attempt to expose a way of formulating problems and solving them which the author employed in his research. With this goal in mind, the text is short on tables of values of material parameters, optical fiber types, optical detector properties, and so on. The reader is given enough numerical examples to develop a feeling for the orders of magnitude of device dimensions, intensities, propagation distances, mode numbers, and so on, that are necessary to achieve certain signal-processing objectives. The main purpose, however, is to develop mathematical models and approaches for the analysis of optical propagation, resonance, and nonlinear optical interactions. These methods are transferrable to other disciplines concerned with analogous phenomena, and in this spirit the present text lays claim to greater generality than its title implies.

The author was greatly helped by discussions with his colleagues and students. The sections on practical applications of optical systems were taken from material made available to the author by Professor M. M. Salour. These are specifically Section 3.8, a portion of Section 4.4, Section 5.7, the experimental part of Section 9.3, and Section 13.7.

The use of a component of the vector potential for the excitation amplitude in the scalar paraxial wave equation was prompted by a discussion with Professor E. P. Ippen. The introductory graduate course "Optics and Optical Electronics," the notes of which developed into this book, was taught jointly by Professor S. Ezekiel and the author. Aside from these specifics, the influence of these and other colleagues pervades the text. Students helped greatly with their thoughtful and penetrating questions. Jay K. Lee and Chris M. Gabriel read the manuscript with care and corrected many errors. Professor Etan Bourkoff of Johns Hopkins also kindly supplied a list of errors. Sching Lih Lin executed Figs. 5.3 and 5.4. Mohammed N. Islam supplied all remaining computer plots and commented penetratingly on the manuscript. The typing of the manuscript was done by Cindy Kopf with inimitable care.

If this text communicates to the readers some of the excitement and sense of discovery the author felt in learning the field and in extending a little further some of its topics, then it will have succeeded.

HERMANN A. HAUS

INTRODUCTION

The invention of the laser revived the research in optics, and caused revolutionary advances in the science and technology of the field that has become known as "quantum optics." Like all successful inventions, the laser provided improvements by orders of magnitude in existing applications of a mature field, and opened up new fields and new applications. We look briefly at the main characteristics of quantum optics that provide the opportunities for new systems, new devices, and new measurement techniques.

Laser radiation is distinguished from thermal (optical) radiation in that it can be made to possess:

1. Spatial coherence over wide apertures
2. Temporal coherence over long times
3. Very high brightness
4. Wide absolute bandwidths

A laser emitting in a single mode has spatial coherence over that mode. The entire power passing a cross section can be focused into a spot of diffraction-limited cross section. This property makes lasers the ideal sources for delivery of large electromagnetic intensities. The spatial coherence property of lasers made holography possible—a process invented by D. Gabor before the invention of the laser but not used practically until laser sources became available. Gabor received the Nobel prize for his invention only after lasers made it practical. Spatial Fourier transformation in "real time" is also made possible by the spatial and temporal coherence of laser radiation. The temporal coherence of lasers allows their use for doppler radar (for clear-air turbulence detection), and helps in holography and spatial Fourier transformation in that differences in travel times between interfering beams are rendered unimportant. It makes possible interferometry and very high resolution spectroscopy. It is essential for optical fiber communications because the narrow-band laser radiation allows for pulse propagation with no pulse-envelope spreading.

The brightness of the laser (large intensity per unit area and solid angle) accounts for materials processing and drilling by laser beams; it makes possible laser annealing; further, it is the requirement for laser fusion through heating and compression of fusion pellets.

The preceding application also utilizes the wide absolute bandwidth. Short pulses can be produced and amplified by laser amplifiers. The reason for this is that a relative bandwidth of a fraction of 1% "translates" into a very large absolute bandwidth because of the (generally) very high carrier frequency. Subpicosecond pulses have been produced by this means. These pulses have opened the way for experiments with very high temporal resolution on solid-state phenomena with relaxation times of the order of 10^{-12} to 10^{-13} s. The potential for the production of picosecond pulses also opens up the possibility of ultra-high-speed instrumentation and communications (encoding and decoding).

The particular properties of laser radiation and quantum optical devices favor particular mathematical approaches. We shall briefly discuss these because they determine the choice of formalisms in this treatise.

An optical wavelength is very short and an optical period is very brief. Within such short distances and small time intervals the gain, nonlinear wave interaction, phase change, and so on, achievable with available materials is small. This permits an analysis of the interactions in terms of first-order differential equations in space and time (as opposed to the second-order wave equation). Interactions may be treated as small "perturbations" of the propagation of a wave packet or the time evolution of a mode. The "coupling of modes formalism" developed for microwave tubes in the 1950s is particularly appropriate for quantum electronic devices. Nonlinear effects are easily incorporated into the equations.

Optical beams are, generally, many wavelengths wide. This enables one to use the results from plane-wave analyses for the description of the main features of any reflection phenomenon. When diffraction effects are important, the paraxial wave equation is adequate. Fresnel diffraction theory results very simply from this approach. Nonlinear effects are easily incorporated into this equation, enabling one to analyze the phenomenon of self-focusing, optical pulse formation, and modelocking.

Optical nonlinearities are "weak." By this we mean that the response of a medium (polarization) to an applied field can be expanded into a Taylor series, of which only a few terms have to be retained. The interference of sources (produced by nonlinearities) at different frequencies is generally destructive unless phase matching is provided, again enabling one to ignore most sidebands produced by the nonlinear interaction. This is of great advantage to the analysis, but poses a challenge to the optical system designer. Indeed, any optical signal-processing, or communication, system requires nonlinear devices of speed compatible with the speeds of the signals to be processed. The weak nonlinearities available from optical media require long propagation distances for their effect to be felt. This means that optical nonlinear devices tend to require "long" delays for their operation. It is in this area that most of the progress in optical devices is to be expected in the future.

This book is intended to serve as an introduction to the mathematical methods, physical ideas, and device concepts of optoelectronics. The emphasis is on waves and fields as derived from Maxwell's equations, with approximations suitable for fields at optical wavelengths. The physics of lasers is not treated. The brief discussion given by Yariv in his remarkable text *Introduction to Optical Electronics* cannot be bettered. The concepts of gain, and gain saturation, are simple enough so that they are introduced ad hoc, and make possible the discussion of laser threshold, power delivered by a laser, and the production of laser pulses by modelocking.

We start with a discussion of Maxwell's equations, and definitions of energy density and power flow, in particular as extended to dispersive media. The concept of group velocity is reviewed. Next, we take up the discussion of transmission and reflection of plane waves at interfaces of two optical media. We then discuss the operation of multilayered media in producing reflection and antireflection layers.

Some properties of the scattering matrix as used in the analysis of microwave circuits are also introduced. The operation of a reflecting surface (mirror) can be treated in general terms leading to the theory of interferometers. Various common types of interferometers are discussed. Because the degree of temporal and spatial coherence can be measured by interferometry the theory of coherence is developed.

Next we take up the propagation of beams of finite cross section, using the paraxial wave equation. Fourier transformation by a pair of lenses is treated. Next, consider the propagation of Gaussian beams and their transformation. Hermite–Gaussian beams are to diffraction optics what sinusoids are to signal theory. Any input condition at an input reference plane can be expressed as a superposition of Hermite–Gaussians. Transformation of a Hermite–Gaussian beam by an optical system then informs us of the general transformation properties of optical beams. Modes in Fabry–Perot optical resonators can be discussed this way. Optical resonators as filters are analyzed and the special properties of the confocal resonator are discussed.

Guidance of optical radiation by a dielectric slab serves as the introduction for a discussion of guidance in optical waveguides and in optical fibers. The modes in a parabolic index fiber are derived and their characteristics are discussed. Propagation along dispersive fibers is studied and a grating pair system is described that enables one to undo the spreading of a pulse propagating along a dispersive fiber. The following chapter is devoted to the formalism of coupling of modes, an extremely powerful approach for the discussion of optical systems incorporating modes that couple to each other in time or space. The transmission characteristic of a Fabry–Perot interferometer is found to be a general property of any resonator with one input "port" and one output "port." The formalism can also be used to derive the behavior of distributed feedback structures that have been developed for integrated optics application. The structures are related to the layered media discussed previously. This offers an opportunity to develop general expressions for the transmission and reflection (filter) characteristics as functions of frequency.

The succeeding chapter uses the paraxial wave equation to describe some nonlinear optical systems by introducing an intensity dependent index: self-focusing, pulse formation in optical fibers, and saturable absorber mode-locking. These are prototypes of optical devices that can be used for nonlinear optical signal processing. We also discuss the operation of a bistable device (with hysteresis) that has been proposed for information storage.

Nonlinear optical materials are, in general, anisotropic. For many nonlinear processes phase matching and proper polarization orientation is necessary. We discuss, therefore, the propagation characteristics of an anisotropic optical medium and introduce the so-called index surface and normal surface. The electro-optic effect is then introduced and phase modulators and amplitude modulators using this effect are discussed. Modulators can also be constructed using the diffraction of optical waves off index "gratings" set up by acoustic standing or traveling waves.

Next we take up the study of nonlinear optical media in more general terms. Frequency doubling and parametric oscillators are introduced as practical examples. Finally, we study briefly optical detectors and the signal-to-noise problems encountered in optical detection.

1

MAXWELL'S EQUATIONS OF ISOTROPIC MEDIA AND SOME IMPORTANT IDENTITIES

We begin by a statement of Maxwell's equations in the time-dependent form. With linear relationships between the polarization density $\boldsymbol{P}$ and the electric field $\boldsymbol{E}$, and the magnetization density $\boldsymbol{M}$ and the magnetic field intensity $\boldsymbol{H}$, one may derive the wave equation and the well-known plane-wave solutions. Poynting's theorem is reviewed and power flow and energy density are identified for linear media. We review complex notation as used for sinusoidally time-dependent vectors and show that the most general polarization of a sinusoidal time-dependent field is elliptical.

Next we study the complex form of Maxwell's equations. In this form the linear media may be time dispersive; that is, the dielectric constant and magnetic susceptibility may be functions of the radial frequency ω. The Poynting theorem derived from these expressions leads to identities that prove useful in the study of resonating systems. We review Fourier series and Fourier integrals. The concept of group velocity is reviewed by means of the analysis of the propagation of a wave packet. The extension of Fourier methods to statistical time functions is introduced. For more details on Maxwell's equations, the reader is referred to the many excellent texts on electromagnetism (e.g., Refs. 1 and 2).

1.1 MAXWELL'S EQUATIONS IN REAL, TIME-DEPENDENT FORM

In a medium of *magnetization density* $\boldsymbol{M}$ and *polarization density* $\boldsymbol{P}$, Maxwell's equations in *mks units* for the electric field $\boldsymbol{E}$ (V/m), and the magnetic field (intensity) $\boldsymbol{H}$ (A/m) are: [1–4]

Faraday's law:

$$\nabla \times \boldsymbol{E} = -\frac{\partial}{\partial t}\mu_0 \boldsymbol{H} - \frac{\partial}{\partial t}\mu_0 \boldsymbol{M} \tag{1.1}$$

Ampère's law:

$$\nabla \times \boldsymbol{H} = \frac{\partial}{\partial t}\varepsilon_0 \boldsymbol{E} + \frac{\partial \boldsymbol{P}}{\partial t} + \boldsymbol{J} \tag{1.2}$$

Gauss's law for the electric field:

$$\nabla \cdot \varepsilon_0 \boldsymbol{E} = -\nabla \cdot \boldsymbol{P} + \rho \tag{1.3}$$

Gauss's law for the magnetic field:

$$\nabla \cdot \mu_0 \boldsymbol{H} = -\nabla \cdot \mu_0 \boldsymbol{M} \tag{1.4}$$

where ρ is the charge density and $\boldsymbol{J}$ is the current density produced by all charges other than those associated with polarization. The constants in Maxwell's equations are

$$\varepsilon_0 \simeq \frac{1}{36\pi} \times 10^{-9} \frac{\text{A} \cdot \text{s}}{\text{V} \cdot \text{m}}$$

the *dielectric constant* of free space, and

$$\mu_0 = 4\pi \times 10^{-7} \frac{\text{V} \cdot \text{s}}{\text{A} \cdot \text{m}}$$

the *magnetic permeability* of free space.

If the medium is linear, isotropic, and *dispersion-free* (i.e., the response of the medium is instantaneous), then $\boldsymbol{P}$ is simply related to $\boldsymbol{E}$ by the *constitutive* law:

$$\boldsymbol{P} = \varepsilon_0 \chi_e \boldsymbol{E} \tag{1.5}$$

and similarly, the *constitutive* law relating $\boldsymbol{M}$ to $\boldsymbol{H}$ is

$$\boldsymbol{M} = \chi_m \boldsymbol{H} \tag{1.6}$$

where χ_e and χ_m are the *electric* and *magnetic* (scalar) *susceptibilities.* We may then write

$$\nabla \times \boldsymbol{E} = -\mu \frac{\partial \boldsymbol{H}}{\partial t} \tag{1.7}$$

$$\nabla \times \boldsymbol{H} = \varepsilon \frac{\partial \boldsymbol{E}}{\partial t} + \boldsymbol{J} \tag{1.8}$$

where $\varepsilon \equiv \varepsilon_0(1 + \chi_e)$ and $\mu \equiv \mu_0(1 + \chi_m)$ are the *dielectric constant* and *per-*

meability of the medium. Further,

$$\nabla \cdot \varepsilon \boldsymbol{E} = \rho \tag{1.9}$$

$$\nabla \cdot \mu \boldsymbol{H} = 0 \tag{1.10}$$

Consider a source-free medium, $\rho = \boldsymbol{J} = 0$. In order to derive a single equation for $\boldsymbol{E}$ from Maxwell's equations, we take the curl of (1.7), and substitute (1.8) with $\boldsymbol{J} = 0$. We obtain the equation

$$\nabla \times (\nabla \times \boldsymbol{E}) = -\mu\varepsilon \frac{\partial^2 \boldsymbol{E}}{\partial t^2} \tag{1.11}$$

This equation is valid in media with spatial variation of ε. With the use of a well-known identity,* we may rewrite this in the form

$$\nabla(\nabla \cdot \boldsymbol{E}) - \nabla^2 \boldsymbol{E} = -\mu\varepsilon \frac{\partial^2 \boldsymbol{E}}{\partial t^2}$$

The *wave equation* results, when $\nabla \cdot \boldsymbol{E} = 0$; for example, when ε is uniform,

$$\nabla^2 \boldsymbol{E} = \mu\varepsilon \frac{\partial^2 \boldsymbol{E}}{\partial t^2} \tag{1.12}$$

One must note that not every vector field that is a solution of the wave equation is a solution of Maxwell's equations. One must require that

$$\nabla \cdot \boldsymbol{E} = 0 \tag{1.13}$$

The relation above does not hold, in general, in spatially nonuniform dielectric media.

One-dimensional plane waves are solutions of the wave equation for uniform media. Indeed, the following expressions can be checked to satisfy the wave equation: [4, 5]

$$\boldsymbol{E} = \hat{\boldsymbol{x}}[f_+(z - vt) + f_-(z + vt)] \tag{1.14}$$

$$\boldsymbol{H} = \sqrt{\frac{\varepsilon}{\mu}}\, \hat{\boldsymbol{y}}[f_+(z - vt) - f_-(z + vt)] \tag{1.15}$$

$\hat{\boldsymbol{x}}$ and $\hat{\boldsymbol{y}}$ are the unit vectors in the x and y directions of a Cartesian coordinate system, and f_+ and f_- are arbitrary functions; $f_+(z - vt)$ travels in the $+z$ direction, $f_-(z + vt)$ in the $-z$ direction. The quantity $\sqrt{\varepsilon/\mu}$ is the *character istic admittance* of the medium. The velocity $v = 1/\sqrt{\mu\varepsilon}$. The electric and magnetic fields are mutually perpendicular and lie in a plane perpendicular to the direction of propagation, so as to satisfy (1.10) and (1.9) with $\rho = 0$. The solutions (1.14) and (1.15) are illustrated in Fig. 1.1 for an assumed rectangular

*Given a vector $\boldsymbol{A}$ that is twice differentiable, the identity holds:

$$\nabla \times (\nabla \times \boldsymbol{A}) = \nabla(\nabla \cdot \boldsymbol{A}) - \nabla^2 \boldsymbol{A}$$

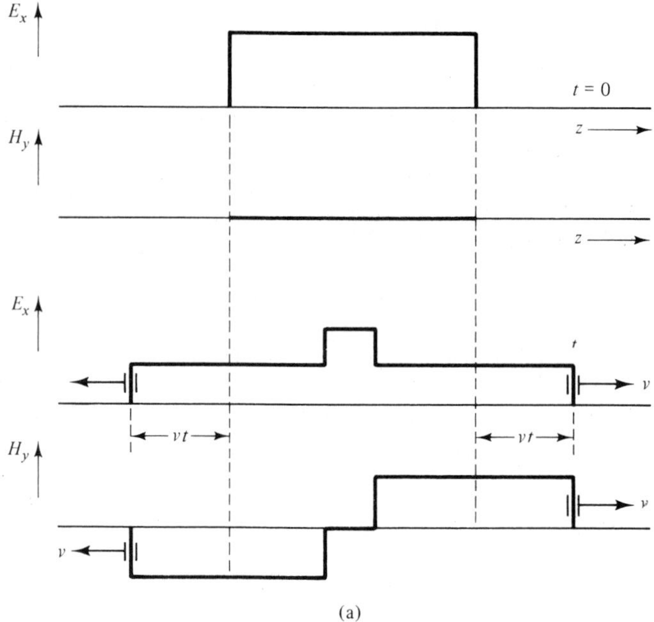

Figure 1.1 (a) and (b) E_x and H_y as functions of z at two different instants of time, with $H_y(z, t = 0) = 0$; (c) E_x and H_y as functions of z and t plotted above the z-t plane. A cut at $t = t_0$ illustrates E_x and H_y at constant t as a function of z.

initial spatial distribution of $E_x(z)$ and with $H_y(z) = 0$ at $t = 0$. Polarization of $\boldsymbol{E}$ in the $\hat{y}$ direction, of $\boldsymbol{H}$ in the $-\hat{x}$ direction, also represents a solution.

Poynting's theorem is the law of power conservation for electromagnetic fields. Poynting's theorem [1, 2, 5] is derived by dot multiplying (1.1) by $\boldsymbol{H}$, (1.2) by $-\boldsymbol{E}$, adding, and using the fact that $\nabla \cdot (\boldsymbol{E} \times \boldsymbol{H}) = (\nabla \times \boldsymbol{E}) \cdot \boldsymbol{H} - (\nabla \times \boldsymbol{H}) \cdot \boldsymbol{E}$.

$$\nabla \cdot (\boldsymbol{E} \times \boldsymbol{H}) + \frac{\partial}{\partial t}\left(\frac{1}{2}\varepsilon_0 \boldsymbol{E}^2\right) + \frac{\partial}{\partial t}\left(\frac{1}{2}\mu_0 \boldsymbol{H}^2\right) + \boldsymbol{E} \cdot \frac{\partial \boldsymbol{P}}{\partial t} + \boldsymbol{H} \cdot \frac{\partial}{\partial t}(\mu_0 \boldsymbol{M}) + \boldsymbol{E} \cdot \boldsymbol{J} = 0 \tag{1.16}$$

$\boldsymbol{E} \cdot \boldsymbol{J}$ is the power per unit volume imparted to $\boldsymbol{J}$. The term

$$\frac{\partial}{\partial t}\left(\frac{1}{2}\varepsilon_0 \boldsymbol{E}^2\right)$$

is the time rate of change of the energy density in the electric field. We arrive at this interpretation by studying the time integral of the power per unit volume represented by this term, that is, the energy delivered over a time interval

$$\int_{-\infty}^{t} dt\, \frac{\partial}{\partial t}\left(\frac{1}{2}\varepsilon_0 \boldsymbol{E}^2\right) = \frac{1}{2}\varepsilon_0 \boldsymbol{E}^2 \bigg|_{-\infty}^{t} = \frac{1}{2}\varepsilon_0 \boldsymbol{E}^2(t) \tag{1.17}$$

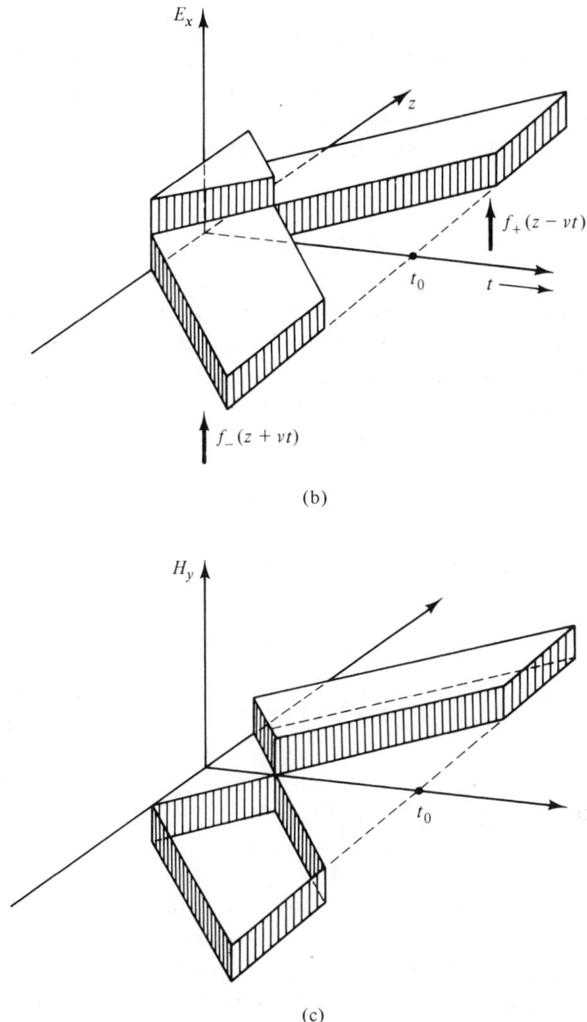

Figure 1.1 *(continued)*

for $\boldsymbol{E} = 0$ at $t = -\infty$. This integral depends only on the final value of $\boldsymbol{E}$, if $\boldsymbol{E} = 0$ at $t = -\infty$, and is equal to the energy per unit volume required to bring $\boldsymbol{E}$ from zero to its value at t, $\frac{1}{2}\varepsilon_0 \boldsymbol{E}^2$. This energy is fully recovered, if the field is made to vanish again at $t = +\infty$:

$$\int_t^{+\infty} dt \, \frac{\partial}{\partial t}\left(\frac{1}{2}\, \varepsilon_0 \, E^2\right) = -\frac{1}{2}\, \varepsilon_0 \, \boldsymbol{E}^2 \bigg|_\infty^t = -\frac{1}{2}\, \varepsilon_0 \, \boldsymbol{E}^2(t)$$

The time integral of the power is negative; the energy has been returned. In a similar way, one interprets $\frac{1}{2}\mu_0 \boldsymbol{H}^2$ as the energy density in the magnetic field.

The term $\boldsymbol{E} \cdot \partial \boldsymbol{P}/\partial t$ represents the power per unit volume delivered to the polarization current density $\partial \boldsymbol{P}/\partial t$. In general, it is not possible to determine whether the time integral of this power is stored, dissipated, or both, but for a given constitutive law such a criterion can be developed. Consider the linear constitutive law (1.5). The power per unit volume delivered to the polarization process is

$$\boldsymbol{E} \cdot \frac{\partial \boldsymbol{P}}{\partial t} = \boldsymbol{E} \cdot \frac{\partial}{\partial t} (\varepsilon_0 \chi_e \boldsymbol{E}) = \frac{\partial}{\partial t} \left(\frac{1}{2} \varepsilon_0 \chi_e \boldsymbol{E}^2 \right)$$

We were able to write the power per unit volume as a time derivative of a positive-definite term. The same argument can be applied as before to identify $\frac{1}{2}\varepsilon_0 \chi_e \boldsymbol{E}^2$ as the energy density stored in the polarization.

We denote by w_e the *electric energy density* stored in the electric field and in the polarization

$$w_e \equiv \tfrac{1}{2}\varepsilon_0(1 + \chi_e)\boldsymbol{E}^2 = \tfrac{1}{2}\varepsilon \boldsymbol{E}^2 \tag{1.18}$$

An analogous argument applied to a linear magnetic material shows that $\frac{1}{2}\mu_0 \chi_m \boldsymbol{H}^2$ is the energy stored in the magnetization. We define by w_m the *magnetic energy density* stored in the magnetic field and in the magnetization

$$w_m \equiv \tfrac{1}{2}\mu_0(1 + \chi_m)\boldsymbol{H}^2 = \tfrac{1}{2}\mu \boldsymbol{H}^2 \tag{1.19}$$

With this notation, the Poynting theorem for a linear isotropic medium assumes the form

$$\nabla \cdot (\boldsymbol{E} \times \boldsymbol{H}) + \frac{\partial}{\partial t}(w_e + w_m) + \boldsymbol{E} \cdot \boldsymbol{J} = 0 \tag{1.20}$$

The power per unit volume imparted to the current density $\boldsymbol{J}$, the time rate of change of energy, and the outflow of power per unit volume of the *Poynting vector* $\boldsymbol{E} \times \boldsymbol{H}$ add to zero. Remember, the divergence of a vector $\boldsymbol{A}$ is indeed the outflow of $\boldsymbol{A}$ per unit volume through the surface S enclosing the volume V (Fig. 1.2) according to the definition of divergence:*

$$\nabla \cdot \boldsymbol{A} \equiv \lim_{V \to 0} \frac{\oint_S \boldsymbol{A} \cdot d\boldsymbol{a}}{V}$$

We pick a general surface S, enclosed by a volume V and integrate (1.20) over the volume, and obtain, using Gauss's theorem,

$$\oint_S \boldsymbol{E} \times \boldsymbol{H} \cdot d\boldsymbol{a} + \frac{d}{dt}\int_V (w_e + w_m)\, dv + \int_V \boldsymbol{E} \cdot \boldsymbol{J}\, dv = 0 \tag{1.21}$$

*Line, surface, and volume integrals will be distinguished by the differential of integration, ds for a line integral, $d\boldsymbol{a}$ for a surface integral, and dv for a volume integral.

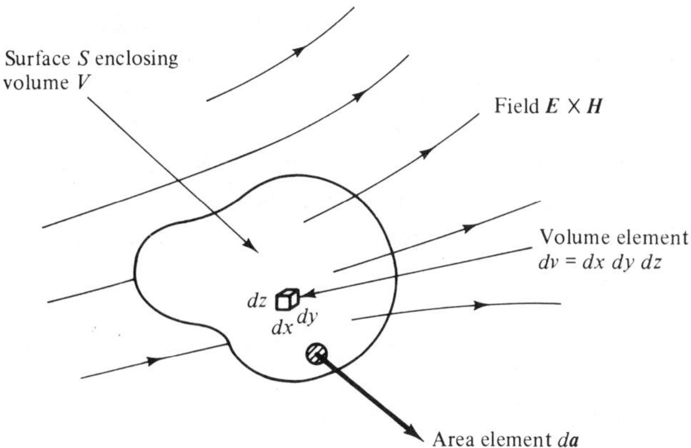

Figure 1.2 Definition of area element and surface.

This is the Poynting theorem in integral form. The flux of $\boldsymbol{E} \times \boldsymbol{H}$ through the surface; that is, the electromagnetic power escaping from the volume, the time rate of change of the energy in the volume, and the power imparted to the current distribution $\boldsymbol{J}$ must add up to zero.

For later use, we introduce the vector potential $\boldsymbol{A}$ and scalar potential Φ. We confine ourselves to nonmagnetic permeabilities, $\mu = \mu_0$. Because $\mu_0 \boldsymbol{H}$ is divergence-free, one may satisfy (1.10) automatically by introducing the vector potential by the definition

$$\mu_0 \boldsymbol{H} = \nabla \times \boldsymbol{A} \tag{1.22}$$

For any particular magnetic field, (1.22) only defines the curl for $\boldsymbol{A}$. Because unique specification of a vector field requires the specification of both its curl and its divergence, we must still specify the divergence of $\boldsymbol{A}$, which is done below. Introducing (1.22) into (1.7) one finds for $\boldsymbol{E}$

$$\boldsymbol{E} = -\frac{\partial \boldsymbol{A}}{\partial t} - \nabla \Phi \tag{1.23}$$

where Φ is the, as yet, unspecified scalar potential. With (1.22) and (1.23) introduced into (1.8), one obtains

$$\nabla \times (\nabla \times \boldsymbol{A}) = -\mu_0 \varepsilon \frac{\partial^2 \boldsymbol{A}}{\partial t^2} - \mu_0 \varepsilon \frac{\partial}{\partial t} \nabla \Phi + \mu_0 \boldsymbol{J} \tag{1.24}$$

Equation (1.9) with (1.23) gives

$$\nabla \cdot \left(\varepsilon \frac{\partial \boldsymbol{A}}{\partial t} + \varepsilon \nabla \Phi \right) = -\rho \tag{1.25}$$

We may now take advantage of the free choice for the divergence of $\boldsymbol{A}$.

We choose

$$\nabla \cdot \boldsymbol{A} + \mu_0 \varepsilon \frac{\partial \Phi}{\partial t} = 0 \tag{1.26}$$

In free space, $\varepsilon = \varepsilon_0$, the foregoing choice of the divergence of $\boldsymbol{A}$ is called the *Lorentz gauge*. It is a convenient choice because, in free space and in a uniform dielectric, it leads to the wave equation for $\boldsymbol{A}$ and Φ. Introducing (1.26) into (1.24), one has

$$\nabla^2 \boldsymbol{A} - \mu_0 \varepsilon \frac{\partial^2 \boldsymbol{A}}{\partial t^2} = -\mu_0 \boldsymbol{J} - \mu_0 (\nabla \varepsilon) \frac{\partial \Phi}{\partial t} \tag{1.27}$$

Similarly, introducing (1.26) into (1.25) gives us

$$\frac{1}{\varepsilon} \nabla \cdot (\varepsilon \nabla \Phi) - \mu_0 \varepsilon \frac{\partial^2 \Phi}{\partial t^2} = -\frac{\rho}{\varepsilon} - \frac{1}{\varepsilon} \nabla \varepsilon \cdot \frac{\partial \boldsymbol{A}}{\partial t} \tag{1.28}$$

The vector potential obeys the wave equation with a source, and a coupling term; Φ couples to $\boldsymbol{A}$ in a nonuniform dielectric medium. In a uniform charge-free medium, Φ obeys the wave equation as well, and Φ and $\boldsymbol{A}$ are uncoupled:

$$\nabla^2 \boldsymbol{A} - \mu_0 \varepsilon \frac{\partial^2 \boldsymbol{A}}{\partial t^2} = 0 \tag{1.29}$$

$$\nabla^2 \Phi - \mu_0 \varepsilon \frac{\partial^2 \Phi}{\partial t^2} = 0 \tag{1.30}$$

1.2 COMPLEX VECTORS [1, 5, 6]

Maxwell's equations in linear media are linear. A sinusoidal excitation (e.g., a sinusoidal current density distribution $\boldsymbol{J}$) at a frequency ω produces a sinusoidal response (i.e., $\boldsymbol{E}$ and $\boldsymbol{H}$ vary sinusoidally with time). Any transient excitation and response can be represented by a superposition of sinusoidal excitations and responses. For this reason, it is important to study sinusoidal steady-state solutions of Maxwell's equations. The sinusoidal steady state is treated conveniently in terms of complex variables. In this section we determine the physical interpretation of complex vectors.

A time-dependent scalar [e.g., a voltage $v(t)$] that is a sinusoidal function of time of frequency ω is written conveniently as the real part of a complex function:

$$v(t) = \text{Re}\,[\mathrm{V} e^{j\omega t}] = |\mathrm{V}| \cos(\omega t + \phi) \tag{1.31}$$

Here $|\mathrm{V}|$ is the magnitude of the complex voltage V and ϕ is its argument, $\phi = \arg[\mathrm{V}]$. We shall use roman type to denote complex excitation amplitudes. A sinusoidally time-dependent vector can be decomposed into three scalar components. Each of the scalar components has an amplitude and phase. The generalization of (1.31) to a vector is

$$A(t) = \text{Re}\,[(\hat{x}\text{A}_x + \hat{y}\text{A}_y + \hat{z}\text{A}_z)e^{j\omega t}] = \hat{x}\,|\text{A}_x|\cos(\omega t + \phi_x) + \hat{y}\,|\text{A}_y|\cos(\omega t + \phi_y) + \hat{z}\,|\text{A}_z|\cos(\omega t + \phi_z) \tag{1.32}$$

Here A_x, A_y, and A_z are complex scalars. The phases ϕ_x, ϕ_y, and ϕ_z are the arguments of the respective complex scalars. We may construct the complex vector

$$\mathbf{A} \equiv \hat{\mathbf{x}}\text{A}_x + \hat{\mathbf{y}}\text{A}_y + \hat{\mathbf{z}}\text{A}_z \tag{1.33}$$

which describes the vector $A(t)$ at all times by

$$A(t) = \text{Re}\,[\mathbf{A}\exp(j\omega t)] = \text{Re}\,[\mathbf{A}]\cos\omega t - \text{Im}\,[\mathbf{A}]\sin\omega t \tag{1.34}$$

where

$$\text{Re}\,[\mathbf{A}] = \hat{\mathbf{x}}\,\text{Re}\,[\text{A}_x] + \hat{\mathbf{y}}\,\text{Re}\,[\text{A}_y] + \hat{\mathbf{z}}\,\text{Re}\,[\text{A}_z]$$

and Im [**A**] is defined analogously in terms of the imaginary parts of the complex vector components. Both Re [**A**] and Im [**A**] are *real* vectors, each defined by three *real* scalars. In combination, they describe the complex vector **A**, which is defined in terms of three *complex* scalars,

$$\mathbf{A} = \text{Re}\,[\mathbf{A}] + j\,\text{Im}\,[\mathbf{A}]$$

The locus of $A(t)$ for all t forms an ellipse (see Fig. 1.3). Indeed, the vector

$$A(t) = \text{Re}\,[\mathbf{A}]\cos\omega t - \text{Im}\,[\mathbf{A}]\sin\omega t$$

must lie in the plane of Re **A** and Im **A**. If we pick a two-dimensional Cartesian coordinate system in that plane, the coordinate vector $\hat{\mathbf{x}}$ along Re **A**, then

$$\text{Im}\,\mathbf{A} = \hat{\mathbf{x}}\,\text{Im}\,\text{A}_x + \hat{\mathbf{y}}\,\text{Im}\,\text{A}_y \tag{1.35}$$

$$A_x(t) = \text{Re}\,\text{A}_x\cos\omega t - \text{Im}\,\text{A}_x\sin\omega t \tag{1.36}$$

$$A_y(t) = -\text{Im}\,\text{A}_y\sin\omega t \tag{1.37}$$

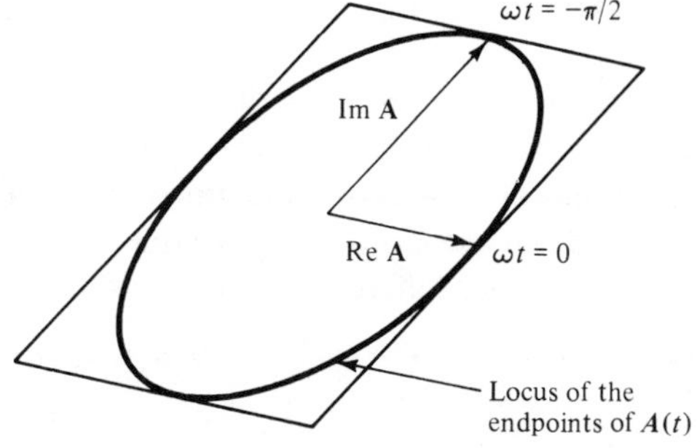

Figure 1.3 Polarization ellipse. The vectors Re **A** and Im **A** are chosen to lie in plane of page. The ellipse is tangent to parallelogram that has sides parallel to Re **A** and Im **A**. These are conjugate axes of the ellipse.

We find from (1.37),

$$\sin \omega t = -\frac{A_y(t)}{\text{Im } \mathrm{A}_y} \tag{1.38}$$

and from (1.36) and (1.38):

$$\cos \omega t = \frac{1}{\text{Re } \mathrm{A}_x} A_x(t) - \frac{\text{Im } \mathrm{A}_x}{\text{Im } \mathrm{A}_y \text{ Re } \mathrm{A}_x} A_y(t) \tag{1.39}$$

By adding the squares of (1.38) and (1.39) we find a quadratic equation for the coordinates $A_x(t)$ and $A_y(t)$ which describes an ellipse in the plane A_x, A_y.

If the electric vector of a one-dimensional plane electromagnetic wave describes an ellipse as a function of time, at any point in space, the wave is said to be *elliptically polarized.* Special cases of elliptical polarization are *linear* and *circular polarizations.* The field of a right-circularly polarized wave rotates in the sense of rotation of a right-handed screw, for advancement of the screw in the same direction as the direction of propagation. Nonplanar waves may have different polarization ellipses at different points.

Time averages. Time averages of products of sinusoidally time-dependent scalars can be evaluated directly from the complex amplitudes. Thus the time-averaged power delivered to a circuit element with a voltage $v(t)$ across its terminals, a current $i(t)$ into the terminals, is $\langle v(t)\, i(t)\rangle = \frac{1}{2} \text{Re } [\mathrm{VI}^*]$. We denote a time average by angular brackets. A similar relation can be proved for vector quantities:

$$\begin{aligned}
\langle \boldsymbol{A}(t) \times \boldsymbol{B}(t)\rangle &= \frac{1}{T}\int_0^T \boldsymbol{A}(t) \times \boldsymbol{B}(t)\, dt \\
&= \frac{1}{T}\int_0^T \frac{1}{4}(\mathbf{A}e^{j\omega t} + \mathbf{A}^* e^{-j\omega t}) \\
&\quad \times (\mathbf{B}e^{j\omega t} + \mathbf{B}^* e^{-j\omega t})\, dt = \frac{1}{2}\text{Re } [\mathbf{A} \times \mathbf{B}^*]
\end{aligned} \tag{1.40}$$

1.3 THE COMPLEX FORM OF MAXWELL'S EQUATIONS

If one introduces into Maxwell's equations for a linear medium, (1.7) and 1.8), complex fields $\mathbf{E}(\boldsymbol{r})e^{j\omega t}$, $\mathbf{H}(\boldsymbol{r})e^{j\omega t}$, and $\mathbf{J}(\boldsymbol{r})e^{j\omega t}$ which are products of complex functions of space and the time function $e^{j\omega t}$, the time dependence can be factored out. Separation of variables is thereby accomplished. Maxwell's equations reduce to differential equations in the spatial variable only.

$$\nabla \times \mathbf{E} = -j\omega\mu\mathbf{H} \tag{1.41}$$

$$\nabla \times \mathbf{H} = j\omega\varepsilon\mathbf{E} + \mathbf{J} \tag{1.42}$$

One obtains real, time-dependent fields, which obey the original equations (1.7) and (1.8), by taking either the real part or the imaginary part of $\mathbf{E}(\boldsymbol{r})e^{j\omega t}$, and so on. By convention the real part is chosen. The two laws of Gauss become

$$\nabla \cdot \varepsilon \mathbf{E} = \rho \tag{1.43}$$

$$\nabla \cdot \mu \mathbf{H} = 0 \tag{1.44}$$

Thus far, complex notation only served to simplify the solution of Maxwell's equations, (1.7)–(1.10), in the sinusoidal steady state. Complex notation has the further advantage that it permits the analysis of media with a noninstantaneous response to $\boldsymbol{E}$ and $\boldsymbol{H}$. In a medium with instantaneous response, a dispersion-free medium, (1.5) holds. In a linear dispersive medium, $\boldsymbol{P}$ is related to $\boldsymbol{E}$ by a linear integrodifferential equation. In the sinusoidal steady state the differential and integral operators operating on the time function are replaced by multiplication or division by $j\omega$ of the complex amplitude. One may still write

$$\mathbf{P} = \varepsilon_0 \chi_e(\omega) \mathbf{E} \tag{1.45}$$

where the dielectric susceptibility χ_e is a function of ω and is in general complex. Similar comments apply to χ_m. The dielectric constant becomes a function of frequency:

$$\varepsilon(\omega) = \varepsilon_0[1 + \chi_e(\omega)] \tag{1.46}$$

and the magnetic permeability as well:

$$\mu(\omega) = \mu_0[1 + \chi_m(\omega)] \tag{1.47}$$

A complex Poynting theorem can be derived from (1.41) and (1.42). Because the time averages of products of sinusoidally time-dependent vectors are evaluated from the products of the complex vector amplitude of one vector with the complex conjugate of the other vector, according to (1.40), the complex Poynting theorem aims at a relation for $\mathbf{E} \times \mathbf{H}^*$ and $\mathbf{E} \cdot \mathbf{J}^*$. Dot multiplication of (1.41) by $\mathbf{H}^*$ and of the complex conjugate of (1.42) by $-\mathbf{E}$ and addition gives

$$\nabla \cdot (\mathbf{E} \times \mathbf{H}^*) + j\omega(\mu \mathbf{H} \cdot \mathbf{H}^* - \varepsilon^* \mathbf{E} \cdot \mathbf{E}^*) + \mathbf{E} \cdot \mathbf{J}^* = 0 \tag{1.48}$$

The term Re $[j\omega(\mu \mathbf{H} \cdot \mathbf{H}^* - \varepsilon^* \mathbf{E} \cdot \mathbf{E}^*)]$ contributes to the real part of the divergence of the complex Poynting vector in the same way as the density of power dissipated by $\mathbf{J}$, Re $[\mathbf{E} \cdot \mathbf{J}^*]$, contributes to it. Therefore, we interpret it as the power per unit volume dissipated in the dielectric and magnetic medium. In a lossless medium, for a real frequency ω, one must have

$$\operatorname{Im} \varepsilon^* \mathbf{E} \cdot \mathbf{E}^* = 0$$

and

$$\operatorname{Im} \mu \mathbf{H} \cdot \mathbf{H}^* = 0$$

Since $\mathbf{E} \cdot \mathbf{E}^*$ and $\mathbf{H} \cdot \mathbf{H}^*$ are real, ε and μ must be real!

$$\text{Im } \varepsilon = 0 \tag{1.49}$$

$$\text{Im } \mu = 0 \tag{1.50}$$

The complex Poynting theorem has many interesting consequences, one of which refers to resonance. Consider a perfectly conducting enclosure containing a source-free and loss-free medium. If the electromagnetic field is resonant, a field may exist in the absence of a source. Integration of (1.48) over the volume contained in the enclosure and use of Gauss's theorem give

$$\oint_S \mathbf{E} \times \mathbf{H}^* \cdot d\boldsymbol{a} + j\omega \int_V (\mu \mathbf{H} \cdot \mathbf{H}^* - \varepsilon \mathbf{E} \cdot \mathbf{E}^*)\, dv = 0$$

The surface integral over the surface S of the enclosure vanishes because $\mathbf{E}$ is parallel to $d\boldsymbol{a}$ on the surface. Therefore, if a field exists inside the volume, the equality must hold:

$$\int_V \varepsilon \mathbf{E} \cdot \mathbf{E}^*\, dv = \int_V \mu \mathbf{H} \cdot \mathbf{H}^*\, dv \tag{1.51}$$

In a dispersion-free medium the equation above implies equality of the time-averaged electric and magnetic energies.

The complex form of Maxwell's equations leads in a simple way to general plane-wave solutions traveling in an arbitrary direction. From the complex Maxwell's equations, with $\boldsymbol{J} = \rho = 0$, and a spatially uniform ε, one may derive the Helmholtz equation for $\mathbf{E}$ in a spatially uniform medium:

$$\nabla^2 \mathbf{E} + \omega^2 \mu \varepsilon \mathbf{E} = 0 \tag{1.52}$$

This equation may handle more complicated cases than the wave equation (1.12) because now μ and ε may be functions of frequency. Again we require that the solution of (1.52) satisfy

$$\nabla \cdot \mathbf{E} = 0 \tag{1.53}$$

in order to be a solution of Maxwell's equations.

It is convenient to describe a plane wave propagating in a general direction without reference to a specific coordinate system. One defines the *propagation vector* $\boldsymbol{k}$ as the vector normal to the phase fronts. One constructs a solution to Maxwell's equations (1.41)–(1.44) by assuming a spatial dependence for $\mathbf{E}$ and $\mathbf{H}$ of the form $\exp(-j\boldsymbol{k} \cdot \boldsymbol{r})$. One sets

$$\mathbf{E} = \mathbf{E}_+ e^{-j\boldsymbol{k}\cdot\boldsymbol{r}} \tag{1.54}$$

and

$$\mathbf{H} = \mathbf{H}_+ e^{-j\boldsymbol{k}\cdot\boldsymbol{r}} \tag{1.55}$$

The phase of $\mathbf{E}$ and $\mathbf{H}$ is constant along planes that satisfy the condition $\boldsymbol{k} \cdot \boldsymbol{r} = \text{const}$ (see Fig. 1.4). The ∇ operator operating on $\mathbf{E}$ or $\mathbf{H}$ may be

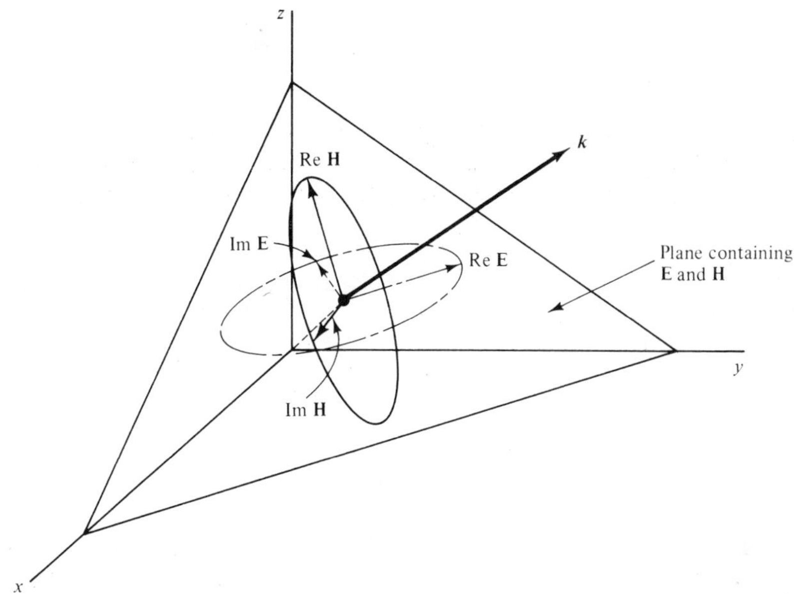

Figure 1.4 Plane of constant phase and the complex field vectors. Re **E**, Re **H**, Im **E**, Im **H**.

replaced by $j\boldsymbol{k}$ because of the assumed form of the spatial dependence. From (1.41) and (1.42)

$$-j\boldsymbol{k} \times \mathbf{E}_+ = -j\omega\mu\mathbf{H}_+ \tag{1.56}$$

$$-j\boldsymbol{k} \times \mathbf{H}_+ = j\omega\varepsilon\mathbf{E}_+ \tag{1.57}$$

Cross-multiplication of (1.56) by $-j\boldsymbol{k}$ and use of (1.57) gives

$$-\boldsymbol{k} \times (\boldsymbol{k} \times \mathbf{E}_+) = -\boldsymbol{k}(\boldsymbol{k} \cdot \mathbf{E}_+) + \boldsymbol{k}^2\mathbf{E}_+ = \omega^2\mu\varepsilon\mathbf{E}_+ \tag{1.58}$$

The triple vector product in (1.58) produces a vector perpendicular to $\boldsymbol{k}$, which is set equal to $\omega^2\mu\varepsilon\mathbf{E}_+$. It follows that $\mathbf{E}_+$ must be perpendicular to $\boldsymbol{k}$. On the other hand, Gauss' law (1.43), in a charge-free medium, predicts

$$j\boldsymbol{k} \cdot \mathbf{E}_+ = 0 \tag{1.59}$$

and therefore, also implies that $\mathbf{E}_+$ is perpendicular to $\boldsymbol{k}$. This coincidence of results is one of many examples in which a particular assumed form of solution introduced into some of Maxwell's equations automatically obeys the remaining laws of Maxwell.

From 1.58, one finds the *dispersion relation* for the magnitude of $\boldsymbol{k}$, k:

$$k^2 = \omega^2\mu\varepsilon \tag{1.60}$$

The solution for **H** follows from (1.56):

$$\mathbf{H}_+ = \frac{1}{\omega\mu} \boldsymbol{k} \times \mathbf{E}_+ = \sqrt{\frac{\varepsilon}{\mu}} \frac{\boldsymbol{k}}{k} \times \mathbf{E}_+ \tag{1.61}$$

The complex magnetic field vector **H** is perpendicular to **E**; $\boldsymbol{k}$, **E**, and **H** define right-handed orthogonal coordinates as shown in Fig. 1.4. The magnitude of $\boldsymbol{k}$ is, according to (1.60) equal to $\omega\sqrt{\mu\varepsilon}$. The phase velocity v_p is the velocity of a plane of constant phase. Because the full space–time dependence is $\exp[j(\omega t - \boldsymbol{k} \cdot \boldsymbol{r})]$, the change $\Delta r_{\parallel}$ parallel to $\boldsymbol{k}$ required to keep the phase fixed as t increases by Δt is

$$\frac{\Delta r_{\parallel}}{\Delta t} = \frac{\omega}{k} = v_p$$

In the case of a plane wave in the medium μ, ε it is equal to

$$v_p = \frac{1}{\sqrt{\mu\varepsilon}} \tag{1.62}$$

Next, apply the complex Poynting theorem to the plane-wave solution. The complex Poynting vector

$$\mathbf{E} \times \mathbf{H}^* = \frac{\boldsymbol{k}}{k} \sqrt{\frac{\varepsilon}{\mu}}\, |\mathbf{E}_+|^2 \tag{1.63}$$

is space independent. Its divergence is zero and hence, from (1.48) for $\boldsymbol{J} = 0$,

$$\varepsilon \mathbf{E} \cdot \mathbf{E}^* = \mu \mathbf{H} \cdot \mathbf{H}^* \tag{1.64}$$

This relation can be confirmed by direct evaluation using (1.61). In a dispersion-free medium the time-averaged electric and magnetic energy densities in a plane wave have to be equal.

Another relation of interest follows from (1.63). Using the dispersion relation (1.60) and (1.64), one may write (1.63) for a dispersion-free medium in the form

$$\frac{1}{2} \mathbf{E} \times \mathbf{H}^* = \frac{\boldsymbol{k}}{k} \frac{1}{\sqrt{\mu\varepsilon}} \langle w_e + w_m \rangle \tag{1.65}$$

where we have employed the expressions for energy densities, (1.18) and (1.19). $\frac{1}{2}\mathbf{E} \times \mathbf{H}^*$ is real for a traveling wave and represents the time-averaged power flow density. According to (1.65), the time-averaged power flow density vector points in the direction of $\boldsymbol{k}$ and is equal in magnitude to the sum of the time-averaged energy densities times the velocity $1/\sqrt{\mu\varepsilon}$. One may imagine that the power flow (density) results from a transport of energy (density) traveling at the velocity $1/\sqrt{\mu\varepsilon}$. We shall see later that dispersive and anisotropic media require a rederivation of energy-density and energy-velocity expressions.

1.4 FOURIER TRANSFORMS AND GROUP VELOCITY

The importance of the sinusoidal steady-state analysis rests on the fact that any *time-dependent process* in a linear system can be treated by a superposition of sinusoidal steady states. A periodic function of time, $f(t)$, of period T can be represented by a Fourier series

$$f(t) = \sum_{n=-\infty}^{\infty} F(n)e^{jn\omega_0 t} \tag{1.66}$$

where $\omega_0 = 2\pi/T$ is the fundamental frequency and the $F(n)$'s are the *Fourier amplitude coefficients.* $F(n)$ is found from the integral

$$F(n) = \frac{1}{T}\int_{-T/2}^{T/2} e^{-jn\omega_0 t} f(t)\, dt \tag{1.67}$$

An aperiodic function calls for the *Fourier integral* transform. An aperiodic function can be treated as a periodic function in the limit as the period T goes to infinity, $T \to \infty$, $2\pi/T \equiv \Delta\omega \to 0$. In this limit, $n\omega_0$ becomes a continuous variable, $n\omega_0 = \omega$. Starting with (1.66), we obtain

$$f(t) = \lim \sum_n F(n)e^{jn\omega_0 t} = \lim_{\Delta\omega\to 0} \sum_n \frac{F(n)}{\Delta\omega} e^{j\omega t} \Delta\omega = \int_{-\infty}^{\infty} F(\omega)e^{j\omega t}\, d\omega$$

where

$$F(\omega) \equiv \lim_{\Delta\omega\to 0} \frac{F(n)}{\Delta\omega}$$

The Fourier transform relation is

$$F(\omega) = \lim_{\Delta\omega\to 0} \frac{F(n)}{\Delta\omega} = \lim_{T\to\infty} \frac{1}{\Delta\omega T}\int_{-T/2}^{T/2} e^{-j\omega t} f(t)\, dt = \frac{1}{2\pi}\int_{-\infty}^{\infty} e^{-j\omega t} f(t)\, dt$$

We restate succinctly the Fourier transform pairs,

$$F(\omega) = \frac{1}{2\pi}\int_{-\infty}^{\infty} f(t)e^{-j\omega t}\, dt \tag{1.68}$$

with the *inverse Fourier transform* relation*

$$f(t) = \int_{-\infty}^{\infty} F(\omega)e^{j\omega t}\, d\omega \tag{1.69}$$

If $f(t)$ is a real function, as any physical variable must be, then the Fourier

*We use the definition of Fourier transform pairs which assigns the factor $1/2\pi$ to the Fourier transform and no factor to the inverse transform.

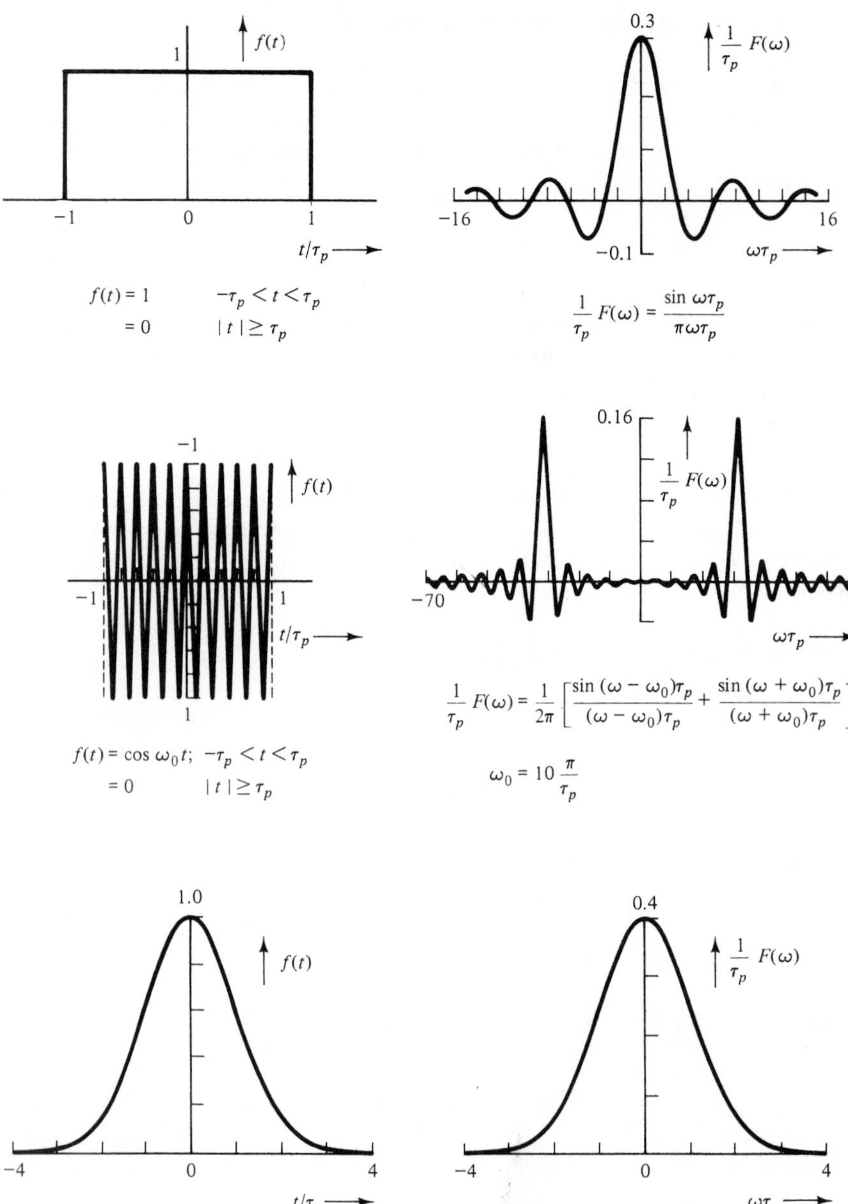

Figure 1.5 Examples of Fourier transform pairs.

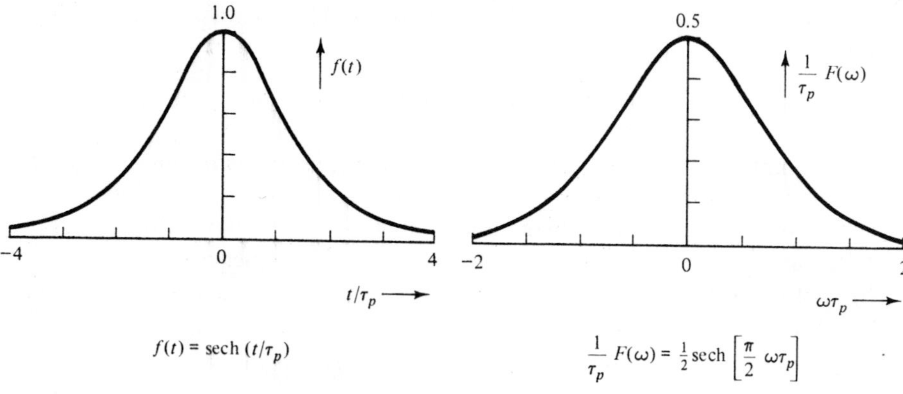

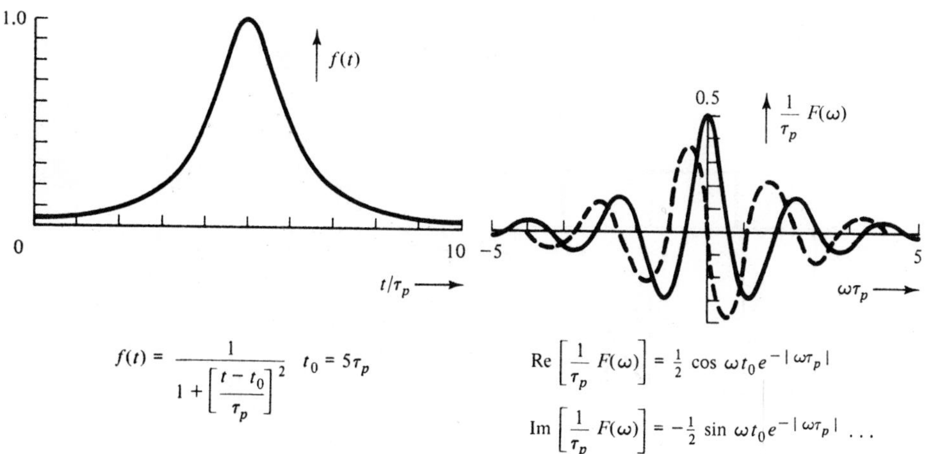

Figure 1.5 *(continued)*

amplitudes $F(\omega)$ are constrained by

$$F(-\omega) = F^*(\omega) \tag{1.70}$$

Figure 1.5 shows several examples of integral Fourier transform pairs.

A convolution of two time functions, $f(t)$ and $g(t)$, is defined symbolically and operationally by

$$g \otimes f = \int_{-\infty}^{\infty} dt'\, g(t - t')f(t') \tag{1.71}$$

The Fourier transform of a convolution is proportional to the product of the Fourier transforms of the two time functions. This is easily demonstrated:

$$\frac{1}{2\pi}\int_{-\infty}^{\infty} g \otimes f e^{-j\omega t}\, dt = \frac{1}{2\pi}\int_{-\infty}^{\infty} dt \int_{-\infty}^{\infty} dt'\, e^{-j\omega(t-t')} g(t - t')f(t')e^{-j\omega t'}$$

Introduce the pair of new variables $t - t'$ and t'. The Jacobian is unity. Therefore,

$$\frac{1}{2\pi}\int_{-\infty}^{\infty} g \otimes f e^{-j\omega t}\, dt = 2\pi G(\omega)F(\omega) \tag{1.72}$$

This completes the proof. Equation (1.72) is the *convolution theorem.* Figure 1.6 shows the steps involved in the convolution equation.

Consider now the propagation of a plane wave, whose Fourier amplitudes are appreciable only in a narrow band of frequencies around $\pm\omega_0$. Specialize to propagation along the z direction $\boldsymbol{k} = \omega\sqrt{\mu\varepsilon}\,\hat{z} \equiv k\hat{z}$. The complex electric field is [compare (1.54)].

$$\mathbf{E}(\omega, z) = \mathbf{E}_{+}(\omega) \exp\left[-jk(\omega)z\right] \tag{1.73}$$

The time-dependent field is obtained from (1.69):

$$E(t, z) = \int_{-\infty}^{\infty} d\omega \exp\{j[\omega t - k(\omega)z]\}\mathbf{E}_{+}(\omega) \tag{1.74}$$

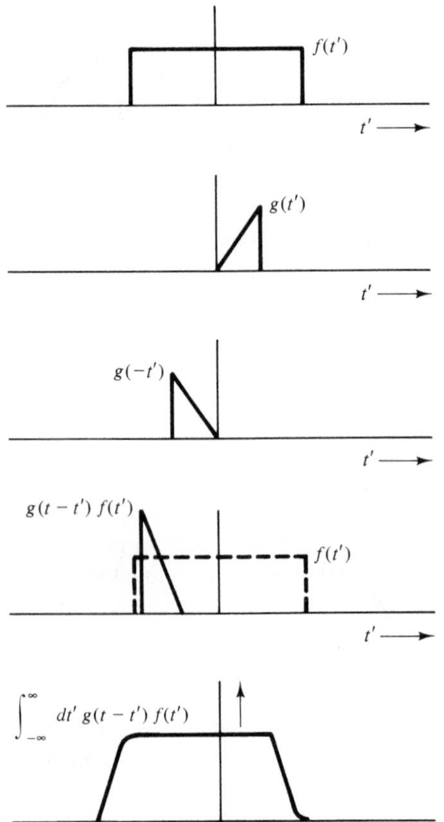

Figure 1.6 Steps involved in convolution.

Thus far we have made no approximation. Suppose that ε is a function of ω and thus k is not simply proportional to ω. We may expand k in the neighborhood of ω_0 to first order in $\Delta\omega = (\omega - \omega_0)$ and ignore higher-order terms, because they contribute negligibly to the integral because of the assumption that $E_+(\omega)$ is nonzero only in a narrow range around ω_0.

$$k(\omega) = k(\omega_0) + \left.\frac{dk}{d\omega}\right|_{\omega 0} \Delta\omega \tag{1.75}$$

Figure 1.7 illustrates a particular *dispersion relation*, k as a function of ω, namely the relation

$$k^2 = \frac{\omega^2}{c^2}\left(1 - \frac{\omega_p^2}{\omega^2}\right) \qquad \text{with } \omega_p^2 \text{ a constant}$$

For this dispersion relation real values of k are obtained only for $|\omega| > \omega_p$. The bandwidth of the Fourier transform $\mathbf{E}(\omega)$ must occupy a narrow range of values around ω_0 for which $k(\omega)$ can be approximated by a straight line in order that the expansion (1.75) be valid. We have from (1.74) and (1.75):

$$\boldsymbol{E}(t, z) = e^{j[\omega_0 t - k(\omega_0)z]} \int_{\text{band}} \mathrm{E}_+(\Delta\omega) e^{j\Delta\omega[t - (dk/d\omega)z]}\, d\Delta\omega + \text{complex conjugate} \tag{1.76}$$

where we have separated out the integrals over positive and negative fre-

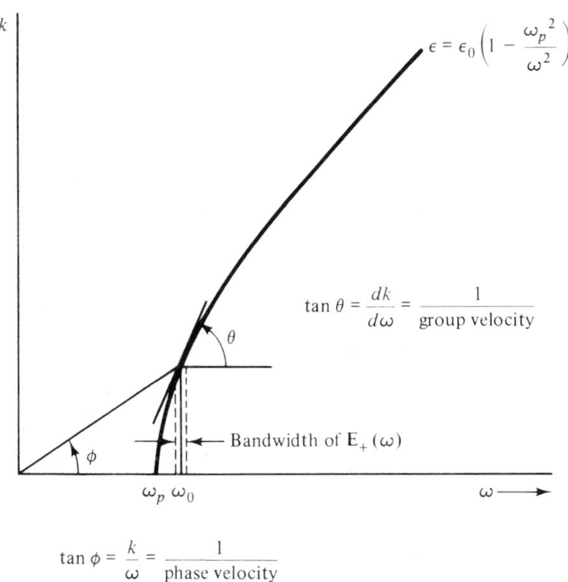

Figure 1.7 Possible dependence of k on ω and the associated phase and group velocities.

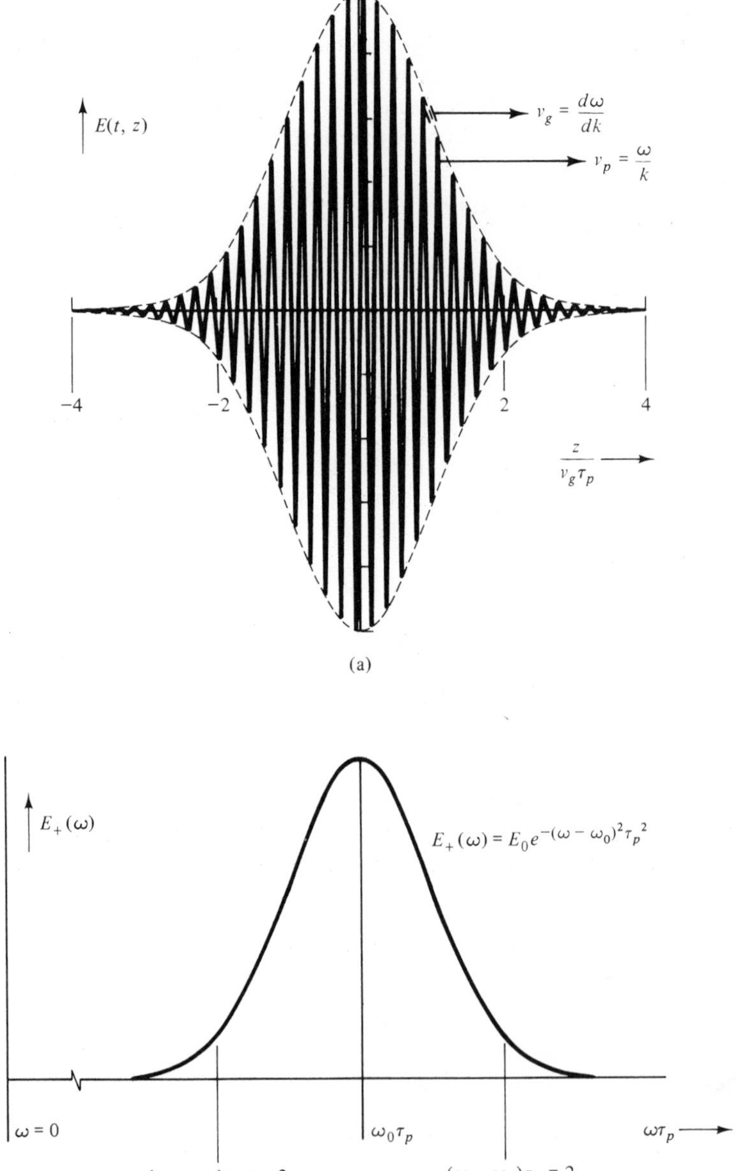

Figure 1.8 Propagation of carrier and envelope: (a) space–time domain; (b) frequency spectrum.

quencies. Equation (1.76) shows that the z dependence of $\boldsymbol{E}(t, z)$ consists of two factors:

1. A rapidly varying term, the carrier, that propagates with the *phase velocity* $\omega_0/k(\omega_0)$
2. A slowly varying envelope that proceeds with the *group velocity* $d\omega/dk = 1/(dk/d\omega)$ (see Fig. 1.8)

For a nondispersive medium, the two velocities are the same, but not for a medium with an ε that depends on frequency, a dispersive medium.

It should be noted that the foregoing derivation of the group velocity is not limited to a plane wave, but applies to any linear process with a spatial dependence $\exp\left[-jk(\omega)z\right]$ at frequency ω. We shall have ample opportunity to use this fact in the discussion of optical fibers.

In the analysis of diffraction we shall need two-dimensional spatial Fourier transforms. Consider a function $f(x, y)$ of the spatial coordinates x and y. The two-dimensional Fourier representation of the function $f(x, y)$ is

$$f(x, y) = \int_{-\infty}^{\infty} dk_x \int_{-\infty}^{\infty} dk_y \; F(k_x, k_y) e^{-j(k_x x + k_y y)} \tag{1.77}$$

where k_x and k_y are the *spatial frequencies.* The Fourier transform relation inverse to (1.77) is

$$F(k_x, k_y) = \left(\frac{1}{2\pi}\right)^2 \int_{-\infty}^{\infty} dx \int_{-\infty}^{\infty} dy \, f(x, y) e^{j(k_x x + k_y y)} \tag{1.78}$$

The convolution of two spatial function $f(x, y)$ and $g(x, y)$ is

$$g \otimes f \equiv \int_{-\infty}^{\infty} dx' \int_{-\infty}^{\infty} dy' \; g(x - x', y - y') f(x', y') \tag{1.79}$$

The Fourier transform of the convolution gives

$$\left(\frac{1}{2\pi}\right)^2 \int_{-\infty}^{\infty} dx \int_{-\infty}^{\infty} dy \; g \otimes f e^{j(k_x x + k_y y)} = (2\pi)^2 G(k_x, k_y) F(k_x, k_y) \tag{1.80}$$

This is the generalization of the convolution theorem to two dimensions.

1.5 POWER SPECTRA AND AUTOCORRELATION FUNCTIONS

Returning to the one-dimensional Fourier transform of a time function, (1.69), we note that its validity is constrained by the requirement that $f(t)$ be square integrable, the integral

$$\int_{-\infty}^{\infty} |f(t)|^2 \, dt$$

must be bounded. This is true for the field amplitudes of pulses of finite energy. Steady-state optical fields are not periodic and, if imagined to have been turned on at $t = -\infty$, not square integrable.* Further, $f(t)$ may be statistical. To treat such a case one uses the following stratagem. One treats the process as if it were periodic, of period T, where T is chosen very large. All quantities of interest are then defined in the limit $T \to \infty$.

Thus suppose that $f(t)$ is a sample function of a steady-state statistical process extending over the time interval $-T/2 \le t \le T/2$. We determine F(n) according to (1.67), treating $f(t)$ as if it were periodic outside the time interval. We define the quantity

$$\lim_{T\to\infty}\left[T\,\frac{\overline{|F(n)|^2}}{2\pi}\right] \equiv \Phi(\omega) \tag{1.81}$$

where ω is identified with $n\omega_0$. The overbar in (1.81) indicates a statistical average over the sample functions obtained from the system under consideration over a sequence of very long (ideally, infinitely long) time intervals. Alternatively, the sample functions may be thought to be obtained from a collection (*ensemble*) of identically prepared systems. The two ways of obtaining sample functions are equivalent if the statistical process is *stationary* and *ergodic.* [7] The function $\Phi(\omega)$ is the *spectral density.* The inverse Fourier transform of the spectral density is equal to the autocorrelation function

$$\int_{-\infty}^{\infty} e^{j\omega\tau}\Phi(\omega)\,d\omega = \overline{f(t)f(t-\tau)} \tag{1.82}$$

where $\overline{f(t)f(t-\tau)}$, the *autocorrelation function,* is the statistical average of the product of $f(t)$ and its replica delayed in time by τ. For a stationary process, the autocorrelation function is independent of t. Relation (1.82) can be proven by introducing (1.81), by replacing the integration over ω by a summation, and using (1.67) to express $F(n)$. With ω_0 of (1.66) and (1.67) replaced by $\Delta\omega = 2\pi/T$ to emphasize its differential character in the limit $T \to \infty$, and with $\omega = n\,\Delta\omega$, we write

$$\begin{aligned}\int_{-\infty}^{\infty} e^{j\omega\tau}\Phi(\omega)\,d\omega &= \lim_{T\to\infty}\sum_n T\,\frac{\overline{|F(n)|^2}}{2\pi}\,\Delta\omega e^{jn\Delta\omega\tau}\\ &= \lim_{T\to\infty}\frac{1}{T}\int_{-T/2}^{T/2} dt\; e^{-jn\Delta\omega t}\,\overline{f(t)F^*(n)}\;e^{jn\Delta\omega\tau}\\ &= \lim_{T\to\infty}\frac{1}{T}\int_{-T/2}^{T/2} dt\;\overline{f(t)f(t-\tau)} = \overline{f(t)f(t-\tau)}\end{aligned} \tag{1.83}$$

because the statistical average is time independent for a stationary process.

*The remainder of this section can be skipped until used in the discussion of coherence, Sections 3.7 and 4.4, and noise in Chapter 14. If these topics are not studied, it can be omitted.

The spectral density, or the autocorrelation function, are the quantities appropriate for the description of a statistical process, as for example an optical field. The autocorrelation function evaluated at $\tau = 0$ gives the mean-square value of $f(t)$.

The definition (1.81) provides a simple recipe for the analysis of a statistical process. One analyzes the system, using Fourier amplitude coefficients $F(n)$ as if they pertained to a sinusoidal, or periodic, process. When the analysis is completed, one takes the absolute square, normalizes by $2\pi/T$, and presumes the average to have been taken. Thus, consider for example a linear system excited by the Fourier component $F(n)$. Suppose that the response is

$$G(n) = H(\omega)F(n)$$

where $\omega = n\,\Delta\omega = 2\pi n/T$. Then the output spectrum $\Phi_g(\omega)$ is related to the input spectrum $\Phi_f(\omega)$ by

$$\Phi_g(\omega) = \lim_{T\to\infty}\left[T\,\frac{\overline{|G(n)|^2}}{2\pi}\right] = \lim_{T\to\infty}\left[|H(\omega)|^2 T\,\frac{\overline{|F(n)|^2}}{2\pi}\right] = |H(\omega)|^2\Phi_f(\omega)$$

In practice, the instruments performing a measurement will take the statistical average automatically.

For optical field amplitudes with spectra within optical frequency bands, it is convenient to introduce inverse Fourier transforms in terms of positive frequencies only. Thus, instead of (1.66) one may define a complex time function for a periodic process by

$$\mathrm{f}(t) = \sqrt{2}\sum_{n=0}^{\infty} F(n)e^{jn\omega_0 t} \tag{1.84}$$

The real function $f(t)$ is related to $\mathrm{f}(t)$ by

$$f(t) = \frac{1}{\sqrt{2}}\,[\mathrm{f}(t) + \mathrm{f}^*(t)] \tag{1.85}$$

The factor $\sqrt{2}$ is used so that time averages of $f^2(t)$ over a few cycles can be evaluated from the absolute value squared of the complex function, $\mathrm{f}(t)\mathrm{f}^*(t)$, with no additional factors. Correspondingly, instead of the full spectrum $F(n)$, $-\infty \le n \le +\infty$ one uses a spectrum for positive frequencies only. If $\mathrm{F}(n)$ is set equal to $\sqrt{2}\,F(n)$ as well, the time average of $f^2(t)$ can be evaluated from

$$\langle f^2(t)\rangle = \sum_{n>0} |\mathrm{F}(n)|^2 \tag{1.86}$$

We shall interpret all optical spectra as single-sided, positive frequencies only, and use the associated complex time function $\mathrm{f}(t)$. The proof (1.83) adapted to this new definition of spectra gives

$$\int_0^\infty e^{j\omega\tau}\Phi(\omega)\,d\omega = \lim_{T\to\infty}\sum_{n>0} T\,\frac{\overline{|F(n)|^2}}{2\pi}\,\Delta\omega\, e^{jn\Delta\omega\tau}$$

$$= \lim_{T\to\infty}\frac{1}{T}\int_{-T/2}^{T/2} dt\; e^{-jn\Delta\omega t}\,\overline{f(t)F^*(n)}\,e^{jn\Delta\omega\tau}$$

$$= \lim_{T\to\infty}\frac{1}{T}\int_{-T/2}^{T/2} dt\;\overline{f(t)f^*(t-\tau)} = \overline{f(t)f^*(t-\tau)} \tag{1.87}$$

The single-sided Fourier transform gives the generalized complex autocorrelation function of the complex time function f(t). We shall use a special symbol for the complex autocorrelation function:

$$\Gamma_{ff}(\tau) \equiv \overline{f(t)f^*(t-\tau)} \tag{1.88}$$

1.6 SUMMARY

The material in the first chapter is mainly intended as a review of concepts previously encountered by the reader and used later in the text. Polarization density $\boldsymbol{P}$, and magnetization density $\boldsymbol{M}$, were introduced and the linear constitutive laws were stated. Complex notation was defined, and the meaning of a complex vector, described in terms of six real scalars, was elaborated. One way of understanding why six real scalars are needed to describe a polarization ellipse is by the following argument: The polarization ellipse requires five numbers for its specification—two for the orientation of the plane of the ellipse, one for the orientation of the major axis of the ellipse in the plane and two for the magnitudes of the major and minor axis. The sixth scalar is required for the time phase. The complex vector is a convenient means for summarizing these data. The real and imaginary parts of the complex vector **A** give the conjugate axes of the ellipse and the ellipse is traversed starting at Re [**A**].

We reviewed the real and complex Poynting theorem. The former is the law of power conservation in a nondispersive medium; the latter shows that the dielectric constant and magnetic permeability of a lossless isotropic medium must be real. The complex Poynting theorem also shows that in a nondispersive medium the time-averaged electric and magnetic energy densities in a traveling wave must be equal to each other, and the time-averaged electric and magnetic energies at resonance are equal. In a dispersive medium $\frac{1}{4}\varepsilon|\mathbf{E}|^2$ and $\frac{1}{4}\mu|\mathbf{H}|^2$ cannot be identified with energy densities, as we shall see in Section 11.5.

We reviewed Fourier transform theory and defined the Fourier series and Fourier integral transforms. In the literature there is no uniform convention for the placement of the factor 2π in the Fourier integral. We have chosen it so that the inverse Fourier transform (1.69) is free of such a factor when written in terms of the angular frequency. We reviewed the convolution theorem and introduced the two-dimensional Fourier transforms. The group velocity concept was introduced in terms of the speed of propagation of the envelope of a wave packet produced by Fourier superposition of amplitudes in a narrow band of frequencies.

We continued with a discussion of power spectra and autocorrelation functions; we showed that they were the Fourier transforms of each other. The power spectrum was interpreted as the limit of a statistical average of the squares of the Fourier amplitudes of a periodic process, with period $T \to \infty$. This interpretation leads to a simple procedure for the treatment of statistical processes in linear systems: They are treated as if they were deterministic, the transformation of the Fourier amplitudes by the linear system is determined, and the result is squared and averaged.

We concluded with the definition of a single-sided spectrum and the associated complex time function. This concept is particularly useful in the analysis of optical signals which possess spectra that do not extend to zero frequency.

PROBLEMS

1.1. Determine the energy densities w_e and w_m, and the Poynting vector $\boldsymbol{E} \times \boldsymbol{H}$ for a plane wave $\hat{x}E_0 \cos(\omega t - kz)$ propagating in free space. Check that Poynting's theorem (1.20) is satisfied.

1.2. Construct the complex vector expression for the electric field of a right-circularly polarized plane wave at frequency ω propagating in free space in the $+z$ direction with peak amplitude E_0 occurring at $z = 0$, $t = 0$ along the x direction. Determine the complex magnetic field and the complex Poynting vector.

1.3. The Shiva laser fusion system at the Lawrence Livermore Laboratory is capable of delivering 10^{13} W of optical power at 1.06 μm wavelength to a pellet of 100 μm diameter.

(a) Find the rms E field in V/m under the assumption of uniform illumination (uniform Poynting vector flux) over the pellet surface (assume linear polarization).

(b) Compare the answer to part (a) to the field produced by a charge of the proton, 1.6×10^{-19} C, at a distance of a Bohr radius 0.5 Å or 0.5×10^{-8} cm.

1.4. A dielectric with one single resonance may be modeled as a distribution of + and − charges, the + charges immobile and the − charges tied to the + charges by a spring constant k. The resonance frequency is then $\omega_0 = \sqrt{k/m}$ and the equation of motion of the − charges is

$$m\left(\frac{d^2}{dt^2} + \sigma \frac{d}{dt} + \omega_0^2\right)\boldsymbol{d} = -q\boldsymbol{E}$$

where m is the mass, σ a phenomenological damping constant, $\boldsymbol{d}$ the displacement of $-$ charges from $+$ charges, and q the magnitude of the charge. Find the polarizability χ_e defined by

$$\mathbf{P} \equiv \varepsilon_0 \chi_e \mathbf{E}$$

where $\mathbf{P} = -qN\mathbf{d}$ with N the number density.

1.5. Find the dispersion relation for transverse electric waves in the medium of Problem 1.4. Sketch the k versus ω curve. Determine the phase velocity $v_p = \omega/k$ and the group velocity $d\omega/dk$ (set the loss coefficient $\sigma = 0$).

1.6. Find the time function $f(t)$ that possesses the Fourier transform

$$F(\omega) = \frac{A}{\omega^2\tau^2 + 1}$$

where A and τ are constants.

1.7. Find the inverse Fourier transform of

$$F(t) = \exp\left[-\frac{(\omega - \omega_0)^2}{2\omega_p^2}\right] + \exp\left[-\frac{(\omega + \omega_0)^2}{2\omega_p^2}\right]$$

where ω_0 and ω_p are constants.

1.8. Determine the convolution of the square pulse $g(t)$, where $g(t) = 1/\tau_p$, $|t| \le \tau_p/2$, $g(t) = 0$ for $|t| > \tau_p/2$, with the time function $f(t) = \cos \omega t$. What is the convolution in the limit $\tau_p \to 0$?

1.9. Consider the time functions

$$f(t) \equiv \cos(\omega_m t)\cos(\omega_0 t) \quad \text{and} \quad f(t) \equiv \frac{1}{1 + \omega_0^2 t^2}$$

For these two functions, determine the complex time functions $\mathrm{f}(t)$ associated with the single-sided spectra (assume that $\omega_m < \omega_0$).

REFERENCES

[1] P. Lorrain and D. Corson, *Electromagnetic Fields and Waves*, W. H. Freeman, San Francisco, 1970.

[2] J. A. Stratton, *Electromagnetic Theory*, McGraw-Hill, New York, 1941.

[3] R. M. Fano, L. J. Chu, and R. B. Adler, *Electromagnetic Fields, Energy, and Forces*, Wiley, New York, 1960.

[4] W. K. H. Panofsky and M. Phillips, *Classical Electricity and Magnetism*, Addison-Wesley, Reading, Mass., 1955.

[5] S. Ramo, J. R. Whinnery, and T. Van Duzer, *Fields and Waves in Communication Electronics*, Wiley, New York, 1965.

[6] R. B. Adler, L. J. Chu, and R. M. Fano, *Electromagnetic Energy Transmission and Radiation*, Wiley, New York, 1960.

[7] J. S. Bendat, *Principles and Applications of Random Noise Theory*, Wiley, New York, 1958.

2

REFLECTION OF PLANE WAVES FROM INTERFACES

Propagation of waves at optical frequencies and their reflection and transmission at interfaces can be treated to a good approximation as the propagation of plane waves because, more often than not, the optical beams have diameters ($2w$) that are large compared with the free-space wavelength (λ). In this limit, diffraction effects can be ignored over propagation distances much less than $\pi w^2/\lambda$ (see Chapters 4 and 5).

In this chapter we determine the reflection coefficient of a plane wave incident upon a plane of discontinuity between two media. In the process we find Snell's law as a consequence of continuity of fields at the interface. The reflection coefficient is a function of the angle of incidence and the polarization of the incident plane wave. We consider the case of total internal reflection at the interface of an optically "dense" medium to one that is less "dense." We derive the "Goos–Hänchen shift," which plays a role in optical waveguides that use the principle of total internal reflection for guidance of optical waves. We introduce the concept of wave impedance which we can employ in the analysis of layered media. Multiple layers can be used to enhance the reflection of a wave from an interface as employed in the fabrication of dielectric mirrors; they can also be chosen so as to prevent reflection (antireflection coating). In the final section we take up the study of reflection from a corrugated perfectly reflecting surface (i.e., grating) and consider the use of gratings in optical spectrometers.

2.1 TRANSVERSE ELECTRIC WAVE REFLECTED FROM BOUNDARY [1, 2]

In Section 1.3 we studied the complex vector formulation for a plane wave, a solution of Maxwell's equations. Consider a plane wave with its electric field polarized parallel to the surface of an interface between two media as shown in Fig. 2.1 (transverse electric, or TE, wave).

$$\mathbf{E}_{\text{incident}} = \hat{y}\mathrm{E}_{+}e^{-j\boldsymbol{k}^{(1)}\cdot\boldsymbol{r}}$$

In general, part of the incident power will be transmitted, part of it will be reflected. The reflection and transmission are determined by satisfying the boundary conditions at the interface. The tangential $\boldsymbol{E}$ and $\boldsymbol{H}$ must be continuous at $z = 0$. But this implies that the x dependences of the reflected and transmitted waves have to be identical with the x dependence of the incident wave (see Fig. 2.1).

$$k_x^{(1)} = k_x^{(2)} = k_x \tag{2.1}$$

Snell's law is the consequence of (2.1):

$$\sqrt{\mu_1\varepsilon_1}\,\sin\theta_1 = \sqrt{\mu_2\varepsilon_2}\,\sin\theta_2 \tag{2.2}$$

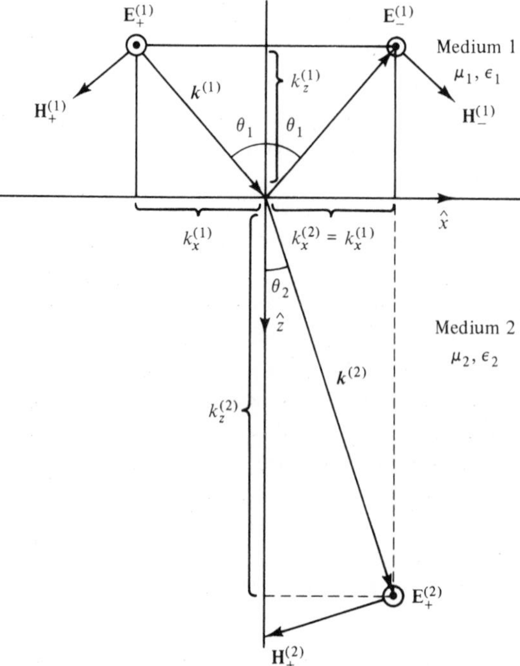

Figure 2.1 TE reflection and transmission at an interface.

The electric field on side 1, $z < 0$, is the superposition of the fields of the incident and reflected waves.

$$E_y = [E_+^{(1)} e^{-jk_z^{(1)}z} + E_-^{(1)} e^{+jk_z^{(1)}z}] e^{-jk_x x} \tag{2.3}$$

The propagation vector of the reflected wave has the same x component as the propagation vector of the incident wave. The x component of the magnetic field is obtained from Faraday's law (1.41):

$$\nabla \times \mathbf{E} = -j\omega\mu\mathbf{H} \tag{2.4}$$

Introducing (2.3) into (2.4), one finds for H_x in the region $z < 0$:

$$H_x = -\frac{k_z^{(1)}}{\omega\mu_1} [E_+^{(1)} e^{-jk_z^{(1)}z} - E_-^{(1)} e^{+jk_z^{(1)}z}] e^{-jk_x x} \tag{2.5}$$

where

$$\frac{k_z^{(1)}}{\omega\mu_1} = \sqrt{\frac{\varepsilon_1}{\mu_1}} \cos\theta_1 \equiv Y_0^{(1)} \tag{2.6}$$

is the *characteristic admittance* $Y_0^{(1)}$ presented by medium 1 to a TE wave at inclination θ_1 with respect to the z direction. Its inverse is the *characteristic impedance* $Z_0^{(1)}$. The transmitted wave is

$$E_y = E_+^{(2)} e^{-jk_z^{(2)}z} e^{-jk_x x} \tag{2.7}$$

with the x component of the $\boldsymbol{H}$ field

$$H_x = -\frac{k_z^{(2)}}{\omega\mu_2} E_+^{(2)} e^{-jk_z^{(2)}z} e^{-jk_x x} \tag{2.8}$$

with the characteristic admittance in medium 2

$$\frac{k_z^{(2)}}{\omega\mu_2} = \sqrt{\frac{\varepsilon_2}{\mu_2}} \cos\theta_2 \equiv Y_0^{(2)}$$

experienced by a TE wave at inclination θ_2 with respect to the z direction. Its inverse is $Z_0^{(2)}$. Continuity of the tangential components of $\boldsymbol{E}$ and $\boldsymbol{H}$ requires that the ratio

$$Z \equiv -\frac{E_y}{H_x} \tag{2.9}$$

be continuous. Z is the *wave impedance* at the interface. One has from (2.3), (2.5), (2.7), and (2.8) at $z = 0$,

$$Z_0^{(1)} \frac{E_+^{(1)} + E_-^{(1)}}{E_+^{(1)} - E_-^{(1)}} = Z_0^{(2)} \tag{2.10}$$

The quantity $\Gamma \equiv E_-/E_+$ is the *reflection coefficient*. One finds $\Gamma^{(1)}$ by solving

(2.10) for $E_-^{(1)}/E_+^{(1)}$:

$$\Gamma^{(1)} = \frac{Z_0^{(2)} - Z_0^{(1)}}{Z_0^{(2)} + Z_0^{(1)}} \tag{2.11}$$

The explicit expression for the reflection coefficient, using Snell's law, is

$$\Gamma^{(1)} = \frac{\sqrt{1 - \sin^2 \theta_1} - \sqrt{1 - \sin^2 \theta_1 \dfrac{\varepsilon_1 \mu_1}{\varepsilon_2 \mu_2}} \sqrt{\dfrac{\varepsilon_2 \mu_1}{\varepsilon_1 \mu_2}}}{\sqrt{1 - \sin^2 \theta_1} + \sqrt{1 - \sin^2 \theta_1 \dfrac{\varepsilon_1 \mu_1}{\varepsilon_2 \mu_2}} \sqrt{\dfrac{\varepsilon_2 \mu_1}{\varepsilon_1 \mu_2}}} \tag{2.12}$$

The density of power flow in the z direction is

$$\tfrac{1}{2} \operatorname{Re} [\mathbf{E} \times \mathbf{H}^*] \cdot \hat{z} = -\tfrac{1}{2} \operatorname{Re} [E_y H_x] = \tfrac{1}{2} Y_0^{(1)} |E_+^{(1)}|^2 (1 - |\Gamma^{(1)}|^2) \tag{2.13}$$

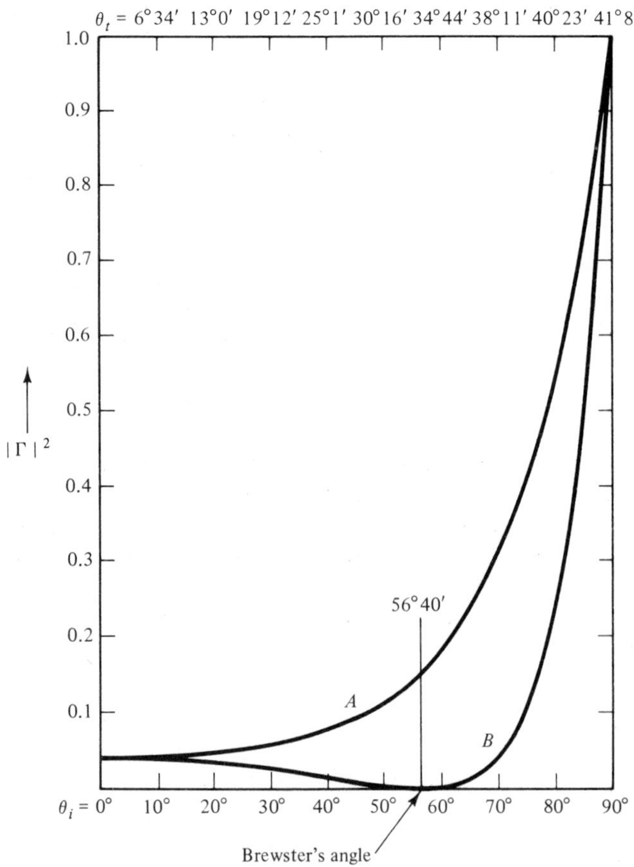

Figure 2.2 Square of reflection coefficient as a function of the angle of incidence: (a) curve A, TE; (b) curve B, TM. $n = 1.52$ index of glass. (After O. D. Chwolson, *Lehrbuch der Physik*, Vol. II, No. 2, 2nd ed., Braunschweig, Vieweg, 1922, p. 716.)

Thus $|\Gamma|^2$ is the ratio of reflected to incident power flow. By continuity of power at the interface, the transmitted power density is obviously equal to (2.13). Figure 2.2 shows $|\Gamma|^2$ as a function of the angle of incidence on a dielectric medium of index $n = \sqrt{\varepsilon/\varepsilon_0} = 1.52$; the index is related to the dielectric constant by $\varepsilon = \varepsilon_0 n^2$. The two nonmagnetic media have $\mu_1 = \mu_2 = \mu_0$.

2.2 TRANSVERSE MAGNETIC WAVE REFLECTED FROM BOUNDARY [1, 2]

Consider a plane wave whose magnetic field is parallel to a boundary between two media (transverse magnetic, or TM, wave) (see Fig. 2.3). As before, there is a transmitted and a reflected wave. At the interface the x dependences are continuous, again leading to Snell's law (2.2). To describe the reflection and transmission, it is more convenient to write the waves in terms of the amplitudes H_+ and H_- of the incident and reflected **H**:

$$H_y = (H_+^{(1)} e^{-jk_z^{(1)}z} + H_-^{(1)} e^{+jk_z^{(1)}z}) e^{-jk_x x} \tag{2.14}$$

The **E** field is obtained from Ampère's law [(1.42) with $\mathbf{J} = 0$]:

$$\nabla \times \mathbf{H} = j\omega\varepsilon\mathbf{E} \tag{2.15}$$

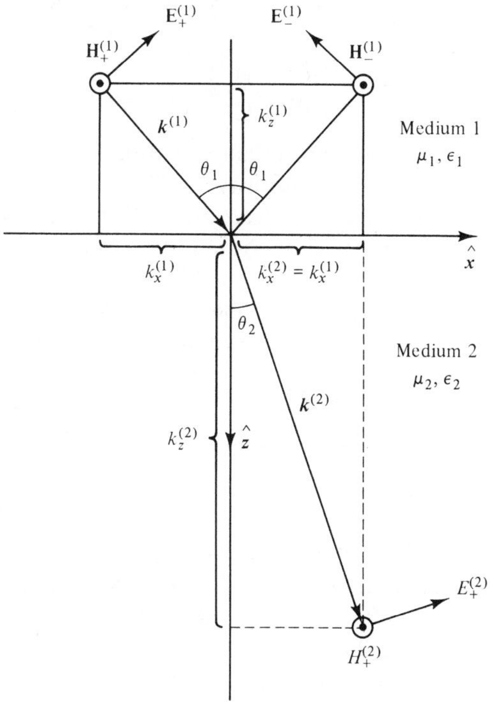

Figure 2.3 TM reflection and transmission at an interface.

Introduction of (2.14) in (2.15) gives for the x component of **E**:

$$\mathrm{E}_x = \frac{k_z^{(1)}}{\omega\varepsilon_1}(\mathrm{H}_+^{(1)}e^{-jk_z^{(1)}z} - \mathrm{H}_-^{(1)}e^{+jk_z^{(1)}z})e^{-jk_x x} \tag{2.16}$$

The transmitted wave is

$$\mathrm{H}_y = \mathrm{H}_+^{(2)}e^{-jk_z^{(2)}z}e^{-jk_x x}$$

$$\mathrm{E}_x = \frac{k_z^{(2)}}{\omega\varepsilon_2}\mathrm{H}_+^{(2)}e^{-jk_z^{(2)}z}e^{-jk_x x}$$

Continuity of the tangential electric fields requires, as before, continuity of the wave impedances $\mathrm{E}_x/\mathrm{H}_y$ at $z = 0$:

$$\frac{k_z^{(1)}}{\omega\varepsilon_1}\frac{\mathrm{H}_+^{(1)} - \mathrm{H}_-^{(1)}}{\mathrm{H}_+^{(1)} + \mathrm{H}_-^{(1)}} = \frac{k_z^{(2)}}{\omega\varepsilon_2} \tag{2.17}$$

The characteristic admittances of the traveling TM waves are

$$Y_0 = \frac{\omega\varepsilon}{k_z} = \sqrt{\frac{\varepsilon}{\mu}}\frac{1}{\cos\theta} \tag{2.18}$$

Note that the cosine of the angle of incidence is now in the denominator, whereas for TE waves it was in the numerator. With the definition of the characteristic admittances, or their inverses, the impedances, one may rewrite (2.17)

$$Z_0^{(1)}\frac{\mathrm{H}_+^{(1)} - \mathrm{H}_-^{(1)}}{\mathrm{H}_+^{(1)} + \mathrm{H}_-^{(1)}} = Z_0^{(2)} \tag{2.19}$$

Comparison with (2.10) shows that the same relation for a reflection coefficient Γ results for a TM wave, as for a TE wave, if Γ is defined as $-\mathrm{H}_-/\mathrm{H}_+$. The reason for this is that, for a given direction of **E**, the magnetic field of the reflected wave points in a direction opposite to that of the magnetic field of the incident wave.

With this definition of Γ, (2.11) applies to a TM wave as well. Use of the expressions for the characteristic impedances and Snell's law gives

$$\Gamma^{(1)} = -\frac{\sqrt{1-\sin^2\theta_1} - \sqrt{1-\sin^2\theta_1\dfrac{\mu_1\varepsilon_1}{\mu_2\varepsilon_2}}\sqrt{\dfrac{\varepsilon_1\mu_2}{\varepsilon_2\mu_1}}}{\sqrt{1-\sin^2\theta_1} + \sqrt{1-\sin^2\theta_1\dfrac{\mu_1\varepsilon_1}{\mu_2\varepsilon_2}}\sqrt{\dfrac{\varepsilon_1\mu_2}{\varepsilon_2\mu_1}}} \tag{2.20}$$

TM waves can be transmitted reflection-free at a dielectric interface. This follows from the fact that Γ of (2.20) vanishes for a real angle when $\mu_1 = \mu_2 = \mu_0$. Indeed, the angle $\theta_1 = \theta_B$ for which this happens, the so-called *Brewster angle*, is from, (2.20) (see Fig. 2.2),

$$\theta_B = \tan^{-1}\sqrt{\frac{\varepsilon_2}{\varepsilon_1}} \tag{2.21}$$

2.3 TOTAL INTERNAL REFLECTION [1, 2]

If medium 1 has a larger value of $\sqrt{\mu\varepsilon}$ (i.e., it is "optically denser") than medium 2, Snell's law (2.2) fails to yield a real angle θ_2 for a certain range of angles of incidence θ_1. At optical frequencies, $\mu \simeq \mu_0$ and only ε can assume values appreciably different from ε_0; $\varepsilon = \varepsilon_0 n^2$, where n is the (refractive) *index*. An optically denser medium has a larger index n. From (2.2), for $\mu_1 = \mu_2 = \mu_0$,

$$\sin\theta_2 = \frac{\sqrt{\varepsilon_1}}{\sqrt{\varepsilon_2}}\sin\theta_1 \tag{2.22}$$

The *critical angle* θ_c is that value of θ_1 beyond which $\sin\theta_2 > 1$, and thus no real solutions exist for θ_2 (see Fig. 2.4)

$$\sin\theta_c = \sqrt{\frac{\varepsilon_2}{\varepsilon_1}} = \frac{n_2}{n_1} \tag{2.23}$$

When no real solutions of θ_2 are found, the assumption that a propagating transmitted wave exists must be reexamined. The continuity of the x dependences,

$$k_x^{(1)} = k_x^{(2)} \tag{2.24}$$

leads to $k_x^{(2)}$ values that cannot be accommodated by a "propagating" wave when $\theta_1 > \theta_c$. The propagation constant must be allowed to become negative imaginary, the field decays into medium 2, $k_z^{(2)} = -j\alpha_z^{(2)}$. In this case

$$[k_x^{(2)}]^2 + [k_z^{(2)}]^2 = [k_x^{(2)}]^2 - [\alpha_z^{(2)}]^2 = \omega^2\mu_0\varepsilon_2$$

and

$$k_x^{(2)} = \sqrt{\omega^2\mu_0\varepsilon_2 + [\alpha_z^{(2)}]^2} \tag{2.25}$$

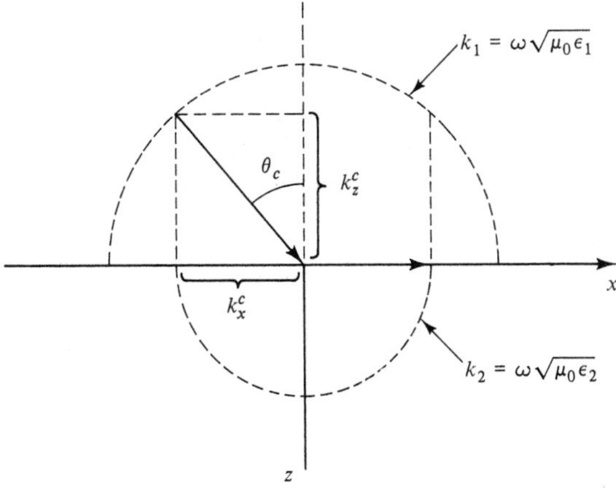

Figure 2.4 Construction of ***k*** vectors for critical angle of incidence.

Thus $k_x^{(2)}$ can be made larger than $\omega\sqrt{\mu_0 \varepsilon_2}$ and the boundary condition (2.24) can be met. In the case of a TE wave, the incident and reflected wave E fields on side 1 are given as before by (2.3) and the x component of $\mathbf{H}$ by (2.5). The transmitted electric field is now

$$\mathrm{E}_y = \mathrm{E}_+^{(2)} e^{-\alpha_z^{(2)} z} e^{-jk_x x} \tag{2.26}$$

and its associated H_x, using Faraday's law (2.4) is

$$-\frac{j}{\omega\mu_0}\frac{\partial \mathrm{E}_y}{\partial z} = \mathrm{H}_x = \frac{j\alpha_z^{(2)}}{\omega\mu_0}\mathrm{E}_+^{(2)} e^{-\alpha_z^{(2)} z} e^{-jk_x x} \tag{2.27}$$

We find by matching the wave impedance $-\mathrm{E}_y/\mathrm{H}_x$ on the two sides of the boundary that

$$\frac{\omega\mu_0}{k_z^{(1)}}\frac{\mathrm{E}_+^{(1)} + \mathrm{E}_-^{(1)}}{\mathrm{E}_+^{(1)} - \mathrm{E}_-^{(1)}} = \frac{j\omega\mu_0}{\alpha_z^{(2)}} = Z_0^{(2)} \tag{2.28}$$

This equation is of the form of (2.10), and the characteristic impedance of medium 2 is now *imaginary*; $Z_0^{(2)} = jX_0^{(2)}$, with $X_0^{(2)}$ real. Solving for $\Gamma \equiv E_-^{(1)}/E_+^{(1)}$, we obtain

$$\Gamma^{(1)} = \frac{\mathrm{E}_-^{(1)}}{\mathrm{E}_+^{(1)}} = \frac{jX_0^{(2)} - Z_0^{(1)}}{jX_0^{(2)} + Z_0^{(1)}} \tag{2.29}$$

The above shows that $|\Gamma^{(1)}| = 1$, and the magnitude of the reflected wave amplitude $\mathrm{E}_-^{(1)}$ is equal to the magnitude of the incident wave amplitude; a standing wave in the z direction is set up in medium 1 with no net power flow

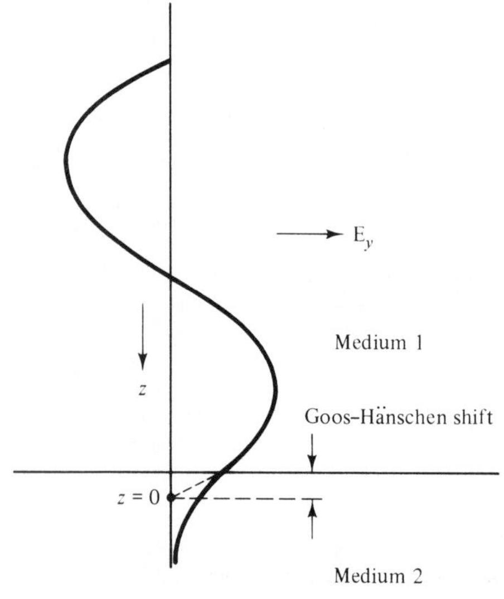

Figure 2.5 The E field for critical reflection of a TE wave. $\mathrm{E}_y(z = 0)$ is taken as real. Then it is real for all z in this cross section.

in this direction. The electric field in medium 1 is of the form

$$\begin{aligned} E_y &= E_+^{(1)}[e^{-jk_z^{(1)}z} + \Gamma^{(1)}e^{+jk_z^{(1)}z}]e^{-jk_x x} \\ &= e^{-j\phi}2E_+^{(1)}\cos(k_z^{(1)}z - \phi)e^{-jk_x x} \end{aligned} \tag{2.30}$$

where $\phi = -\frac{1}{2}\arg(\Gamma^{(1)})$.

Figure 2.5 shows E_y at $x = 0$, as a function of z, assuming that the phase of $E_+^{(2)}$ is equal to 0. Note that $\partial E_y/\partial z$ is continuous at the boundary because $\partial E_y/\partial z$ is proportional to H_x, which must be continuous.

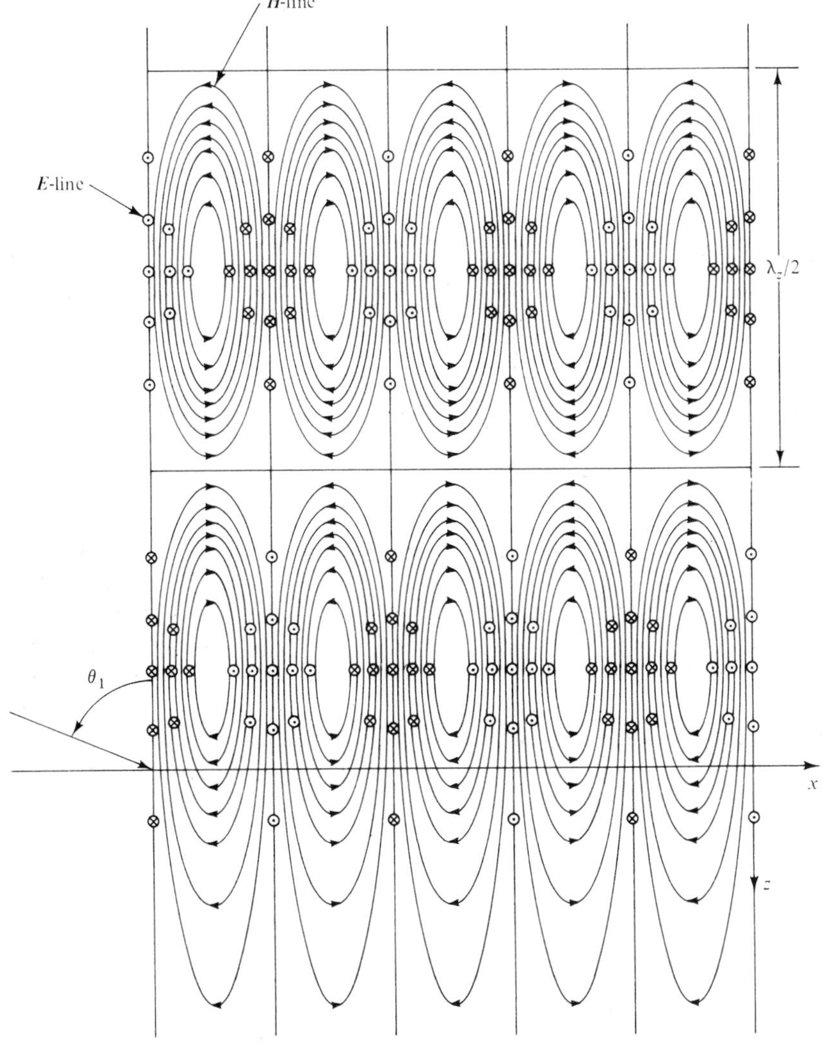

Figure 2.6 TE wave reflected from interface; $\varepsilon_1/\varepsilon_2 = 1.2$. The pattern travels to the right with the phase velocity ω/k_x.

Figure 2.5 shows the null of the standing wave of E extended into the region of medium 2. The null occurs beyond the interface and its distance from the interface is called the Goos–Hänchen shift. [3, 4] Figure 2.6 presents a plot of the electric and magnetic fields of a TE wave at one instant of time. The construction of these graphs is explained in Chapter 6.

The reader is encouraged to derive expressions for the tangential $\boldsymbol{E}$ and $\boldsymbol{H}$ fields at total internal reflection for an assumed TM wave of incidence. Figure 2.7 shows the TM fields at one instant of time.

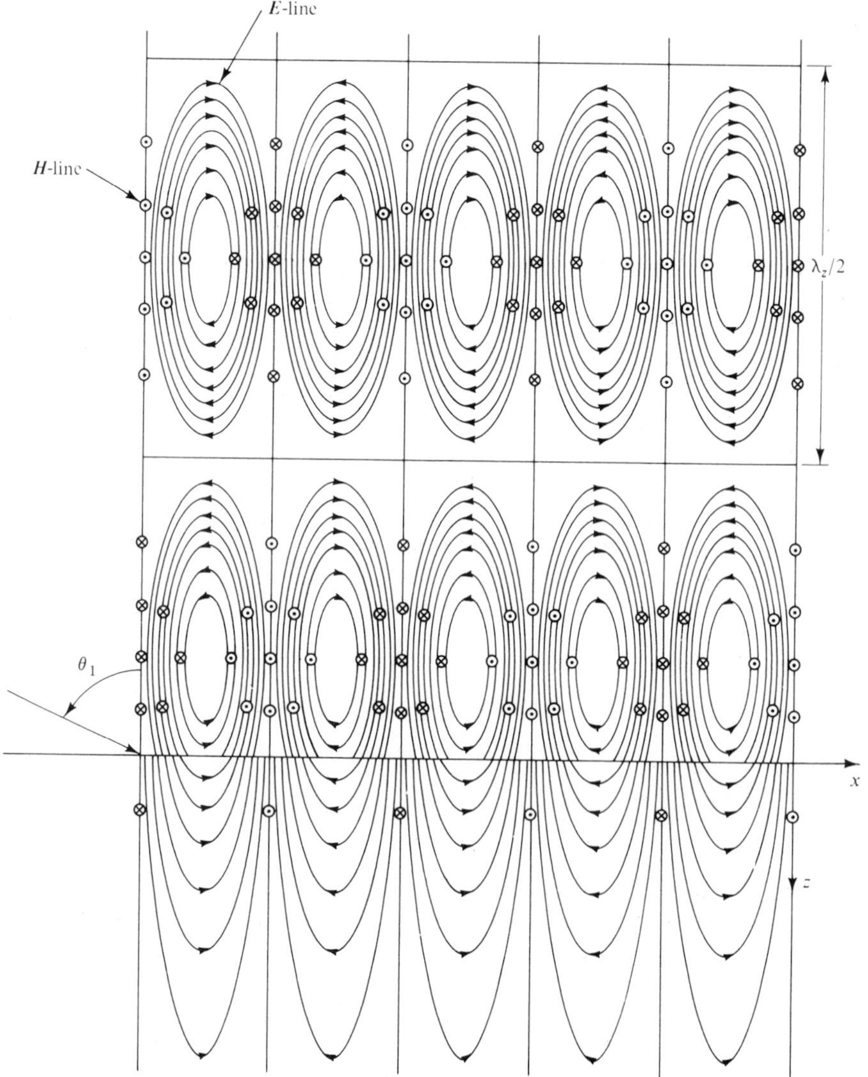

Figure 2.7 TM wave reflected from interface; $\varepsilon_1/\varepsilon_2 = 1.2$. The pattern travels to the right with the phase velocity ω/k_x.

Total internal reflection can be used to confine electromagnetic radiation to the interior of optical slab waveguides or tapered index optical fibers as described in Chapter 6.

2.4 TRANSFORMATION OF WAVE IMPEDANCE AND REFLECTION COEFFICIENT

In the preceding sections it was convenient to use the concept of wave impedance at an interface. The concept is useful in a more general sense; it can be associated with a wave at an arbitrary cross section. For example, the wave impedance $Z(z) = -E_y/H_x$ of a TE wave in medium 1 at any position z, according to (2.3) and (2.5), is

$$Z(z) = -\frac{E_y}{H_x} = Z_0 \frac{1 + \Gamma e^{2jk_z z}}{1 - \Gamma e^{2jk_z z}} \tag{2.31}$$

We omit a specific reference to medium 1, because in this form the equation applies to a plane wave in any medium (e.g., medium 2). Equation (2.31) can be used to evaluate the wave impedance $Z(z)$ at any cross section z, given the wave impedance $Z(0)$ at $z = 0$. First, apply the equation to $z = 0$:

$$Z(0) = Z_0 \frac{1 + \Gamma}{1 - \Gamma} \tag{2.32}$$

solve for Γ in terms of $Z(0)$, and substitute in (2.31). After some manipulation, one obtains

$$Z(z) = Z_0 \frac{Z(0) - jZ_0 \tan k_z z}{Z_0 - jZ(0) \tan k_z z} \tag{2.33}$$

This is a useful formula. We shall employ it right away to show that the wave impedance $Z(z)$ at $z = -[(2m + 1)/2k_z]\pi$, with m an integer, is related to $Z(0)$ by

$$Z\left(-\frac{2m + 1}{2k_z}\pi\right) = \frac{Z_0^2}{Z(0)} \tag{2.34}$$

This result follows from (2.33) because $\tan k_z z = \infty$ at these positions. The normalized impedance $Z(z)/Z_0$ for $z = -(2m + 1)\pi/2k_z$ is the inverse of its value at $z = 0$. These positions are at a distance from the $z = 0$ plane equal to an odd-integer multiple of quarter-wavelengths with the wavelength defined by (see Figs. 2.6 and 2.7)

$$\lambda_z = \frac{2\pi}{k_z} \tag{2.35}$$

The wavelength measures the spacing along the z axis of planes of equal phase:

$$\lambda_z = \frac{\lambda}{\cos\theta} \tag{2.36}$$

where $\lambda = 2\pi/\omega\sqrt{\mu\varepsilon}$ is the wavelength measured along the direction of propagation and θ is the angle of incidence. It is worthwhile to generalize the reflection coefficient to an arbitrary position z. The ratio of the electric fields of

TABLE 2.1. Summary of Key Formulas

TE	TM
$\Gamma(z) = \frac{E_-}{E_+} e^{+j2k_z z}$	$\Gamma(z) = -\frac{H_-}{H_+} e^{+j2k_z z}$
$Z_0 = \sqrt{\frac{\mu}{\varepsilon}} \frac{1}{\cos\theta}$	$Z_0 = \sqrt{\frac{\mu}{\varepsilon}} \cos\theta$
$Z(z) = -\frac{E_y}{H_x}$	$Z(z) = \frac{E_x}{H_y}$

$$\frac{Z(z)}{Z_0} = \frac{1 + \Gamma(z)}{1 - \Gamma(z)}$$

$$\Gamma(z) = \frac{Z(z) - Z_0}{Z(z) + Z_0}$$

$$Z(z) = Z_0 \frac{Z(0) - jZ_0 \tan k_z z}{Z_0 - jZ(0) \tan k_z z}$$

Corresponding formulas can be written in terms of the inverse of the impedance, the admittance

$$Y \equiv \frac{1}{Z}$$

TE	TM
$Y_0 = \sqrt{\frac{\varepsilon}{\mu}} \cos\theta$	$Y_0 = \sqrt{\frac{\varepsilon}{\mu}} \frac{1}{\cos\theta}$
$Y(z) = -\frac{H_x}{E_y}$	$Y(z) = \frac{H_y}{E_x}$

$$\frac{Y(z)}{Y_0} = \frac{1 - \Gamma(z)}{1 + \Gamma(z)}$$

$$\Gamma(z) = \frac{Y_0 - Y(z)}{Y_0 + Y(z)}$$

$$Y(z) = Y_0 \frac{Y(0) - jY_0 \tan k_z z}{Y_0 - jY(0) \tan k_z z}$$

the reflected and incident waves at any position z is equivalently

$$\Gamma e^{2jk_z z} \equiv \Gamma(z) \tag{2.37}$$

With these generalizations we have a pair of relations between the wave impedance and reflection coefficient at any cross section

$$Z(z) = Z_0 \frac{1 + \Gamma(z)}{1 - \Gamma(z)} \tag{2.38}$$

The inverse relation, expressing $\Gamma(z)$ in terms of $Z(z)$, is

$$\Gamma(z) = \frac{Z(z) - Z_0}{Z(z) + Z_0} \tag{2.39}$$

Completely analogous relations may be derived for TM waves (see Table 2.1).

The transformation (2.37) of the reflection coefficient shows that the reflected wave is in phase with the incident wave, $\Gamma(z)$ is real and positive, at positions z spaced by $k_z \, \Delta z = \pi$. These are separated by integer multiples of $\lambda_z/2$. Halfway between are planes at which the incident and reflected waves are in antiphase.

We shall use relation (2.33) in Section 6.4 for the analysis of an optical guiding layer, and relation (2.38) in the next chapter for the characterization of a Fabry-Perot interferometer.

2.5 QUARTER-WAVE DIELECTRIC LAYERS

Reflection of a plane wave from a dielectric interface may be eliminated by coating the interface with a layer of dielectric of different dielectric constant, a quarter wavelength thick at a particular wavelength and angle of incidence. Consider the case illustrated in Fig. 2.8. The substrate is a medium described by μ_0, $\varepsilon = \varepsilon_0 n^2$. A TE wave incident from the top as shown produces a transmitted wave in the substrate. The wave impedance presented by the wave in the substrate is $Z(0) = \sqrt{\mu_0/\varepsilon}/\cos\theta = \sqrt{\mu_0/\varepsilon_0}/(n \cos\theta)$.

This impedance is transformed according to (2.34). The impedance is real if the layer is a quarter wavelength thick as measured in medium 1 at angle θ. Picking the value of d so that

$$d = \frac{\lambda_z^{(1)}}{4} = \frac{\lambda^{(1)}}{4} \frac{1}{\cos\theta_1} \tag{2.40}$$

we have at $z = -d$, according to (2.34),

$$Z(-d) = \frac{Z_{01}^2}{Z(0)} = \frac{n \cos\theta}{\sqrt{\mu_0/\varepsilon_0}} \frac{\mu_0/\varepsilon_0}{n_1^2 \cos^2\theta_1} \tag{2.41}$$

If the substrate is to be matched, $Z(-d)$ must be made equal to the characteristic impedance in the air space above, $Z_0 = \sqrt{\mu_0/\varepsilon_0}(1/\cos\theta_0)$. This is achieved

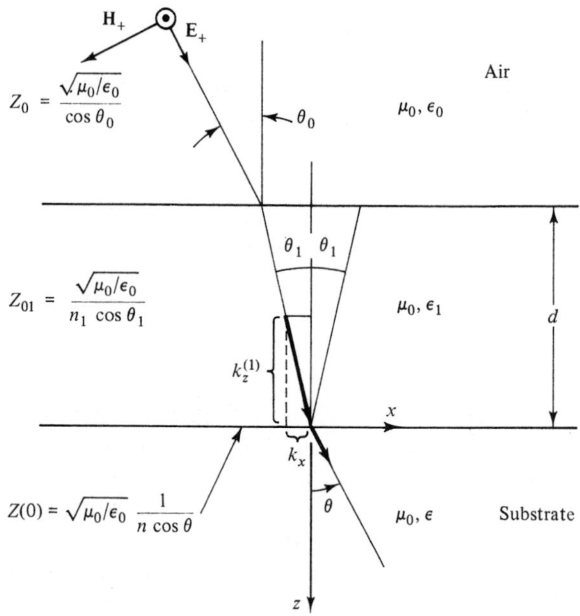

Figure 2.8 Field-impedance transformation by "quarter-wave layer."

when

$$n_1^2 \cos^2 \theta_1 = n \cos \theta \cos \theta_0 \tag{2.42}$$

The angles θ and θ_1 are fixed by Snell's law for a given θ_0. Equation (2.42) gives the index of the quarter-wave layer required to *match* the substrate, eliminate the reflected wave. A coating applied for this purpose is an *anti-reflection coating.* In particular for $\theta_0 = 0$,

$$n_1^2 = n$$

The matching condition is a function of angle. If met for normal incidence, the matching is imperfect for other angles of incidence. Figure 2.9 shows the square of the reflection coefficient $|\Gamma|^2$ for TE and TM reflection from a surface with $\sqrt{\varepsilon_1/\varepsilon_0} = n_1 = 3.65$ as a function of angle of incidence.

It may be difficult to find a dielectric material with an index n_1 that satisfies (2.42) for a given n and, in addition, adheres well to the substrate. For this reason one usually applies several quarter-wave layers. Consider the system of alternating layers of Fig. 2.10. The wave impedance at the interface marked 1 has been evaluated in (2.41). Application of (2.34) to determine the impedance at the interface 2 gives

$$Z(z_2) = \sqrt{\frac{\mu_0}{\varepsilon_0}} \frac{1}{n \cos \theta} \left(\frac{n_1 \cos \theta_1}{n_2 \cos \theta_2} \right)^2 \tag{2.43}$$

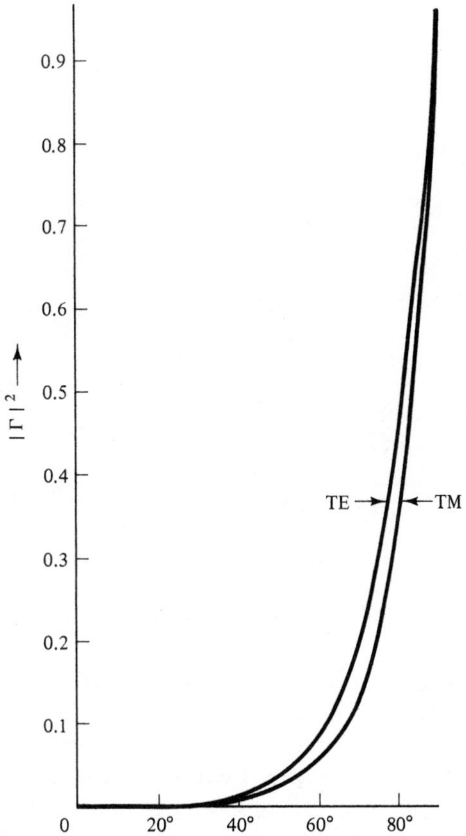

Figure 2.9 Reflection from surface "matched" with quarter wave layer for $\theta = 0$, as function of angle.

A pair of layers transforms the impedance by multiplying it by the factor $n_1^2 \cos^2 \theta_1 / n_2^2 \cos^2 \theta_2$. If m pairs of layers are applied to the substrate, the wave impedance seen at the "input plane" in Fig. 2.10 is

$$Z(z_i) = \sqrt{\frac{\mu_0}{\varepsilon_0}} \frac{1}{n \cos \theta} \left(\frac{n_1 \cos \theta_1}{n_2 \cos \theta_2} \right)^{2m} \tag{2.44}$$

Even for a ratio $n_1 \cos \theta / n_2 \cos \theta_2$ close to unity, the multiplier can be made large by choosing a large m. The multiple layers act as an antireflection coating, if $(n_1 \cos \theta_1 / n_2 \cos \theta_2)$ is chosen so that $Z(z_i)$ is equal to the characteristic impedance of the wave in the "input" medium i:

$$Z(z_i) = Z_0^{(i)} = \sqrt{\frac{\mu_0}{\varepsilon_0}} \frac{1}{n_i \cos \theta_i} \tag{2.45}$$

One may choose the indices n_1 and n_2 so that $Z(z_i)$ is greatly different from $Z_0^{(i)}$. Then one achieves strong reflection. The coating is a *reflection coating*.

$$Z(z_1) = \sqrt{\frac{\mu_0}{\epsilon_0}}\,\frac{n\cos\theta}{n_1^2\cos^2\theta_1}$$

$$Z(z_2) = \sqrt{\frac{\mu_0}{\epsilon_0}}\,\frac{1}{n\cos\theta}\,\frac{n_1^2\cos^2\theta_1}{n_2^2\cos^2\theta_2}$$

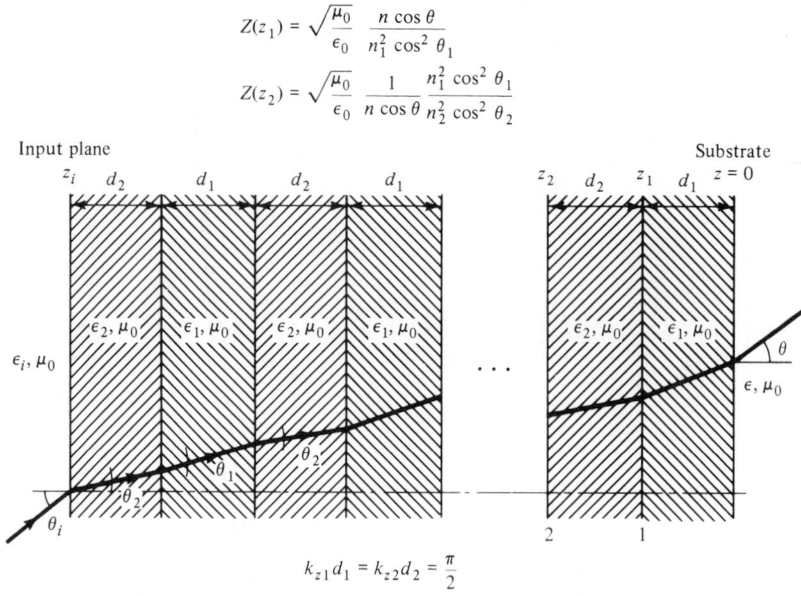

Figure 2.10 Matching with m pairs of layers.

Highly reflecting mirrors of low loss are made by reflection coatings consisting of many dielectric layers, rather than with metallic coatings, because the latter possess, generally, higher losses. Of course, the antireflection or reflection property of a coating is frequency dependent. The layers are a quarter-wavelength thick only at one specific frequency. A change of frequency leads to a change of reflectivity or transmissivity. The analysis of the properties of multiple layers that are not of quarter-wavelength thickness is considerably more complicated and will not be pursued further here. In Chapter 8 we shall use a simplified approximate model by which we shall analyze the frequency dependence of the reflectivity of multiple layers.

2.6 REFLECTION GRATINGS [5]

A reflection grating is formed by a periodically "corrugated" reflecting surface. An incident single-frequency plane wave is separated, upon reflection, into different *orders* (i.e., the reflected plane waves come off at different angles). The angles depend on the frequency of the incident wave. In this way, spectral separation of incident waves of different frequencies is accomplished. We develop a construction for the angles of the reflected waves produced by a reflection grating. The reflected waves are evaluated for the case of a grating with a sinusoidal reflecting surface. Gratings with strong reflection in the -1 order and one of their applications are discussed.

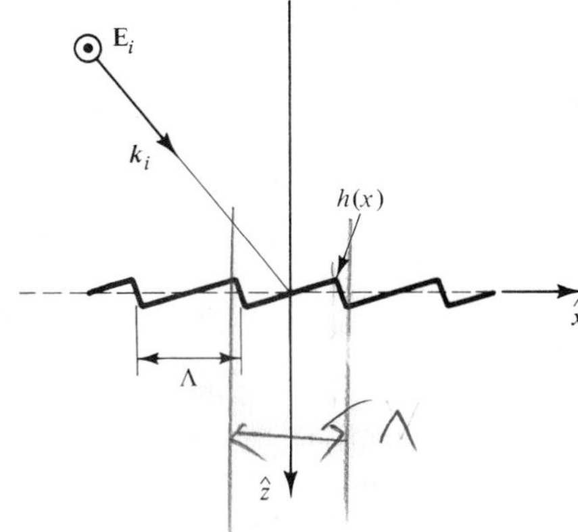

Figure 2.11 Schematic of grating surface.

Figure 2.11 shows the schematic of a grating surface. We idealize it as perfectly reflecting. The spatial period of the grating is Λ. Reflected waves are excited so as to satisfy the boundary condition of zero tangential E field on the surface of the grating. We start with polarization normal to the paper (y-directed). If the incident wave is $\hat{\boldsymbol{y}}\mathrm{E}_i \exp(-j\boldsymbol{k}_i \cdot \boldsymbol{r})$, then the superposition of incident and reflected waves $\hat{\boldsymbol{y}}\mathrm{E}_R$ will have to satisfy the condition

$$\mathrm{E}_i \exp\left[-(jk_{ix}x + jk_{iz}z)\right]\Big|_{z=h(x)} + \mathrm{E}_R(x, z)\Big|_{z=h(x)} = 0 \tag{2.46}$$

The field of the incident wave on the perturbed surface is of the form

$$\exp(-jk_{ix}x)\exp\left[-jk_{iz}h(x)\right]$$

Since $h(x)$ is a periodic function of x with period Λ, $\exp[-jk_{iz}h(x)]$ is also a periodic function with the same period. The reflected field must cancel the incident field at $z = h(x)$. The reflected field can accomplish this if, and only if, it is of the form

$$\mathrm{E}_R(x, z) = \exp(-jk_{ix}x)g(x, z) \tag{2.47}$$

where $g(x, z)$ is a periodic function of x with period Λ. Indeed, $g(x, h(x))$ is a periodic function of x as well, and thus the boundary condition (2.46) can be met.

The reflected wave must be a solution of Maxwell's equations. A function of the form (2.47) can be constructed out of plane waves, solutions of Maxwell's equations, if one sets

$$\mathrm{E}_R = \exp(-jk_{ix}x)\sum_m \mathrm{R}_m \exp(jk_{Rz}^{(m)}z)\exp\left(-j\frac{2\pi m}{\Lambda}x\right) \tag{2.48}$$

where the R_m's are constants and the $k_{Rz}^{(m)}$'s obey the constraint imposed by (1.60).

$$\left(k_{ix} + \frac{2\pi}{\Lambda}m\right)^2 + k_{Rz}^{(m)2} = \omega^2\mu_0\varepsilon_0 \tag{2.49}$$

The reflected electric field is composed of an infinite sum of plane waves propagating at different angles with respect to the z axis. Figure 2.12 shows the construction for the $\boldsymbol{k}$ vectors of the plane waves.

The angle of reflection of mth order may be related to the angle of incidence noting that $k_{ix}/\omega\sqrt{\mu_0 \varepsilon_0} = \sin\theta_i$ and $[k_{ix} + (2\pi/\Lambda)m]/\omega\sqrt{\mu_0 \varepsilon_0} = \sin\theta_R^{(m)}$ and $\omega\sqrt{\mu_0 \varepsilon_0} = 2\pi/\lambda$

$$\sin\theta_R^{(m)} = \sin\theta_i + \frac{m\lambda}{\Lambda} \tag{2.50}$$

This is the grating reflection law. Note the dependence of $\theta_R^{(m)}$ on λ. This dependence accounts for the use of gratings for spectral analysis.

To show how the coefficients R_m in (2.48) can be evaluated, we consider the simple case of a small sinusoidal corrugation.

$$h(x) = h_0 \cos\frac{2\pi}{\Lambda}x \qquad \left(\frac{h_0}{\Lambda} \ll 1\right) \tag{2.51}$$

For a small corrugation one may use a perturbation approach, a Taylor expansion in terms of h_0/Λ. The main effect of the reflecting surface is to produce a reflected wave in the same way as a mirror would produce a reflected wave. The field due to the incident wave and reflected wave is canceled on the corrugated surface by waves of order $m \neq 0$ that are by a factor h_0/Λ smaller in amplitude than the incident wave. The field of the higher-order waves on the corrugated surface is approximately equal to their field at $z = 0$, the deviation is of order $(h_0/\Lambda)^2$. The field of a planar perfect reflector for an incident wave $E_i \exp(-jk_{ix}x)\exp(-jk_{iz}z)$ is

$$E_i \exp(-jk_{ix}x)[\exp(-jk_{iz}z) - \exp(jk_{iz}z)] \tag{2.52}$$

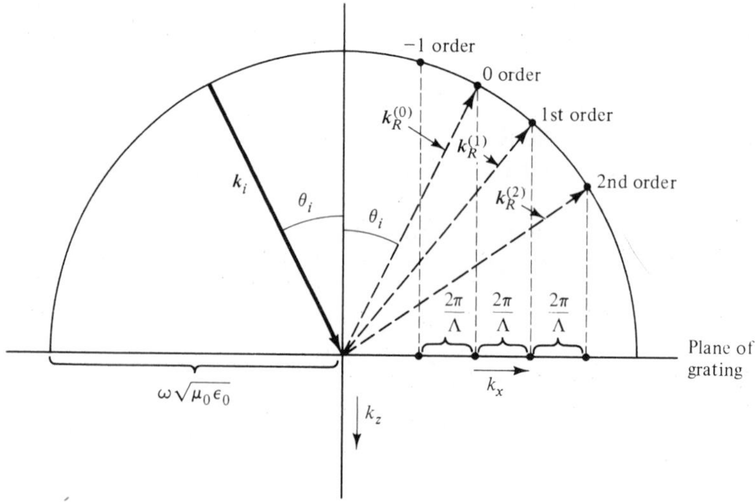

Figure 2.12 Construction of angles for higher-order reflections from grating.

On the perturbed surface, $z = h(x)$, this field is

$$\mathrm{E}_i \exp(-jk_{ix}x)\left[\exp\left\{-j\left(k_{iz}h_0 \cos\frac{2\pi}{\Lambda}x\right)\right\} - \exp\left\{j\left(k_{iz}h_0 \cos\frac{2\pi}{\Lambda}x\right)\right\}\right]$$

$$\simeq -jk_{iz}\Lambda\mathrm{E}_i \exp(-jk_{ix}x)\frac{h_0}{\Lambda}\left[\left(\exp j\frac{2\pi}{\Lambda}x\right) + \exp\left(-j\frac{2\pi}{\Lambda}x\right)\right] \qquad (2.53)$$

to first order in h_0/Λ. This field has to be canceled by the higher-order waves. Because this amplitude at $z = h(x)$ is approximately equal to the amplitude at $z = 0$, we find from (2.46), (2.48), and (2.53):

$$\sum_m \mathrm{R}_m \exp\left(-j\frac{2\pi m}{\Lambda}x\right) - jk_{iz}\Lambda E_i \frac{h_0}{\Lambda} \cdot \left(\exp j\frac{2\pi}{\Lambda}x + \exp -j\frac{2\pi}{\Lambda}x\right) = 0 \qquad (2.54)$$

Therefore,

$$\mathrm{R}_{+1} = \mathrm{R}_{-1} = jk_{iz}h_0\mathrm{E}_i \qquad (2.55)$$

and all other R_m's are zero.

We shall not study further the analysis of gratings with nonsinusoidal $h(x)$, or deeper grooves, or "rulings," greater values of h_0/Λ. Instead, we shall look at a case where the solution can be obtained by inspection.

Gratings are often used in lasers as the "frequency-sensitive mirror" at one end of the optical resonator so as to select one single mode of the resonator for "lasing." In such applications the power in the first-order term can be made to be as much as 95% of the incident power, if the angle of incidence is the blaze angle. A simple intuitive picture of such a "grating reflector" is presented in Fig. 2.13. The grating is made of reflecting steps that are perpendicular to the incident radiation. The reflection in the same direction will be very strong if the step height H is a multiple of half wavelengths. Since the

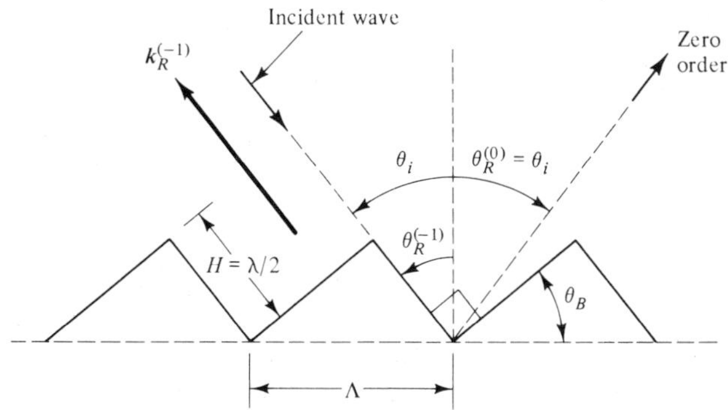

Figure 2.13 Maximum reflection from step grating.

wavelengths are usually so small that tolerances are hard to obey, only the case $H = \lambda/2$ is of practical interest. Consider the condition of reflection of the -1 order into the direction of incidence. From the grating reflection law (2.50)

$$-\sin\theta_R^{(-1)} = \sin\theta_i = \frac{\lambda}{\Lambda} - \sin\theta_i$$

Therefore,

$$\frac{\lambda}{2\Lambda} = \sin\theta_i$$

But $\lambda/2\Lambda$ is equal to $\sin\theta_B$, where θ_B, the *blaze angle*, is shown in Fig. 2.13. Thus the grating reflection is consistent with the argument based on the coherent superposition of the reflections from different steps.

The efficiency of a grating depends on the polarization of the incident wave. If a grating is illuminated at $\theta_i = \theta_B$ by a wave with the $\boldsymbol{E}$ field polarized in the plane of the paper in Fig. 2.13, all boundary conditions are matched with the nodal planes of the standing waves parallel to, and coincident with, the grating steps. The grating reflects only into the -1 order. If the polarization is normal to the plane of the paper, the boundary conditions are not matched this simply. Reflection into other orders occurs.

2.7 GRATING SPECTROMETERS [5]

Figure 2.14 shows the schematic of a grating spectrometer. The light source whose spectrum is to be analyzed is focused on a narrow slit. The light that passes the slit is collimated, frequently with a focusing mirror rather than a lens. The grating disperses the light (θ_R is a function of λ), the dispersed light is then brought to a focus by a second lens or mirror. Generally, the two focus-

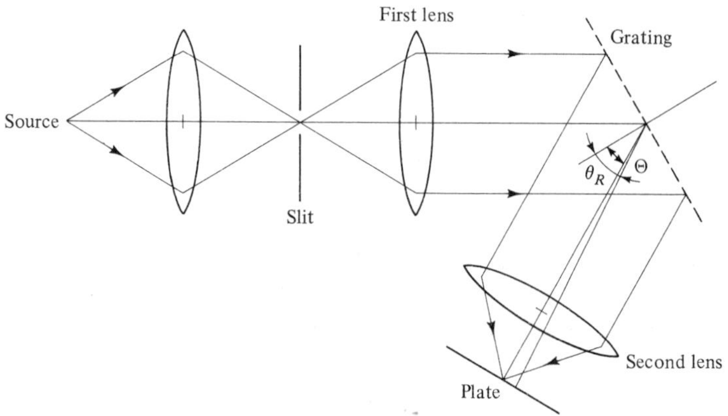

Figure 2.14 Plane-grating spectrometer.

ing lenses or mirrors are identical and, for best results, the slit width is adjusted to produce a grating illumination consistent with the theoretical resolution limit of the grating (see Section 4.4).

In some instruments, a photographic plate is placed in the focal plane of the second lens. Different wavelengths record at different positions on the plate. In other instruments a slit is located in the focal point, and the grating is rotated in a known fashion, thereby directing different wavelengths through the slit.

Chromatic Resolving Power

The grating reflection law (2.50) applies to infinite parallel plane waves incident upon an infinitely wide grating. The finite width limits the ability to resolve waves that differ in wavelengths by less than the chromatic resolving power. The angle θ_R of the grating reflection law determines the central angle of the grating diffraction pattern [Section 4.2, Eq. (4.31)]. The wave reflected from the grating at the nominal angle θ_R has an amplitude pattern

$$I(\theta) = \frac{\sin^2\left[\dfrac{N\pi}{\lambda}\Lambda(\sin\theta_R - \sin\theta)\right]}{\sin^2\left[\dfrac{\pi}{\lambda}\Lambda(\sin\theta_R - \sin\theta)\right]} \tag{2.56}$$

where N is the total number of grooves. A principal maximum occurs when both numerator and denominator are zero. When N is large, the numerator varies much more quickly than the denominator, so it alone determines the angular width of the maximum.

At the principal maximum, $\sin\theta = \sin\theta_R$. If we increase or decrease the argument slightly, the value of $I(\theta)$ falls sharply, and when the numerator reaches zero again, the denominator is no longer zero. Thus the intensity of the diffracted light falls to zero when θ is increased by a value of $\delta\theta$ given by

$$\frac{N\pi}{\lambda}\Lambda\,\delta(\sin\theta) = \pi \tag{2.57}$$

or

$$\delta\theta = \frac{\lambda}{N\Lambda\cos\theta} \tag{2.58}$$

This is also approximately the full angular width (at half-maximum) of the diffracted beam.

Next, estimate the smallest resolvable wavelength difference $\delta\lambda$ between two spectral lines. Two spectral lines begin to be resolved when the maximum of one wavelength coincides with the first zero of the other. The reflection angle θ_R is a function of wavelength λ. The maximum is followed when θ, the angle of observation, changes according to the constraint

$$\delta\theta \cos\theta = \delta\theta_R \cos\theta_R \tag{2.59}$$

We relate $\delta\theta_R$ to $\delta\lambda$ by differentiating the grating equation (2.50) to find that

$$\delta\lambda = \frac{\Lambda}{m} \cos\theta_R \, \delta\theta_R \tag{2.60}$$

and introduce this relation along with (2.59) into (2.58), from which we find that

$$\frac{\lambda}{\delta\lambda} = mN \tag{2.61}$$

where $\lambda/\delta\lambda$ is known as the *chromatic resolving power*.

For resolution on the order of 1 nm or less, the product mN must be several thousand. Because m is generally 1, the total number of grooves, or rulings, on the grating must be sizable. Typical gratings are made with 6000 rulings/per centimeter or more, and a large, high-quality diffraction grating has tens of thousands of rulings.

2.8 SUMMARY

Reflection and transmission of optical waves at an interface is described by Snell's law, which results from matching of the spatial dependences of the fields on the two sides of the interface, and by the reflection coefficient Γ, which is evaluated by matching the tangential electric and magnetic fields. Snell's law is independent of polarization; the reflection coefficient is not. In particular, TM waves can traverse an interface between two dielectrics with no reflection, at the Brewster angle. Total internal reflection occurs between two dielectric media when the wave is incident upon the interface from the medium of greater dielectric constant at an angle greater than the critical angle. This total internal reflection can be utilized for the guidance of optical waves as in optical fibers and optical waveguides to be discussed in Chapter 6.

The continuity of tangential electric magnetic fields at interfaces is expressed conveniently in terms of the wave impedance, or its inverse, the wave admittance. The wave impedance is the ratio of the tangential electric and magnetic fields. The transformation of the wave impedance as a function of distance from the interface, (2.33), is a central result which we employed for the analysis of reflection from multiple layers, antireflection coatings, and reflection coatings. The use of multiple layers extends the range of achievable values of index discontinuities that can be matched with available grating materials and enables one to achieve highly reflecting surfaces.

The reflection from a "corrugated" grating is an example of a boundary value problem in which a single reflected wave is insufficient to satisfy the boundary conditions. The angles of the higher-order reflections are frequency (or wavelength) dependent. The use of gratings in spectrometers for frequency resolution is based on this property of grating reflection. The chromatic resolving power of a grating spectrometer was defined as $\lambda/\delta\lambda$, where $\delta\lambda$ is the wavelength range resolved by the instrument. It was found to be equal to mN, where m is the order of the reflection and N is the number of grooves in the grating.

PROBLEMS

2.1. The reflection at the Brewster angle θ_B is zero. Evaluate the sensitivity of reflection to angular deviation from the Brewster angle. For this purpose expand Γ to first order in $\theta_1 - \theta_B$, and square. If $|\Gamma|^2$ is to be $\leq 10^{-2}$, how big an angular deviation is allowed? Use $n = 1.7$ for the medium, interfacing with air.

2.2. Derive equations analogous to (2.28), (2.29), and (2.30) for a TM wave at total internal reflection.

2.3. Mirrors for the He–Ne laser wavelength ($\lambda = 6328$ Å) are constructed from multiple layers of ZnS and ThF_2. The refractive indices are 2.5 and 1.5, respectively. The substrate is glass, with $n = 1.5$. What is the minimum number of layers to produce a reflection of the incident power greater than 99.5%? Assume normal incidence.

2.4. When a TE wave is totally reflected from a dielectric interface, the field on side 1 forms a perfect standing wave (Fig. P2.4). The maximum of the wave has phase ϕ with respect to the interface; $\theta_G \equiv 90° - \phi$ is the Goos–Hänchen shift.

(a) Solve the boundary value problem in the case when $\theta_1 > \theta_c$. Show that $|E_+| = |E_-|$.

(b) Develop an expression for $\cot \theta_G$ in terms of θ_1 and θ_c (θ critical). For $\theta_c = 45°$, sketch θ_G versus θ_1.

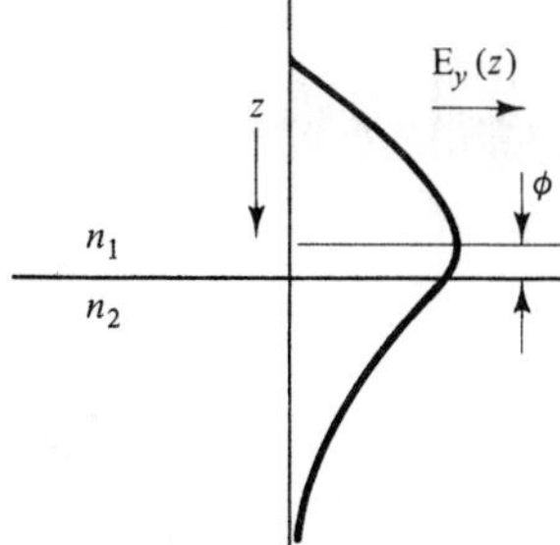

Figure P2.4

2.5. (a) Show that the impedance $Z(z)$ of (2.33) is pure imaginary for every value of z, if $Z(0)$ is imaginary. How does this relate to power conservation?

(b) Show that $|\Gamma(z)| = 1$, when $Z(z)$ is pure imaginary, and relate this fact to power conservation.

2.6. To first order in a Taylor expansion in terms of h_0/Λ, find the amplitudes of the waves reflected from a perfectly conducting grating with a square wave profile of height h_0, period Λ, for an incident wave $\mathbf{E} = \hat{\mathbf{y}}\, E_i \exp - jk_z z \exp - jk_x x$, with the grooves parallel to $\hat{\mathbf{y}}$.

2.7. Determine the wavelength dependence of the -1 order reflection from a grating at $\theta_i = \theta_B$: $\lambda\, |d\theta_R^{(-1)}/d\lambda|$. Write it in terms of the blaze angle.

REFERENCES

[1] S. Ramo, J. R. Whinnery, and T. Van Duzer, *Fields and Waves in Communication Electronics*, Wiley, New York, 1965.

[2] R. B. Adler, L. J. Chu, and R. M. Fano, *Electromagnetic Energy Transmission and Radiation*, Wiley, New York, 1960.

[3] N. S. Kapany and J. J. Burke, *Optical Waveguides*, Academic Press, New York, 1972, p. 7.

[4] H. Kogelnik and H. P. Weber, "Rays, stored energy, and power flow in dielectric waveguides," *J. Opt. Soc.*, *64*, no. 2, 174–185, Feb. 1974.

[5] M. Born and E. Wolf, *Principles of Optics*, Macmillan, New York, 1964.

3

MIRRORS AND INTERFEROMETERS

A plane mirror partially reflects and partially transmits the wave incident upon it. The action of the mirror may be described in terms of a linear set of relationships between the incident, transmitted, and reflected wave amplitudes. Since a wave may be incident from either side, there are two incident wave amplitudes and two reflected amplitudes. The linear relationship defines a second-order matrix, the so-called scattering matrix. In this format the scattering matrix describes the operation of any linear system operating on a pair of incident waves and reflected waves. Properties of the scattering matrix are restricted by the principle of reciprocity, which holds for any isotropic material, and may be further restricted by the requirement of losslessness if the material of which the mirror is made can be idealized as such.

In Section 3.1 we derive the reciprocity principle from Maxwell's equations. In Section 3.2 we present the scattering matrix formalism for any linear network. Section 3.3 states the properties of the scattering matrix if the network is reciprocal and/or lossless. In Section 3.4 we apply the scattering matrix formalism to a partially transmitting mirror.

Two partially transmitting mirrors placed a distance l apart, transmit maximally an optical wave incident upon them, when the frequency of the wave is equal to some multiple of the inverse travel time between the mirrors. Thus the two mirrors in cascade can be used as a passband filter with peak transmission occurring at a set of characteristic frequencies. Such a system is called the "Fabry–Perot" interferometer. The same system provides the "resonator" required for the construction of an optical oscillator, a laser. The Michelson interferometer discussed next serves a different purpose. In a Michelson interferometer an incoming wave is separated into two parts by a "half-silvered" mirror. The two parts of the wave are recombined after a suitable delay introduced in each. Measurement of the interference pattern for different delays of the two beams measures the degree of *temporal coherence.* We describe, finally, the Twyman–Green interferometer which is used to test nominally flat optical components for their "true parallelism."

3.1 RECIPROCITY PRINCIPLE [1]

In this section we derive the field-theoretical form of the reciprocity principle for an isotropic medium. We start with one solution of Maxwell's equations, denoted by a superscript (a), which satisfies the equations

$$\nabla \times \mathbf{E}^{(a)} = -j\omega\mu\mathbf{H}^{(a)} \tag{3.1}$$

$$\nabla \times \mathbf{H}^{(a)} = j\omega\varepsilon\mathbf{E}^{(a)} \tag{3.2}$$

We set up another solution (b) in the same manner. We dot-multiply (3.1) by $\mathbf{H}^{(b)}$, (3.2) by $\mathbf{E}^{(b)}$ and add

$$(\nabla \times \mathbf{E}^{(a)}) \cdot \mathbf{H}^{(b)} + (\nabla \times \mathbf{H}^{(a)}) \cdot \mathbf{E}^{(b)} = -j\omega[\mu\mathbf{H}^{(b)} \cdot \mathbf{H}^{(a)} - \varepsilon\mathbf{E}^{(b)} \cdot \mathbf{E}^{(a)}] \tag{3.3}$$

Next we interchange the superscripts in (3.3) in the equation above and subtract. The right-hand sides cancel. The left-hand sides produce a divergence

$$\nabla \cdot (\mathbf{E}^{(a)} \times \mathbf{H}^{(b)} - \mathbf{E}^{(b)} \times \mathbf{H}^{(a)}) = 0 \tag{3.4}$$

We integrate (3.4) over a given volume V, enclosed by a surface S. Use of Gauss's theorem yields

$$\oint_S (\mathbf{E}^{(a)} \times \mathbf{H}^{(b)}) \cdot d\boldsymbol{a} = \oint_S (\mathbf{E}^{(b)} \times \mathbf{H}^{(a)}) \cdot d\boldsymbol{a} \tag{3.5}$$

This is the field-theoretical form of the *reciprocity theorem.* It imposes a constraint on the solutions of Maxwell's equations in a medium described by a scalar dielectric constant and magnetic permeability.

3.2 SCATTERING MATRIX FORMALISM [2, 3, 4]

We develop the scattering matrix formalism for plane waves passing through slabs of optical media. Suppose that we have a plane wave impinging on, transmitted through, and reflected from such a system (see Fig. 3.1). If we pick one polarization, then the amplitudes of the incident and reflected waves at the input and output planes constitute a full description of the action of the system—a total of four *wave amplitudes.* A linear system relates linearly the reflected wave amplitudes to the incident wave amplitudes so that only two wave amplitudes are independently specifiable. Denote *incident wave amplitudes* by a and *reflected wave amplitudes* by b.

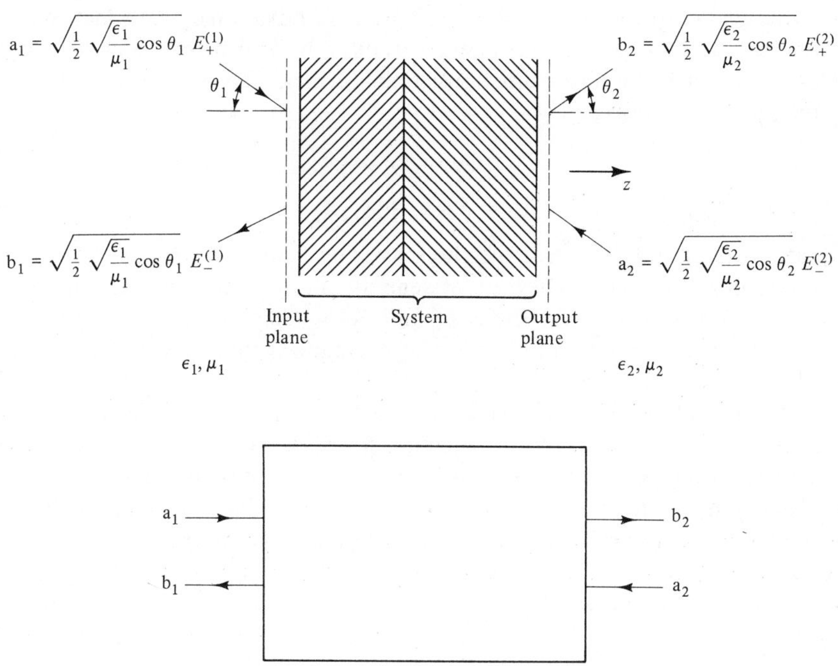

Figure 3.1 TE wave incident upon system of slabs. *Incidence need not be normal.*

We normalize the amplitudes a_1 and b_1 so that power per unit area incident from the left is given by

$$\frac{1}{2}\,\text{Re}\,[\mathbf{E}\times\mathbf{H}^*\cdot\hat{z}] = |a_1|^2 - |b_1|^2 \tag{3.6}$$

An analogous normalization is used for the amplitudes a_2 and b_2. The identification of the right side of (3.6) with power is consistent with the results of Section 2.1 for a TE-wave. There, we saw that E_y can be written

$$E_y = \sqrt{2/Y_0^{(1)}}\,[a_1 + b_1]e^{-jk_x x}$$

and

$$H_x = \sqrt{2Y_0^{(1)}}\,[a_1 - b_1]e^{-jk_x x}$$

where

$$a_1 \equiv \sqrt{Y_0^{(1)}/2}\; E_+^{(1)} e^{-jk_z{}^{(1)}z}$$
$$b_1 \equiv \sqrt{Y_0^{(1)}/2}\; E_-^{(1)} e^{+jk_z{}^{(1)}z} \tag{3.7}$$

An analogous identification can be made for the TM waves of Section 2.2. In fact, the interpretation (3.6) is quite general and applies, in modified form, to optical waveguide modes as we shall have the opportunity to see in Section 6.8.

Returning to the system of Fig. 3.1 we take advantage of the fact that a_1 and a_2 can be chosen as independent variables, b_1 and b_2 as dependent variables, the latter being linear functions of the former. The system is describable by the scattering matrix

$$b_1 = S_{11}a_1 + S_{12}a_2 \tag{3.8}$$

$$b_2 = S_{21}a_1 + S_{22}a_2 \tag{3.9}$$

Equations (3.8) and (3.9) describe the scattering from a "two-port," a system with two ports of access. They are of general validity; they may be applied to waveguide ports or optical beams of finite cross section—not only to plane waves. In this case $|a_i|^2$ and $|b_i|^2$ are so normalized that they give the power in the waves.

The reflection of a TE wave from an interface (Section 2.1) is an example of a particular excitation of a particular two-port. To understand better the formalism developed in this section, we relate the present notation to that problem. The normalized wave amplitudes are given by (3.7) with $z = 0$. The elements of the scattering matrix determined in Section 2.1 are

$$S_{11} = \Gamma^{(1)} = \frac{Y_0^{(1)} - Y_0^{(2)}}{Y_0^{(1)} + Y_0^{(2)}} \tag{3.10}$$

$$\begin{aligned} S_{12} = S_{21} &= \sqrt{Y_0^{(2)}}\, E_+^{(2)} / \sqrt{Y_0^{(1)}}\, E_+^{(1)} = (1 + \Gamma^{(1)}) \sqrt{\frac{Y_0^{(2)}}{Y_0^{(1)}}} \\ &= \frac{2Y_0^{(1)}}{Y_0^{(1)} + Y_0^{(2)}} \sqrt{\frac{Y_0^{(2)}}{Y_0^{(1)}}} \end{aligned} \tag{3.11}$$

The replacement of $E_+^{(2)}/E_+^{(1)}$ by $(1 + \Gamma^{(1)})$ follows from continuity of the tangential **E** at the interface. S_{22} is obtained from (3.10) by an interchange of indices 1 and 2.

$$S_{22} = -S_{11} \tag{3.12}$$

The total electric field in medium 1 at the interface is related to a_1 and b_1 by

$$E_y^{(1)} = \sqrt{\frac{2}{Y_0^{(1)}}}\,(a_1 + b_1)e^{-jk_x x} \tag{3.13}$$

The tangential magnetic field is

$$H_x^{(1)} = -\sqrt{\frac{Y_0^{(1)}}{2}}\,(a_1 - b_1)e^{-jk_x x} \tag{3.14}$$

Note that the time-averaged power flow density is equal to

$$-\tfrac{1}{2}\,\mathrm{Re}\,[E_y^{(1)}H_x^{(1)*}] = |a_1|^2 - |b_1|^2 \tag{3.15}$$

and does not contain the characteristic admittance when expressed in terms of a_1 and b_1. Similar expressions hold for TM waves.

It is convenient to cast (3.8) and (3.9) into matrix form by defining the two column matrices

$$\mathbf{a} \equiv \begin{bmatrix} a_1 \\ a_2 \end{bmatrix}, \qquad \mathbf{b} \equiv \begin{bmatrix} b_1 \\ b_2 \end{bmatrix}$$

and the matrix of second rank

$$\mathbf{S} \equiv \begin{bmatrix} S_{11} & S_{12} \\ S_{21} & S_{22} \end{bmatrix}$$

With these definitions the compact expression

$$\mathbf{b} = \mathbf{S}\mathbf{a} \tag{3.16}$$

replaces the two equations (3.8) and (3.9).

3.3 PROPERTIES OF SCATTERING MATRIX [2–4]

If the linear system described by the scattering matrix **S** is reciprocal, the elements of the scattering matrix have to obey a *reciprocity condition.* If the system is lossless, another *losslessness condition* is imposed on **S**. Finally, *time reversibility*, a property obeyed by solutions of Maxwell's equations for a linear lossless system, imposes a constraint of its own. In this section we shall study all of these conditions. They are useful in the characterization of a linear system in terms of a scattering matrix, because they serve as a check on the analysis, and/or enable one to determine some elements of the scattering matrix from the knowledge of other elements.

First, we consider the reciprocity principle and rewrite it in terms of wave amplitudes **a** and **b**.

The surface S in (3.5) is to be interpreted as a cylindrical surface enclosing the system, with end planes at the input and output planes. In the analysis of waves in systems of finite cross section, the choice of the cylinder surface presents no problems, it is chosen in the field-free region (Fig. 3.2a). When the reciprocity principle is applied to infinite parallel plane waves inclined with respect to the reference planes, two problems arise.

(a) The fields have an $\exp(-jk_x x)$ dependence (see Fig. 3.2b) and if (3.5) is applied to wave solutions with the same inclination the integrands in (3.5) have an $\exp(-2jk_x x)$ dependence.

(b) There are contributions to the integrands from the cylinder surface in Fig. 3.2b.

Both these difficulties are removed if one chooses for the solutions (a) and (b) waves with opposite angles of incidence so that $\exp(-jk_x^{(a)} x) = \exp(jk_x^{(b)} x)$. Note that the scattering matrices for the two solutions in a plane

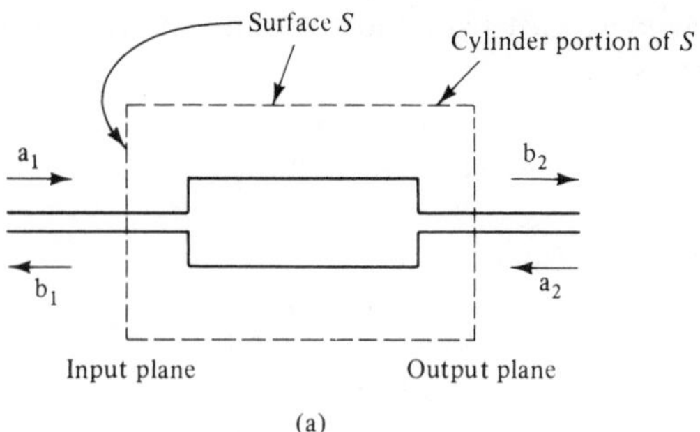

(a)

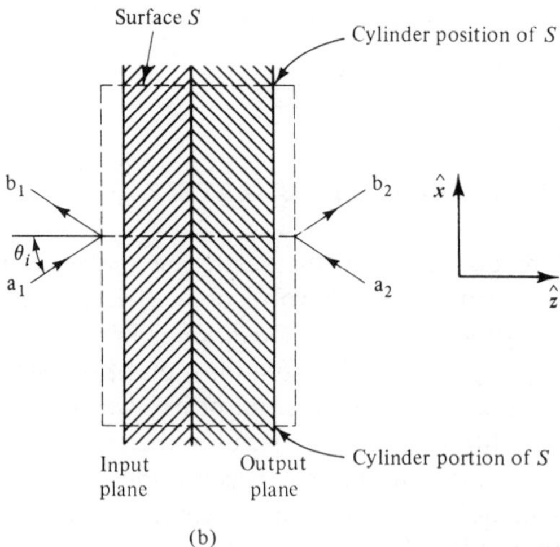

(b)

Figure 3.2 (a) Schematic of linear system connected to two guides. Fields are zero on cylinder portion of surface. (b) Schematic of plane-wave reflection and transmission from layered medium.

parallel system are identical, by symmetry. The contributions of $\int \mathbf{E}^{(a)} \times \mathbf{H}^{(b)} \cdot d\mathbf{a}$ over the cylinder portion of S shown in Fig. 3.2b cancel as one can convince oneself by drawing the associated fluxes of $\mathbf{E}^{(a)} \times \mathbf{H}^{(b)}$. The surface integrals in (3.5) reduce to integrals over the end planes. Expressed in terms of the wave amplitudes one of the integrals becomes [compare (3.15)]

$$\int_S \mathbf{E}^{(a)} \times \mathbf{H}^{(b)} \cdot d\mathbf{a} \rightarrow -(a_1^{(a)} + b_1^{(a)})(a_1^{(b)} - b_1^{(b)}) - (a_2^{(a)} + b_2^{(a)})(a_2^{(b)} - b_2^{(b)}) \tag{3.17}$$

An analogous expression holds for $\int_S \mathbf{E}^{(b)} \times \mathbf{H}^{(a)} \cdot d\mathbf{a}$ with superscripts (a) and (b) interchanged in (3.17). The lengthy expression on the right in (3.17) can be streamlined somewhat by writing it in matrix notation. The transpose of the matrix $\mathbf{S}$ indicated by the subscript t, is the matrix $\mathbf{S}$ mirrored around the diagonal. In the case of a matrix of second rank:

$$\mathbf{S}_t = \begin{bmatrix} S_{11} & S_{21} \\ S_{12} & S_{22} \end{bmatrix} \tag{3.18}$$

The transpose $\mathbf{a}_t$ of the column matrix $\mathbf{a}$, is a row matrix

$$\mathbf{a}_t = [\mathrm{a}_1, \mathrm{a}_2] \tag{3.19}$$

With these definitions one may write for (3.17)

$$\int_S \mathbf{E}^{(a)} \times \mathbf{H}^{(b)} \cdot d\boldsymbol{a} = -(\mathbf{a}_t^{(a)} + \mathbf{b}_t^{(a)})(\mathbf{a}^{(b)} - \mathbf{b}^{(b)})$$

The reciprocity principle (3.5) then gives

$$\mathbf{b}_t^{(a)}\, \mathbf{a}^{(b)} = \mathbf{a}_t^{(a)}\, \mathbf{b}^{(b)} \tag{3.20}$$

where we have taken advantage of the fact that $\mathbf{b}_t\, \mathbf{a} = \mathbf{a}_t\, \mathbf{b}$ by definition of the transpose of a column matrix and canceled terms on both sides of the equation.

Introducing (3.16) and noting that

$$\mathbf{b}_t = \mathbf{a}_t\, \mathbf{S}_t,$$

by definition of the transpose, we obtain for (3.20)

$$\mathbf{a}_t^{(a)}\, \mathbf{S}_t\, \mathbf{a}^{(b)} = \mathbf{a}_t^{(a)}\, \mathbf{S}\, \mathbf{a}^{(b)} \tag{3.21}$$

Because $\mathbf{a}^{(a)}$ and $\mathbf{a}^{(b)}$ can be assigned arbitrary phases and amplitudes, (3.21) requires that

$$\mathbf{S}_t = \mathbf{S} \tag{3.22}$$

The scattering matrix of a linear reciprocal system is symmetric. In the case of a two-port, (3.22) implies

$$S_{12} = S_{21} \tag{3.23}$$

Power Conservation

If the system is lossless, the net power flowing into the system must be zero:

$$|\mathrm{a}_2|^2 - |\mathrm{b}_2|^2 + |\mathrm{a}_1|^2 - |\mathrm{b}_1|^2 = 0$$

This relation can be written in matrix form:

$$\mathbf{a}^+[\mathbf{1} - \mathbf{S}^+\mathbf{S}]\mathbf{a} = 0 \tag{3.24}$$

where $\mathbf{1}$ is the identity matrix and the operation $+$ complex conjugates and transposes the matrix on which it appears as a superscript:

$$[\mathbf{S}^+]_{ij} = S_{ji}^*$$

Because the excitation "vector" **a** is arbitrary, (3.24) implies that

$$\mathbf{S}^{+}\mathbf{S} = \mathbf{1}$$

$\mathbf{S}^{+}\mathbf{S}$ is equal to the identity matrix, or

$$\mathbf{S}^{+} = \mathbf{S}^{-1} \tag{3.25}$$

The scattering matrix of a lossless system is *unitary*. The derivation above, once cast into matrix form, is not restricted to a two-port but applies to a system of any number of ports. In the case of a two-port, (3.25) written out in component form gives

$$|S_{11}|^2 + |S_{21}|^2 = 1 \tag{3.26}$$

$$|S_{22}|^2 + |S_{12}|^2 = 1 \tag{3.27}$$

$$S_{11}^{*}S_{12} + S_{21}^{*}S_{22} = 0 \tag{3.28}$$

The scattering matrix of a two-port has four complex components. The reciprocity relation (3.23) eliminates one of them, leaving three complex components—six real numbers—as parameters of the system. Equations (3.26), (3.27), and (3.28), with the constraint (3.23), constitute three independent real equations. A lossless reciprocal two-port is, therefore, described in terms of three real parameters.

Time Reversal

Maxwell's equations (1.7) and (1.8), in the absence of a source, $\boldsymbol{J} = 0$, remain unchanged under a replacement of t by $-t$ and $\boldsymbol{H}$ by $-\boldsymbol{H}$. A similar property holds more generally for isotropic dispersive media with the dielectric constant and magnetic permeability functions of frequency. If one takes the complex conjugate of (1.41) and (1.42) for real ω, the form of the equations remains unchanged, with $\mathbf{E}^*$ and $-\mathbf{H}^*$ as the new fields provided $\varepsilon(\omega) = \varepsilon^*(\omega)$ and $\mu(\omega) = \mu^*(\omega)$. This holds for a lossless medium. A wave traveling in the $+z$ direction with the dependence $\exp(-j\beta z)$ is converted into a wave with the dependence $\exp(j\beta z)$. The time dependence, chosen to be $\exp(j\omega t)$ by definition, combined with the new spatial dependence $\exp(j\beta z)$ describes a backward traveling wave. The new fields **E** and **H** are the complex conjugates of the old fields with the sign of **H** reversed.

Time reversibility leads to constraints on the scattering matrix coefficients that are similar to, but not identical with, those imposed by power conservation. To show this, start with (3.16) and take the complex conjugate

$$\mathbf{b}^* = \mathbf{S}^*\mathbf{a}^* \tag{3.29}$$

We may interpret $\mathbf{b}^*$ as time-reversed amplitudes of reflected waves (i.e., as incident waves, $\mathbf{b}^* \Rightarrow \mathbf{a}$). Similarly, $\mathbf{a}^* \Rightarrow \mathbf{b}$, (see Fig. 3.3) and we obtain from (3.29)

$$\mathbf{a} = \mathbf{S}^*\mathbf{b}$$

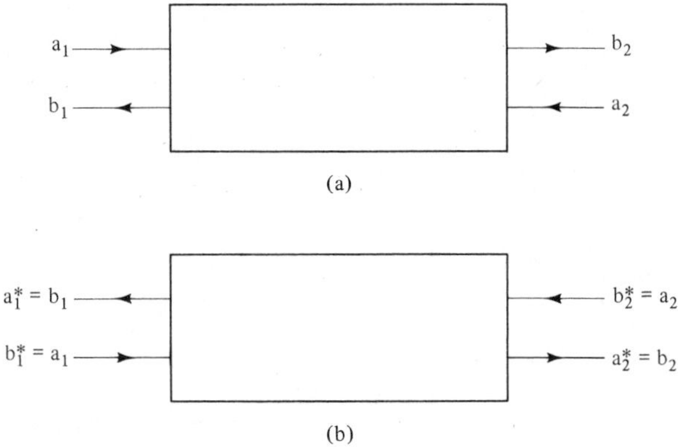

Figure 3.3 An example of an excitation and it's time reversed version.

or

$$\mathbf{b} = [\mathbf{S}^*]^{-1}\mathbf{a}$$

Because the equation above describes the same two-port as (3.16), one must have

$$[\mathbf{S}^*]^{-1} = \mathbf{S}$$

or

$$\mathbf{S}^* = \mathbf{S}^{-1} \tag{3.30}$$

This is not the same condition as the one derived from power conservation, (3.25). The two are made consistent only by setting

$$\mathbf{S}^* = \mathbf{S}^+$$

The complex conjugate of the scattering matrix is equal to the complex conjugate transpose; in other words, **S** is a symmetric matrix. Thus time reversibility and power conservation imply reciprocity.

3.4 SCATTERING MATRIX OF PARTIALLY TRANSMITTING MIRROR

We now apply the scattering matrix formalism to a lossless reflecting and transmitting system—the general case of a lossless partially transmitting mirror. We may choose the position of reference plane 1 so that the reflected wave (see Fig. 3.4) is in antiphase with the incident wave. (Note comment in

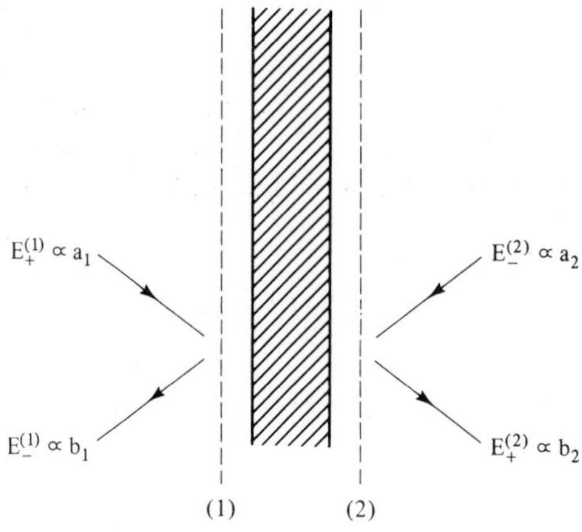

Figure 3.4 Partially transmitting mirror.

connection with the transformation properties of the reflection coefficient, (2.37).)

$$E_-^{(1)} = -r_1 E_+^{(1)} \tag{3.31}$$

or

$$b_1 = -r_1 a_1 \tag{3.32}$$

where r_1 is real and positive. The same may be done with reference plane 2. Thus

$$S_{11} = -r_1 \tag{3.33}$$

$$S_{22} = -r_2 \tag{3.34}$$

Reciprocity requires a symmetric **S** matrix according to (3.23). If the system is lossless, we must have from (3.26), (3.27), (3.33), and (3.34):

$$|S_{21}|^2 = 1 - |S_{11}|^2 = 1 - r_1^2 \tag{3.35}$$

$$|S_{12}|^2 = 1 - |S_{22}|^2 = 1 - r_2^2 = |S_{21}|^2 \tag{3.36}$$

Thus the matrix elements $S_{11} = -r_1$ and $S_{22} = -r_2$ are equal, $r_1 = r_2 = r$, even if the system is *not symmetric*. We find from (3.28)

$$S_{12} = -S_{21}^* \frac{S_{22}}{S_{11}^*} \tag{3.37}$$

For the special choice of reference planes, S_{22}/S_{11}^* is equal to unity; thus S_{12} is a pure imaginary quantity. We choose

$$S_{12} = jt \tag{3.38}$$

where t is the "transmissivity"

$$t = \sqrt{1 - r^2} \tag{3.39}$$

The transmissivity t need not be positive; both signs of the square root are permissible. Through a particular choice of reference planes the **S** matrix of a lossless system has been cast into the form

$$\mathbf{S} = \left[\begin{array}{c|c} -r & jt \\ \hline jt & -r \end{array}\right] \tag{3.40}$$

where

$$t = \sqrt{1 - r^2} \tag{3.41}$$

The **S** matrix contains only one real adjustable parameter, the reflectivity r.

The present general derivation of the mirror **S** matrix is independent of the angle and polarization of the incident wave; it can be TE, TM, or a combination of the two. Of course, the value of r may be a function of the polarization and angle of incidence.

The **S** matrix (3.40) is by necessity complex if it represents a loss-free network. The phases of the S parameters are adjustable by proper choice of the reference planes. Once the reference planes have been fixed, and **S** has been cast into the form (3.40) at one particular frequency ω_0, the same form of **S** cannot be maintained over all frequencies: (1) because the choice of the reference planes is frequency dependent, (2) because r (and t) may be frequency dependent, (3) because at the negative frequency $-\omega_0$, the **S** matrix must be replaced by its complex conjugate, the transmission coefficient must change sign. The last statement shows that a *frequency independent* **S** matrix of the form (3.40) is unachievable, in principle. Some systems permit choice of reference planes so that the **S** matrix is frequency independent; then, however, it is not of the form (3.40) [see (3.10, 3.11, and 3.12)].

3.5 FABRY–PEROT INTERFEROMETER [5–7]

The Fabry–Perot interferometer has many applications: It serves as a narrowband transmission filter, it forms the resonator structure for lasers, it can be used as an optical spectrum analyzer. In this section we present the analysis of the Fabry–Perot interferometer, derive its scattering matrix and study briefly some of its applications.

Two parallel partially transmitting mirrors separated by a distance l form a Fabry–Perot interferometer. The distance l is, in general, large compared with a wavelength. In this section we study the transmission characteristic of a Fabry–Perot interferometer.

A wave of frequency ω is incident from the left on mirror 1 of Fig. 3.5a, is partially transmitted with amplitude $jt_1 \mathrm{a}_1$, and partially reflected with amplitude $-r_1 \mathrm{a}_1$. The transmitted part continues to the second mirror, where it arrives phase shifted by $(\omega n/c) \cos \theta l \equiv \delta/2$. Here a portion $-r_2(jt_1)e^{-j\delta/2}\mathrm{a}_1$ is reflected and the portion $(jt_2)(jt_1)e^{-j\delta/2}\mathrm{a}_1$ is transmitted. The reflected part is

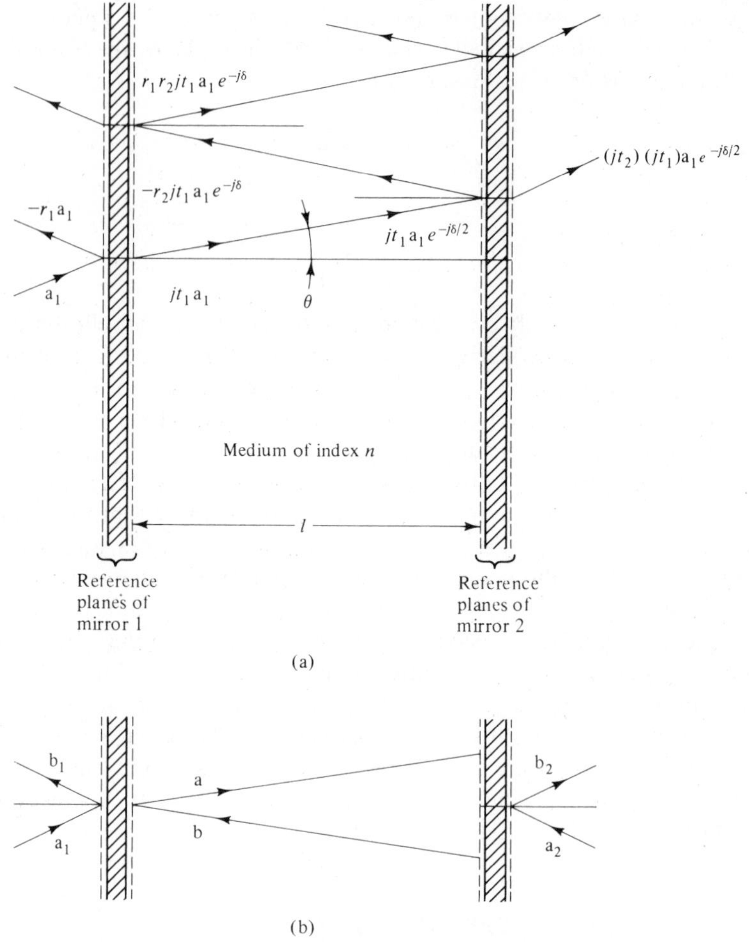

Figure 3.5 (a) Schematic of reflections and transmission in Fabry–Perot interferometer. (b) Definition of incident, reflected, and internal waves.

returned to mirror 1 with an additional phase shift $\delta/2$ and then reflected: $r_1 r_2 (jt_1)e^{-j\delta}a_1$. This amplitude, reflected from mirror 1, goes through the same process as the initally transmitted wave, $(jt_1)a_1$. Each repetition of the process amounts to multiplication by $r_1 r_2 e^{-j\delta}$. The total wave leaving mirror 1 and traveling to the right, at the right-hand reference plane of mirror 1 (Fig. 3.5b), is

$$a = \sum_{m=0}^{\infty} (r_1 r_2 e^{-j\delta})^m jt_1 a_1 = \frac{jt_1}{1 - r_1 r_2 e^{-j\delta}} a_1 \tag{3.42}$$

From a and a_1 one may solve for the reflected wave, b_1, and for the total internal wave, b (Fig. 3.5b), that travels away from mirror 1, using the scattering matrix elements of mirror 1 which give the two relations

$$b_1 = -r_1 a_1 + jt_1 b \tag{3.43}$$

$$a = jt_1 a_1 - r_1 b \tag{3.44}$$

Solved for b_1, with the use of (3.41), these two equations give

$$b_1 = -\frac{r_1 - r_2 e^{-j\delta}}{1 - r_1 r_2 e^{-j\delta}} a_1 \tag{3.45}$$

The transmitted wave b_2 follows from a of (3.42) delayed by exp $(-j\delta/2)$, and transmitted through mirror 2:

$$b_2 = -\frac{t_1 t_2 e^{-j\delta/2}}{1 - r_1 r_2 e^{-j\delta}} a_1 \tag{3.46}$$

Equations (3.45) and (3.46) provide the scattering matrix elements S_{11} and S_{21} of the Fabry–Perot interferometer. By reciprocity, $S_{21} = S_{12}$, and S_{22} is obtained from S_{11} by interchange of the subscripts 1 and 2. Thus the complete scattering matrix of the Fabry–Perot interferometer is

$$\mathbf{S} = \frac{1}{1 - r_1 r_2 e^{-j\delta}} \left[\begin{array}{c|c} -(r_1 - r_2 e^{-j\delta}) & -t_1 t_2 e^{-j\delta/2} \\ \hline -t_1 t_2 e^{-j\delta/2} & -(r_2 - r_1 e^{-j\delta}) \end{array} \right] \tag{3.47}$$

In problem 3.3 the reader is asked to check the **S**-matrix above for power conservation. To complete the analysis, we solve for b from (3.44) using (3.42):

$$b = -\frac{jt_1 r_2 e^{-j\delta}}{1 - r_1 r_2 e^{-j\delta}} a_1 \tag{3.48}$$

The important property of a Fabry–Perot interferometer is its frequency-dependent transmission of power. The transmitted power (per unit area) is

$$|b_2|^2 = |S_{21}|^2 |a_1|^2 = \frac{t_1^2 t_2^2 |a_1|^2}{(1 - r_1 r_2)^2 + 4 r_1 r_2 \sin^2(\delta/2)} \tag{3.49}$$

In this form, the expression for the transmitted power is equally valid for mirrors with loss; then r_1, r_2, t_1 and t_2 may be complex and do not obey (3.41). The expression simplifies when $t_1^2 = t_2^2 = 1 - r_1^2$, $r_1^2 = r_2^2 = r^2 \equiv R$ and $t_1^2 - t_2^2 = 1 - R$. Then

$$|b_2|^2 = |a_1|^2 \frac{(1 - R)^2}{(1 - R)^2 + 4R \sin^2(\delta/2)} \tag{3.50}$$

Here, R is the reflectivity of the mirrors expressing the reflected *power* per unit incident power. The transmission is shown in Fig. 3.6. The transmission peaks occur when $\delta/2 = (\omega n/c) \cos \theta l = m\pi$, with m an integer, when the frequency

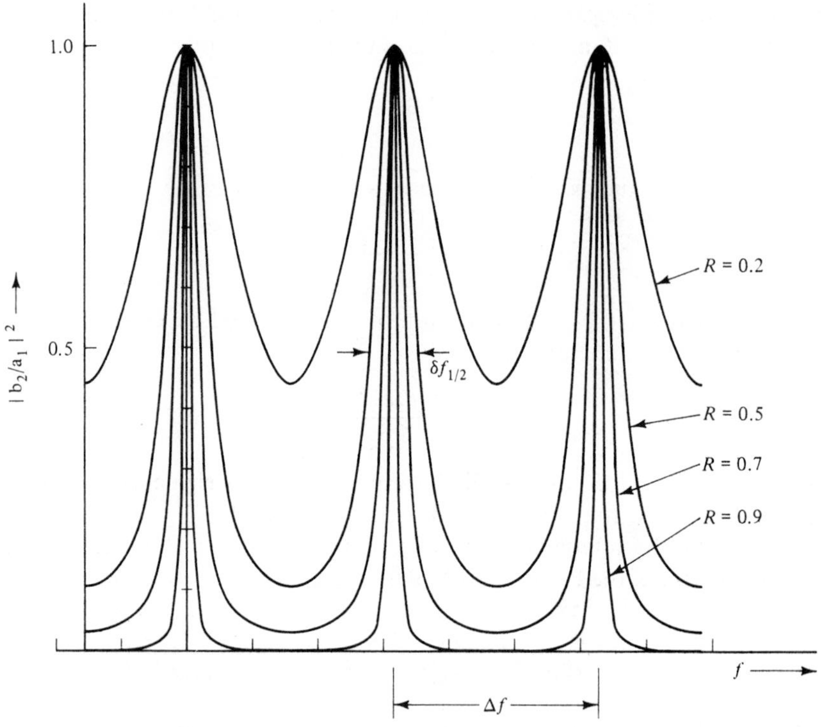

Figure 3.6 Transmission of a lossless Fabry–Perot interferometer. $r_1 = r_2$.

$f = \omega/2\pi$ is equal to one of a set of characteristic values

$$f_m = \frac{mc}{2nl \cos \theta} \tag{3.51}$$

The spacing l of the inner reference planes of the mirrors is m half-wavelengths as measured in terms of the inclined wave vector with angle θ with respect to the mirror normal. The frequency separation Δf of the peaks is given by

$$f_{m+1} - f_m = \Delta f = \frac{c}{2nl \cos \theta} \tag{3.52}$$

The *free spectral range* of a Fabry–Perot interferometer is defined as the frequency separation of transmission peaks, expressed in (free-space) wavelength, $\Delta\lambda$:

$$|\Delta\lambda| = \left(\frac{|\Delta f|}{f}\right)\lambda = \frac{\lambda^2}{2nl} \cos \theta \tag{3.53}$$

The "full width at half-maximum" (FWHM) of the transmission peaks is obtained from (3.50) approximately, when $1 - R \ll 1$ (see Fig. 3.6):

$$\delta f_{1/2} = \frac{(1 - R)c}{2\pi\sqrt{R}\, nl \cos\theta} \tag{3.54}$$

The quantity

$$\frac{\pi\sqrt{R}}{1 - R} \equiv F = \frac{\Delta f}{\delta f_{1/2}} \tag{3.55}$$

is the *finesse* of the interferometer.

Next consider the wavelength or frequency resolution of the Fabry–Perot interferometer. We adopt the criterion that two frequencies are resolved when their maxima are separated by the FWHM, $\delta f_{1/2}$. The chromatic resolving power of the Fabry–Perot interferometer is

$$\frac{f}{\delta f_{1/2}} = \left|\frac{\lambda}{\delta\lambda_{1/2}}\right| = F\,\frac{2\cos\theta l}{\lambda} \tag{3.56}$$

This should be compared with the chromatic resolving power for the grating spectrometer (2.61). The role of mN is played by the parameter $F(2\cos\theta l)/\lambda$. The finesse F may be made as high as 30 or 50 before other factors, such as alignment of the mirrors, begin to influence resolution. The mirrors may be separated by 1 m or more, so that the parameter is typically several hundred thousand for visible light. Thus the chromatic resolving power of the Fabry–Perot interferometer can be much larger than that of the grating spectrometer.

The so-called "scanning Fabry–Perot resonator" followed by a detector is an instrument that can be made equivalent to a spectrum analyzer. To see this, consider the distance over which one of the mirrors has to be moved in order to shift the frequency of the *mode* with m half-wavelengths between the mirrors by one *mode spacing* $c/(2l\cos\theta)$ (in free space, $n = 1$). In the sequel, we shall assume that $\theta = 0$, $\cos\theta = 1$; the beam to be studied is collimated into a beam that approximates an infinite parallel plane wave normal to the mirrors. We have for the frequency f_m of the mth mode from (3.51) ($n = 1$):

$$f_m l = \frac{mc}{2} \tag{3.57}$$

and thus the change of length Δl is related to the frequency shift Δf_m by

$$\frac{\Delta f_m}{f_m} = -\frac{\Delta l}{l} \tag{3.58}$$

When Δf_m is equal to the mode spacing $c/2l$, we find from (3.57) and (3.58):

$$\Delta l = \frac{\lambda}{2}$$

When one of the mirrors is moved by $\lambda/2$, the entire transmission characteristic of Fig. 3.6 is translated by one of the transmission-peak spacings.

Suppose that the mirror position is scanned as a function of time over a distance $\lambda/2$. If the system is illuminated by a parallel plane wave composed of different frequency components occupying a frequency range less than, or equal to Δf, then the transmitted optical intensity follows the spectral distribution $|a_1(\omega)|^2$ as a function of time, provided that the half-power width of the mode is much smaller than the spectral width of $|a_1(\omega)|^2$. The detector output displayed against mirror position gives the spectrum $\overline{|a_1(\omega)|^2}$.

Fabry–Perot structures are also used as laser resonators. The analysis of a laser resonator, to be done in greater detail in Chapter 5, starts with a *resonant mode* in a Fabry–Perot-like structure in the limit of perfect reflectivity (zero loss). The analysis presented in this section provides a means for defining such resonant modes. For this purpose return to (3.42), which gives the total wave a "leaving" mirror 1 inside the Fabry–Perot interferometer as produced by an excitation a_1. When $r_1 = r_2 = 1$ and θ becomes zero, the expression becomes infinite at a frequency such that $\delta = 2m\pi$, for $a_1 \neq 0$. Or a finite wave amplitude can exist inside the resonator, for zero input, $a_1 \rightarrow 0$. This corresponds to a *resonance*, in which a plane wave of the proper frequency bounces back and forth between the perfectly reflecting mirrors, forming the *resonant mode* of the Fabry–Perot. Note that the power transmitted by the Fabry–Perot transmission resonator "peaks" at the frequencies of the resonant modes of the closed Fabry–Perot resonator. The wave b inside the resonator at mirror 1 (see Fig. 3.5b) is related to a by [compare (3.42) and (3.48)]:

$$b = -a$$

Because $a \propto E_+$, the **E** field of the forward traveling wave in the resonator, and $b \propto E_-$, the **E** field of the backward traveling wave, the **E** field is zero at the right-hand reference plane of mirror 1; the reference plane is a nodal plane of the field. The same conclusion holds for the left-hand reference plane of mirror 2.

Interference Filters

The most common interference filters are designed to transmit a narrow band of wavelengths and block all wavelengths outside the band. These are thin-film Fabry–Perot interferometers made of metal films on the two sides of a dielectric slab.

The two metal films are very thin and partially transparent. They serve as the Fabry–Perot mirrors whose spacing is determined by the layer of dielectric. The optical thickness of this layer is determined by the relation $m = 2nd$, where n is the index of the dielectric slab and m is an integer, almost always chosen to be 1 or 2. Unwanted transmittance maxima occur at wavelengths greatly different from the design wavelength and are blocked, for example, with a colored-glass substrate.

Interference filters may be designed to pass very narrow frequency bands, but usually sacrifice peak transmittance in that case. A typical interference filter may have a 7-nm passband with 70% peak transmission.

3.6 THE MICHELSON INTERFEROMETER [6]

The scanning Fabry–Perot interferometer was shown to measure the spectrum of an incident optical wave. The Michelson interferometer on the other hand can measure the autocorrelation function, the Fourier transform of the spectrum. Whereas the Fabry–Perot is ideally suited to the analysis of narrow band processes, the Michelson is well adapted to the measurement of broad band optical waveforms. We derive the scattering matrix of the Michelson interferometer in this section and study the measurement of the autocorrelation function with a Michelson interferometer in the following section.

The Michelson interferometer is shown in Fig. 3.7. An incident wave a_1 is split by the "half-silvered" mirror ($r = 1/\sqrt{2}$, $t = 1/\sqrt{2}$) into two compo-

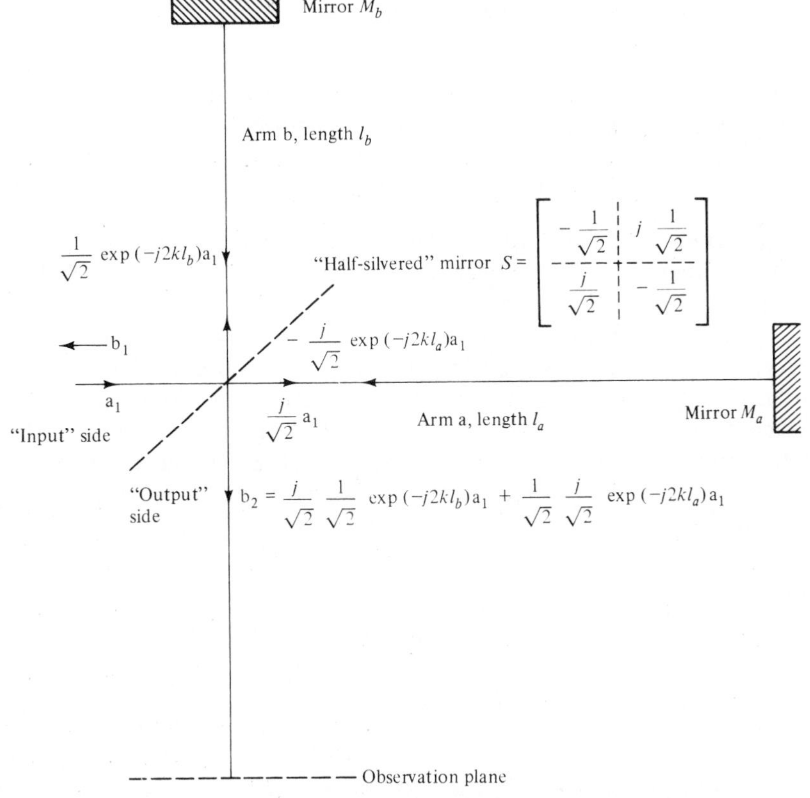

Figure 3.7 Michelson interferometer.

nents. The one shown horizontally in arm a in the figure, $j/\sqrt{2}\,\mathrm{a}_1$, travels to the end mirror and is returned with the phase factor $-\exp(-j2kl_a)$, where $k = \omega/c$. The component shown traveling vertically in arm b, $-1/\sqrt{2}\,\mathrm{a}_1$, is returned with the phase factor $-\exp(-j2kl_b)$. They are recombined at the "output" side of the mirror to produce b_2:

$$\begin{aligned}\mathrm{b}_2 &= \frac{j}{2}\exp(-j2kl_a)\mathrm{a}_1 + \frac{j}{2}\exp(-j2kl_b)\mathrm{a}_1 \\ &= j\exp[-jk(l_a + l_b)]\cos k(l_a - l_b)\mathrm{a}_1 = S_{21}\mathrm{a}_1 \end{aligned} \tag{3.59}$$

The reflected wave amplitude at the input reference plane is

$$\begin{aligned}\mathrm{b}_1 &= \tfrac{1}{2}[\exp(-j2kl_a) - \exp(-j2kl_b)]\mathrm{a}_1 \\ &= -j\exp[-jk(l_a + l_b)]\sin k(l_a - l_b)\mathrm{a}_1 = S_{11}\mathrm{a}_1 \end{aligned} \tag{3.60}$$

The scattering matrix of the Michelson interferometer is

$$\mathbf{S} = \begin{bmatrix} -\sin k(l_a - l_b) & \cos k(l_a - l_b) \\ \cos k(l_a - l_b) & \sin k(l_a - l_b) \end{bmatrix} j\exp[-jk(l_a + l_b)] \tag{3.61}$$

The element S_{22} is obtained from S_{11} by interchange of l_a and l_b. This follows directly from the layout of the interferometer, Fig. 3.7, or can be derived from the losslessness condition (3.28). The transmitted power is

$$|\mathrm{b}_2|^2 = |S_{21}|^2|\mathrm{a}_1|^2 = \cos^2 k(l_a - l_b)|\mathrm{a}_1|^2 \tag{3.62}$$

Thus, in contrast to the Fabry–Perot interferometer, the transmission as a function of frequency of the Michelson interferometer is not narrowly peaked around the transmission maxima.

If the interferometer is illuminated by a source of well-defined frequency, a screen is held in the observation plane (Fig. 3.7), and one of the mirrors is translated, bright illumination alternates with zero illumination on the screen for each change of $(l_a - l_b)$ by one quarter-wavelength. If mirror M_a is tilted slightly through a small angle $\theta/2$, as shown in Fig. 3.8, the wave returning from the mirror and reflected by the half-silvered mirror propagates at an angle θ relative to the wave from arm b. For waves polarized along y in Fig. 3.8, the electric field in the wave b_2 is [compare (3.59)]

$$\mathrm{E}_y = \sqrt[4]{\frac{\mu_0}{\varepsilon_0}}\frac{j}{\sqrt{2}}\left\{\frac{1}{\cos\theta}\exp[-jkl_a(1 + \cos\theta)]e^{-jk(\cos\theta z - \sin\theta x)} + \exp(-2jkl_b)e^{-jkz}\right\}\mathrm{a}_1$$

θ is a small angle in order to obtain observable fringes, $\cos\theta \simeq 1$, $\sin\theta \simeq \theta$. The Poynting flux (intensity) as a function of position x, in a plane of constant

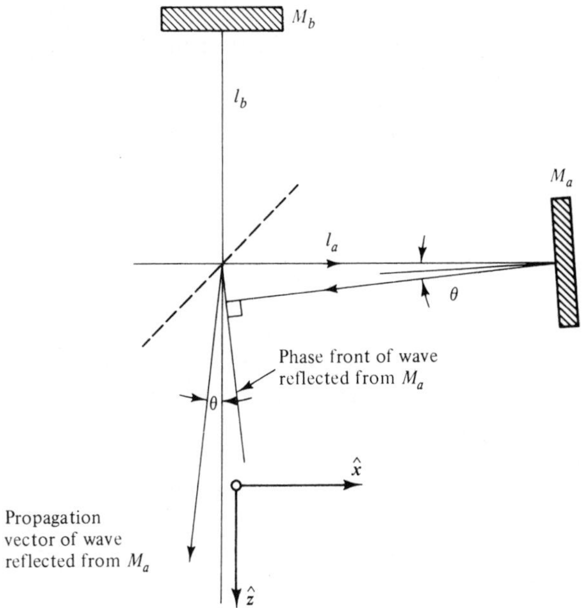

Figure 3.8 Michelson interferometer with a tilted mirror.

z, is

$$\frac{1}{2}\sqrt{\frac{\varepsilon_0}{\mu_0}}\,|\mathrm{E}_y|^2 = \cos^2 k\left(\frac{\theta}{2}\,x + l_b - l_a\,\frac{1+\cos\theta}{2}\right)|a_1|^2$$

The intensity pattern shows fringes of width λ/θ. If the angle θ is small, the fringe widths can be made much larger than a wavelength.

3.7 COHERENCE [6]

The Michelson interferometer superimposes two waves which interfere. Thus far we have not considered the possibility that the two waves may be statistical in nature and that the interference, therefore, may be imperfect.

Equation (3.59) is written in terms of a complex amplitude a_1 that has the implied time dependence $e^{j\omega t}$. If one treats a periodic time-dependent process, one has to interpret (3.59) as an equation relating the Fourier coefficients $\mathrm{a}_1(n)$ and $\mathrm{b}_2(n)$ of frequency $\omega_n = n\,\Delta\omega$, where $\Delta\omega = 2\pi/T$, with T the period of the process (compare Section 1.4):

$$\mathrm{a}_1(t) = \sum_n \mathrm{a}_1(n)e^{j\omega_n t} \tag{3.63}$$

A statistical process may be treated as the limit of a periodic process with the period T made to approach infinity.

Equation (3.59) interpreted as the relation for one Fourier coefficient of a periodic process can be cast into the time domain by multiplying it by $e^{jn\Delta\omega t}$ and summing over all n. With $k = \omega/c$:

$$\mathrm{b}_2(t) = \frac{j}{2}\,[\mathrm{a}_1(t - \tau_a) + \mathrm{a}_1(t - \tau_b)] \tag{3.64}$$

where

$$\tau_a = \frac{2l_a}{c}, \qquad \tau_b = \frac{2l_b}{c}$$

We do not distinguish quantities representing time functions or Fourier components by a change of symbol, but indicate them by the argument in parentheses. The factor j in front of (3.64) is surprising because it leads to imaginary amplitudes $\mathrm{b}_2(t)$. Our assumptions have to be reexamined. We have mentioned earlier that the **S** matrix of the mirror is not frequency independent—at negative frequencies the **S** matrix is the complex conjugate of the same matrix evaluated at positive frequencies. By assuming that $r = 1/\sqrt{2}$ and $t = j/\sqrt{2}$ for all frequencies, we have violated the requirement that t be replaced by $-t$ for negative frequencies, aside from the fact that r (and with it t) itself may be frequency dependent. If the optical signal is sufficiently narrow-band, the ω dependences of r and t can be ignored—the sign reversal of t, however, cannot. We resort here to the redefinition of Fourier transform pairs of Section 1.4 based on the single-sided Fourier transform. It is commonly used in quantum optics and avoids the preceding difficulty.

We interpret $\mathrm{a}_1(t)$ and $\mathrm{b}_2(t)$ as the complex time functions associated with the single-sided spectrum of positive frequencies. Then the transmission coefficient of the half-silvered mirror $j/\sqrt{2}$ for positive frequencies is changed in sign automatically in the expression for $\mathrm{b}_2^*(t)$ associated with the spectrum of negative frequencies.

The time-averaged power transmitted through the Michelson interferometer is

$$\langle|\mathrm{b}_2(t)|^2\rangle = \tfrac{1}{4}\{2\langle|\mathrm{a}_1(t)|^2\rangle + \langle\mathrm{a}_1(t - \tau_a)\mathrm{a}_1^*(t - \tau_b)\rangle + \langle\mathrm{a}_1^*(t - \tau_a)\mathrm{a}_1(t - \tau_b)\rangle\} \tag{3.65}$$

where we have taken into account that $\langle|a_1(t)|^2\rangle$ is time independent. For a stationary statistical process the quantity

$$\langle\mathrm{a}_1(t - \tau_a)\mathrm{a}_1^*(t - \tau_b)\rangle = \langle\mathrm{a}_1(t)\mathrm{a}_1^*(t - \tau)\rangle \tag{3.66}$$

is the complex autocorrelation function $\Gamma_{11}(\tau)$ of a_1, [6] a function of τ, where $\tau = \tau_b - \tau_a$. A detector that responds to time-averaged power mounted at the output of the interferometer detects

$$\langle|\mathrm{b}_2(t)|^2\rangle = \tfrac{1}{2}\{\Gamma_{11}(0) + \tfrac{1}{2}[\Gamma_{11}(\tau) + \Gamma_{11}^*(\tau)]\} \tag{3.67}$$

Translation of one of the mirrors traces out the real part of the autocorrelation function. If the illuminating radiation is of narrow bandwidth centered around the frequency ω_0, then it is convenient to write

$$a_1(t) = A_1(t)e^{j\omega_0 t} \tag{3.68}$$

where $A_1(t)$ varies with t much more slowly than the exponential factor; $A_1(t)$ is the envelope of $a_1(t)$.

The correlation function is

$$\Gamma_{11}(\tau) = \langle A_1(t)A_1^*(t-\tau)\rangle\, e^{j\omega_0 \tau} \tag{3.69}$$

The correlation function $\langle A_1(t)A_1^*(t-\tau)\rangle$ goes to zero when τ is much longer than the coherence time τ_c, which is defined as that value of τ for which the correlation function has decreased to $1/e$ of its value at $\tau = 0$. The dependence of $\Gamma_{11}(\tau) + \Gamma_{11}^*(\tau)$ on τ is

$$\tfrac{1}{2}(\Gamma_{11}(\tau) + \Gamma_{11}^*(\tau)) = |\langle A_1(t)A_1^*(t-\tau)\rangle| \cos(\omega_0 \tau + \phi) \tag{3.70}$$

where ϕ is the argument of $\Gamma_{11}(\tau)$. As one of the interferometer mirrors is moved, the detector output goes through maxima and minima with the ratio

$$\frac{\langle |A_1(t)|^2\rangle + |\langle A_1(t)A_1^*(t-\tau)\rangle|}{\langle |A_1(t)|^2\rangle - |\langle A_1(t)A_1^*(t-\tau)\rangle|}$$

In this way the magnitude of the normalized autocorrelation function

$$\frac{|\langle A_1(t)A_1^*(t-\tau)\rangle|}{\langle |A_1(t)|^2\rangle}$$

can be determined, and along with it the coherence time. The Michelson interferometer is well suited to the measurement of *temporal coherence*. Figure 3.9 shows the "output" $\langle |b_2(t)|^2\rangle$ as a function of τ for two statistical proc-

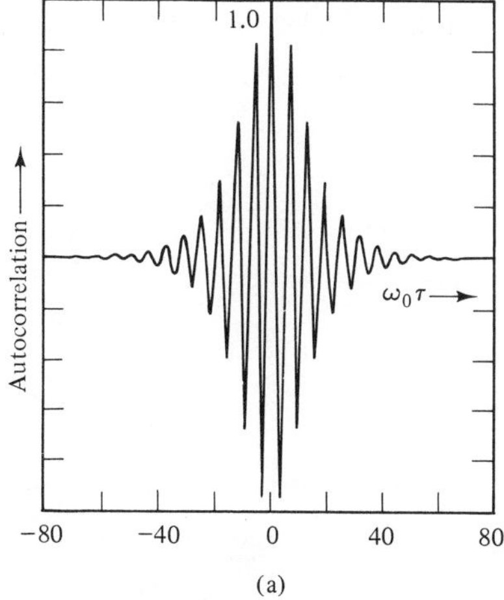

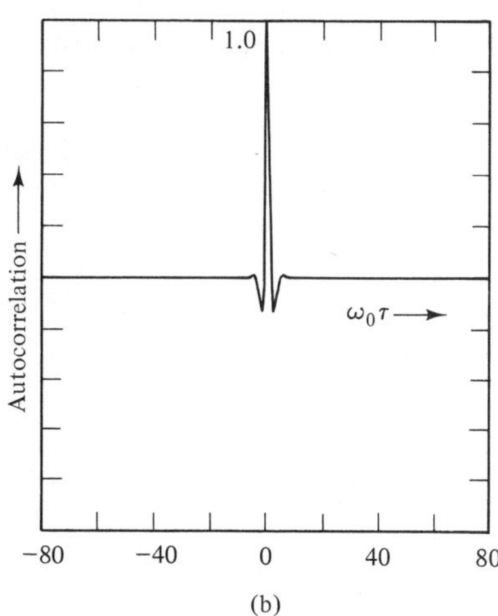

Figure 3.9 Time-averaged intensity as a function of $\omega_0\tau$: (a) $\omega_0\tau_c = 16.6$; (b) $\omega_0\tau_c = 1.66$.

esses with different coherence times. For the purpose of illustration we have picked $\langle A_1(t)A_1^*(t + \tau)\rangle = 1/\cosh \alpha\tau$ with α a constant. The coherence time τ_c is the value of τ for a decrease of this function by a factor of $1/e$; $\tau_c = 1.657/\alpha$.

The input to the Michelson interferometer was assumed to be describable in terms of a plane wave of amplitude a that was uniform across the wavefront. When this is not the case, and the amplitude a varies statistically across the wavefront, the radiation is said to be spatially incoherent (or partially coherent). Spatial coherence is measured conveniently by diffraction and will be described in the next chapter.

3.8 TWYMAN–GREEN INTERFEROMETER AND ITS USE [6]

This instrument is closely related to the Michelson interferometer (Fig. 3.10). It is used to test nominally flat optical windows and other optics whose transmission (as opposed to reflection) is important.

The Twyman–Green interferometer is set up with collimated light; the mirrors are adjusted sufficiently parallel that a single fringe covers the entire field. A test piece, say an optical "flat," is inserted in one arm. Any fringes that appear as a result of the flat's presence represent optical-path variations within the flat. For example, if the flat is slightly thicker on one side than on the other, it is said to have wedge. The nonuniform phase delay through the flat makes the mirror appear slightly tilted and nearly straight fringes appear in the observation plane. Similarly, if the surfaces of the flat are slightly spherical and not parallel, the fringes appear circular.

The Twyman–Green interferometer is also used for testing lenses. One mirror is replaced by a small, reflecting sphere, and the lens is positioned so that the center of the sphere coincides with the focal point of the lens. The beam returning through the lens is thus collimated if the lens is of high quality.

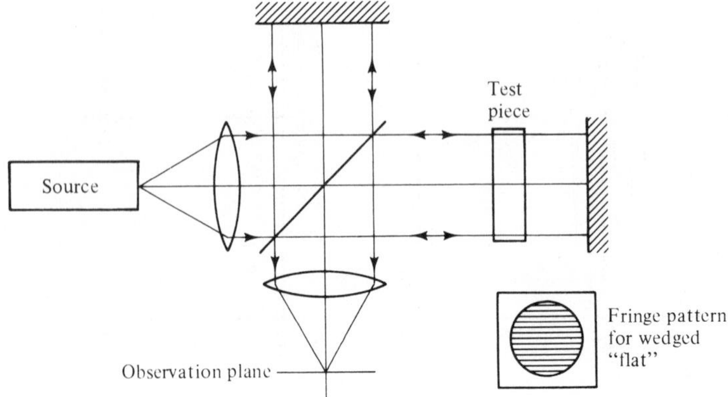

Figure 3.10 Twyman–Green interferometer.

Aberrations or other defects in the lens cause the returning wavefronts to deviate from planes and result in a fringe pattern that can be used to evaluate the performance of the lens.

3.9 SUMMARY

We introduced the concept of a linear two-port described in terms of linear relationships between the two incident and the two reflected waves. The scattering matrix of the two-port, a matrix of second rank, comprises four complex coefficients, in general functions of frequency. The reciprocity principle imposes the restriction of symmetry upon the scattering matrix. If the two-port is lossless, there are three further real relations that constrain the coefficients leaving three real parameters for the full characterization. Choice of the two reference planes disposes of two, leaving only one—the reflectivity. The scattering matrix of a partially transmitting mirror was obtained in this manner. The choice of reference planes is, in general, frequency dependent.

The Fabry–Perot interferometer was analyzed by adding the multiple reflections between the two mirrors and its scattering matrix was obtained. Because the spacing between the mirrors is usually many wavelengths long, and maximum transmission occurs for an integer number of half wavelengths between the mirrors, the frequency dependence of the transmission of power through the interferometer is usually a very rapid function of frequency. The scanning Fabry–Perot can resolve the spectrum of an incident wave if it occupies a frequency range less than the mode spacing. The Fabry–Perot also provides the basic resonator structure for laser oscillators. A standing-wave solution exists inside a Fabry–Perot interferometer in the ideal limit of perfectly reflecting mirrors for every frequency at which the spacing of the mirror reference planes is an integer multiple of half wavelengths. The field patterns pertaining to these solutions form the Fabry–Perot resonator *modes*.

The scattering matrix of the Michelson interferometer has a dependence on frequency that is more gradual than that of the Fabry–Perot. The Michelson interferometer can be used to measure the temporal coherence of light incident upon it by the ratio of the maxima and minima of the transmitted intensity. The autocorrelation function of the field can be determined in this way.

According to Section 1.5, the spectral density and the autocorrelation function are related by a Fourier transform. Thus both the Fabry–Perot interferometer and the Michelson interferometer provide basically the same information, but in different format with different ease of use. The scanning Fabry–Perot can resolve unambiguously spectra of width less than the mode spacing, whereas the Michelson interferometer is limited to coherence times less than the difference of travel times between the two arms of the interferometer. The Michelson interferometer in its Twyman–Green modification is a useful instrument for the determination of quality of optical flats.

PROBLEMS

3.1. Using time reversibility, show that zero reflection at the Brewster angle θ_B is achieved both by passage of a plane wave from the low index medium to the high index medium as well as from the high index medium to the low index medium. Show that a slab of medium onto which a plane wave is incident at the Brewster angle provides perfect transmission with no reflection at either of the two interfaces.

3.2. Check for power conservation the **S**-matrix coefficients (3.10), (3.11), and (3.12) for a TE wave incident upon and reflected from a dielectric interface.

3.3. Check the **S** matrix of the Fabry–Perot interferometer, (3.47), for power conservation.

3.4. Two-ports may be described by matrices other than scattering matrices. If one uses a_1 and b_1 as independent variables (Fig. P3.4), one may write

$$b_2 = T_{ba} a_1 + T_{bb} b_1$$

$$a_2 = T_{aa} a_1 + T_{ab} b_1$$

(a) Find the elements of the **T** matrix in terms of the scattering matrix coefficients. Show that det $[\mathbf{T}] = 1$ when $S_{12} = S_{21}$.

(b) Find the **T** matrix for a lossless mirror of reflectivity r.

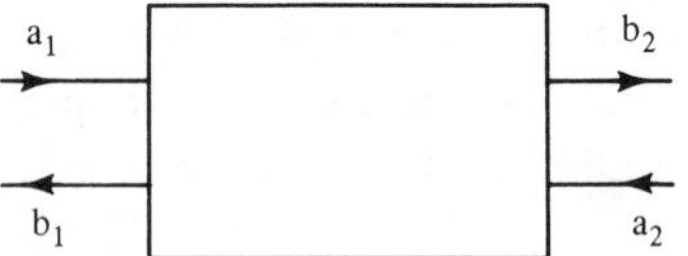

Figure P3.4

3.5. The **T** matrix of Problem 3.4 is particularly well suited to analyze a cascade of two-ports. Consider the two-ports α and β in Fig. P3.5. The output of two-port α is the input of two-port β. $\mathbf{v}_\alpha = \mathbf{T}_\alpha \mathbf{u}_\alpha$, $\mathbf{v}_\beta = \mathbf{T}_\beta \mathbf{u}_\beta = \mathbf{T}_\beta \mathbf{v}_\alpha = \mathbf{T}_\beta \mathbf{T}_\alpha \mathbf{u}_\alpha$. The **T** matrix of a cascade of two-ports is the matrix product of the **T** matrices of the individual two-ports.

$$\mathbf{u}_\alpha = \begin{bmatrix} a_1^{(\alpha)} \\ b_1^{(\alpha)} \end{bmatrix} \qquad \mathbf{v}_\alpha = \begin{bmatrix} b_2^{(\alpha)} \\ a_2^{(\alpha)} \end{bmatrix} = \begin{bmatrix} a_1^{(\beta)} \\ b_1^{(\beta)} \end{bmatrix} = \mathbf{u}_\beta$$

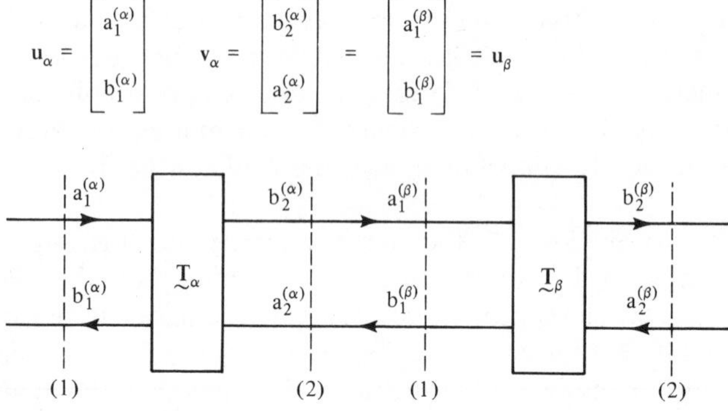

Figure P3.5 Transformation by a cascade of two-ports.

(a) Find the **T** matrix for propagation of a forward wave and a backward wave with $\boldsymbol{k}$ vectors inclined by θ with respect to the z axis, between reference planes at $z = 0$ and $z = l$.

(b) Derive the T matrix for a Fabry–Perot by cascading a mirror, a free-space section of length l, and a mirror. Confirm that $1/T_{ab}$ of the cascade is equal to S_{21} of the scattering matrix (3.47).

3.6. Suppose that the loss per single pass in a symmetric Fabry–Perot is $L(\ll 1)$. You may incorporate the loss into the transmission matrix of each of the mirrors by setting

$$t^2 = (1 - r^2 - L)$$

Using this treatment in (3.49), determine the transmitted power as a function of L, for given r.

3.7. For the Fabry–Perot with $r_1 \neq r_2$, determine the finesse, that is, the mode spacing (in frequency) divided by the FWHM. Determine the power transmitted at resonance as a function of $R_1 \equiv r_1^2$ and $R_2 \equiv r_2^2$. Show that it is unity for $R_1 = R_2$ and less than unity otherwise.

3.8. The frequencies emitted by a laser in a Fabry–Perot configuration can be evaluated from the Fabry–Perot transmission formula (3.50). The laser medium has gain, so that $k = (\omega/c)n$ must be replaced by $(\omega/c)n + j\alpha_g$, where $\alpha_g(>0)$ is the spatial rate of growth of the amplitude. This makes δ complex! (Assume that $r_1 = r_2$.)

(a) Find an expression for the ratio of transmitted power to incident power in a symmetric Fabry–Perot interferometer for normal incidence ($\theta = 0$) as a function of $\alpha_g l$ at resonance $[(\omega/c)nl = m\pi]$.

(b) What gain factor $e^{2\alpha_g l}$ is required to make the ratio infinite? At this value of $a_g l$ the Fabry–Perot has an output without any input—it lases.

3.9. An He-Ne laser emitting around a "frequency" measured in wavelength, $\lambda =$ 6328 Å, can oscillate at several different frequencies corresponding to its resonator modes (Fabry–Perot modes) spaced $\Delta f = c/2l$ apart [compare 3.52)]. Assume that three modes produce output waves $\mathrm{a}^{(k)}$ ($k = 1, 2, 3$) of equal amplitude but phases varying randomly with time. Suppose that the wave $\mathrm{a}_1 = \sum_k \mathrm{a}^{(k)}$ is the "input" to a Michelson interferometer. Predict the output intensity $\langle |\mathrm{b}_2|^2 \rangle$ as a function of $l_a - l_b$.

3.10. The Michelson interferometer produces no output wave b_2 when the mirror spacing is adjusted properly. The purpose of the problem is to check for the reflected wave b_1 in the input arm as a function of arm lengths l_a and l_b. What standing-wave ratio

$$\frac{|\mathrm{a}_1| + |\mathrm{b}_1|}{|\mathrm{a}_1| - |\mathrm{b}_1|}$$

is observed in the "input" as a function of l_a and l_b? Can you comment on power conservation?

REFERENCES

[1] K. Kurokawa, *An Introduction to the Theory of Microwave Circuits*, Academic Press, New York, 1969, p. 219.

[2] R. E. Collins, *Foundations for Microwave Engineering*, McGraw-Hill, New York, 1966.

[3] J. C. Slater, *Microwave Circuits*, McGraw-Hill, New York, 1950.

[4] R. N. Ghose, *Microwave Circuit Theory and Analysis*, McGraw-Hill, New York, 1963.

[5] C. Fabry and A. Perot, "Théorie et applications d'une nouvelle méthode de spectroscopie interférentielle," *Ann. Chim. Phys.*, *16*, 115, 1899.

[6] M. Born and E. Wolf, *Principles of Optics*, Macmillan, New York, 1964.

[7] A. Yariv, *Introduction to Optical Electronics*, Holt, Rinehart and Winston, New York, 1976.

4

FRESNEL DIFFRACTION IN PARAXIAL LIMIT

The preceding chapters were concerned with the propagation, reflection, and transmission of plane waves and with systems utilizing these phenomena. Any physical optical beam is of finite transverse cross section. Beams of finite cross section may be described in terms of a superposition of plane waves, analogous to the representation of a time function of finite duration in terms of a Fourier superposition of sinusoids. The paraxial approximation simplifies the analysis of the propagation of beams of finite cross section carried out in this chapter. We use a superposition of plane waves to construct such a beam and derive its propagation properties in terms of the paraxial Fresnel diffraction integral. We study Fraunhofer diffraction, which is the far-field limit of Fresnel diffraction. The Fraunhofer diffraction pattern of a rectangular aperture and an array of such apertures is derived. Then we study the effect of a thin lens upon a beam passing through it in preparation for the spatial Fourier transform operation of lenses.

We conclude by an alternative approach to the analysis of optical beams, the paraxial wave equation, and show how the Fresnel diffraction integral relates to this equation.

4.1 THE FRESNEL DIFFRACTION INTEGRAL [1, 2]

The Fresnel diffraction integral expresses the amplitude distribution of a scalar wave (like a pressure wave) at any cross section of constant z, in terms of a given distribution at $z = 0$. To apply the scalar diffraction theory to an electromagnetic field, one has to replace the wave equation of the electromagnetic field by the scalar wave equation.

In Chapter 2 we studied the propagation of the complex electric field associated with a plane electromagnetic wave of frequency ω and propagation vector $\boldsymbol{k}$. The field obeys the vector wave equation (1.52). The scalar wave equation results from the vector wave equation if one looks for solutions of the form $\hat{\boldsymbol{n}}\psi(x, y, z)$, where $\hat{\boldsymbol{n}}$ is a unit vector in some direction and $\psi(x, y, z)$ is a scalar function. We have mentioned that an electric field must be divergence-free in free space and thus the substitution of $\hat{\boldsymbol{n}}\psi(x, y, z)$ for the electric field $\mathbf{E}$ is not legitimate, in general.

In free space, the vector potential also obeys the vector wave equation, (1.29), and $\nabla \cdot \mathbf{A}$ need not be zero; it simply defines the scalar potential Φ by (1.26). Thus if we assume the functional dependence

$$\mathbf{A}(\boldsymbol{r}, t) = \hat{\boldsymbol{n}}\psi(x, y, z)e^{j\omega t} \tag{4.1}$$

in (1.29), we obtain

$$\nabla^2\psi + k^2\psi = 0 \tag{4.2}$$

The amplitude $\psi(x, y, z)$ of the vector potential obeys a scalar wave equation and no other constraints are imposed on ψ. It is possible to restore the vector nature of the vector potential to the solutions of the scalar wave equation through proper identification of the field polarization.

A general plane-wave solution of the scalar wave equation in Cartesian coordinates is of the form

$$e^{-jk_x x}e^{-jk_y y}e^{-jk_z z}$$

with

$$k_x^2 + k_y^2 + k_z^2 = k^2 = \frac{\omega^2}{c^2} = \left(\frac{2\pi}{\lambda}\right)^2 \tag{4.3}$$

If the propagation vector $\boldsymbol{k}$ is inclined by a small angle with respect to the z axis (see Fig. 4.1), then the wave vector is *paraxial*, and

$$k_z = \sqrt{k^2 - k_x^2 - k_y^2} \simeq k - \frac{k_x^2 + k_y^2}{2k} \tag{4.4}$$

This is the paraxial approximation for the z component of $\boldsymbol{k}$. We shall limit ourselves to fields that are composed of waves that obey this restriction. Henceforth we shall not carry along explicitly the phase factor $\exp(-jkz)$. We shall define

$$\psi(x, y, z) = u(x, y, z) \exp(-jkz) \tag{4.5}$$

and use $u(x, y, z)$ as the amplitude distribution. Whenever the complete field solution is sought, the factor $\exp(-jkz)$ will be restored.

Let us build up an amplitude distribution $u(x, y, z)$ by superposition of

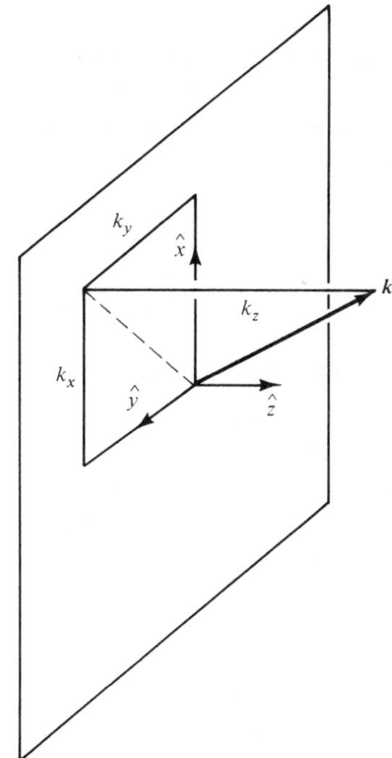

Figure 4.1 The $\boldsymbol{k}$-vector decomposition.

plane waves [compare (1.77)]:

$$u(x, y, z) = \int_{-\infty}^{\infty} dk_x \int_{-\infty}^{\infty} dk_y \; U_0(k_x, k_y) e^{-j(k_x x + k_y y)} e^{[j(k_x^2 + k_y^2)/2k]z} \tag{4.6}$$

$U_0(k_x, k_y)$ is the amplitude of the plane wave solution with particular transverse components of $\boldsymbol{k}$, k_x, and k_y. In order that the paraxial approximation be valid, the function $U_0(k_x, k_y)$ must vanish for arguments that lie outside the range

$$\frac{\sqrt{k_x^2 + k_y^2}}{k} \ll 1$$

The meaning of $U_0(k_x, k_y)$ becomes clear by noting that, at $z = 0$, the amplitude distribution $u_0(x, y)$ is, according to (4.6),

$$u_0(x, y) = \int_{-\infty}^{\infty} dk_x \int_{-\infty}^{\infty} dk_y \; U_0(k_x, k_y) e^{-j(k_x x + k_y y)} \tag{4.7}$$

The wave amplitude function $U_0(k_x, k_y)$ is the Fourier transform of the amplitude distribution at $z = 0$, $u_0(x, y)$. Expressing $U_0(k_x, k_y)$ in terms of $u_0(x, y)$, using (1.78) one has the relation

$$U_0(k_x, k_y) = \left(\frac{1}{2\pi}\right)^2 \int_{-\infty}^{\infty} dx_0 \int_{-\infty}^{\infty} dy_0 \; u_0(x_0, y_0) e^{j(k_x x_0 + k_y y_0)} \tag{4.8}$$

In the paraxial approximation, which is valid when the field is known to be expressible in terms of a superposition of waves with *paraxial* wave vectors, one is able to express the solution to the scalar wave equation $u(x, y, z)$, in terms of the known distribution $u_0(x, y)$, at $z = 0$. Combining (4.6) and (4.8), one has

$$u(x, y, z) = \int_{-\infty}^{\infty} dx_0 \int_{-\infty}^{\infty} dy_0\, u_0(x_0, y_0) \left(\frac{1}{2\pi}\right)^2 \int_{-\infty}^{\infty} dk_x \int_{-\infty}^{\infty} dk_y \cdot e^{-j[k_x(x-x_0)+k_y(y-y_0)]} e^{j[(k_x^2+k_y^2)/2k]z} \tag{4.9}$$

This expression is the convolution of $u_0(x, y)$ with the *Fresnel kernel*

$$h(x, y, z) = \left(\frac{1}{2\pi}\right)^2 \int_{-\infty}^{\infty} dk_x \int_{-\infty}^{\infty} dk_y\, e^{-j(k_x x + k_y y)} e^{j[(k_x^2+k_y^2)/2k]z} \tag{4.10}$$

The integral in (4.10) can be carried out by completion of the square in the exponent. Consider the integral over k_x:

$$\begin{aligned}\int_{-\infty}^{\infty} dk_x\, e^{-jk_x x} e^{j(k_x^2/2k)z} &= e^{-j(kx^2/2z)} \int_{-\infty}^{\infty} dk_x\, e^{j[k_x - k(x/z)]^2 z/2k} \\ &= \sqrt{\frac{jk}{z}}\, e^{-j(kx^2/2z)} \int_{-\infty}^{\infty} d\xi\, e^{-\xi^2/2} = \sqrt{\frac{2\pi jk}{z}}\, e^{-j(kx^2/2z)}\end{aligned}$$

where

$$\xi^2 \equiv -j\frac{[k_x - k(x/z)]^2 z}{k}$$

An analogous procedure is applied to the integral over k_y. The result for the Fresnel kernel is

$$h(x, y, z) = \frac{j}{\lambda z} e^{-jk[(x^2+y^2)/2z]} \tag{4.11}$$

Equation (4.11) introduced in (4.9) gives the *Fresnel diffraction integral* in the paraxial approximation

$$\begin{aligned}u(x, y, z) &= \frac{j}{\lambda z} \int_{-\infty}^{\infty} dx_0 \int_{-\infty}^{\infty} dy_0\, u_0(x_0, y_0) e^{-j(k/2z)[(x-x_0)^2+(y-y_0)^2]} \\ &= h \otimes u_0\end{aligned} \tag{4.12}$$

This integral is useful, because one can often make intelligent guesses about the distribution $u_0(x_0, y_0)$. If, for example, a plane wave is incident upon a thin absorbing screen with a hole as shown in Fig. 4.2, the wave is, presumably, absorbed except in the hole through which it proceeds with minimal disturbance if the hole diameter is of the order of many wavelengths λ. The distribution of the vector potential at $z = 0$ to the right of the screen for a

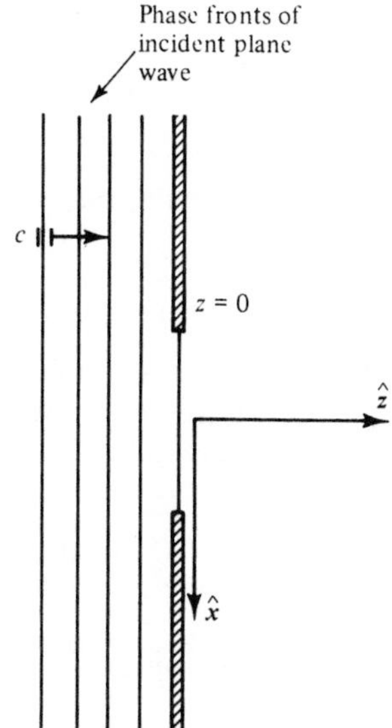

Figure 4.2 Absorbing screen with hole.

y-polarized electric field is

$$\mathbf{A} = \frac{j}{\omega}\hat{y}\mathrm{E}_+$$

and $u(x_0, y_0)$ is uniform across the hole, zero outside.

The effect on $u_0(x_0, y_0)$ of propagation through free space is expressed by the Fresnel diffraction integral. The function $h(x, y, z)$ in (4.12) is the spatial "impulse response" of diffraction optics to an excitation at $x = y = z = 0$. This fact follows from the observation that (4.12) must yield $u_0(x_0, y_0)$ when the limit $z \to 0$ is approached. The way in which h acts as an impulse function in this limit may be visualized by considering the behavior of h as given in (4.11) in the limit of small z. The phase of h as a function of x and y has zero slope at $x = 0$, $y = 0$ (is stationary) but then varies infinitely fast in the limit $z \to 0$, as x and y are increased by finite amounts. Convolution of a function $u_0(x_0, y_0)$ with $\lim_{z\to 0} h(x, y, z)$ reproduces the function, because u contributes to the integral only at the point of stationary phase of h.

The integral (4.12) further shows that $u(x, y, z)$ will remain roughly equal to $u_0(x_0, y_0)$ over those distances z over which h has not "broadened" significantly; the first nulls of h near $x = y = 0$ must occupy a range much smaller than the range over which u_0 varies appreciably. Suppose that u varies appreciably only over a scale $\Delta x \geq d_x$, $\Delta y \geq d_y$. As long as

$$\frac{k}{2z} d_x^2 \gg \pi, \qquad \frac{k}{2z} d_y^2 \gg \pi \tag{4.13}$$

$u(x, y, z)$ is essentially independent of z. The wave of finite transverse extent retains its profile as it propagates in the "near-field region" defined by

$$z \ll \frac{d_x^2}{\lambda}, \qquad \frac{d_y^2}{\lambda}$$

We may further show that, within the near-field region, a profile with an inclined phase front

$$u_0(x_0, y_0) = f_0(x_0, y_0)e^{-jk_x x_0}$$

is undistorted and shifts in the transverse direction as it propagates along z. Here $f_0(x_0, y_0)$ is assumed to vary with x_0 much less rapidly than $\exp(-jk_x x_0)$. Equation (4.12) gives:

$$\begin{aligned} h \otimes u_0 &= \frac{j}{\lambda z}\int_{-\infty}^{\infty} dx_0 \int_{-\infty}^{\infty} dy_0\, f_0(x_0, y_0)e^{-jk_x x_0}e^{-j(k/2z)[(x-x_0)^2+(y-y_0)^2]} \\ &\simeq \sqrt{\frac{j}{\lambda z}}\int_{-\infty}^{\infty} dx_0\, f_0(x_0, y_0)e^{-j(k/2z)[x-(k_x/k)z-x_0]^2}e^{-jk_x x}e^{j(k/2)(k_x/k)^2 z} \\ &\simeq f_0\left(x - \frac{k_x}{k}z, y\right)e^{-jk_x x}e^{j(k/2)(k_x/k)^2 z} \end{aligned} \tag{4.14}$$

where we have used the unit impulse nature of

$$\sqrt{\frac{j}{\lambda z}}\exp\left[-j\frac{k}{2z}\left(x - \frac{k_x}{k}z - x_0\right)^2\right] \quad \text{and} \quad \sqrt{\frac{j}{\lambda z}}\exp\left[-j\frac{k}{2z}(y - y_0)^2\right]$$

in the limit of small z.

We clearly see the shift of the profile in the argument $x - (k_x z)/k$. Further there is the inclination of the wavefront as expressed by the factor $e^{-jk_x x}$. Finally, there is the change of the phase delay due to the inclination angle k_x/k of the $\boldsymbol{k}$-vector with respect to the z-axis represented by the last factor.

In Fourier transform coordinates [i.e., in terms of the amplitude function $U_0(k_x, k_y)$], the propagation through free space is expressed much more simply than in (4.12). From (4.6) we see that $U(k_x, k_y, z)$, the Fourier transform of $u(x, y, z)$, is given by

$$U(k_x, k_y, z) = U_0(k_x, k_y)e^{j[(k_x^2+k_y^2)/2k]z} \tag{4.15}$$

We define the function

$$H(k_x, k_y, z) = \frac{1}{(2\pi)^2}e^{j[(k_x^2+k_y^2)/2k]z} \tag{4.16}$$

It is the Fourier transform of the function $h(x, y, z)$. In this shorthand notation, (4.15) becomes

$$U(k_x, k_y, z) = (2\pi)^2 H(k_x, k_y, z)U_0(k_x, k_y) \tag{4.17}$$

In Fourier transform space, the convolution of the two functions is replaced by the product of their Fourier transforms in accordance with the convolution theorem.

In all our investigations of wave propagation in the z direction, the direction of the vector potential $\hat{\boldsymbol{n}}$ will be picked in the x-y plane. The electric field $\mathbf{E}$ will be, then, polarized mainly in the x-y plane, although small longitudinal components along z are predicted as required in order to satisfy the condition of $\nabla \cdot \mathbf{E} = 0$. This follows from (1.23) and (1.26) applied to the complex form of the vector potential. Indeed, from (1.23),

$$\mathbf{E} = -j\omega\mathbf{A} - \nabla\Phi \tag{4.18}$$

and the scalar potential is

$$\Phi = \frac{j}{\omega\mu_0\,\varepsilon_0}\,\nabla \cdot \mathbf{A} \tag{4.19}$$

Thus

$$\mathbf{E} = -j\omega\left(\mathbf{A} + \frac{1}{\omega^2\mu_0\,\varepsilon_0}\,\nabla\nabla \cdot \mathbf{A}\right) \tag{4.20}$$

If $\mathbf{A} = \hat{\boldsymbol{n}}\, u(x, y, z) \exp(-jkz)$ is transverse to $\hat{z}$, the divergence involves derivatives with respect to the (much slower) transverse variation. The scalar potential contributes a longitudinal component of the electric field that is much smaller than the transverse component and gives an even smaller contribution to the transverse component.

The power flow density along the z axis is approximately equal to

$$\frac{1}{2}\,\mathrm{Re}\,[\mathbf{E} \times \mathbf{H}^* \cdot \hat{z}] \simeq \frac{1}{2}\sqrt{\frac{\varepsilon_0}{\mu_0}}\,\omega^2\,|u|^2 \tag{4.21}$$

In Section 4.5 we shall prove the correctness of this interpretation.

4.2 FRAUNHOFER DIFFRACTION [1]

Fraunhofer diffraction is the limit of Fresnel diffraction for large distances between the input plane at $z = 0$, at which $u_0(x_0, y_0)$ is specified, and the observation plane at z. In this limit, the *far-field limit*, one approximates the argument of the exponential in (4.12):

$$\frac{k}{z}\,[(x - x_0)^2 + (y - y_0)^2] \simeq \frac{k}{z}\,[(x^2 + y^2) - 2xx_0 - 2yy_0] \tag{4.22}$$

and ignores the term $k(x_0^2 + y_0^2)/z$. Thus the Fraunhofer approximation is valid if the amplitude distribution in the input plane extends over a transverse dimension d such that

$$d \ll \sqrt{\frac{z}{k}}$$

In this limit

$$u(x, y, z) = \frac{j}{\lambda z} e^{-j[k(x^2+y^2)/2z]} \int_{-\infty}^{\infty} dx_0 \int_{-\infty}^{\infty} dy_0 \, u_0(x_0, y_0) \cdot \exp\left[\frac{jk}{z}(xx_0 + yy_0)\right] \tag{4.23}$$

The integral over x_0 and y_0 produces the Fourier transform of $u_0(x_0, y_0)$ [compare (4.8)]. Denote it by U_0. The arguments of U_0 are kx/z and ky/z:

$$u(x, y, z) = j\frac{(2\pi)^2}{\lambda z} \exp\left[-j\frac{k(x^2+y^2)}{2z}\right] U_0\left(\frac{kx}{z}, \frac{ky}{z}\right) \tag{4.24}$$

Note, however, that there is a phase factor multiplying the amplitude distribution: $\psi(x, y, z) = u(x, y, z) \exp(-jkz)$, where

$$\text{phase factor} = \exp\left\{-j\left[\frac{k(x^2+y^2)}{2z} + kz\right]\right\}$$

indicating that the phase front is curved. To derive the curvature of the phase front on axis, we explore the surface of constant phase ϕ for small deviations $\sqrt{x^2+y^2}$ from the axis. On the axis, $x^2 + y^2 = 0$ and $kz = \phi$. For small angles (paraxial approximation) $\sqrt{x^2+y^2} \ll z$ and z in the denominator in the exponent of the phase factor can be replaced by a constant $z \simeq \phi/k$. Therefore, the equation of constant phase is

$$\frac{k^2(x^2+y^2)}{2\phi} + kz = \phi$$

This is the equation of a paraboloid. The parabola produced by intersection of the x-z plane with the paraboloid ($y = 0$) has the radius of curvature R at $x = 0$ given by

$$\frac{1}{R} = \frac{-d^2z/dx^2}{\sqrt{1 + (dz/dx)^{2^3}}} = \frac{k}{\phi} = \frac{1}{z} \tag{4.25}$$

The Fraunhofer diffraction pattern has the radius of curvature that would be produced at z by a point source at the center of the aperture. The radius of curvature is defined positive if the phase front is convex when viewed from the right (larger z).

Consider as an example the uniform illumination of a slit at $z = 0$:

$$u_0(x_0, y_0) = \begin{cases} 1, & |x_0| < d_x/2; \quad 0 < |y_0| < d_y/2 \\ 0, & d_x/2 \le |x_0|; \quad d_y/2 \le |y_0| \end{cases}$$

Then

$$u(x, y, z) = \frac{j}{\lambda z} \exp\left[-\frac{jk(x^2+y^2)}{2z}\right] d_x \, d_y \frac{\sin(kd_x x/2z) \sin(kd_y y/2z)}{(kd_x x/2z)(kd_y y/2z)} \tag{4.26}$$

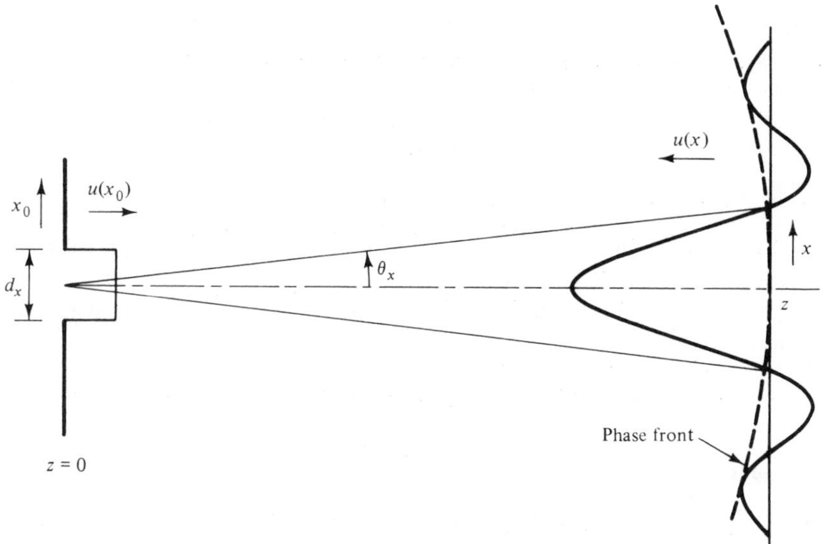

Figure 4.3 Fraunhofer diffraction pattern of uniformly irradiated rectangular slit. Amplitude is plotted without phase factor.

The widths of the diffraction patterns in the x and y directions are characterized by the first nulls at

$$x = \frac{2\pi z}{kd_x} \qquad \text{and} \qquad y = \frac{2\pi z}{kd_y}$$

The angles subtended from the center of the slit to the first nodal lines of the diffraction pattern (see Fig. 4.3) are the diffraction angles,

$$\theta_x = \frac{2\pi}{kd_x} = \frac{\lambda}{d_x}, \qquad \theta_y = \frac{2\pi}{kd_y} = \frac{\lambda}{d_y} \tag{4.27}$$

Figure 4.4 shows a three-dimensional plot of (4.26) with the phase factor omitted.

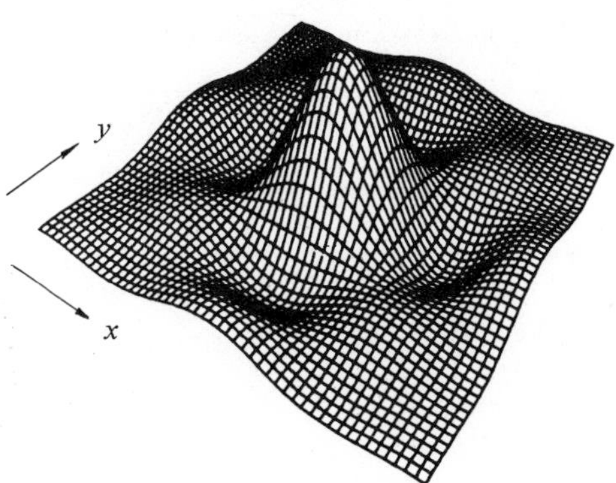

Figure 4.4 Plot of (4.26) without phase factor. Origin is at center, the directions of x and y are as shown.

Next, we consider an array of slits, N in number, as shown in Fig. 4.5. They are excited uniformly and in phase. This array of slits can be the model of a grating as discussed in Section 2.5. Each slit has its own excitation pattern, $u_n(x_0, y_0)$. To derive the Fraunhofer diffraction pattern one starts with (4.23) and replaces $u_0(x_0, y_0)$ by

$$\sum_n u_0(x_0 - n\Lambda, y_0)$$

a summation over the spatially shifted amplitude distributions of the $n = 0$ slit. A change of variable in each term of the sum, $x_0 \to x_0' + n\Lambda$, $y_0 \to y_0'$ makes all terms in the integral (4.23) identical in form. The exponential associated with each term changes, however.

$$\exp\left[\frac{jk}{z}(xx_0 + yy_0)\right] \to \exp\left[\frac{jk}{z}(xx_0' + yy_0')\right]\exp\left(nj\frac{kx}{z}\Lambda\right)$$

The integral can be rewritten

$$u(x, y, z) = \frac{j}{\lambda z}e^{-[jk(x^2+y^2)/2z]}\left\{\int_{-\infty}^{\infty}dx_0'\int_{-\infty}^{\infty}dy_0'\,u_0(x_0', y_0')\right.$$
$$\left.\cdot\exp\left[\frac{jk}{z}(xx_0' + yy_0')\right]\right\}\sum_{n=-[(N-1)/2]}^{(N-1)/2}\exp\left(nj\frac{kx}{z}\Lambda\right) \tag{4.28}$$

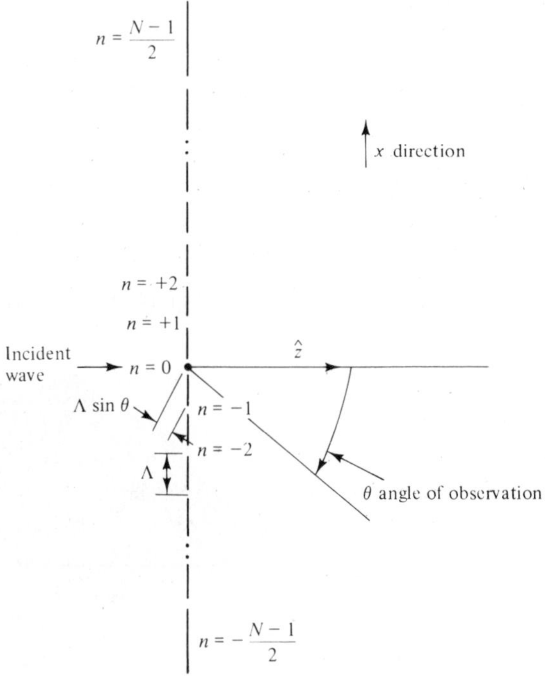

Figure 4.5 "Array" of slits.

The diffraction pattern is that of a single slit multiplied by the array factor

$$A(\theta, \phi) \equiv \sum_{n=-[(N-1)/2]}^{(N-1)/2} \exp(jnk\Lambda \sin\theta \cos\phi) \tag{4.29}$$

where we have introduced spherical coordinates and set

$$\frac{x}{z} = \sin\theta \cos\phi$$

The sum can be carried out in closed form:

$$A(\theta, \phi) = \frac{\sin\left(\dfrac{Nk\Lambda \sin\theta \cos\phi}{2}\right)}{\sin\left(\dfrac{k\Lambda \sin\theta \cos\phi}{2}\right)} \tag{4.30}$$

Figure 4.6 shows the array factor for different N and Λ values as a function of θ, for $\phi = 0$.

The present analysis can be adapted to treat the resolution limit of a grating spectrometer as discussed in Section 2.7. For this purpose one has to treat oblique incidence and/or reflection. The array factor is easily adapted to this more general case. As Fig. 4.7 shows, the nth slit of the array of slits is illuminated obliquely by an incident wave with its $\boldsymbol{k}$ vector in the x-z plane at an angle θ_i with respect to the z axis. The amplitude pattern at the center of the nth slit is phase advanced by $k \sin\theta_i n\Lambda$. In the far field, at the angle θ of observation, the radiation emanating from the nth slit is phase delayed by $k \sin\theta \cos\phi\, n\Lambda$. The array factor $A(\theta)$ is thus

$$A(\theta) = \frac{\sin[(Nk\Lambda/2)(\sin\theta \cos\phi - \sin\theta_i)]}{\sin[(k\Lambda/2)(\sin\theta \cos\phi - \sin\theta_i)]} \tag{4.31}$$

If the array is excited by reflection, θ_i is to be replaced by θ_R, where θ_R is the reflection angle. The array factor (4.31) is a more general result than Snell's law for a grating, (2.50). As N, the number of slits (or grooves) in (4.31), approaches infinity, the array factor splits into a sequence of impulse functions centered at

$$\sin\theta_R \cos\phi = \sin\theta_i + \frac{2\pi m}{k\Lambda}$$

In the plane of incidence, $\cos\phi = 1$, the above reduces to (2.50). The array factor contains in addition the width of the "orders" of transmission or reflection as determined by the finite width of the grating. The width of the pattern for $\sin\theta_i = 0$ between the first two nulls is

$$\delta\theta \simeq \frac{4\pi}{Nk\Lambda} = \frac{2\lambda}{N\Lambda}$$

which checks with (4.27) for $d_x = N\Lambda$.

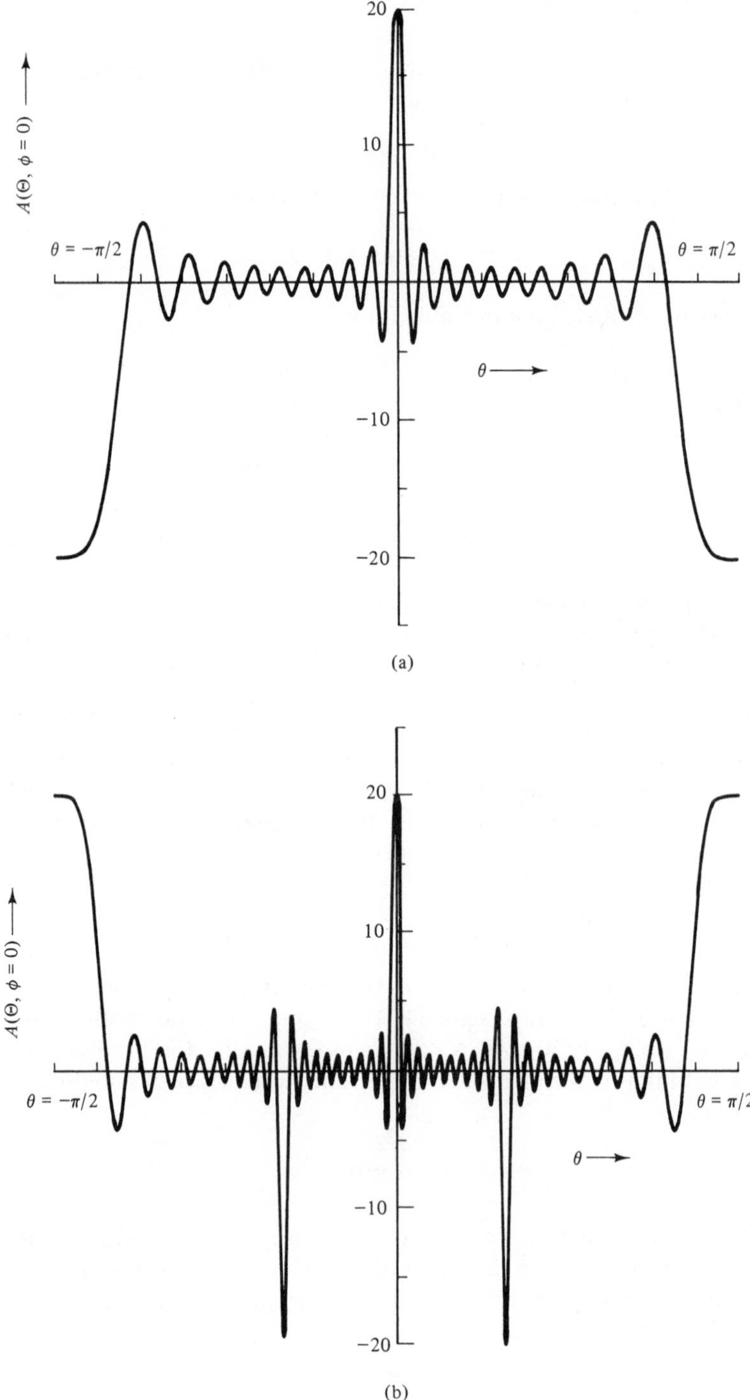

Figure 4.6 Array factor for array of slits: (a) $N = 20$, $k\Lambda = 2\pi$: (b) $N = 20$, $k\Lambda = 4\pi$.

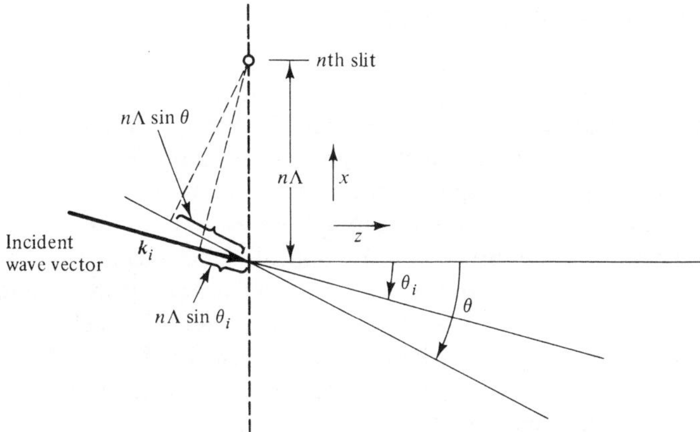

Figure 4.7 Array of slits with incident wave at angle θ_i.

The array factor (4.31) was used in Section 2.7 to determine the resolution of the grating spectrometer.

Spatial Coherence

Fraunhofer diffraction of two slits provides a means for measuring spatial coherence. Equation (4.28) applied to two slits separated by a distance Λ and illuminated by wave amplitudes a(0) and a(Λ) gives, in the paraxial limit,

$$u = u_0(x, y, z)[\mathrm{a}(0) + \mathrm{a}(\Lambda) \exp(-jk\Lambda\theta)]$$

where $u_0(x, y, z)$ is the diffraction pattern of the slit centered at $x_0 = y_0 = 0$. The time-averaged intensity pattern is then

$$\langle |u|^2 \rangle = |u_0|^2[\langle |\mathrm{a}(0)|^2 \rangle + \langle |\mathrm{a}(\Lambda)|^2 \rangle + \langle \mathrm{a}(0)\mathrm{a}^*(\Lambda) \rangle \exp(jk\Lambda\theta) + \langle \mathrm{a}^*(0)\mathrm{a}(\Lambda) \rangle \exp(-jk\Lambda\theta)] \tag{4.32}$$

When Λ is very much smaller than the *coherence length* λ_c, and both slits have the same illumination, so that

$$\langle |\mathrm{a}(0)|^2 \rangle = \langle |\mathrm{a}(\Lambda)|^2 \rangle$$

then

$$\langle |\mathrm{a}(0)\mathrm{a}^*(\Lambda)| \rangle \simeq \langle |\mathrm{a}(0)|^2 \rangle$$

and the observed fringe pattern as a function of θ has perfect contrast. When Λ is much larger than λ_c, then $\langle \mathrm{a}(0)\mathrm{a}^*(\Lambda) \rangle \simeq 0$ and no interference fringes are seen. Thus observation of the fringe pattern of a pair of slits as a function of slit separation yields information about the coherence length λ_c. Of course, the fringe pattern becomes finer and finer with increasing Λ and may have to be magnified by a microscope when Λ becomes too large.

4.3 THE ACTION OF A THIN LENS

Thus far we have studied the propagation of an optical beam of finite cross section through free space. We have derived the Fresnel diffraction integral in the paraxial approximation, which expresses the amplitude pattern (of the vector potential) $u(x, y, z) \exp(-jkz)$ at any position x, y, z in terms of the amplitude pattern $u_0(x_0, y_0)$ at the "input cross section" $z = 0$.

The beam can be processed by optical components such as lenses, prisms, gratings. We have studied the action of a grating in Chapter 2; the grating splits the incident waves into different orders, to each of which the Fresnel diffraction integral can be applied in turn. In this and the next sections, we shall be concerned with the action of a thin lens on an optical beam in the paraxial approximation.

Consider a wave with the complex amplitude profile $u(x, y)$ incident upon a doubly convex *converging* lens of focal distance f as shown in Fig. 4.8. The portion of the wavefront near the axis of the lens travels through a thicker section of the optically dense material of the lens than the portion farther away from the axis. The phase delay experienced by the center portion of the wavefront is larger than that of the off-axis portion, the phase delay is a monotonically decreasing function of distance from the lens axis. If the lens is *thin*, then the output profile $u'(x, y)$ has the same amplitude as the input profile; the lens only introduces an (x, y)-dependent phase delay $\phi(x, y)$. An ideal thin lens produces the phase delay

$$\phi(x, y) = \phi_0 - \frac{k}{2f}(x^2 + y^2) \tag{4.33}$$

where ϕ_0 is the delay at the center, and f is the focal distance. Ignoring the constant phase, the action of the lens is represented by the output (u')–input (u)

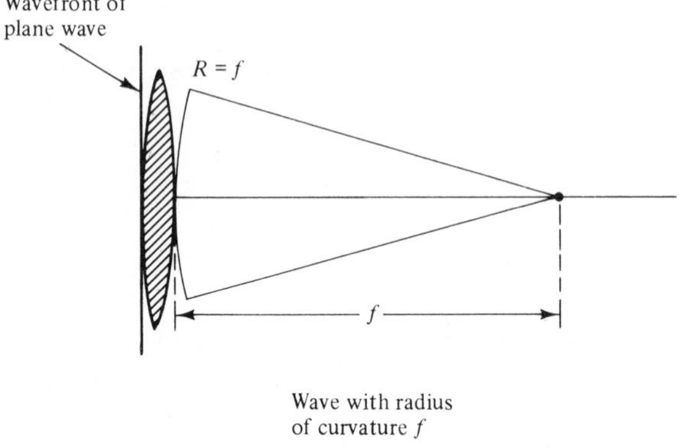

Figure 4.8 Plane wave focused by a lens of focal distance f.

relation:

$$u'(x, y) = u(x, y) \exp\left[j \frac{k}{2f}(x^2 + y^2)\right] \equiv u(x, y)l(x, y) \tag{4.34}$$

The focal distance f characterizes the lens. It is familiar as the distance from the lens at which rays of an incident plane wave meet after passage through the lens (the focus).

We show next that a plane wave acted on by a lens according to (4.33) comes to a focus a distance f behind the lens. Consider the full-phase factor of a plane wave after passage through the lens:

$$\text{phase factor} = \exp\left\{-j\left[-\frac{k(x^2 + y^2)}{2f} + kz\right]\right\} \tag{4.35}$$

We have shown in the preceding section that a phase factor of the form (4.35) implies a radius of curvature of the phase front of $-f$, the minus sign indicating a phase front that is concave as seen from the right, in contrast to the previous case, where it was convex (compare Fig. 4.3). The wave will come to a focus at a distance f from the lens, as originally claimed. Note, however, that a beam of finite transverse cross section does not produce a focal spot of zero transverse dimension as we shall note in Chapter 5. There we shall also find that the focus (position of minimum transverse extent) of a beam of finite transverse cross section occurs at a distance from the lens that differs slightly from f.

4.4 FOURIER TRANSFORMATION BY A LENS [2]

Fraunhofer diffraction was shown to produce a Fourier transform of the input amplitude distribution in the far field, within a phase factor representing a curved phase front. A lens system can produce a Fourier transform without causing a curved wavefront and, further, the transform need not be in the far-field. Thus a lens system is ideal to perform a spatial Fourier transform on a given amplitude distribution once, or repeatedly. A simple realization of the system is shown in Fig. 4.9. The input amplitude distribution is produced in the focal plane in front of the lens. It propagates a distance f through free space, is transformed by the lens, and then propagates by another distance f to be observed in the second focal plane of the lens. The Fourier transform of the input appears in this plane.

The mathematical derivation of the Fourier transformation by a lens is rather formal. The input amplitude distribution is transformed onto the front face of the lens, is carried through the lens via the transformation (4.34), and then is transformed once more by free space propagation through the distance f. This approach is pursued in the appendix. Another approach is followed

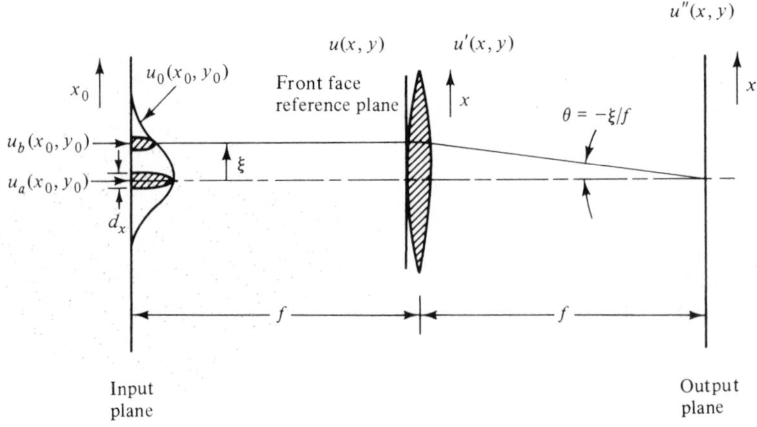

Figure 4.9 Transformation by lens between focal planes.

here which has merit of its own. It separates the input amplitude distribution into a superposition of illuminated sections covering areas $d_x d_y$ with d_x and d_y small enough so that the lens is in the far field of each of these sections, $d_x^2 \ll \lambda f$, $d_y^2 \ll \lambda f$. [Fig. 4.9]. Each of these sections produces its own Fourier transform at the front face of the lens with a curved phase front. The phase front curvature is removed by passage through the lens, so that a superposition of plane wave Fourier transforms appears behind the lens. This superposition is not yet the Fourier transform of the superposition at the input plane, the phase factors of the transforms are not the proper ones. Propagation of the plane through the distance f corrects the phase factors so that a Fourier transform is produced in the output plane, $z = f$. We present the details of this approach in this section.

Consider a general illumination over the input plane, $u_0(x_0, y_0)$ [Fig. 4.9]. If $u_0(x_0, y_0)$ is of very small transverse extent d_x, d_y, so that

$$\frac{d_x^2}{\lambda f} \ll 1, \qquad \frac{d_y^2}{\lambda f} \ll 1$$

then the lens is in the far-field and the excitation at the front face reference plane of the lens is according to (4.24):

$$u(x, y) = \frac{j(2\pi)^2}{\lambda f} e^{-[jk(x^2+y^2)/2f]} U(kx/f, ky/f) \tag{4.36}$$

Transmission through the lens removes the exponential factor, according to the transformation by a lens, (4.34):

$$u'(x, y) = \frac{j(2\pi)^2}{\lambda f} U(kx/f, ky/f) \tag{4.37}$$

The perfect Fourier transform has been taken; the lens removed the curvature of the phase front. However, we required that the lens be in the far field of the object (input) plane. Also, we found that the Fourier transform appears anywhere beyond the lens, not in the focal plane as stated at the outset. In order to adapt the analysis to an extended illumination for which the lens is not in the far field, consider two samples of the extended illumination $u_0(x_0, y_0)$, the samples $u_a(x_0, y_0)$ and $u_b(x_0 - \xi, y_0)$ as shown in Fig. 4.9. The samples are assumed to be of sufficiently small transverse extent, so that the lens is in their far field. The displacement ξ, however, does not obey the requirement $\xi^2/\lambda f \ll 1$. The Fourier transform of the sum of $u_a(x_0, y_0)$ and $u_b(x_0 - \xi, y_0)$ is

$$\text{F.T. } \{u_a(x_0, y_0) + u_b(x_0 - \xi, y_0)\} = U_a(k_x, k_y) + U_b(k_x, k_y)e^{jk_x\xi} \tag{4.38}$$

This is the relationship we want to find in order to prove the Fourier transformation by a lens. Consider the physical transformation wrought by propagation up to the front reference plane of the lens and transformation by the lens of the illumination

$$u(x_0, y_0) = u_a(x_0, y_0) + u_b(x_0 - \xi, y_0) \tag{4.39}$$

We have in correspondence with (4.34), at the front face reference plane of the lens

$$u(x, y) = \frac{j(2\pi)^2}{\lambda f}\{U_a(kx/f, ky/f)e^{-jk[(x^2+y^2)/2f]} + U_b(k(x-\xi)/f, ky/f)e^{-jk[(x-\xi)^2+y^2]/2f}\}$$

Passage through the lens provides the multiplier $\exp\,[jk(x^2 + y^2)/2f]$

$$u(x', y') = \frac{j(2\pi)^2}{\lambda f}\,[U_a(kx/f, ky/f) + U_b(k(x-\xi)/f, ky/f)e^{jkx\xi/f}e^{-jk\xi^2/2f}] \tag{4.40}$$

This is not what is desired [i.e., an expression of the form (4.38)]. But now study what happens as one follows the propagation from the back-face plane to the output plane.

As a consequence of the smallness of the samples the second focal plane is in the near field of the amplitude distribution (4.40). Indeed U_a and U_b experience appreciable variation when the arguments $k_x = kx/f$ and $k_y = ky/f$ change by $\Delta k_x \sim 1/d_x$ and $\Delta k_y \sim 1/d_y$ where d_x and d_y are the scale of variation of u_a and u_b. The focal plane at the distance f is in the near field, because using (4.13) for the change Δx over which an appreciable change of $U_a(kx/f)$ occurs:

$$\frac{k(\Delta x)^2}{f} = \frac{k}{f}\left(\frac{\Delta k_x f}{k}\right)^2 = \frac{f}{kd_x^2} = \frac{\lambda f}{2\pi d_x^2} \gg 1$$

if d_x is small enough. An analogous argument holds for Δy. Because the second focal plane is in the near field, the propagation takes place with no diffraction, or distortion, of the amplitude profiles. The amplitude distribution $U_a(kx/f, ky/f)$ gets phase shifted by $\exp(-jkf)$, but, according to convention, this factor is dropped. The excitation b, is a wave inclined by the angle $\theta = -\xi/f$

with respect to the z axis and shifts its center position by ξ according to (4.14). The phase delay is

$$k_z f = \left(k - \frac{k_x^2}{2k}\right) f = kf - \frac{k\xi^2}{2f}$$

(i.e., an advance with respect to u_a of $k\xi^2/2f$). Thus the amplitude u_b at the output plane is

$$\frac{j(2\pi)^2}{\lambda f} U_b(kx/f, ky/f) e^{j(kx\xi)/f}$$

as required by the Fourier transform condition (4.38). Thus the transformation of (4.36) from the input reference plane to the output reference plane has produced the Fourier transform with $k_x \equiv kx/f$ and $k_y \equiv ky/f$. Extension of the argument to a superposition of a complete set of samples of the input amplitude distribution of a general illumination $u_0(x_0, y_0)$ at the input plane, with Fourier transform $U_0(k_x, k_y)$, gives

$$u''(x, y) = j \frac{(2\pi)^2}{\lambda f} U_0(kx/f, ky/f) \tag{4.41}$$

The fact that a lens performs a Fourier transform of a two-dimensional complex amplitude distribution is of great practical importance in image processing. Even with the fast Fourier transform algorithm, digital Fourier transformation requires a large amount of computation. Usually, image processing does not require the accuracy inherent in a digital Fourier transformation. The analog method available with optical signal processing can be of sufficient accuracy and can be much faster, limited in speed solely by detector response, and in accuracy by the spatial resolution of the lens system and the detector array.

Figure 4.10 illustrates an optical system used for coherent optical data processing. Light from a He–Ne laser is brought to a point focus at a pinhole by a lens of short focal length. This lens/pinhole arrangement is indicated in Fig. 4.10 as a "spatial filter." Another lens, L_1, of focal length f_1, is placed at a distance f_1 from the plane of the pinhole to collimate the laser light. The input

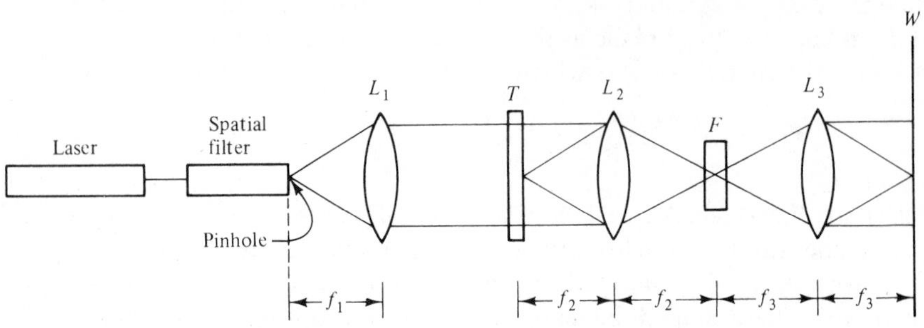

Figure 4.10 Optical spatial filtering arrangement.

information, T (e.g., a photographic transparency) is placed at a distance equal to the focal length f_2 from the converging lens, L_2. This lens produces the Fourier transform of the input amplitude distribution in its back focal plane, F. The diffracted light is recombined by the imaging lens L_2 (which takes the inverse transform), to form the image of the input in the output plane, W. Spatial filtering experiments are performed by inserting stops in the transform plane, F, to remove certain components from the spatial frequency spectrum of the input. The image in the output plane, W, is altered thereby.

Before concluding this section we will make one important observation. The paraxial Fresnel diffraction formalism predicts a perfect Fourier transform by one lens. Repetition of the process by one more lens recovers a *perfect image* in the output plane of the cascade of two lenses. The paraxial Fresnel diffraction theory predicts perfect imaging because the limits in the integrals extend from $-\infty$ to $+\infty$ by assumption. In practice, *apertures* limit the transverse dimensions of an optical system, and diffraction of the radiation passing the aperture prevents the formation of perfect images. In the next chapter, where we study the propagation of optical beams of finite cross section, we shall find the limits on the resolution of an optical system imposed by its finite cross section. In this context, note further that the paraxial assumption on p. 82 limits the Fourier decomposition of the object amplitude distribution to a finite range of k_x and k_y. An amplitude distribution with Fourier components limited to a finite "bandwidth" of "spatial frequencies" k_x and k_y cannot exhibit appreciable variation over ranges $\Delta x_0 \leq 1/k_x$, $\Delta y_0 \leq 1/k_y$. This, in itself imposes a restriction on the resolution of the object.

4.5 THE PARAXIAL WAVE EQUATION [3]

There are two equivalent circuit-theoretical descriptions of the response (output) of a linear system to an input excitation. One starts with the differential equation of the system and looks for particular and homogeneous solutions of these equations. The response is written as a linear superposition of these solutions. The other approach starts with the impulse response of the system and writes the output in terms of the convolution of the impulse response and excitation. The Fresnel diffraction integral can be viewed as the convolution of the input excitation $[u_0(x, y)]$ at $z = 0$ with the impulse response $h(x, y, z)$ to give the output response at the plane z. One might expect that there exists an alternative approach to the paraxial diffraction problem, namely the statement of a differential equation with characteristic solutions, the superposition of which gives the response. We shall now show that such an alternative formulation exists. It has its own advantages that will become evident later.

The paraxial formulation of diffraction theory starts with a superposition of approximate solutions to the wave equation in which the angle between the $\boldsymbol{k}$ vector and the z axis has to be small. An equivalent statement is that the x-y

dependence of the function of $\psi(x, y, z)$ must be much less rapid than its z dependence. This suggests that we write the function $\psi(x, y, z)$ as a product of $\exp(-jkz)$ times a factor $u(x, y, z)$, (4.5):

$$\psi(x, y, z) = u(x, y, z)e^{-jkz} \tag{4.42}$$

where the x, y, and z dependences of u are much less rapid than that of the factor $\exp(-jkz)$. When (4.42) is introduced in the wave equation (4.2), one obtains

$$\nabla_T^2 u + \frac{\partial^2}{\partial z^2} u - 2jk \frac{\partial u}{\partial z} = 0 \tag{4.43}$$

where

$$\nabla_T \equiv \hat{x} \frac{\partial}{\partial x} + \hat{y} \frac{\partial}{\partial y} \tag{4.44}$$

Because $|(\partial/\partial z)u| \ll ku$ we may neglect $\partial^2 u/\partial z^2$ compared with $k(\partial u/\partial z)$ and obtain the *paraxial wave equation*

$$\nabla_T^2 u - 2jk \frac{\partial u}{\partial z} = 0 \tag{4.45}$$

which is an approximate form of the wave equation.

It is easy to confirm by differentiation that $h(x, y, z)$ of (4.11) is an exact solution of the paraxial wave equation, as an impulse response must be. Thus the paraxial wave equation and its characteristic solutions must lead to a description equivalent to the Fresnel diffraction theory. The advantage of the paraxial wave equation is that it is easily generalizable to treat problems of propagation in nonuniform dielectrics and nonlinear dielectrics, a generalization not easily achieved with the Fresnel diffraction integral.

The impulse response $h(x, y, z)$ can be derived directly from the point source solution $\exp(-jkr)/r$ of the scalar wave equation (4.2), where $r = \sqrt{x^2 + y^2 + z^2}$. If one limits the analysis to points near the z axis, then

$$r \simeq z + \frac{x^2 + y^2}{2z} \tag{4.46}$$

and the *impulse response* reduces to

$$\frac{e^{-jkr}}{r} \simeq e^{-jkz} \frac{1}{z} e^{-j[k(x^2+y^2)/2z]} \tag{4.47}$$

where we have approximated r in the denominator by z, whereas in the sensitive exponent the full expression (4.46) is used. Note that, aside from the factor $\exp(-jkz)$, the right-hand side of (4.47) is identical in form with (4.11), the Fresnel kernel. This derivation serves as a simple mnemonic rule for the quick reconstruction of the Fresnel kernel.

The paraxial wave equation gives a simple picture of the incipient effect of diffraction, at small distances from the input plane positioned, say, at $z = 0$. From (4.45)

$$\Delta u = -\frac{j}{2k}\,\Delta z\;\nabla_T^2 u \tag{4.48}$$

Suppose that we start with a plane wavefront, real u. Then the initial effect of diffraction is to add a quadrature component to u. The phase delay $\Delta\phi$ imparted to u is, from (4.48),

$$\Delta\phi = \frac{\Delta z}{2k}\,\frac{\nabla_T^2 u}{u} \tag{4.49}$$

If u is initially *Gaussian* as shown in Fig. 4.11, the phase imparted to u is as shown. The center of the beam is phase delayed, the outside is phase advanced; the initially plane-phase front is curved. Points on the surface of constant phase farther away from the axis have smaller z values. The curved phase front initiates the diffraction of the beam—the beam diameter increases as a function of distance. We shall see later, in Chapters 6 and 10, how diffraction can be prevented if the phase change (4.49) is counteracted by a transverse index profile which produces a phase change of opposite sign.

The paraxial wave equation describes the behavior of the vector potential $\mathbf{A} = \hat{\boldsymbol{n}}u(x, y, z)\exp(-jkz)$ of optical beams in the limit when the transverse extent of the beam is much greater than a wavelength [the transverse

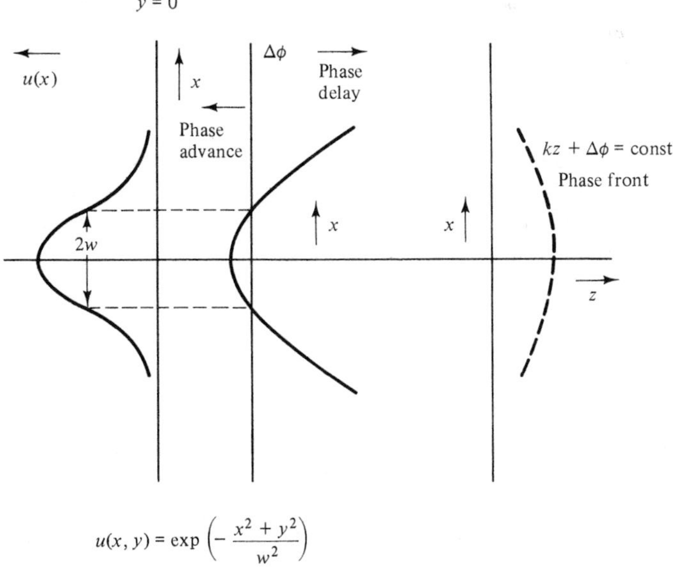

Figure 4.11 Phase $\Delta\phi$ acquired by initial Gaussian by propagation through small distance Δz.

dependence of $u(x, y, z)$ is much less rapid than the z dependence of $\exp(-jkz)$]. The approximations inherent in the paraxial wave equation imply attendant approximations in the power flow expression. The question arises as to the proper interpretation of the z component of the Poynting vector. We shall now show that the proper interpretation is the one of Equation (4.21) identifying the transverse electric field $\mathbf{E}_{\mathrm{T}}$ with

$$\mathbf{E}_{\mathrm{T}} = -j\omega\hat{\boldsymbol{n}}ue^{-jkz} \tag{4.50}$$

and the transverse magnetic field as

$$\mathbf{H}_{\mathrm{T}} = \sqrt{\varepsilon_0/\mu_0}\,\hat{z} \times \mathbf{E}_{\mathrm{T}}. \tag{4.51}$$

To prove the assertion above, we study the conservation of the quantity

$$\int_{-\infty}^{\infty}\int_{-\infty}^{\infty} |u|^2\, dx\, dy$$

which by our hypothesis, is proportional to the power passing a given cross section and must be independent of z. One must have

$$\frac{\partial}{\partial z}\int_{-\infty}^{\infty}\int_{-\infty}^{\infty} |u|^2\, dx\, dy = 0 \tag{4.52}$$

To prove this, we use (4.45):

$$\frac{\partial}{\partial z}\int_{-\infty}^{\infty}\int_{-\infty}^{\infty} |u|^2\, dx\, dy = \frac{1}{2jk}\int_{-\infty}^{\infty}\int_{-\infty}^{\infty}[u^*\nabla_{\mathrm{T}}^2 u - u\nabla_{\mathrm{T}}^2 u^*]\, dx\, dy$$

The right-hand side can be written

$$\frac{1}{2jk}\int_{-\infty}^{\infty}\int_{-\infty}^{\infty}\nabla_{\mathrm{T}}\cdot[u^*\nabla_T u - u\nabla_T u^*]\, dx\, dy = \frac{1}{2jk}\oint ds\,\hat{\imath}\cdot[u^*\nabla_T u - u\nabla_T u]$$

where the line integral is over the contour at ∞, with $\hat{\imath}$ the unit normal to the contour in the x-y plane. Because u vanishes at infinity the integral is zero and (4.52) is proven. The pattern of the power flow density (*intensity*) is thus correctly interpreted as equal to $\frac{1}{2}\sqrt{\varepsilon_0/\mu_0}\,\omega^2|u|^2$.

4.6 SUMMARY

Plane waves with a $\boldsymbol{k}$-vector inclined by a small angle with respect to the z axis have the component k_z of the propagation vector

$$k_z \simeq k - \frac{k_x^2 + k_y^2}{2k}$$

This fact explains the form of the spatial Fourier transform of the impulse response function $h(x, y, z)\exp(-jkz)$ of Fresnel diffraction theory in the paraxial approximation, which is proportional to

$$\exp\left[-j\left(k_x x + k_y y - \frac{k_x^2 + k_y^2}{2k} z\right)\right] \exp(-jkz)$$

The impulse response applied to the amplitude distribution $u_0(x, y)$ at the reference plane $z = 0$ gives the response at z as the convolution of u_0 with $h(x, y, z)$. This convolution is the Fresnel diffraction integral in the paraxial approximation. We applied it in the Fraunhofer limit and showed that the far-field pattern is the Fourier transform of the input amplitude, except for a phase factor that expresses the curvature of the phase front. We studied the diffraction by a rectangular aperture and an array of such apertures. In the latter case, the diffraction pattern is the product of an array factor and the diffraction pattern of a single aperture. Measurement of the diffraction pattern gives information on the spatial coherence of the radiation illuminating the slit array.

We determined the action of a thin lens on a wave transmitted through the lens. Next we showed that an amplitude distribution at one of the focal planes of a thin lens appears Fourier transformed at the other focal plane. Here no additional phase factor appears, as in the Fraunhofer diffraction, and thus Fourier transforms can be performed repeatedly.

We showed that a system of thin lenses, with no aperturing, transforms an object illumination into a perfect image. The diffraction limitation of imaging systems is due to the finite transverse extent of the (apertures in) imaging systems as will become apparent in the next chapter. Finally, we introduced the paraxial wave equation. The impulse response developed by plane wave superposition was shown to be a solution of the paraxial wave equation. This equation provides an alternative route to the analysis of the diffraction problem—a treatment in terms of a superposition of eigensolutions, a topic taken up in the next chapter.

APPENDIX 4A

Formal Derivation of Fourier Transformation by a Lens

We have found that free-space propagation through a distance z is described by the simple multiplier $(2\pi)^2 H(k_x, k_y, z)$ on the Fourier transform $U_0(k_x, k_y)$, whereas it is expressed as the Fresnel convolution integral on the input amplitude distribution $u_0(x_0, y_0)$. Since the system of Fig. 4.9 involves two free-space propagations, it is best to describe the action of the system in Fourier transform space in which the free-space propagation is expressed more simply. The transformation of the lens is described by (4.34). In Fourier transform space:

$$U'(k_x, k_y) = \int dk'_x \int dk'_y \; L(k_x - k'_x, k_y - k'_y) U(k'_x, k'_y) = L \otimes U \tag{4A.1}$$

where

$$L(k_x, k_y) \equiv \left(\frac{1}{2\pi}\right)^2 \int_{-\infty}^{\infty} dx \int_{-\infty}^{\infty} dy \; l(x, y) e^{j(k_x x + k_y y)}$$

$$= \frac{j\lambda f}{(2\pi)^2} \exp\left[-\frac{jf}{2k}(k_x^2 + k_y^2)\right] \tag{4A.2}$$

[compare (4.10)].

The Fourier transform of the amplitude distribution in the plane at the front face of the lens U is related to the Fourier transform U_0 in the "input plane" located at the focal distance

$$U = (2\pi)^2 H(f) U_0 \tag{4A.3}$$

The Fourier transform at the output plane is transformed by the lens followed by free-space propagation:

$$U'' = (2\pi)^4 H(f)\{L \otimes H(f) U_0\} \tag{4A.4}$$

or written out explicitly:

$$U''(k_x, k_y) = \frac{j\lambda f}{(2\pi)^2} \int_{-\infty}^{\infty} dk'_x \int_{-\infty}^{\infty} dk'_y \exp\left\{-\frac{jf}{2k}[(k_x - k'_x)^2 + (k_y - k'_y)^2]\right\} U_0(k'_x, k'_y) \exp\left[\frac{jf}{2k}(k'^2_x + k'^2_y)\right] \cdot \exp\left[\frac{jf}{2k}(k_x^2 + k_y^2)\right]$$

$$= \frac{j\lambda f}{(2\pi)^2} \int dk'_x \int dk'_y \; U_0(k'_x, k'_y) \exp\left[\frac{jf}{k}(k_x k'_x + k_y k'_y)\right] \tag{4A.5}$$

This relation shows that $U''(k_x, k_y)$ is proportional to the inverse Fourier transform of $U_0(k_x, k_y)$:

$$U''(k_x, k_y) = \frac{j\lambda f}{(2\pi)^2} u_0\left(-\frac{fk_x}{k}, -\frac{fk_y}{k}\right) \tag{4A.6}$$

The Fourier transform of the output amplitude distribution is proportional to the input amplitude distribution. The system of Fig. 4.9 "takes" a spatial Fourier transform.

The Fourier transform of (4A.6) gives

$$u''(x, y) = j\frac{(2\pi)^2}{\lambda f} U_0\left(\frac{kx}{f}, \frac{ky}{f}\right) \tag{4A.7}$$

which agrees with (4.41) of the text.

PROBLEMS

4.1. Find the Fraunhofer diffraction pattern of a pair of identical slits of widths and heights d_x, d_y, respectively, spaced along x by $L > d_x$. Sketch the intensity pattern in normalized variables for $y = 0$, with $L = 2d_x$.

4.2. An optical beam with a Gaussian cross section

$$u(x, y) = e^{-(x^2+y^2)/w_0^2}$$

with w_0 a real parameter, impinges upon a thin lens of focal distance f (Fig. P4.2). Evaluate $u(x, y, z)$ beyond the lens from the Fresnel diffraction integral. Find the minimum diameter of the beam (focal spot size). Note that the minimum spot size does not occur precisely at the distance f—a consequence of the finite beam cross section.

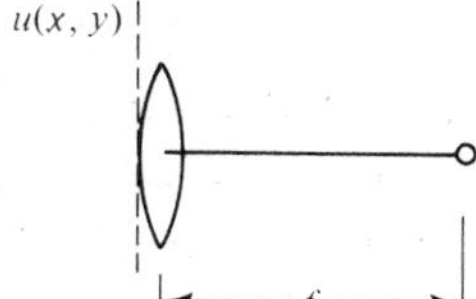

Figure P4.2

4.3. Consider a Gaussian beam with its waist at plane 1 that gets displaced laterally as shown in Fig. P4.3.

$$u_0(x_0, y_0) = A \exp\left[-\frac{(x_0 - x_d)^2 + y_0^2}{w^2}\right]$$

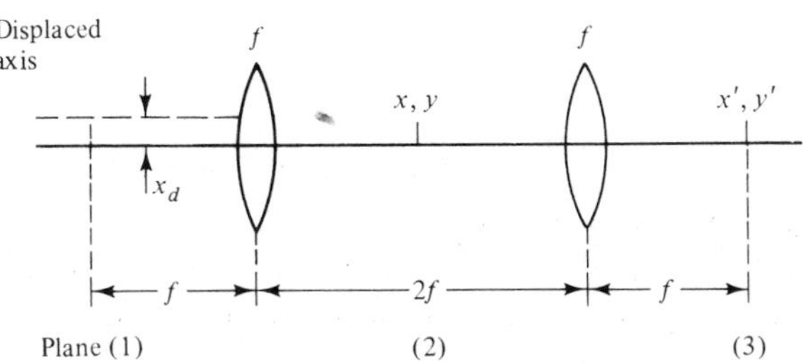

Figure P4.3

(a) What is the expression for the beam in plane 2 as a function of x and y? Can you put an iris in plane 2 that will act on the beam amplitude independent of its displacement x_d?

(b) Is there a ray-optical explanation of what happens?

(c) What is the emerging beam in plane 3? Is there a scale change?

(d) Suppose that we had a general illumination function $u_0(x_0, y_0)$ at plane 1. Without necessarily repeating the algebra, can you state $u(x', y')$ in plane 3?

4.4. **(a)** Obtain the far-field diffraction pattern $u(x, y, z)$ for the *amplitude* produced by a square hole illuminated by a normally incident plane wave centered with respect to the origin of a Cartesian coordinate system (Fig. P4.4a).

(b) Obtain the *intensity* diffraction pattern of the aperture shown (Fig. P4.4b) for the same illumination as in part (a).

(c) Sketch the intensity pattern along the line $x = y$.

(d) Screen out the center portion of the aperture of part (a) over a square area $(a/2)$ (Fig. P4.4c). Find the far-field diffraction pattern.

(e) Sketch the absolute square of the pattern at $y = 0$ as a function of x.

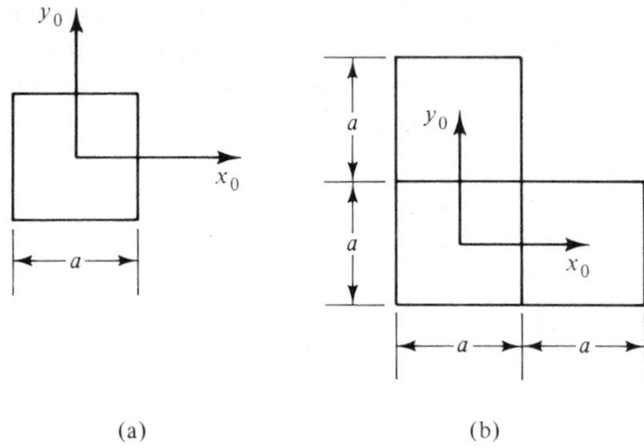

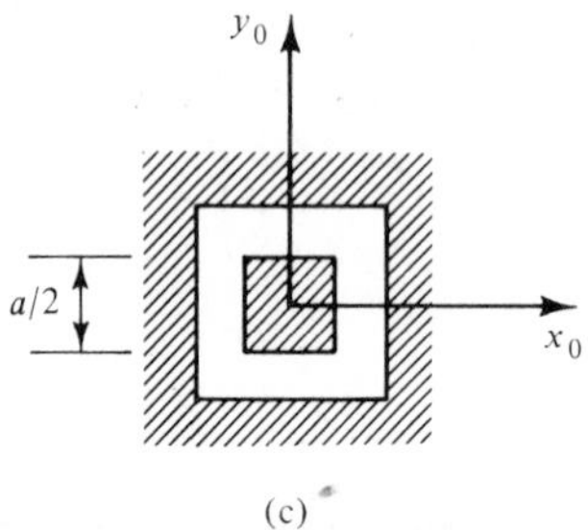

Figure P4.4

4.5. An amplitude distribution

$$u_0(x_0, y_0) = A\left(1 + \cos\frac{2\pi}{\Lambda}x_0\right)$$

appears at plane 1 of Fig. P4.3. If a disk of radius

$$R < \frac{\pi}{\Lambda}\frac{f}{k}$$

is introduced on axis in plane 2, what is the intensity pattern in plane 3?

4.6. Consider the "longitudinal" vector potential

$$\mathbf{A} = \hat{z}\exp\left[-\frac{jk(x^2 + y^2)}{2(jb + z)}\right]e^{-jkz}$$

Show that a negligible electric field is associated with this potential in the paraxial approximation.

4.7. Show that the Fresnel diffraction integral in two dimensions, x and z, is given by

$$u(x, z) = \sqrt{\frac{j}{\lambda z}} \left\{ \int dx_0 \, u_0(x_0) \exp\left[-j \frac{k(x - x_0)^2}{2z} \right] \right\}$$

Show that the kernel obeys the paraxial wave equation in two dimensions:

$$\frac{\partial^2}{\partial x^2} u - 2jk \frac{\partial u}{\partial z} = 0$$

REFERENCES

[1] M. Born and E. Wolf, *Principles of Optics*, Macmillan, New York, 1964.

[2] J. W. Goodman, *Introduction to Fourier Optics*, McGraw-Hill, New York, 1968.

[3] H. Kogelnik, "On the propagation of Gaussian beams of light through lenslike media including those with a loss and gain variation," *Appl. Opt.*, *4*, 1562, 1965.

5

HERMITE–GAUSSIAN BEAMS AND THEIR TRANSFORMATIONS

The Fresnel diffraction theory developed in the preceding chapter expresses the excitation of the vector potential at any arbitrary cross section of constant z as the convolution of the specified amplitude distribution at the cross section $z = 0$ with the impulse response. Instead of a description of wave propagation in terms of an impulse response, one may study propagation of characteristic solutions of the wave equation (in the paraxial approximation) and their transformation by optical systems. The second approach has its own advantages. In particular when constructing interferometers or resonators, one is interested in the modes and the resonance frequencies of these structures. The Hermite–Gaussian beam solutions of the paraxial wave equation are well suited for this purpose.

In this chapter we develop the theory of Gaussian beams, their transformation by optical systems, and the study of the higher-order Hermite–Gaussian beams. We deviate from the traditional approach to this problem, taking advantage of unpublished work by Kogelnik. We start with the impulse response of the paraxial wave equation developed in the preceding chapter. Translation of the coordinate system in the z direction leaves the paraxial wave equation invariant. Therefore, the impulse response remains a solution of the paraxial wave equation under such a translation of the coordinate system. This fact is not changed if the translation distance is made imaginary. The Gaussian beam solution is constructed by such a translation. The magnitude of the translation determines the minimum diameter of the Gaussian beam. In Section 5.2 we construct standing-wave solutions by superposition of two counter-traveling Gaussian beams. At the nodal surfaces of the standing-wave solutions, which are spherical, one may mount spherical mirrors without perturbing the solution. In this way we obtain modes of spherical mirror resonators and the conditions under which modes exist for mirrors of given radii of curvature and spacing.

In Section 5.3 we develop the complete set of Hermite–Gaussian solutions in terms of which one may expand any excitation at an "input plane." The spatial development of these solutions contains all the physics that is described by Fresnel diffraction theory. Then we look at the complete set of modes of a curved-mirror Fabry–Perot resonator and determine their natural frequencies.

We consider next the transformation of Gaussian and Hermite–Gaussian beams by optical systems and introduce the so-called q parameter. Aside from the fact that the q parameter permits a simple description of the effect of an optical system upon a Gaussian beam, it also allows for an alternative derivation of the criterion for the existence of modes in a curved-mirror Fabry–Perot resonator. This is done in Section 5.6 after a brief look at the ray-optical description of optical systems and the establishment of the connection between ray optics and the q-parameter characterization of Gaussian beams.

In the last section we consider the influence of diffraction on formation of an image and show various methods for the determination of "nonideality" of optical components, in particular, aberrations.

5.1 GAUSSIAN BEAMS

We have shown that the exact solution of the scalar wave equation reduces to an exact solution of the paraxial wave equation if r is expanded to first order in powers of $(x^2 + y^2)/z^2$. This solution is proportional to the impulse response function

$$h(x, y, z) = \frac{j}{\lambda z} \exp\left(-jk \frac{x^2 + y^2}{2z}\right)$$

The paraxial wave equation is invariant with respect to a translation of the coordinate z, $z \to z - z_0$. Thus another solution of the equation is $h(x, y, z - z_0)$.

A very interesting solution of the paraxial wave equation can be constructed by making z_0 imaginary, $z_0 = -jb$. [1] This removes the singularity of the solution on the real z axis. One obtains a solution of the scalar paraxial wave equation which we denote, for reasons that will become self-evident later, by subscripts "00":

$$u_{00}(x, y, z) = \frac{j}{\lambda(z + jb)} \exp\left[-jk \frac{x^2 + y^2}{2(z + jb)}\right] \tag{5.1}$$

It is convenient to normalize $u_{00}(x, y, z)$ through multiplication by a constant, so that

$$\int_{-\infty}^{\infty} dx \int_{-\infty}^{\infty} dy \, |u_{00}(x, y, z)|^2 = 1$$

Normalization at one cross section, say $z = 0$, assures the same normalization for any z because of power conservation, (4.52). When the integral is carried out, it is found that the required constant is $\sqrt{2b\lambda}$. Thus

$$u_{00}(x, y, z) = j\sqrt{\frac{kb}{\pi}}\left(\frac{1}{z + jb}\right) \exp\left[-jk\frac{x^2 + y^2}{2(z + jb)}\right] \tag{5.2}$$

Separation of the argument of the exponential into real and imaginary parts, and of $(z + jb)$ into magnitude and phase, puts u_{00} into the form

$$u_{00}(x, y, z) = \frac{\sqrt{2}}{\sqrt{\pi}\, w} \exp(j\phi) \exp\left(-\frac{x^2 + y^2}{w^2}\right) \exp\left[-\frac{jk}{2R}(x^2 + y^2)\right] \tag{5.3}$$

where

$$w^2(z) = \frac{2b}{k}\left(1 + \frac{z^2}{b^2}\right) \tag{5.4}$$

$$\frac{1}{R(z)} = \frac{z}{z^2 + b^2} \tag{5.5}$$

$$\tan\phi = \frac{z}{b} \tag{5.6}$$

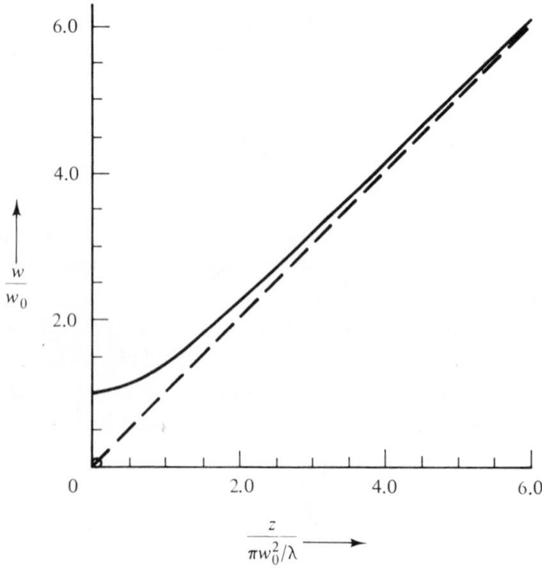

Figure 5.1 Beam radius as a function of distance.

The solution (5.3) is a wave traveling in the $+z$ direction with a Gaussian amplitude profile (*Gaussian beam*) and curved phase fronts of radius $R(z)$ (compare Section 4.2). The beam radius, at which the amplitude decreases to $1/e$ of its value on axis, is w. We see that the minimum beam radius w_0 is

$$w_0 = \sqrt{\frac{2b}{k}} \tag{5.7}$$

or, conversely, that the choice of the displacement of the origin of z by an imaginary distance $-jb$ has fixed the minimum radius w_0. The parameter $b = kw_0^2/2 = \pi w_0^2/\lambda$ is the *confocal parameter*.

We are usually interested in Gaussian beams of given minimum radius w_0, so it is convenient to write the parameters in (5.4) through (5.6) in terms of w_0:

$$w^2 = w_0^2\left[1 + \left(\frac{\lambda z}{\pi w_0^2}\right)^2\right] \tag{5.8}$$

$$\frac{1}{R} = \frac{z}{z^2 + (\pi w_0^2/\lambda)^2} \tag{5.9}$$

$$\tan\phi = \frac{z}{\pi w_0^2/\lambda} \tag{5.10}$$

Figures 5.1 and 5.2 give w and $1/R$ as functions of z. Equation (5.8) shows that a beam of waist size w_0 expands asymptotically in a cone of cone angle θ,

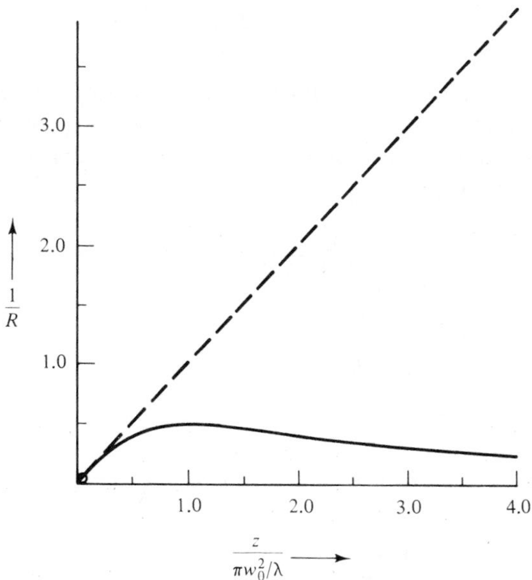

Figure 5.2 Inverse radius of curvature of phase front as function of distance.

where

$$\tan\theta = \frac{w}{z} \simeq \theta = \frac{\lambda}{\pi w_0} \tag{5.11}$$

The angle of expansion is known as the diffraction angle in microwave antenna theory. An antenna of size d produces a radiation pattern limited to an angle of the order of λ/d. A similar relation has been encountered in (4.27) for the *diffraction angle* of an aperture.

Phase Velocity of Gaussian Beam

The phase velocity of the Gaussian beam solution is not $c = 1/\sqrt{\mu_0 \varepsilon_0}$, as in free space. We have for the effective propagation constant, from (5.3) with the multiplier $\exp(-jkz)$ restored,

$$\int_0^z k_{\text{eff}}\, dz = kz - \phi \tag{5.12}$$

and thus

$$k_{\text{eff}} = k - \frac{d\phi}{dz} = \frac{\omega}{c} - \frac{b}{b^2 + z^2} < \frac{\omega}{c} \tag{5.13}$$

The phase velocity is greater than the speed of light. At $z = 0$, in particular, the effective propagation constant is

$$k_{\text{eff}} = k - \frac{2}{k w_0^2} \tag{5.14}$$

This effective propagation constant can be explained by the fact that the field is composed of a superposition of plane waves so as to produce a beam of finite transverse extent. The typical x and y components of the propagation vector $\boldsymbol{k}$ of these waves are

$$k_x = k_y \simeq \frac{\sqrt{2}}{w_0} \tag{5.15}$$

Thus

$$k_z^2 + k_x^2 + k_y^2 = k^2$$

gives

$$k_z = k - \frac{k_x^2 + k_y^2}{2k} = k - \frac{2}{k w_0^2} \tag{5.16}$$

in agreement with k_{eff} of (5.14).

Paraxial Approximation

The Gaussian beam solution (5.2) can be used to ascertain the approximations made in the derivation of the paraxial wave equation. We have ignored $|\partial^2 u/\partial z^2|$ compared with $k|(\partial u/\partial z)|$, or, equivalently, $|\partial u/\partial z|$ compared with $k|u|$. The derivative $\partial u/\partial z$ of the Gaussian solution (5.1) is

$$\frac{\partial u_{00}}{\partial z} = -\left[\frac{1}{z+jb} - j\frac{k(x^2+y^2)}{2(z+jb)^2}\right]u_{00} \tag{5.17}$$

The omission of this term compared with ku implies that $1/b \ll k = 2\pi/\lambda$, or

$$\frac{1}{bk} = \frac{\lambda^2}{2\pi^2 w_0^2} \ll 1 \tag{5.18}$$

The beam diameter must be large compared with a wavelength. Further

$$\frac{x^2+y^2}{z^2+b^2} \ll 1 \tag{5.19}$$

for the range of values of x and y over which the amplitude distribution extends. Since $x^2 + y^2$ is of order w^2 and $w^2 = w_0^2[1 + (z^2/b^2)]$, (5.19) implies that $w_0^2/b^2 \ll 1$, which is the same condition as (5.18).

The Electric and Magnetic Fields of Gaussian Beam

The magnetic field **H** and the electric field **E** follow from the complex vector potential **A** using (1.22) and (1.23):

$$\mu_0 \mathbf{H} = \nabla \times \mathbf{A} \tag{5.20}$$

$$\mathbf{E} = -j\omega\mathbf{A} - \nabla\Phi \tag{5.21}$$

where Φ, according to (1.26) is, in free space

$$\Phi = \frac{j}{\omega\mu_0\varepsilon_0}\nabla\cdot\mathbf{A} \tag{5.22}$$

Suppose that **A** is polarized along $\hat{\boldsymbol{x}}$:

$$\mathbf{A} = \hat{\boldsymbol{x}} u_{00}(x, y, z)e^{-jkz} \tag{5.23}$$

Then

$$\mu_0 \mathbf{H} = \nabla \times [\hat{\boldsymbol{x}} u_{00}(x, y, z)e^{-jkz}] = -jk\left[\hat{\boldsymbol{y}} u_{00} - j\hat{\boldsymbol{z}}\frac{\partial u_{00}}{k\,\partial y}\right]e^{-jkz} \tag{5.24}$$

where we have ignored $\partial u_{00}/\partial z$ compared with ku_{00} as determined above. The magnetic field has a small z component in addition to the y component as

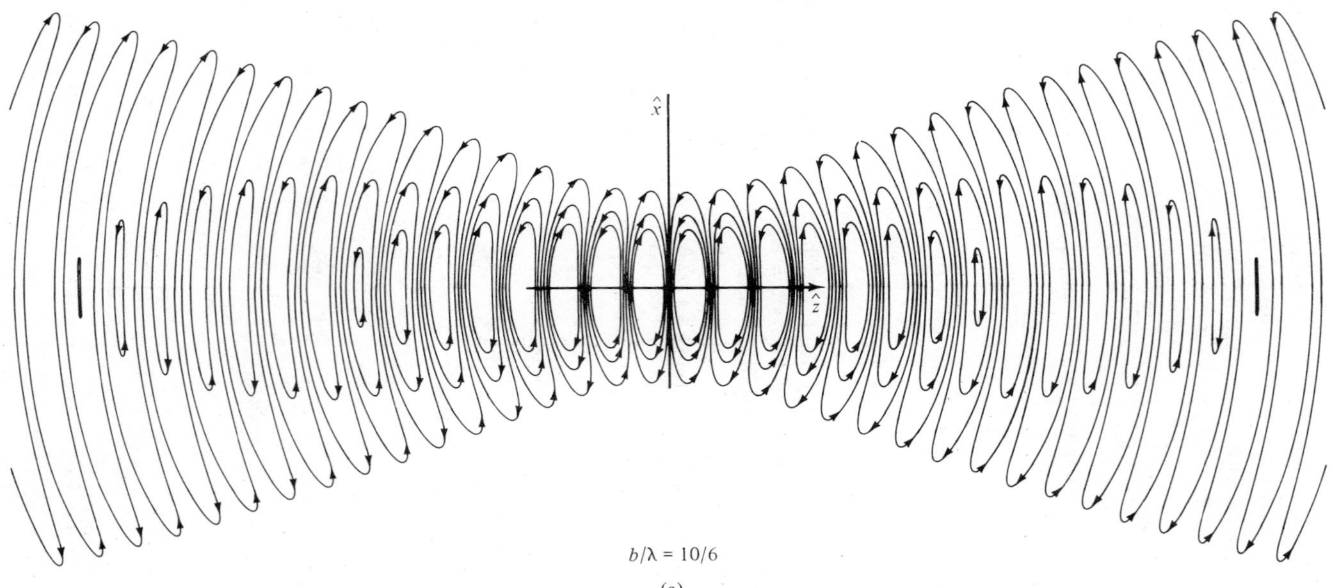

$b/\lambda = 10/6$

(a)

Figure 5.3 Electric field of Gaussian beam in x-z plane at one instant of time. Pattern moves to right as a function of time. Note that spacing between nodes decreases with distance from $z = 0$. This signifies change of phase velocity which approaches c as $z \to \infty$. (a) $b/\lambda = 10/6$; (b) $b/\lambda = 10/2.75$.

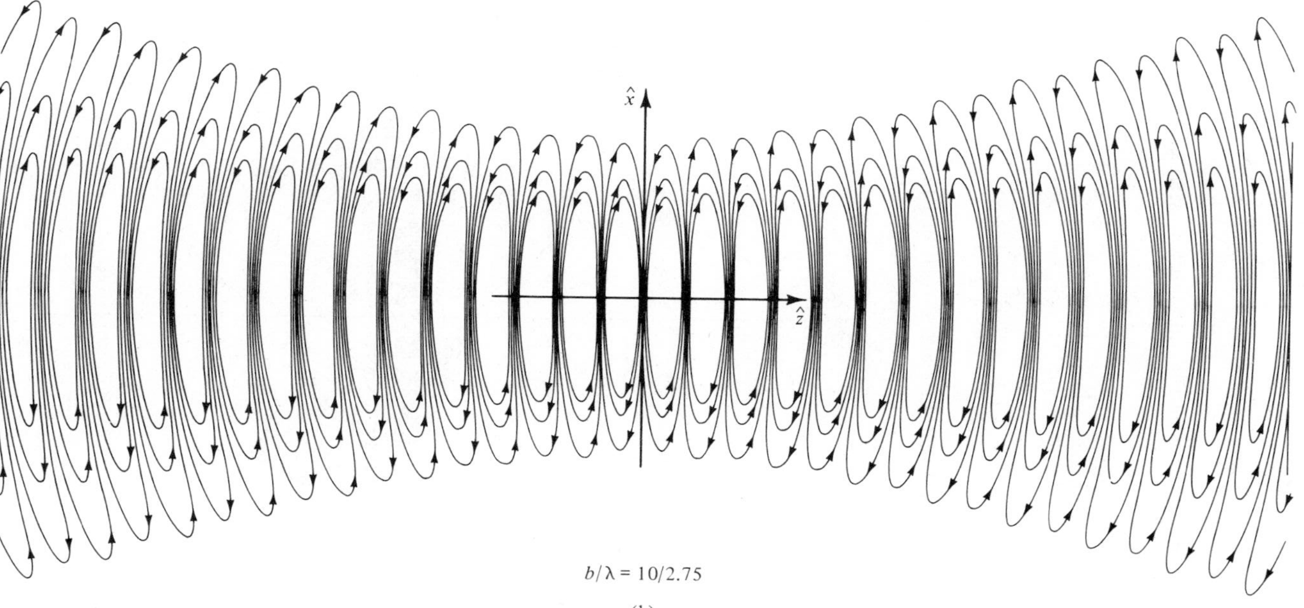

$b/\lambda = 10/2.75$

(b)

Figure 5.3 *(continued)*

required by Gauss's law. The electric field is, to the same degree of approximation,

$$\mathbf{E} = -j\omega\left[\hat{\mathbf{x}}u_{00} - j\hat{\mathbf{z}}\frac{\partial u_{00}}{k\,\partial x}\right]e^{-jkz} \tag{5.25}$$

The electric field of the Gaussian beam solution, written in the form of (5.2), is found from (5.25) to be

$$\mathbf{E} = \sqrt{\frac{kb}{\pi}}\left(\frac{\omega}{z+jb}\right)\exp\left[-jk\frac{x^2+y^2}{2(z+jb)}\right]\exp(-jkz) \cdot\left[\left(\hat{\mathbf{x}} - \hat{\mathbf{z}}\frac{x}{R}\right) + j\hat{\mathbf{z}}\frac{xb}{z^2+b^2}\right] \tag{5.26}$$

Two plots of $\boldsymbol{E}$ at constant t are shown in Fig. 5.3. As an aid in plotting one notes that the divergence-free electric field must be the curl of a vector

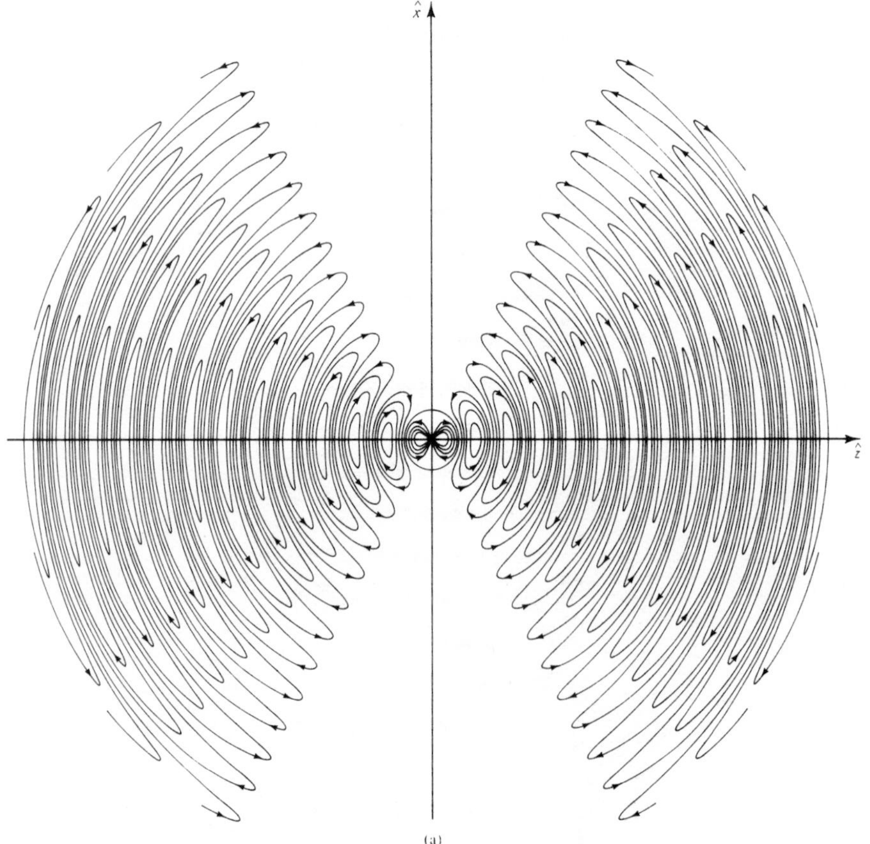

Figure 5.4 Electric field of radiating dipole with vector potential $A_x = \exp(-jkr)/r$ at two instants of time separated by $\Delta t = \pi/2\omega$.

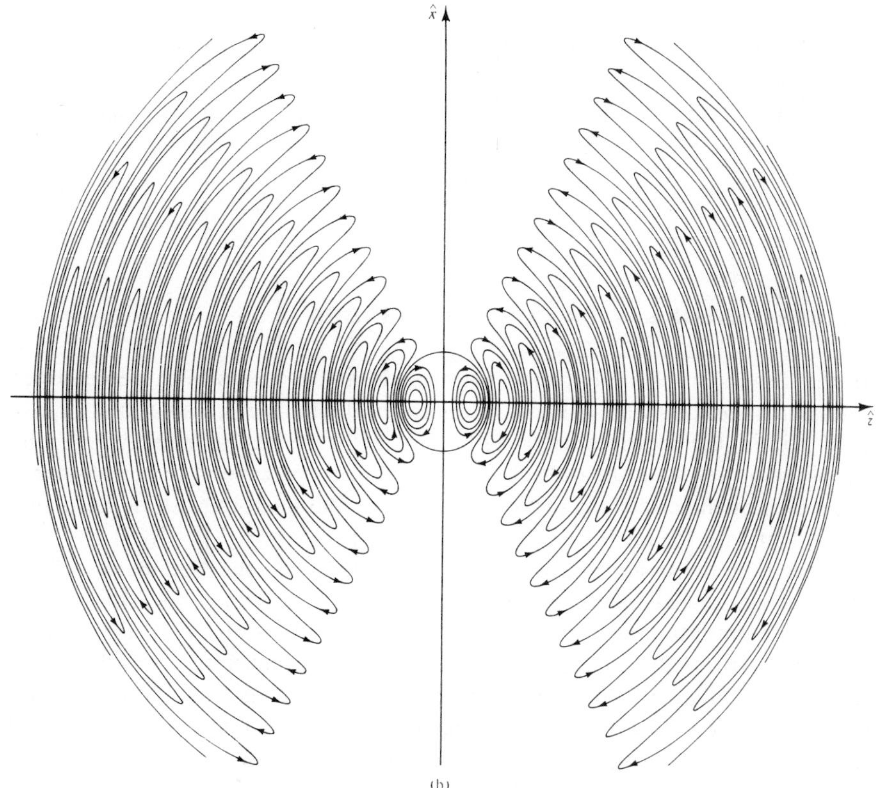

Figure 5.4 (*continued*)

potential which we denote by $\mathbf{C}(x, y, z)$. It is easily confirmed that the vector potential appropriate for (5.25) is within the paraxial approximation

$$\mathbf{C} = -\hat{y}\,\frac{\omega}{k}\,u_{00}(x, y, z)e^{j(\omega t - kz)}$$

The field lines are perpendicular to the lines of steepest descent of the two-dimensional scalar potential $C_y(x, y = \text{const}, z)$; thus the field lines are the lines of equal height (isohypses) of the potential $C_y(x, y = \text{const}, z)$. The density of field lines is proportional to the field strength. In this way a simple algorithm for a computer plot is obtained. Figure 5.3 shows the expansion of two Gaussian beams of different diameters for exagerated ratios of λ/w_0.

For comparison with Fig. 5.3, Fig. 5.4 shows the electric field lines of the vector potential solution $A_x = e^{-jkr}/r$ of equation (4.2), the radiating dipole. The two plots show the field at two time instants, separated in time by $\Delta t = \pi/2\omega$. The resemblance to the Gaussian beam solution is unmistakable. Of course, this is no surprise because the Gaussian beam solution is obtained from the radiating dipole solution by removing the singularity at the origin by an imaginary translation of the source and by making the paraxial approximation.

The power in the Gaussian beam of unit amplitude is

$$\frac{1}{2}\int_{-\infty}^{\infty} dx \int_{-\infty}^{\infty} dy \ \mathrm{Re}\ [\mathbf{E} \times \mathbf{H}^*]_z = \frac{1}{2}\omega^2 \sqrt{\frac{\varepsilon_0}{\mu_0}} \int_{-\infty}^{\infty} dx \int_{-\infty}^{\infty} dy \ |u_{00}|^2$$

$$= \frac{1}{2}\omega^2 \sqrt{\frac{\varepsilon_0}{\mu_0}} \tag{5.27}$$

5.2 RESONATORS WITH CURVED MIRRORS [2, 3]

In the discussion of the Fabry–Perot interferometer, Section 3.5, we described *resonant modes* of a Fabry–Perot resonator in the limit when an excitation can exist within the interferometer with mirrors of perfect reflectivity, with no supply of power from the outside. The resonances correspond to perfect standing waves of the electromagnetic field between the mirrors, with nodal planes coincident with the mirror reference planes.

In that discussion we were not concerned with the boundaries of the mirrors; we treated them as infinite in the transverse dimensions. In fact, the transverse amplitude distribution of a mode in a physical Fabry–Perot interferometer with planar mirrors of finite transverse dimensions is controlled by the diffraction of waves at the mirror boundaries (causing diffraction loss). Curved-mirror Fabry–Perot interferometers practically eliminate the effect of mirror boundaries on the mode amplitude distribution and the associated diffraction loss. This is described below.

Consider the Gaussian beam solution (5.2):

$$u_{00}(x, y, z)e^{-jkz} = \frac{j\sqrt{kb/\pi}}{z + jb}\, e^{-jkz} e^{-[jk(x^2+y^2)/2(z+jb)]} \tag{5.28}$$

This solution corresponds to a wave of Gaussian amplitude profile traveling in the $+z$ direction. If we replace z by $-z$, we produce a solution of the paraxial wave equation

$$\nabla_T^2 u + 2jk\frac{\partial u}{\partial z} = 0$$

which describes waves propagating in the $-z$ direction:

$$u_{00}(x, y, -z)e^{jkz} = \frac{-j\sqrt{kb/\pi}}{z - jb}\, e^{jkz} e^{+[jk(x^2+y^2)/2(z-jb)]} \tag{5.29}$$

Superposition of solutions (5.28) and (5.29) produces standing waves with nodal surfaces of the electric field parallel to the phase fronts with radii of curvature R given by

$$\frac{1}{R} = \frac{z}{z^2 + b^2} \tag{5.30}$$

If mirror 1 is placed at position z_1, with the radius of curvature

$$\frac{1}{R_1} = \frac{z_1}{z_1^2 + b^2} \tag{5.31}$$

and mirror 2 at position z_2 with the radius of curvature

$$\frac{1}{R_2} = \frac{z_2}{z_2^2 + b^2} \tag{5.32}$$

(see Fig. 5.5), one may "capture" this standing wave (i.e., satisfy the boundary conditions called for by the standing-wave solution). Of course, the spacing between the mirrors has to correspond precisely to an integer number of nodes of the standing wave between the two mirrors. If the mirror diameters are chosen much larger than the Gaussian beam diameter, the fact that the mirror is not infinite is of no practical significance because the field of the Gaussian beam at the mirror edges is negligible.

We attacked the problem as one in which a system is constructed to fit a particular solution. In practice, the problem is reversed. Given a set of mirrors of radii of curvature R_1 and R_2 and their spacing d, the resonant mode appropriate for this configuration is sought. This means that (5.31) and (5.32) must be solved for b under the constraint $z_2 - z_1 = d$. Note also that it is customary to distinguish convex from concave mirrors by denoting the former by a negative radius of curvature. Thus the problem corresponding to Fig. 5.5 reads as follows: Solve for b, given that

$$z_2 - z_1 = d \tag{5.33}$$

$$\frac{1}{R_2} = \frac{z_2}{z_2^2 + b^2} \tag{5.34}$$

and

$$-\frac{1}{R_1} = \frac{z_1}{z_1^2 + b^2} \tag{5.35}$$

where R_1 and R_2 are the mirror radii taken positive for concave mirrors, negative for convex mirrors.

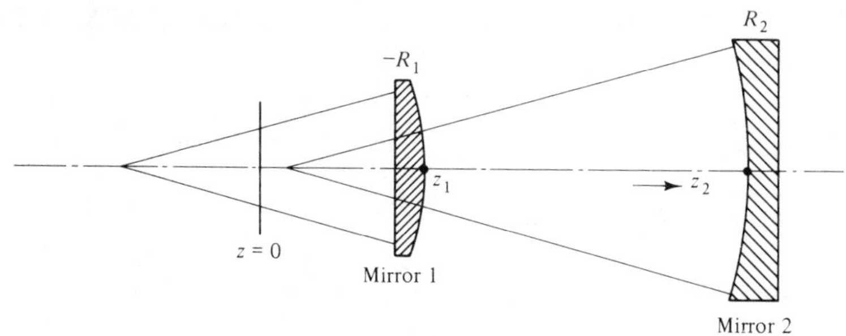

Figure 5.5 Resonator geometry.

Case 1. One plane, one concave mirror. The simplest case is when $R_1 = \infty$, $R_2 = R_0$ (see Fig. 5.6). Then $z_1 = 0$ and $z_2 = d$. We find from (5.34)

$$b = \frac{\pi w_0^2}{\lambda} = \sqrt{d(R_0 - d)} \tag{5.36}$$

The minimum beam radius of the resonant mode occurs at the flat mirror. Solutions are found provided that $R_0 > d$. No solutions are found for $R_0 < d$. In fact, the beam diameter goes to zero at the critical value $R_0 = d$. Here one must question the validity of the paraxial wave analysis, because a beam diameter much smaller than a wavelength violates the assumptions that went into the derivation of the paraxial wave equation.

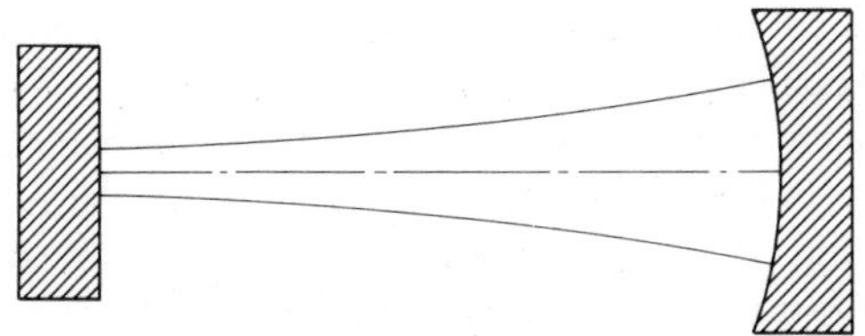

Figure 5.6 Resonator of (5.36).

Case 2. The symmetric case. The case when $R_1 = R_2 = R_0$ can be constructed simply by setting the system of Fig. 5.6 back to back with its mirror image (see Fig. 5.7). The distance d of (5.36) must now be interpreted as $d/2$ and one has, from (5.36),

$$b = \frac{\pi w_0^2}{\lambda} = \sqrt{\frac{d}{2}\left(R_0 - \frac{d}{2}\right)} \tag{5.37}$$

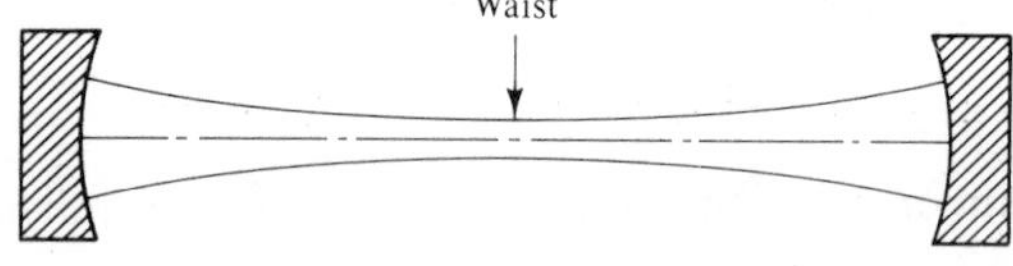

Figure 5.7 Symmetric resonator.

Case 3. The general case. In the general case when $R_1 \neq R_2$ (see Fig. 5.5), one finds from (5.33), (5.34), and (5.35) that

$$\frac{R_1}{2} + \frac{R_2}{2} \pm \sqrt{\frac{R_1^2}{4} - b^2} \pm \sqrt{\frac{R_2^2}{4} - b^2} = d \tag{5.38}$$

By squaring, one may obtain an equation for b^2:

$$4b^2 = \frac{R_1^2 R_2^2}{4} \frac{1 - [(2d/R_1 R_2)(d - R_1 - R_2) + 1]^2}{[d - (R_1 + R_2)/2]^2} \tag{5.39}$$

This equation yields real solutions for b only when the numerator is positive.

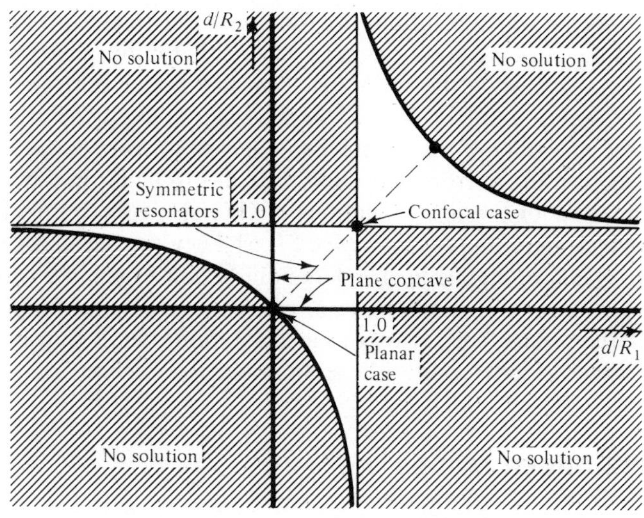

Figure 5.8 The d/R_1–d/R_2 plane and regions in which resonator solutions exist.

Simple manipulation shows that this is the case when

$$0 \leq \left(1 - \frac{d}{R_1}\right)\left(1 - \frac{d}{R_2}\right) \leq 1 \tag{5.40}$$

Figure 5.8 gives the diagram which shows the range of values of d/R_1 and d/R_2 for which solutions are found. [3, 4]

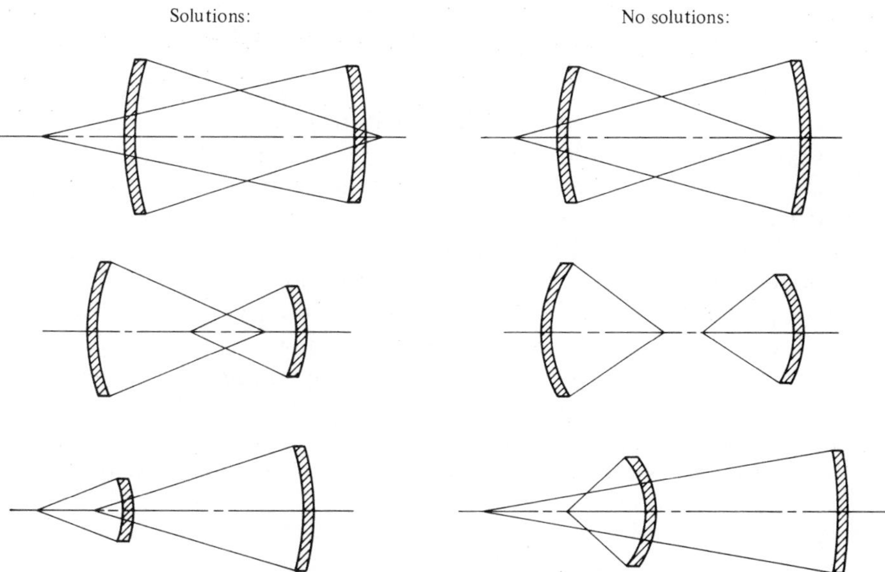

Figure 5.9 Resonator configurations with and without mode solutions.

The cases for which solutions are found are illustrated in Fig. 5.9. When both mirrors are concave, solutions exist when either both centers of curvature of the mirrors lie outside the space between mirrors, or both are inside and closer to the opposite mirror. When one mirror is convex, the other concave, the centers of curvature of the convex and concave mirrors have to lie outside the space between the mirrors and the former has to lie farther away than the latter. Resonators for which no solutions of Gaussian modes exist that reproduce themselves are *unstable resonators.* [4, 5] They are useful for laser oscillators using media of high gain. The large diffraction losses of unstable resonators help select a stable mode pattern, whereas a medium of high gain in a stable resonator, which has Gaussian mode solutions, tends to produce power outputs in higher-order Hermite–Gaussian modes (see Section 5.3).

5.3 HIGHER-ORDER MODES [2, 3, 6–8]

We have found one solution to the paraxial wave equation, the Gaussian beam solution u_{00}. For the representation of a beam with an arbitrary amplitude distribution at an input plane, an infinite set of solutions of the paraxial wave equation is required. There exists an orthogonal set of functions with the Gaussian as its lowest-order member, namely functions formed of products of Hermite–Gaussians (Appendices 5A and 5B).

A Hermite–Gaussian of order m of the independent variable ξ is defined as the product of a Hermite polynomial of order m, $H_m(\xi)$, and the Gaussian $\exp(-\xi^2/2)$. The lowest-order Hermite polynomials are (see Appendix 5C)

$$H_0(\xi) = 1, \qquad H_1(\xi) = 2\xi, \qquad H_2(\xi) = 4\xi^2 - 2$$

The Hermite–Gaussians are their own Fourier transforms (Appendix 5C). This means that Hermite–Gaussian amplitude distributions at an input plane ($z = 0$) have the same functional dependence in the far-field (Fraunhofer diffraction). But do they remain Hermite–Gaussians in the near field, in Fresnel diffraction? The Fresnel diffraction integral is the convolution of the input pattern with a Gaussian Fresnel kernel [(4.11)]. It will be shown that the convolution of a Hermite–Gaussian with a Gaussian yields the product of a Hermite polynomial with a Gaussian as it propagates through space. This is a generalization of the mode concept known from waveguide theory. [9] A waveguide wave of a particular mode retains strictly its transverse pattern as it propagates. In free space, diffraction spreads the beam, so that the pattern scales to larger and larger dimensions.

We use the abbreviated symbol $\psi_m(\xi)$ for the Hermite–Gaussian of mth order:

$$\psi_m(\xi) \equiv H_m(\xi) e^{-\xi^2/2} \tag{5.41}$$

We define the (two-dimensional) Hermite–Gaussian forward traveling wave of the mode of order m, n at $z = 0$, $u_{mn}(x_0, y_0)$, by

$$u_{mn}(x_0, y_0) = C_{mn}\,\psi_m\!\left(\frac{\sqrt{2}\,x_0}{w_0}\right)\psi_n\!\left(\frac{\sqrt{2}\,y_0}{w_0}\right) \tag{5.42}$$

Figure 5.10 shows some of the lowest-order Hermite–Gaussians. The Gaussian mode of the preceding section is a special case with $m = n = 0$. The forward traveling wave of the mode u_{mn} evaluated at z via the Fresnel diffraction

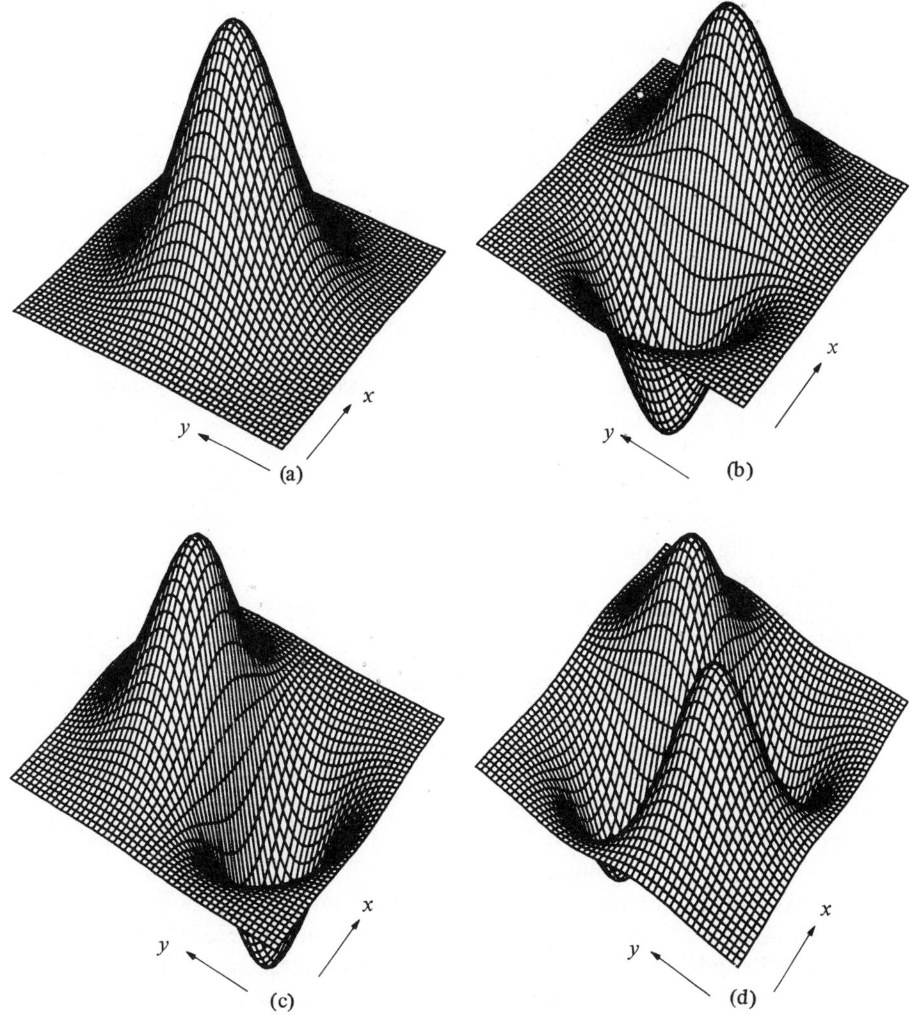

Figure 5.10 Plots of lowest-order Hermite–Gaussians. Origins of x-y coordinates are at centers of the plots. x axis goes from lower left to upper right, y axis from lower right to upper left. (a) $m = 0$, $n = 0$; (b) $m = 1$, $n = 0$; (c) $m = 0$, $n = 1$; (d) $m = 1$, $n = 1$.

integral is

$$u_{mn}(x, y, z) = \frac{jC_{mn}}{\lambda z} \int_{-\infty}^{\infty} dx_0 \int_{-\infty}^{\infty} dy_0 \, \psi_m\left(\frac{\sqrt{2}\,x_0}{w_0}\right)\psi_n\left(\frac{\sqrt{2}\,y_0}{w_0}\right)$$
$$\cdot \exp\left\{-\frac{jk}{2z}\left[(x - x_0)^2 + (y - y_0)^2\right]\right\} \tag{5.43}$$

This is a convolution of the Hermite–Gaussian with a Gaussian. The convolution carried out in one dimension yields, as shown in Appendix 5C:

$$\int_{-\infty}^{\infty} d\xi_0 \, \psi_n(\xi_0) e^{-(a/2)(\xi - \xi_0)^2}$$
$$= \sqrt{\frac{2\pi}{a + 1}} \left(\frac{a - 1}{a + 1}\right)^{n/2} \exp\left[\frac{a\xi^2}{2(a^2 - 1)}\right] \psi_n\left(\frac{a}{\sqrt{a^2 - 1)}}\,\xi\right) \tag{5.44}$$

This identity is to be applied twice, once for the integration over x_0, and once for the integration over y_0, with ξ_0 and ξ defined by

$$\xi_0 \equiv \frac{\sqrt{2}\,x_0}{w_0}, \qquad \xi = \frac{\sqrt{2}\,x}{w_0} \tag{5.45}$$

and a corresponding identification for the integration over y_0. We note that a in (5.44) has to be identified with

$$a = \frac{jkw_0^2}{2z} = j\,\frac{b}{z} \tag{5.46}$$

where b is the previously defined confocal parameter, $b = \pi w_0^2/\lambda$. The same normalization applies to the integral over y_0.

The double integral (5.43) thus leads to the result

$$u_{mn}(x, y, z) = \frac{C_{mn}}{\sqrt{1 + z^2/b^2}} \psi_m\left(\frac{\sqrt{2}\,x}{w}\right)\psi_n\left(\frac{\sqrt{2}\,y}{w}\right)$$
$$\cdot \exp\left[-\frac{jk}{2R}(x^2 + y^2)\right] e^{j(m+n+1)\phi} \tag{5.47}$$

where w, R, and ϕ, functions of z, were defined previously in (5.4) through (5.6) for the Gaussian beam solution. Equation (5.47) shows that the forward traveling wave of the mode m, n with the amplitude distribution $u_{mn}(x, y, z)$ expands in diameter as it propagates along z. A curvature of the phase front sets in. All modes have the same radius of curvature and same basic scale parameter w as functions of z! The only difference in behavior of the modes of different order as a function of distance along z appears in the phase factor $\exp\,[j(m + n + 1)\phi]$, the traveling waves of higher-order modes experience greater phase advances. This follows from the fact that the traveling waves of higher-order modes are made up of larger transverse components k_x, k_y of propagation vectors and thus have greater phase velocities in the z direction (smaller k_z). Note, further, that the field patterns as referred to the curved phase fronts [i.e., the patterns with the factor $\exp(-jk/2R)(x^2+y^2)$ omitted] remain Hermite–Gaussians as they propagate. The scale changes by the factor w/w_0.

The normalization constant C_{mn} can be chosen so that

$$\int_{-\infty}^{\infty} dx_0 \int_{-\infty}^{\infty} dy_0 \, |u_{mn}(x_0, y_0)|^2 = 1 \tag{5.48}$$

Using the identity developed in Appendix 5C,

$$\int_{-\infty}^{\infty} H_n^2(\xi) e^{-\xi^2} \, d\xi = \sqrt{\pi}\, 2^n n! \tag{5.49}$$

we find that

$$C_{mn} = \left(\frac{2}{w_0^2 \pi \cdot 2^{m+n} m! n!} \right)^{1/2} \tag{5.50}$$

Power conservation assures that the normalization is maintained as a function of z.

Fresnel Diffraction Theory Derived from Hermite–Gaussian Modes

The complete set of Hermite–Gaussians can be used for the expansion of an arbitrary input amplitude function $u_0(x_0, y_0)$ at $z = 0$. The simple propagation characteristics of the forward traveling waves of these modes may then be used to predict the amplitude function $u(x, y, z)$ at any cross section z. Thus if we set

$$u_0(x_0, y_0) = \sum A_{mn} u_{mn}(x_0, y_0, z = 0)$$
$$= \sum_{m,n} \frac{A_{mn}\sqrt{2}}{\sqrt{w_0^2 \pi \cdot 2^{m+n} m! n!}} \psi_m\!\left(\frac{\sqrt{2}\, x_0}{w_0} \right) \psi_n\!\left(\frac{\sqrt{2}\, y_0}{w_0} \right) \tag{5.51}$$

we may evaluate the expansion coefficients A_{mn} using the orthogonality condition of the Hermite–Gaussian [Appendix 5C, Eqs. (5C.15) and (5C.16):

$$A_{mn} = \frac{\sqrt{2}}{\sqrt{w_0^2 \pi \cdot 2^{m+n} m! n!}}^{1/2} \int_{-\infty}^{\infty} dx_0 \int_{-\infty}^{\infty} dy_0 \; u_0(x_0, y_0) \psi_m\!\left(\frac{\sqrt{2}\, x_0}{w_0} \right) \psi_n\!\left(\frac{\sqrt{2}\, y_0}{w_0} \right) \tag{5.52}$$

The spatial behavior of $u(x, y, z)$ is then

$$u(x, y, z) = \sum_{m,n} A_{mn} u_{mn}(x, y, z) \tag{5.53}$$

where $u_{mn}(x, y, z)$ is given by (5.47). We have mentioned that the z dependence of the different modes is different only in the phase factor $\exp[j(m+n+1)\phi]$. In the limit when z is large, $z/b \gg 1$, $\phi = \pi/2$. Thus

$$\lim_{z \text{ large}} u(x, y, z) = \sum_{m,n} (j)^{(m+n+1)} \frac{b}{z} \frac{A_{mn}\sqrt{2}}{\sqrt{w_0^2 \pi \cdot 2^{m+n} m! n!}} \cdot \psi_m\left(\frac{\sqrt{2}\,x}{w}\right)\psi_n\left(\frac{\sqrt{2}\,y}{w}\right) \exp\left[-\frac{jk}{2z}(x^2+y^2)\right] \tag{5.54}$$

We note that (5.54) differs from (5.51) in the following respects:

1. The beam diameter w_0 has been scaled up to $w = w_0(z/b)$.
2. The amplitude has been changed by $b/z = w_0/w$.
3. The phase front is curved with curvature $1/R \sim 1/z$.
4. The phase $(m+n+1)(\pi/2)$ has been introduced in each term.

These four modifications of the input amplitude distribution by free-space propagation are a rederivation of Fraunhofer diffraction. Indeed, the Hermite–Gaussian of order m is equal to its Fourier transform within the factor $(1/\sqrt{2\pi})(j)^m$ (Appendix 5C).

For z not much greater than b, the phase factor is more complicated, and thus the mode pattern at z is not the Fourier transform of the pattern at $z = 0$. Yet all intricacies of the paraxial Fresnel diffraction theory are contained merely in the phase factors $\exp[j(m+n+1)\phi]$.

Higher-Order Modes of Fabry–Perot Resonator

The electric field $\mathbf{E}$ and the magnetic field $\mathbf{H}$ of the forward traveling wave of the mode of order m, n polarized along the x direction are derived in the same way as for $m = n = 0$, (5.25) and (5.24),

$$\mathbf{E} = -j\omega\left(\hat{x}u_{mn} - j\hat{z}\frac{\partial u_{mn}}{k\,\partial x}\right)e^{-jkz} \tag{5.55}$$

$$\mu_0 \mathbf{H} = -jk\left(\hat{y}u_{mn} - j\hat{z}\frac{\partial u_{mn}}{k\,\partial y}\right)e^{-jkz} \tag{5.56}$$

The electric field can be written in two parts:

$$\mathbf{E} = \sqrt{\frac{kb}{\pi \cdot 2^{m+n} m! n!}}\frac{\omega}{z+jb} e^{j(m+n)\phi} \exp\left[-\frac{jk}{2R}(x^2+y^2)\right]\exp(-jkz) \left[\left(\hat{x} - \hat{z}\frac{x}{R}\right)\psi_m\psi_n - \hat{z}j\frac{\sqrt{2}}{kw}\psi'_m\psi_n\right] \tag{5.57}$$

The part $(\hat{x} - \hat{z}\,x/R)\psi_m\psi_n$ is parallel to the phase front; the quadrature component is perpendicular to it (within the paraxial approximation). The field patterns for polarization in the y direction are derived analogously to (5.55) and (5.56).

Two counter-traveling waves of equal amplitude of the mode m, n produce nodal surfaces of u_{mn} over the phase fronts; the electric field tangential to the phase fronts vanishes halfway between the surfaces. Thus, as before for the Gaussian resonant mode of the Fabry–Perot resonator, we construct the higher-order modes by superposition of forward and backward traveling waves of Hermite–Gaussian modes. To the solution $u_{mn}(x, y, z) \exp(-jkz)$ we add another solution obtained replacing z by $-z$. Nodal surfaces of the tangential electric field are produced in this manner over the curved phase fronts with radii of curvature $R = (z^2 + b^2)/z$ functions of z. The nodal surfaces are separated by half-wavelengths measured in terms of the effective propagation constant k_{eff} [compare (5.12)]

$$\int_0^z k_{\text{eff}}(z)\, dz = kz - (m + n + 1)\phi$$

Because of the phase advance $(m + n + 1)\phi$ the different Hermite–Gaussian modes resonate at different frequencies from those of the fundamental mode $m = n = 0$. Consider the symmetric resonator of Fig. 5.7 as an example. In order to satisfy the boundary conditions on the mirrors (or rather mirror reference planes defined as nodal surfaces of the tangential electric field) the phase shift between the mirrors must be an integer multiple of π, say $p\pi$, with p an integer. On the other hand, the phase advance of a Hermite–Gaussian of order m, n is $(m + n + 1)\phi$, where $\phi = \tan^{-1}(z/b)$, from the beam waist at $z = 0$ to the position z. In the symmetric resonator, the mirror is at the distance $z = d/2$ from the waist. The phase shift from mirror to mirror is twice that, $2(m + n + 1)\phi$ with $\phi = \tan^{-1}(d/2b)$.

$$\int_{-d/2}^{d/2} k_{\text{eff}}(z)\, dz = p\pi = kd - 2(m + n + 1)\tan^{-1}(d/2b)$$

Since $k = 2\pi f/c$, we have for the frequency f_{mnp} of the resonant mode of order m, n, p (the subscripts indicate the variations of the field pattern in the x, y, and z directions, respectively):

$$f_{mnp} = \frac{c}{2d}\left[p + \frac{2(m + n + 1)}{\pi}\tan^{-1}\frac{d}{2b}\right] \tag{5.58}$$

The resonance frequencies of the modes of Section 5.2 are the special case $m = 0$, $n = 0$, p. The confocal parameter b is evaluated from (5.37) for any choice of d and mirror radii R_0.

There is one special practical case in which the frequency separation of the resonant modes is particularly simple, when $2b = d$ (i.e., when $R_0 = d$). This is the *confocal* case because in this case the foci of the mirrors coincide. (The foci f are a distance $R/2$ in front of the mirror—see Section 5.4.) In this

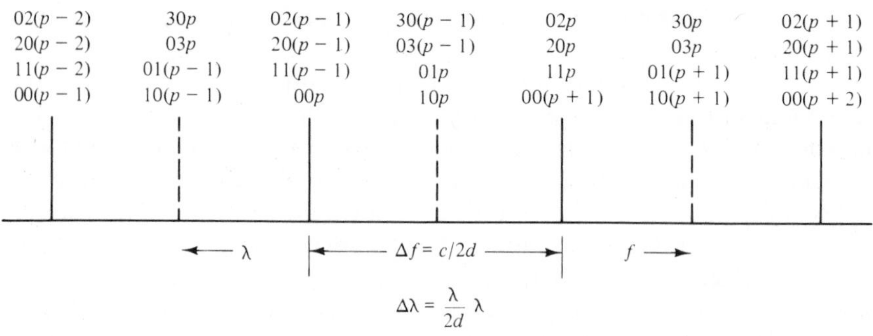

Figure 5.11 Schematic of resonance frequencies of confocal Fabry–Perot.

case,

$$f_{mnp} = \frac{c}{2d}\left[p + \frac{1}{2}(m + n + 1)\right] \tag{5.59}$$

All even-ordered modes ($m + n$ is an even integer) occupy the same *comb of resonance frequencies*, and all odd ones occupy a comb halfway between the even combs (see Fig. 5.11). This fact has an important application for scanning Fabry–Perot transmission interferometers that we are now ready to discuss. When we first studied the scanning Fabry–Perot interferometer we implied that only a single value of θ, the angle of the incident wave, was used for the input plane wave. Any beam of finite size is made up of a superposition of $\boldsymbol{k}$ vectors. Thus different $\boldsymbol{k}$ vector components at the same frequency excite different Hermite–Gaussian modes that resonate at different mirror spacings and produce "false" signals at the output. The confocal mirror Fabry–Perot avoids this difficulty.

5.4 THE q PARAMETER OF A GAUSSIAN BEAM AND ITS TRANSFORMATION [3, 5, 7]

The resonator analysis in Section 5.2 was carried out by fitting reflecting boundaries to nodal planes of a standing-wave solution. Similar results can be obtained from the so-called q-parameter description of Gaussian beams. This description has the additional advantage that it extends to higher-order Hermite–Gaussian modes and gives a concise description of their transformation by lenses, mirrors, and other optical components.

A Gaussian beam is completely described by the complex parameter $z + jb$. The real part gives the distance from the position of the minimum radius, the imaginary part gives the radius.

$$\frac{1}{z + jb} = \frac{z - jb}{z^2 + b^2} = \frac{1}{R} - j\frac{\lambda}{\pi w^2} \equiv \frac{1}{q} \tag{5.60}$$

Thus, if we know how q transforms, we know how to transform Gaussian beams. Because all higher-order Hermite–Gaussian modes are described by the same R and w, the same q parameter, their transformation is governed by the same law, except of course in so far as the phase change $(m + n + 1)\phi$ is concerned. We look now at the transformation of the q parameter by three basic optical system components.

Section of free space of length d. At a plane z, the q parameter is $z + jb$; at a plane z' at distance d from the plane z (see Fig. 5.12):

$$q' = z' + jb = z + d + jb$$

and thus the q parameter changes from q to q', where

$$q' = q + d \tag{5.61}$$

A thin lens of focal distance f. If a Gaussian beam of the form

$$\frac{\sqrt{2}}{\sqrt{\pi w^2}} e^{j\phi} e^{-(x^2+y^2)/w^2} e^{-jk(x^2+y^2)/2R} \tag{5.62}$$

is incident on a thin lens from the left (Fig. 5.13), it does not change the radius w as it passes through the lens, but experiences a phase advance $\phi(x, y)$, (4.33). The exponent $-jk(x^2 + y^2)/2R$ acquires a different x-y dependence:

$$-\frac{jk(x^2 + y^2)}{2R} + \frac{jk(x^2 + y^2)}{2f} = -\frac{jk(x^2 + y^2)}{2R'} \tag{5.63}$$

The beam has a new radius of curvature of the phase front, R':

$$\frac{1}{R'} = \frac{1}{R} - \frac{1}{f} \tag{5.64}$$

Using the definition of the q-parameter (5.60), one concludes that the q-parameter transformation is,

$$\frac{1}{q'} = \frac{1}{q} - \frac{1}{f} \tag{5.65}$$

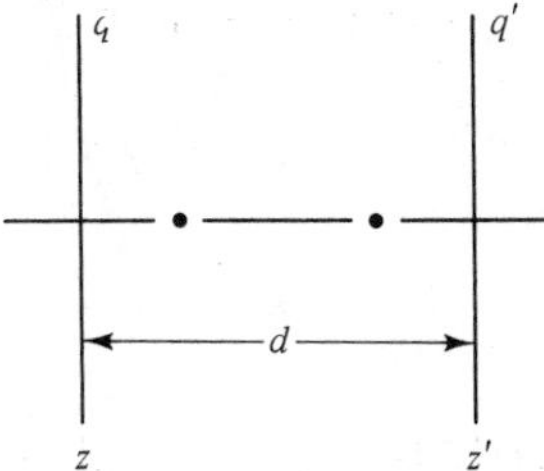

Figure 5.12 Transformation of q parameter by propagation over distance d.

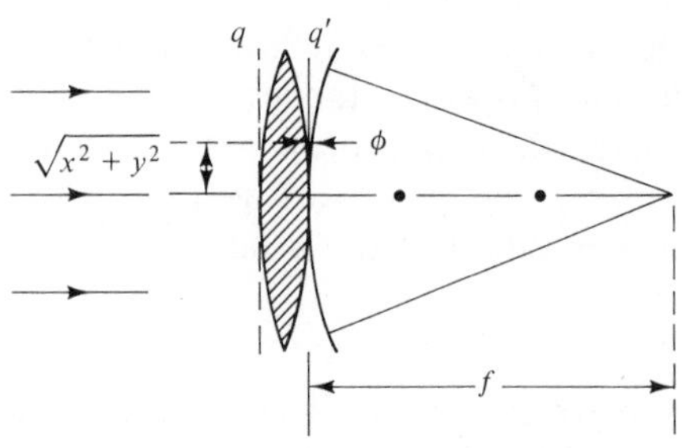

Figure 5.13 Transformation of radius of curvature of phase front by lens.

or

$$q' = \frac{q}{-(q/f) + 1} \tag{5.66}$$

A mirror of radius R_0. A mirror of radius R_0 reflects the beam and changes the radius of curvature of the phase front. If we unfold the beam as shown in Fig. 5.14, we find that the incident phase delay $k(x^2 + y^2)/2R$ is advanced by $2[k(x^2 + y^2)/2R_0]$ because the path is shortened twice, once upon incidence and once upon reflection. Thus R' of the "unfolded" reflected beam is given by

$$\frac{1}{R'} = \frac{1}{R} - \frac{2}{R_0} \tag{5.67}$$

A mirror acts like a lens with a focal distance $f = R_0/2$.

We find that the transformations of q by any of the three optical system

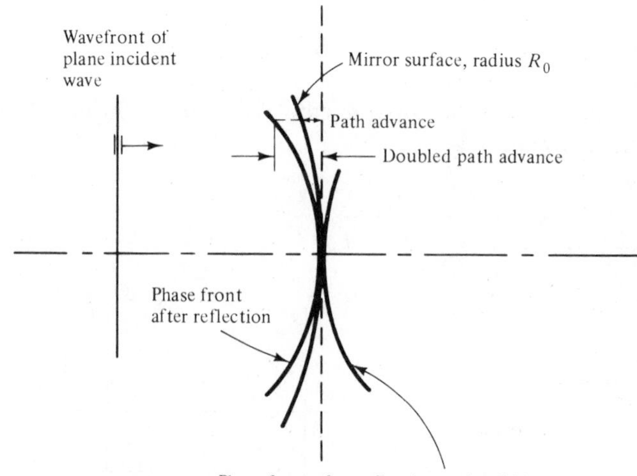

Figure 5.14 Reflection from curved mirror and its unfolding.

components may be expressed as a *bilinear transformation*

$$q' = \frac{Aq + B}{Cq + D} \tag{5.68}$$

with the transformation matrix $\begin{bmatrix} A & B \\ C & D \end{bmatrix}$. For free space

$$\begin{bmatrix} A & B \\ C & D \end{bmatrix} = \begin{bmatrix} 1 & d \\ 0 & 1 \end{bmatrix} \tag{5.69}$$

For a thin lens of focal distance f:

$$\begin{bmatrix} A & B \\ C & D \end{bmatrix} = \begin{bmatrix} 1 & 0 \\ -1/f & 1 \end{bmatrix} \tag{5.70}$$

For a mirror of radius R_0 one gets the lens transformation formula with $f = R_0/2$. One may show with a little algebra that the transformation produced by a cascade of two optical systems yields (see Fig. 5.15)

$$q_2 = \frac{Aq_0 + B}{Cq_0 + D} \tag{5.71}$$

where

$$\begin{aligned} \begin{bmatrix} A & B \\ C & D \end{bmatrix} &= \begin{bmatrix} A_2 & B_2 \\ C_2 & D_2 \end{bmatrix} \begin{bmatrix} A_1 & B_1 \\ C_1 & D_1 \end{bmatrix} \\ &= \begin{bmatrix} A_2 A_1 + B_2 C_1 & A_2 B_1 + B_2 D_1 \\ C_2 A_1 + D_2 C_1 & C_2 B_1 + D_2 D_1 \end{bmatrix} \end{aligned} \tag{5.72}$$

is the matrix product of the matrices of the individual transformations.

For all transformation matrices found thus far, which relate the input q parameter in free space (air) to the output q' parameter in free space:

$$\det \begin{bmatrix} A & B \\ C & D \end{bmatrix} = 1 \tag{5.73}$$

Since the determinant of the product of matrices is equal to the product of determinants, this holds for any cascade.

The multiplication of the individual matrices of systems in a cascade has been encountered in problems 3.4 and 3.5 in connection with the **T** matrix of a two-port. The reflection coefficient $\Gamma = b/a$ obeys a bilinear transformation in which the **T** matrix elements appear as coefficients. One may raise the question

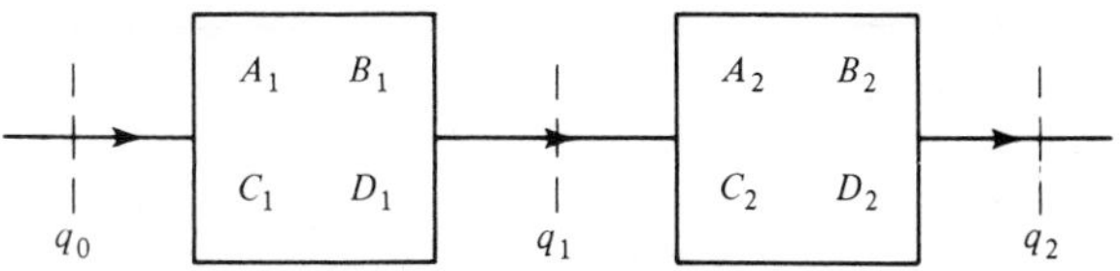

Figure 5.15 Cascade of optical systems.

of whether the q parameter is analogous to a reflection coefficient, that is, is equal to the ratio of two excitation amplitudes that transforms like a reflection coefficient. This question can be answered in the affirmative. The transformation of the q parameter is in one-to-one correspondence with a ray-optical ratio as we shall show in the next section. The proof also shows the correspondence of the *ABCD* matrices derived in this section, with the *ABCD* matrices derived from ray-optical considerations. In many cases it is easier to derive the *ABCD* matrices from ray optics rather than directly from the transformation of the q parameter.

5.5 THE *ABCD* MATRIX IN RAY OPTICS

The formation of an image by an optical system is described in ray optics in terms of the transformation of the rays emerging from a point on the object so as to intersect in a point of the image. This description is not at variance with paraxial Fresnel diffraction theory. Indeed, the analysis in Section 4.4 has shown that a lens produces a perfect image if one allows for apertures of infinite diameter (i.e., diffraction effects are introduced by apertures of finite size). Hence one expects a correspondence between relationships derived from ray optics and relationships of Fresnel diffraction theory. The impulse response function underlying diffraction theory is the solution of the wave equation of a point source with spherical wavefronts. The $\boldsymbol{k}$ vector at any point of the spherical wavefront may be described as a ray emerging from the source. Transformation of the wavefront by an optical system may be described alternately in terms of bending of the rays.

In this section we derive the *ABCD* matrix of ray optics. We then show that this matrix is in one-to-one correspondence with the *ABCD* matrix encountered in the transformation of the q parameter.

Consider the object shown as a vertical arrow in Fig. 5.16 imaged by an optical system. If an image is formed, rays emerging from any point of the object must meet at the corresponding point on the image. Pick any two

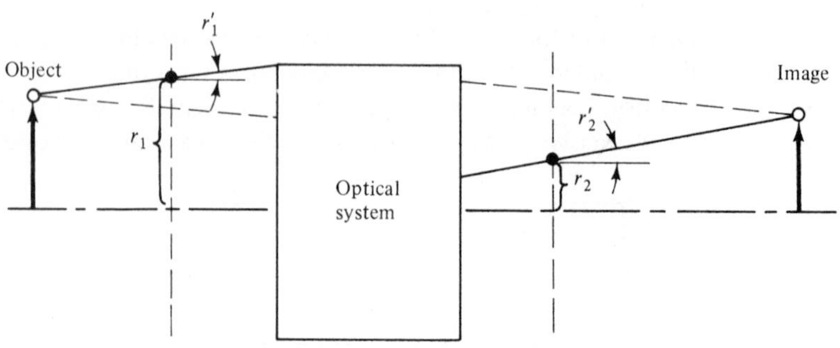

Figure 5.16 Ray transformation of an optical system.

reference planes somewhere between the object and system, image and system. Rays "pierce" the reference planes at distances r_1 and r_2 from the axis, with slopes r'_1 and r'_2 as shown. The optical system relates r_2 and r'_2 to r_1 and r'_1 by a transformation law

$$r_2 = f(r_1, r'_1) \tag{5.74}$$

$$r'_2 = g(r_1, r'_1) \tag{5.75}$$

where f and g are functions of r_1 and r'_1. If the optical system is to form a perfect image in the image plane, the transformation laws (5.74) and (5.75) must be linear. We demonstrate this statement with the aid of Fig. 5.17. The top of the object arrow has one ray which passes reference plane 1 at r_1 with slope r'_1. The transformed ray goes, by definition, through the top of the image arrow. It passes reference plane 2 at r_2 with slope r'_2. A point at a decreased height, ξ times the arrow height, must be imaged at ξ times the image height. The corresponding rays must pass at ξ times the original heights r_1 and r_2, and must have ξ times the original angles. This shows that the relationship between r_2, r'_2 and r_1, r'_1 must be a linear relationship of the general form

$$r_2 = Ar_1 + Br'_1 \tag{5.76}$$

$$r'_2 = Cr_1 + Dr'_1 \tag{5.77}$$

The relation can be written in matrix form by defining the column matrics $\begin{bmatrix} r_1 \\ r'_1 \end{bmatrix}$ and $\begin{bmatrix} r_2 \\ r'_2 \end{bmatrix}$:

$$\begin{bmatrix} r_2 \\ r'_2 \end{bmatrix} = \begin{bmatrix} A & B \\ C & D \end{bmatrix} \begin{bmatrix} r_1 \\ r'_1 \end{bmatrix} \tag{5.78}$$

A cascade of optical systems has an *ABCD* matrix that is the product of the matrices of the component systems. Consider first the transformation law (5.76) and (5.77) for a thin lens with reference planes as shown in Fig. 5.18. Because a thin lens does not change the distance r from the axis of a ray passing through it, $r_2 = r_1$ and thus $A = 1$, $B = 0$. The slope r'_2 is unchanged for a ray passing through the center of the lens, $r_1 = 0$. Hence $D = 1$. Finally, a

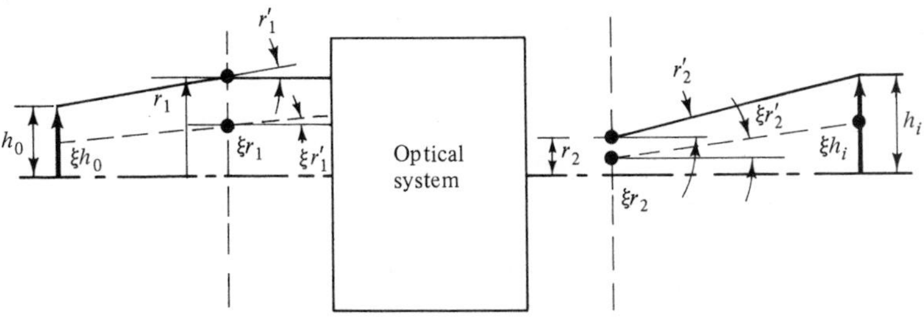

Figure 5.17 Linearity of transformation by an ideal optical system.

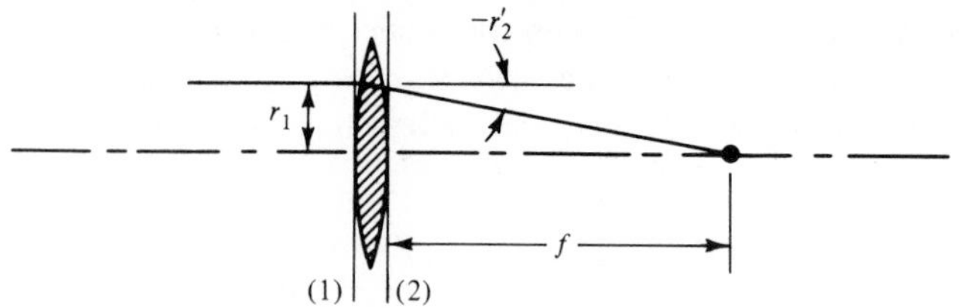

Figure 5.18 Thin lens with reference planes adjacent to it.

ray parallel to the axis, $r'_1 = 0$, goes through the focal point

$$r'_2 = -\frac{r_1}{f}\bigg|_{r'_1=0}$$

and thus

$$C = -\frac{1}{f}$$

The $ABCD$ matrix of a thin lens is

$$\begin{bmatrix} A & B \\ C & D \end{bmatrix} = \begin{bmatrix} 1 & 0 \\ -1/f & 1 \end{bmatrix} \tag{5.79}$$

This is the same matrix as obtained from the q-parameter transformation.

Next, consider the transformation matrix of a slab of dielectric of index n, of thickness d. For $n = 1$ this reduces to a section of free space. At the interface between air and the medium, the ray is deflected according to Snell's law in its paraxial form (see Fig. 5.19).

$$r'_2 = \frac{n_1}{n_2} r'_1 = \frac{1}{n} r'_1$$

and

$$r_2 = r_1$$

Thus the $ABCD$ matrix of the interface is

$$\begin{bmatrix} A & B \\ C & D \end{bmatrix} = \begin{bmatrix} 1 & 0 \\ 0 & 1/n \end{bmatrix} \tag{5.80}$$

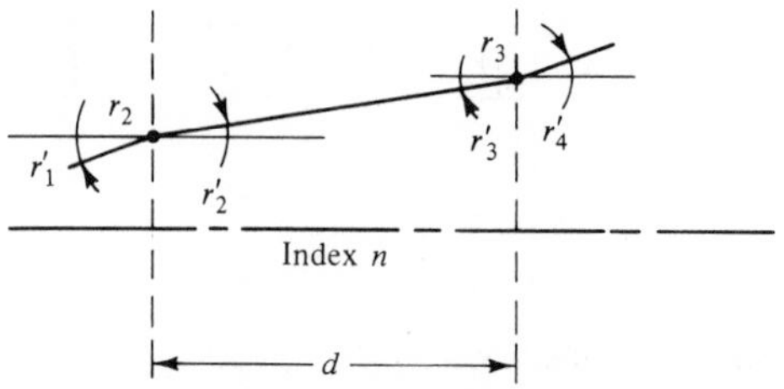

Figure 5.19 A section of free space.

The travel distance of length d gives the transformation

$$r_3 = r_2 + dr'_2$$
$$r'_3 = r'_2$$

With the $ABCD$ matrix

$$\begin{bmatrix} A & B \\ C & D \end{bmatrix} = \begin{bmatrix} 1 & d \\ 0 & 1 \end{bmatrix} \tag{5.81}$$

This is followed by the second interface which has the $ABCD$ matrix of the form (5.80) except that n is replaced by $1/n$. The cascade of the three regions gives the product matrix

$$\begin{bmatrix} A & B \\ C & D \end{bmatrix} = \begin{bmatrix} 1 & d/n \\ 0 & 1 \end{bmatrix} \tag{5.82}$$

For $n = 1$ this reduces to the $ABCD$ matrix obtained previously for the q-parameter transformation by a section of free space of length d. We have generalized it to take into account a change of index. We have shown that the $ABCD$ matrix of a slab of index n, thickness d, is the same as that of a section of free space of length d/n.

Consider as an example a thin lens with reference planes at distances l_1 in front, and l_2 in back, of the lens. The $ABCD$ matrix of the entire system is the product of the $ABCD$ matrices of the three "components":

$$\begin{bmatrix} A & B \\ C & D \end{bmatrix} = \begin{bmatrix} 1 - \dfrac{l_2}{f} & l_1 + l_2\left(1 - \dfrac{l_1}{f}\right) \\ -\dfrac{1}{f} & 1 - \dfrac{l_1}{f} \end{bmatrix} \tag{5.83}$$

If the object is positioned at l_1, and l_2 is identified as the image position, we must have

$$r_2 = Mr_1 \tag{5.84}$$

where M is the magnification, and r_2 is independent of r'_1. Thus

$$B = l_1 + l_2\left(1 - \frac{l_1}{f}\right) = 0$$

or

$$\frac{1}{l_1} + \frac{1}{l_2} = \frac{1}{f} \tag{5.85}$$

This is the well-known relation between the positions of object and image planes. Further, the magnification is equal to $|A|$:

$$M = |A| = \left|1 - \frac{l_2}{f}\right| = \left|\frac{l_2}{l_1}\right| \tag{5.86}$$

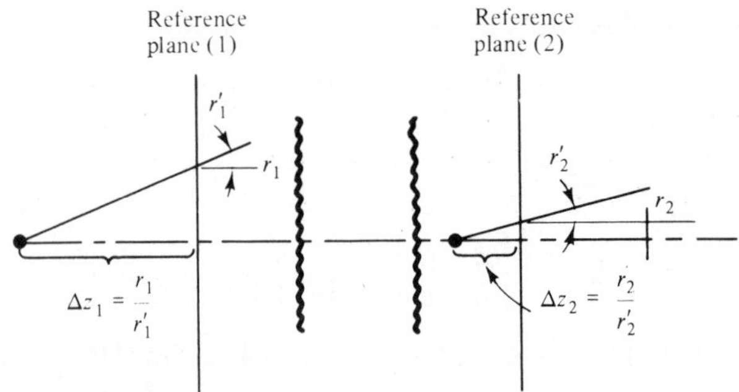

Figure 5.20 Interpretation of the ratio r/r'.

Returning to the general transformation law (5.78), consider the transformation law for the ratio r/r',

$$\frac{r_2}{r_2'} = \frac{A(r_1/r_1') + B}{C(r_1/r_1') + D} \tag{5.87}$$

This is a bilinear transformation law identical in form with the transformation law of the q parameter. Note that r_1/r_1' and r_2/r_2' are the distances Δz_1 and Δz_2 in Fig. 5.20, that is, the distances between the reference planes and the intersection points of the rays with the axis of the system. We have noted that a Gaussian beam may be considered to be the paraxial limit of the point-source solution shifted by an imaginary distance $z_0 = -jb$. The distance $z + jb = q$ measures the "complex distance" from the reference plane to the intersection point with the axis of a "complex ray" pertaining to the Gaussian mode. This is why q obeys the same transformation law as $\Delta z = r/r'$.

The most general $ABCD$ matrix has four adjustable parameters, its determinant is not equal to 1, like that of (5.80). In Appendix 5D we study the general optical system represented by such a matrix.

5.6 APPLICATIONS OF THE *ABCD* MATRIX

In Section 5.2 we described the design of Fabry–Perot resonators using mirrors of given radii of curvature spaced a distance d apart. We matched the nodal surfaces of the tangential electric field with mirror surfaces and obtained criteria for mirror spacings for which resonator modes exist.

The same results can be obtained using the $ABCD$ formalism. A resonator consisting of two curved mirrors may be unfolded as shown in Fig. 5.21. The requirement for the existence of a mode is that the beam repeat itself after $2d$. We have for the $\begin{bmatrix} A & B \\ C & D \end{bmatrix}$ matrix of the cascade consisting of a lens of focal distance f_1, followed by free-space propagation over a distance d, a lens of

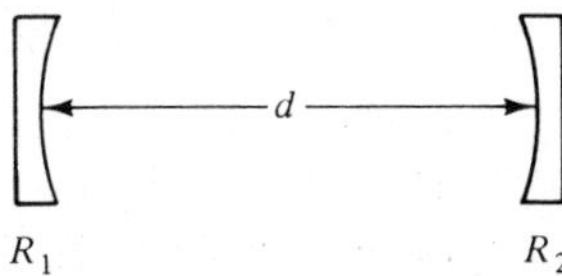

Unfolded

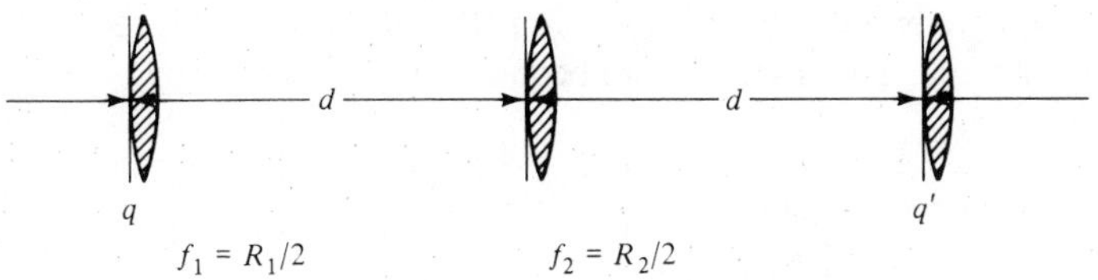

Figure 5.21 Unfolding of resonator.

focal distance f_2, again followed by free space of the same distance:

$$\begin{bmatrix} A & B \\ C & D \end{bmatrix} = \begin{bmatrix} \left(1 - \frac{d}{f_1}\right)\left(1 - \frac{d}{f_2}\right) - \frac{d}{f_1} & d\left(1 - \frac{d}{f_2}\right) + d \\ -\frac{1}{f_2}\left(1 - \frac{d}{f_1}\right) - \frac{1}{f_1} & 1 - \frac{d}{f_2} \end{bmatrix} \tag{5.88}$$

The q parameter transforms into q':

$$q' = \frac{Aq + B}{Cq + D} \tag{5.89}$$

If a mode is to exist, q must repeat after one full round trip: $q' = q$. We get the following equation for q:

$$Cq^2 + (D - A)q - B = 0 \tag{5.90}$$

or

$$q = \frac{A - D}{2C} \pm \sqrt{\left(\frac{A + D}{2C}\right)^2 - \frac{1}{C^2}} \tag{5.91}$$

where we have used the fact that

$$AD - CB = 1 \tag{5.92}$$

A real beam radius is obtained, according to (5.60), when Im $q \neq 0$. In order that q have an imaginary part, (5.91) requires that the following condition be met:

$$\left(\frac{A + D}{2}\right)^2 \leq 1 \tag{5.93}$$

Introducing (5.88), we get

$$0 < \left(1 - \frac{d}{2f_1}\right)\left(1 - \frac{d}{2f_2}\right) < 1 \tag{5.94}$$

This is the same criterion as that derived in (5.40).

Let us study in greater detail the matrix (5.88) in the special case when $1/f_1 = 0$. This is the case of one flat mirror and one curved mirror forming the resonator. The reference plane is picked at the flat mirror (Fig. 5.21) where the beam waist occurs. The matrix (5.88) becomes, replacing f_2 by f:

$$\begin{bmatrix} A & B \\ C & D \end{bmatrix} = \begin{bmatrix} 1 - \dfrac{d}{f} & 2d - \dfrac{d^2}{f} \\ -\dfrac{1}{f} & 1 - \dfrac{d}{f} \end{bmatrix} \tag{5.95}$$

We have found that $A = D$. This is as it must be, because q of (5.91) must be imaginary, since the beam waist occurs at the reference plane. Solving for q, one finds that

$$q = j\sqrt{d(2f - d)} \tag{5.96}$$

the same result as (5.36) with $2f = R_0$.

Matrices of the form (5.95) have some special simple properties; in particular it is easy to evaluate the matrix of a cascade of m systems, each of which is described by the matrix (5.95). Such cascades are of interest because they reconstruct a Gaussian beam of plane wavefront at every reference plane. The diffraction that the beam experiences over the propagation distance $2d$ is counteracted by the lens. In order to develop the expression for the matrix of a cascade we note that (5.95) can be written in the "angular" form

$$\begin{bmatrix} A & B \\ C & D \end{bmatrix} = \begin{bmatrix} \cos\theta & \chi\sin\theta \\ -\dfrac{1}{\chi}\sin\theta & \cos\theta \end{bmatrix} \tag{5.97}$$

where the angle θ is defined by

$$\cos\theta = 1 - \frac{d}{f} \tag{5.98}$$

and the parameter χ is

$$\chi = \sqrt{d(2f - d)} \tag{5.99}$$

If we cascade m systems described by the $ABCD$ matrix (5.97), the resulting matrix is

$$\begin{bmatrix} A & B \\ C & D \end{bmatrix}^m = \begin{bmatrix} \cos\theta & \chi\sin\theta \\ -\dfrac{1}{\chi}\sin\theta & \cos\theta \end{bmatrix}^m = \begin{bmatrix} \cos m\theta & \chi\sin m\theta \\ -\dfrac{1}{\chi}\sin m\theta & \cos m\theta \end{bmatrix} \tag{5.100}$$

Equation (5.100) can be proven by induction, by assuming the relation above to be true for the power m, and then by proving it to be correct for the power $m + 1$.

The Graded Refractive Index (GRIN) Lens

An important special case of a cascade of identical optical systems (of differential length) is the case of a dielectric rod of parabolic index profile as discussed in the next chapter—an example of an optical fiber. It is also the description of the Graded Refractive Index lenses that are commercially available under the name Selfoc lenses. They are thin rods (of the order of 1 mm diameter) that act as lenses and are convenient in controlling beams emerging from semiconductor diode lasers and optical fibers. Consider a dielectric rod with the optical index profile

$$n = n_0\left(1 - \frac{x^2 + y^2}{2h^2}\right) \tag{5.101}$$

where n_0 is the index at the center of the rod and $1/h$ is a measure of the parabolic dependence (rate of decrease) of the index n with increasing distance from the axis. A differential length Δz of rod acts on a beam in the following way:

1. There is the action of the index profile curving the phase front. This effect is equivalent to the action of a lens of focal distance f that produces a radius-dependent phase advance $-\phi$ [compare (4.33)]

   $$-\phi = \frac{k}{2f}(x^2 + y^2) = \frac{1}{2}\frac{\omega}{c}\frac{n_0}{h^2}(x^2 + y^2)\,\Delta z$$

   from which we may gather that

   $$\frac{1}{f} = \frac{n_0}{h^2}\Delta z \tag{5.102}$$

2. There is the action of the rod as a slab of thickness Δz of index n_0. This effect may be represented by the action of two slabs of thickness $\Delta z/2$ and index n_0 preceding and following the lens. The net *ABCD* matrix is given by (5.95) with $d \to \Delta z/2n_0$ [compare (5.82)] and $1/f$ given by (5.102):

$$\begin{bmatrix} A & B \\ C & D \end{bmatrix} = \begin{bmatrix} 1 - \dfrac{\Delta z^2}{2h^2} & \dfrac{\Delta z}{n_0} - \dfrac{(\Delta z)^3}{4h^2 n_0} \\ -\dfrac{n_0}{h^2}\Delta z & 1 - \dfrac{\Delta z^2}{2h^2} \end{bmatrix} \simeq \begin{bmatrix} 1 - \dfrac{\Delta z^2}{2h^2} & \dfrac{\Delta z}{n_0} \\ -\dfrac{n_0 \Delta z}{h^2} & 1 - \dfrac{\Delta z^2}{2h^2} \end{bmatrix} \tag{5.103}$$

The angle θ of the differential optical system is obtained by comparing (5.103) with (5.97):

$$\theta = \frac{\Delta z}{h} \tag{5.104}$$

and the parameter χ is

$$\chi = \frac{h}{n_0} \tag{5.105}$$

It is now an easy matter to determine the $ABCD$ matrix of a rod of length l. The net angle $m\theta$ in (5.100) for a propagation distance $l = m\,\Delta z$ is $m\theta = l/h$, and the $ABCD$ matrix is according to (5.104):

$$\begin{bmatrix} A & B \\ C & D \end{bmatrix} = \begin{bmatrix} \cos\dfrac{l}{h} & \dfrac{h}{n_0}\sin\dfrac{l}{h} \\ -\dfrac{n_0}{h}\sin\dfrac{l}{h} & \cos\dfrac{l}{h} \end{bmatrix} \tag{5.106}$$

A rod described by the matrix above maintains a Gaussian beam with a planar wavefront with a (imaginary) q parameter evaluated from (5.91) using the expressions for the matrix elements (5.106):

$$q = \frac{jh}{n_0} = j\,\frac{\pi w^2}{\lambda} \quad \text{or} \quad w^2 = \frac{h\lambda}{\pi n_0} \tag{5.107}$$

The lens-like performance of a cylindrical rod with a parabolic index profile is perceived very simply from (5.106) in the special case when $l/h = \pi/2$ (see Fig. 5.22). The resulting $ABCD$ matrix is identical with that of a thin lens of focal distance h/n_0 and with reference planes picked in its two focal planes [compare (5.106) for $l/h = \pi/2$ with (5.95) with $d = f$].

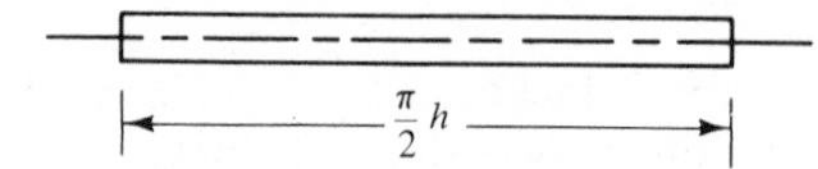

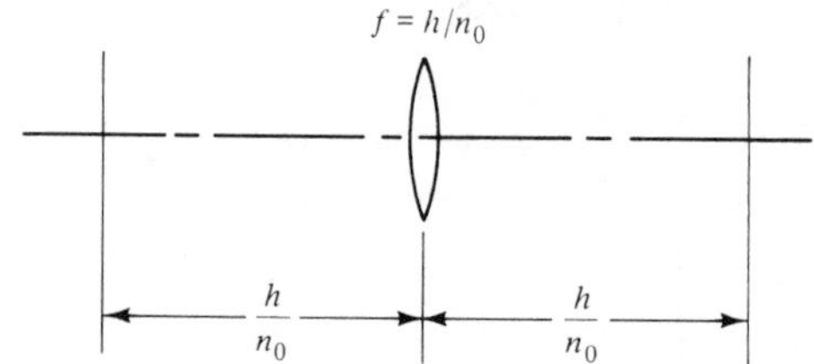

Figure 5.22 GRIN lens of "length" $\pi/2$ and its equivalent.

5.7 METHODS OF TESTING OPTICAL SYSTEMS [10–12]

Thus far, we have discussed ideal optical systems, systems that transform spherical wavefronts into spherical wavefronts and that form perfect images of objects. In reality, images are imperfect partly due to diffraction effects, partly because of imperfections in the optical system. Diffraction effects are caused by finite apertures. Ray optics cannot describe such effects; diffraction optics can. The Gaussian beam analysis gives a simple qualitative picture of the influence of diffraction by a finite-size aperture. If the optical beam has to pass through a lens of aperture radius w and is then focused into a minimum spot size w_0, the spot size w_0 must obey, roughly, the transformation law of a Gaussian beam (see Fig. 5.23).

We find that

$$w_0 = \frac{\lambda f}{\pi w} \tag{5.108}$$

For a finite wavelength, w_0 cannot be made equal to zero. For $f/\pi w \to 1$ one can make w_0 approach λ. This case corresponds roughly to an F number of the lens of unity (the F number is defined as the ratio of focal length to lens diameter).

Diffraction effects are imposed by the nature of light and are unavoidable (although poor system design can make them worse than necessary). Optical components deviate from the ideal behavior assumed in the theory discussed thus far. There are many tests devised to ascertain deviation from ideal behavior.

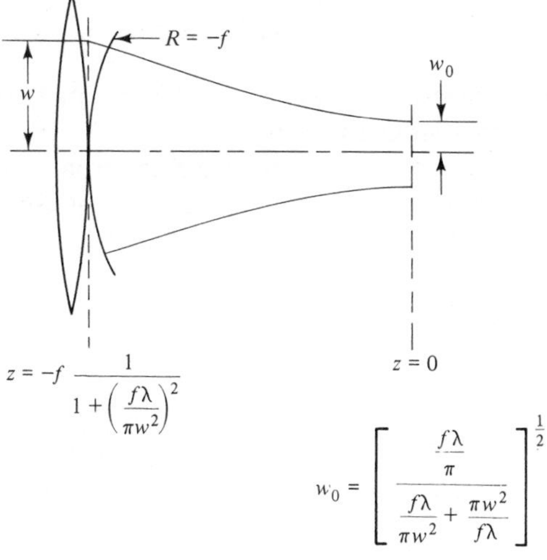

Figure 5.23 Finite spot size due to finite aperture.

Aberrations

Abberations can be defined in terms of the departure of initially spherical wavefronts from sphericity. By comparing an aberrated (i.e., nonspherical) wavefront with a truly spherical wavefront, we can obtain the wavefront aberration. This can be done in a variety of interferometers, in which the aberrated wavefront is made to interfere with a truly spherical wavefront. If the comparison wavefronts are made planar, the interference is as in Fig. 5.24, where Σ represents the aberrated wavefront and S_0 the comparison or true wavefront. To find the resulting fringe pattern, draw construction wavefronts $S_1, S_2, \ldots$ at 1, 2, ... wavelengths in advance of S_0 and similarly $S_{-1}, S_{-2}, \ldots$ in retardation. Bright fringes are formed where Σ is intersected by any of these surfaces, as indicated. Thus the fringe system is a plot of contours or levels of Σ with respect to S at wavelength intervals. Some realizations of this principle are described in the following paragraphs.

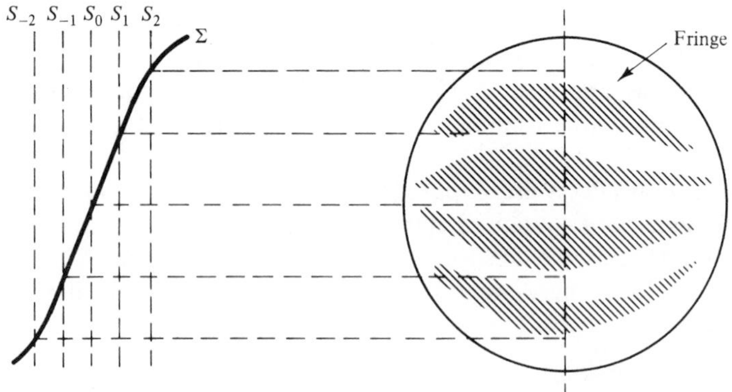

Figure 5.24 Interference between true and aberrated wavefronts.

The Fizeau interferometer. The Fizeau interferometer is used for testing nominally plane-parallel transparent plates. A fringe pattern is formed due to the interference of the reflections from the front and back surfaces of the plate. The interference fringes indicate the loci of constant $2nt$ at wavelength intervals (Fig. 5.25).

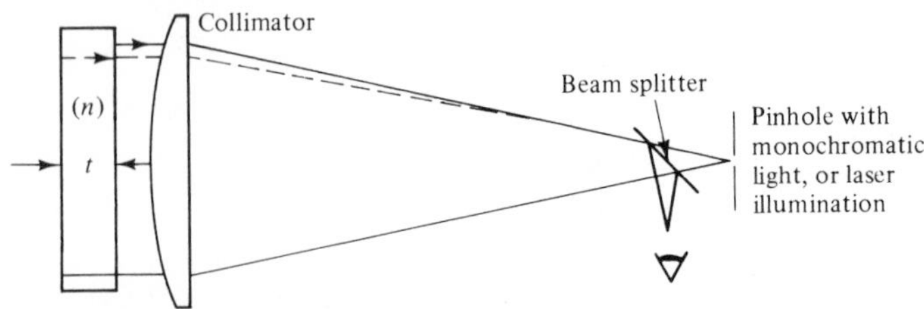

Figure 5.25 Fizeau interferometer for testing a plane-parallel plate of index n.

The Twyman–Green interferometer. In the Twyman–Green interferometer (Fig. 5.26) an incoming beam is divided and recombined at a beam splitter. The two mirrors are suitably tilted, until wedge fringes are seen. Then the mirrors are adjusted until the phase fronts emerging at the bottom of Fig. 5.26 are parallel, and the fringes disappear. The nominally plane-parallel plate to be tested is introduced in the side arm as shown in Fig. 5.27. The resulting fringes are contours at wavelength intervals of constant $2(n - 1)t$.

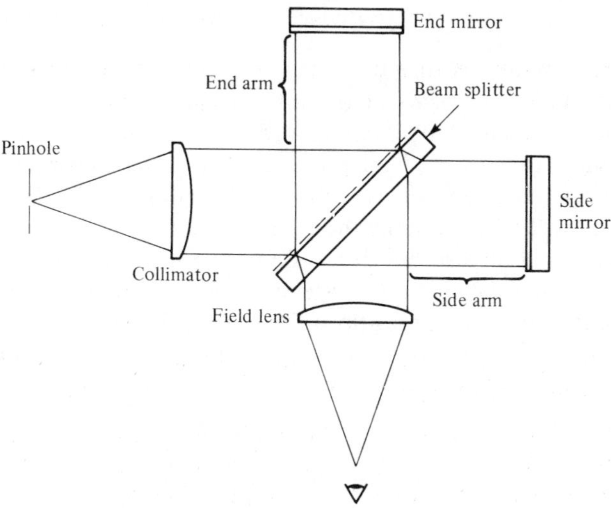

Figure 5.26 Twyman–Green interferometer.

Figure 5.28 shows a test of a dispersing prism; the fringes indicate the overall deviation from flatness in the wavefront. A lens test is set up as in Fig. 5.29, with a convex mirror that has its center of curvature at F', the nominal focus of the lens. In the space to the right of the lens the convex mirror sets up a reference spherical wavefront. If the lens has different foci for rays in two perpendicular planes through the lens axis, they may be determined by suitable displacements of the center of curvature of the curved mirror relative to F'.

The light source used with the Twyman–Green and similar interferometers is usually a low-pressure mercury lamp, filtered to give the green line,

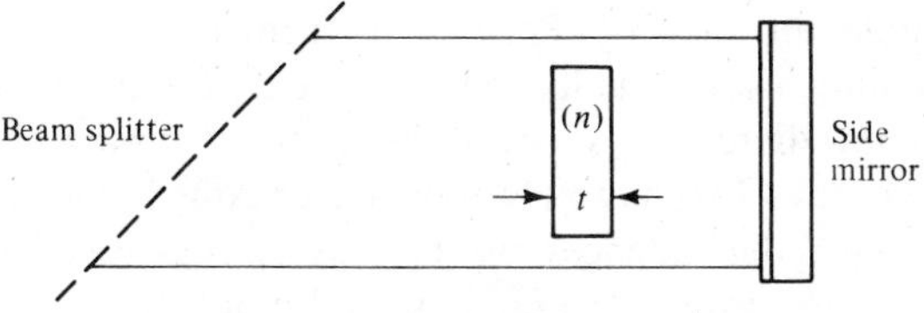

Figure 5.27 Testing a plane-parallel plate.

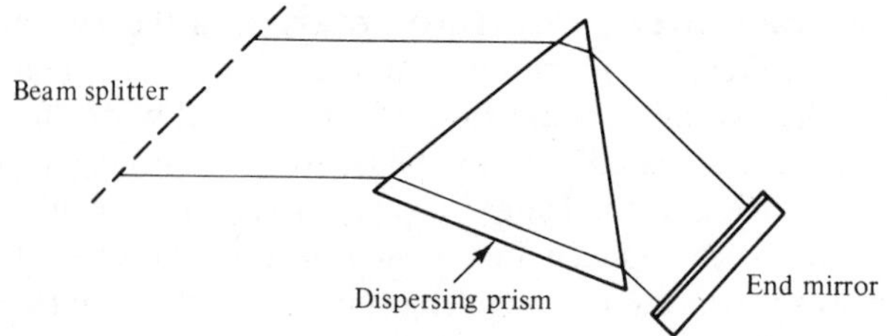

Figure 5.28 Testing a dispersing prism in the Twyman–Green interferometer.

0.546 μm wavelength. Because this line has a finite spectral width the optical path lengths in the two arms of the interferometer must be made approximately equal (to within 10 to 50 mm) in order to get fringes of good contrast.

If a laser is used as the light source the foregoing constraint is removed. Also it becomes possible to devise other modifications of the Twyman–Green interferometer with, for example, the beam splitter in convergent light. However, laser illumination also has disadvantages; a laser has such a long coherence length that extraneous fringe systems are formed involving every beam of scattered light and every parasitic reflection in the system; also, the laser wavelength may not be the one for which a lens system has been designed, so that the performance may be different at the design frequency.

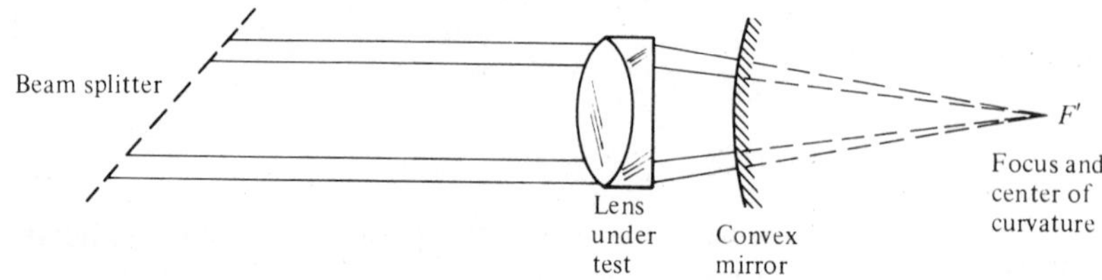

Figure 5.29 Testing a lens.

5.8 SUMMARY

The Gaussian beam is a solution of the paraxial wave equation. This solution demonstrates the simplest features of diffraction, the spreading angle of a beam of given minimum diameter and the dependence of phase-front curvature upon distance. Standing-wave solutions constructed of Gaussian beams illustrate the fundamental modes in a Fabry–Perot interferometer with curved mirrors of finite transverse dimensions. Such modes exist only for certain mirror spacings and the minimum diameter is controllable by the mirror geometry. The higher-order Hermite–Gaussian solutions are obtained from the Fresnel integral; they possess z dependences of the beam cross section and radius of curvature identical to those of the fundamental mode. If diffraction is analyzed as a superposition of Hermite–Gaussian modes, all the intricacies of Fresnel diffraction are contained in the different phase dependences of the different Hermite–Gaussian modes. The higher-order modes of the Fabry–Perot resonator can be constructed from Hermite–Gaussian standing waves and the confocal resonator is found to be the one with simplest mode structure for a Fabry–Perot transmission resonator.

The q parameter is a succinct characterization of the radius of curvature and beam diameter of a Gaussian beam, and as an extension, of any Hermite–Gaussian beam. The q parameter obeys a bilinear transformation law involving the coefficients of the *ABCD* matrix. This matrix also appears in a ray-optical analysis of an optical system. The distance from the reference plane of the intersection of a ray with the system axis also obeys the same bilinear transformation law. This fact is made plausible by the argument that the q parameter is the (complex) distance from the reference plane of the source generating the Gaussian beam.

The *ABCD* formalism leads to the well-known relations for the formation of an image from an object by a lens. We analyzed a cascade of differential lenses—a description of the GRIN lens.

With an understanding of diffraction optics and wavefront transformation we were ready to study methods for the testing of aberration by optical systems.

APPENDIX 5A

The Defining Equation of Hermite–Gaussians

The Hermite–Gaussian functions are best understood with the aid of the differential equation for which they form a complete set of solutions, the Schrödinger equation of the one-dimensional quantum mechanical harmonic oscillator, [13,14,15] which is in normalized form:

$$\frac{d^2\psi}{d\xi^2} + (\lambda - \xi^2)\psi = 0 \tag{5A.1}$$

An orthogonal set of functions $\psi_m(\xi)$ of a single independent variable ξ is generated by this differential equation in the sense that

$$\int_{-\infty}^{\infty} \psi_m(\xi)\psi_n^*(\xi)\, d\xi = 0 \qquad \text{for } m \neq n \tag{5A.2}$$

To prove this, and gain further understanding of the solutions of (5A.1), we study the geometric interpretation of (5A.1) by means of Fig. 5A.1. Because the coefficients of equation (5A.1) are symmetric with respect to ξ, the solutions must be either symmetric or antisymmetric. A symmetric solution that starts out from the center, $\xi = 0$, with zero slope going to the right has a prescribed slope and curvature for ever after, as determined by the second-

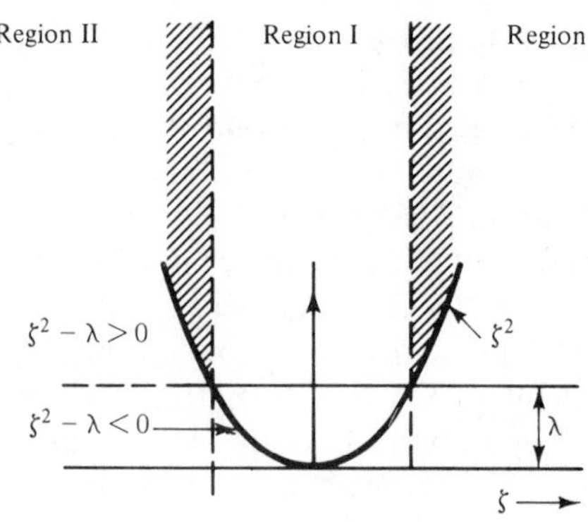

Figure 5A.1

order differential equation. It will be concave toward the ξ axis in region I, convex in region II. It will shoot off toward $\pm\infty$ at $\xi \rightarrow \infty$, unless λ is carefully chosen—at the so-called eigenvalue of λ corresponding to a bounded solution. The lowest-order solution has the lowest curvature, the lowest eigenvalue, and only one extremum. It is the Gaussian $e^{-\xi^2/2}$ with $\lambda = 1$. Each higher eigenvalue belongs to a solution with one more extremum. The next solution is antisymmetric with two extrema, the following one is symmetric with three extrema, and so forth.

Next, we investigate how the higher-order solutions are related to the lower-order ones. For this purpose it is convenient to introduce the "creation" and "annihilation" operators, or "raising" and "lowering" operators $d/d\xi \mp \xi$, where the $-$ sign goes with raising, the $+$ sign with lowering. Consider a function $\psi(\xi)$ which is assumed to obey (5A.1) and which vanishes for $\xi \rightarrow \pm\infty$. Operate on (5A.1) with $d/d\xi \mp \xi$ and rearrange the terms so that $(d/d\xi \mp \xi)$ is brought to the right of $d^2/d\xi^2$ and ξ^2. For this purpose we note that

$$\left(\frac{d}{d\xi} \mp \xi\right)\frac{d^2\psi}{d\xi^2} = \frac{d^2}{d\xi^2}\left[\left(\frac{d}{d\xi} \mp \xi\right)\psi\right] \pm 2\frac{d\psi}{d\xi} \tag{5A.3}$$

$$\left(\frac{d}{d\xi} \mp \xi\right)\xi^2\psi = \xi^2\left(\frac{d}{d\xi} \mp \xi\right)\psi + 2\xi\psi \tag{5A.4}$$

Using (5A.3) and (5A.4) in (5A.1) operated on by $d/d\xi \mp \xi$, we obtain

$$\frac{d^2}{d\xi^2}\left[\left(\frac{d}{d\xi} \mp \xi\right)\psi\right] + [(\lambda \pm 2) - \xi^2]\left[\left(\frac{d}{d\xi} \mp \xi\right)\psi\right] = 0 \tag{5A.5}$$

We have recovered the original equation, where the new solution $(d/d\xi \mp \xi)\psi$ has the eigenvalue $\lambda \pm 2$. Consider the lowest-order solution $\exp(-\xi^2/2)$ with $\lambda = 1$. It has the lowest possible negative curvature in the range where $(\lambda - \xi^2)$ is positive, and hence the lowest possible value of λ. The next solution obtained by operating with the raising operator, $(d/d\xi - \xi)\exp(-\xi^2/2) = -2\xi\exp(-\xi^2/2)$ has two extrema. Each successive application produces one

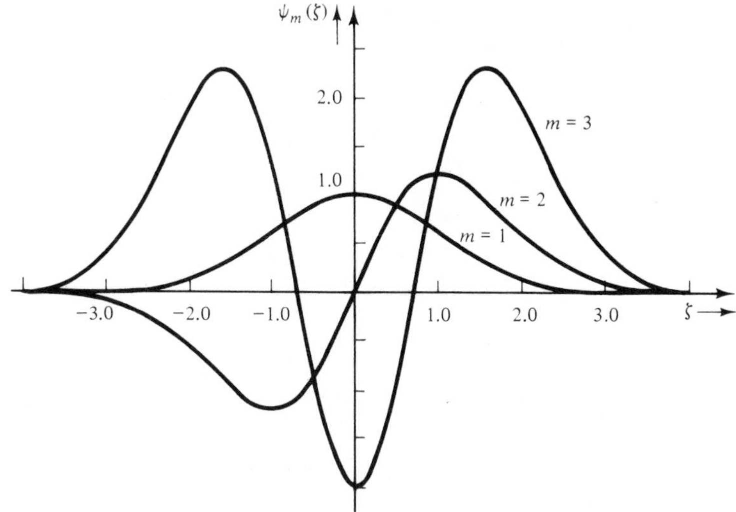

Figure 5A.2 The three lowest-order Hermite–Gaussians.

more extremum. Hence we collect *all* possible solutions by successive application of the raising operator. The mth eigenvalue λ_m is given by $\lambda_m = 2(m + \frac{1}{2})$. Some eigenfunctions are shown in Fig. 5A.2.

Conversely, operation by the lowering operator produces a lower-order solution from a higher-order one by "climbing down" the eigenvalue "ladder" in increments of 2, producing a solution on one "lower rung" of the "ladder." The solutions of (5A.1) for the different discrete eigenvalues are the *Hermite–Gaussians.*

APPENDIX 5B

Orthogonality Property of Hermite–Gaussian Modes

The Hermite–Gaussians $\psi_m(\xi)$ are orthogonal in the sense that [13, 14]

$$\int_{-\infty}^{\infty} d\xi \; \psi_m(\xi)\psi_n(\xi) = 0 \tag{5B.1}$$

if $m \neq n$. To show this one uses their defining equation (5A.1):

$$\frac{d^2\psi_m}{d\xi^2} + \lambda_m \psi_m - \xi^2\psi_m = 0 \tag{5B.2}$$

where

$$\psi_m \equiv H_m(\xi)e^{-\xi^2/2} \tag{5B.3}$$

Multiplying (5B.2) by ψ_n and subtracting an equation such as (5B.2) applied to ψ_n multiplied by ψ_m, one finds

$$(\lambda_m - \lambda_n) \int_{-\infty}^{\infty} \psi_m \psi_n \, d\xi = \int_{-\infty}^{\infty} \frac{d}{d\xi} \left(\psi_m \frac{\partial \psi_n}{\partial \xi} - \psi_n \frac{\partial \psi_m}{\partial \xi} \right) d\xi = 0 \qquad \text{(5B.4)}$$

because ψ_m and ψ_n vanish at $\xi = \pm\infty$. Thus

$$\int_{-\infty}^{\infty} \psi_m \psi_n \, d\xi = 0 \qquad \text{(5B.5)}$$

when

$$\lambda_m \neq \lambda_n$$

Equation (5B.4) applied to the integral in (5B.1) with the use of (5B.5) gives the orthogonality condition (5B.1).

When (5B.3) is introduced into (5B.2) we obtain the differential equation obeyed by the Hermite polynomials

$$\frac{d^2 H_m}{d\xi^2} - 2\xi \frac{dH_m}{d\xi} + 2mH_m = 0 \qquad \text{(5B.6)}$$

APPENDIX 5C

The Generating Function and Convolution of Hermite–Gaussians

The generating function of the Hermite–Gaussians is (as we prove below)

$$F(s, \xi) \equiv \exp\left(-s^2 + 2s\xi - \frac{\xi^2}{2}\right) = \sum_{n=0}^{\infty} \frac{s^n}{n!} H_n(\xi) e^{-\xi^2/2}$$

$$= \sum_{n=0}^{\infty} \frac{s^n}{n!} \psi_n(\xi) \qquad \text{(5C.1)}$$

where we use the sumbol $\psi_n(\xi)$ for the Hermite–Gaussians. The Hermite–Gaussians are the "coefficients" of the Taylor expansion in s of $F(s, \xi)$. Comparison of the two sides of (5C.1) for $s = 0$, gives $\psi_0(\xi) = \exp(-\xi^2/2)$. We shall now show, through application of the lowering operator $(\partial/\partial\xi + \xi)$ to both sides of (5C.1) that all remaining terms in the series are solutions of (5A.1). We use a partial derivative symbol because F is a function of two variables, s and ξ. We obtain

$$\left(\frac{\partial}{\partial \xi} + \xi\right) F(s, \xi) = 2sF(s, \xi) = 2 \sum_{n=0}^{\infty} \frac{s^{n+1}}{n!} \psi_n(\xi)$$

$$= \sum_{n=0}^{\infty} \frac{s^n}{n!} \left(\frac{\partial}{\partial \xi} + \xi\right) \psi_n(\xi) \qquad \text{(5C.2)}$$

where we have differentiated the exponential function $F(s, \xi)$ directly, observing that the operator $(\partial/\partial\xi + \xi)$ operating on $F(s, q)$ is equivalent to multiplication by $2s$. Then we replace $F(s, \xi)$ by its defining expansion, and finally equate the result to the operation of $(\partial/\partial\xi + \xi)$ on the defining expansion. By comparing equal powers of s, we obtain

$$\left(\frac{d}{d\xi} + \xi\right)\psi_{n+1}(\xi) = 2(n + 1)\psi_n(\xi) \tag{5C.3}$$

The lowering operator transforms the $(n + 1)$st function $\psi_{n+1}(\xi)$ into the nth function $\psi_n(\xi)$. Because the function $\psi_0(\xi)$ is the pure Gaussian, the function $\psi_1(\xi)$ must be the first higher-order solution of the differential equation (5A.1). The remaining eigenfunctions along the "ladder" may be identified by induction.

One may use the generating function to evaluate $\psi_n(\xi) \equiv H_n(\xi)e^{-\xi^2/2}$. Expanding $F(s, \xi)$ in powers of its exponent in s, and equating terms in (5C.1), one has

$$H_0(\xi) = 1 \tag{5C.4}$$

$$H_1(\xi) = 2\xi \tag{5C.5}$$

$$H_2(\xi) = 4\xi^2 - 2 \tag{5C.6}$$

The generating function can be used to relate $dH_n/d\xi$ to H_{n-1}. This is accomplished by taking a derivative with respect to ξ of (5C.1) and rewriting the result, $(2s - \xi)\, F(s, \xi)$, in terms of the defining sums. Equating terms of the same powers of s one obtains

$$\frac{d}{d\xi} H_n = 2nH_{n-1} \tag{5C.7}$$

If one differentiates the above, uses the differential equation obeyed by H_n, (5B.6), and (5C.7) to eliminate the derivatives, one obtains the recursion formula

$$H_{n+1} - 2\xi H_n + 2nH_{n-1} = 0 \tag{5C.8}$$

Another very important use of the generating function is evaluation of convolutions and Fourier transforms of the eigenfunctions $\psi_n(\xi)$. Consider first the Fourier transform of the Taylor expansion of the generating function

$$\begin{aligned}
&\frac{1}{2\pi}\int_{-\infty}^{\infty} \sum_{n=0}^{\infty} \frac{s^n}{n!} H_n(\xi)e^{-s^2/2}e^{jk\xi}\, d\xi \\
&= \frac{1}{2\pi}\int_{-\infty}^{\infty} \exp\left(-s^2 + 2s\xi - \frac{\xi^2}{2} + jk\xi\right) d\xi \\
&= \frac{1}{2\pi}\int_{-\infty}^{\infty} \exp\left(-\frac{\xi^2}{2} + (2s + jk)\xi - \frac{1}{2}(2s + jk)^2\right) d\xi \, \exp\left(s^2 + 2jsk - \frac{1}{2}k^2\right)
\end{aligned}$$

The integral evaluates to $\sqrt{2\pi}$ and we recognize the factor in the last expression to be the generating function $F(js, k)$. Thus

$$\text{F.T.}\left[\sum_{n=0}^{\infty} \frac{s^n}{n!} H_n(\xi) e^{-\xi^2/2}\right] = \frac{1}{\sqrt{2\pi}} \sum_{n=0}^{\infty} \frac{(js)^n}{n!} H_n(k) e^{-k^2/2} \tag{5C.9}$$

The Fourier transform of $H_n(\xi)e^{-\xi^2/2}$ is $(1/\sqrt{2\pi})\,(j)^n$ times the same function of k.

Consider next the convolution of the Hermite–Gaussians $\psi_n(\xi) \equiv H_n(\xi)e^{-\xi^2/2}$ with another Gaussian [15]:

$$\begin{aligned}
&\sum_{n=0}^{\infty} \frac{s^n}{n!} \int_{-\infty}^{\infty} \psi_n(\xi_0) e^{-(a/2)(\xi-\xi_0)^2}\, d\xi_0 \\
&= \int_{-\infty}^{\infty} d\xi_0 \exp\left[-s^2 + 2s\xi_0 - \frac{\xi_0^2}{2} - \frac{a}{2}(\xi - \xi_0)^2\right] \\
&= \int_{-\infty}^{\infty} d\xi_0 \exp\left[-\frac{a+1}{2}\xi_0^2 + 2\left(s + \frac{a}{2}\xi\right)\xi_0 - \frac{2}{a+1}\left(s + \frac{a}{2}\xi\right)^2\right] \\
&\quad \cdot \exp\left[-\left(s\sqrt{\frac{a-1}{a+1}}\right)^2 + 2\left(s\sqrt{\frac{a-1}{a+1}}\right)\frac{a\xi}{\sqrt{a^2-1}} - \frac{a}{2(a+1)}\xi^2\right]
\end{aligned} \tag{5C.10}$$

In the first step we convolve the entire series of $\psi_n(\xi_0)$ functions, equate them to the convolution of the generating function and then evaluate the convolution of the latter by completion of the square. The integral evaluates to $\sqrt{2\pi/(a+1)}$. The remaining exponential factor is the generating function of a Hermite-Gaussian. Equating the first expression in (5C.10) to the last one expanded as a series of ψ_n functions one obtains

$$\begin{aligned}
&\sum_{n=0}^{\infty} \frac{s^n}{n!} \int_{-\infty}^{\infty} \psi_n(\xi_0) e^{-(a/2)(\xi-\xi_0)^2}\, d\xi_0 \\
&= \sqrt{\frac{2\pi}{a+1}} \sum_{n=0}^{\infty} \frac{1}{n!}\left(s\sqrt{\frac{a-1}{a+1}}\right)^n \psi_n\left(\frac{a\xi}{\sqrt{a^2-1}}\right) \exp\left[\frac{a\xi^2}{2(a^2-1)}\right]
\end{aligned} \tag{5C.11}$$

Term-by-term identification gives

$$\begin{aligned}
&\int_{-\infty}^{\infty} \psi_n(\xi_0) e^{-(a/2)(\xi-\xi_0)^2}\, d\xi_0 \\
&= \sqrt{\frac{2\pi}{a+1}} \left(\frac{a-1}{a+1}\right)^{n/2} \psi_n\left(\frac{a\xi}{\sqrt{a^2-1}}\right) \exp\left[\frac{a\xi^2}{2(a^2-1)}\right]
\end{aligned} \tag{5C.12}$$

This is the formula used in the text. Note that the square root in the argument of ψ_n has to be interpreted so as to yield solutions decaying with increasing

$|\xi|$. This same interpretation has to be given to

$$\sqrt{\frac{a-1}{a+1}} = \frac{a-1}{\sqrt{a^2-1}}$$

Finally, consider the product of two generating functions for the purpose of evaluating the normalization integral:

$$\int_{-\infty}^{\infty} e^{s^2+2s\xi-\xi^2/2} e^{-p^2+2p\xi-\xi^2/2}\, d\xi$$

$$= \sum_{m=0}^{\infty} \sum_{n=0}^{\infty} \frac{s^m p^n}{m!n!} \int_{-\infty}^{\infty} H_m(\xi) H_n(\xi) e^{-\xi^2}\, d\xi \tag{5C.13}$$

The left-hand side is easily evaluated to give

$$\sqrt{\pi}\, e^{2sp} = \sqrt{\pi} \sum_n \frac{(2sp)^n}{n!} \tag{5C.14}$$

If equal powers of s and p are equated after substitution of the value of the integral (5C.14) into (5C.13) one obtains

$$\int_{-\infty}^{\infty} H_n^2(\xi) e^{-\xi^2}\, d\xi = \sqrt{\pi}\, 2^n n! \tag{5C.15}$$

and the orthogonality condition:

$$\int_{-\infty}^{\infty} H_m(\xi) H_n(\xi) e^{-\xi^2}\, d\xi = 0 \tag{5C.16}$$

APPENDIX 5D

The General Paraxial Optical System and Its *ABCD* Matrix

The *ABCD* matrix of a thin lens between arbitrarily positioned reference planes is given in (5.83). The elements are real and the determinant of the matrix is unity. Therefore, there are three arbitrary parameters that describe the matrix (three degrees of freedom). The most general real *ABCD* matrix has four degrees of freedom. We show that the construction describing a thick lens, with two different focal distances leads to the most general *ABCD* matrix. Different focal distances occur in systems in which the media at the entry and exit of the optical system are different.

A thick lens is characterized by the positions of the so-called principal planes and the two foci [12] as shown in Fig. 5D.1. The ray construction

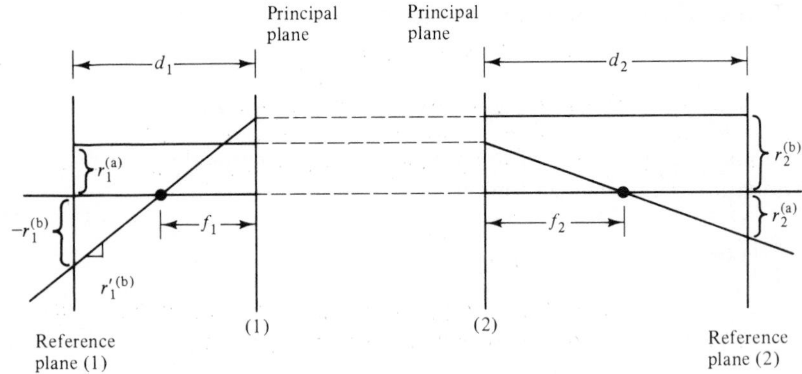

Figure 5D.1 Ray-optical construction for thick lens.

proceeds as follows:

(a) For a ray parallel to the axis on side 1, with given $r_1^{(a)}$ and $r_1'^{(a)} = 0$, one proceeds parallel to the axis from principal plane 1 to principal plane 2 and then makes the ray pass through the focus 2. This determines the values of

$$r_2^{(a)} = r_1^{(a)}\left(1 - \frac{d_2}{f_2}\right), \qquad r_2'^{(a)} = -\frac{1}{f_2}\,r_1^{(a)}$$

From (a) one finds two elements of the $ABCD$ matrix:

$$A = 1 - \frac{d_2}{f_2} \tag{5D.1}$$

$$C = -\frac{1}{f_2} \tag{5D.2}$$

(b) For a ray that passes through the focus 1 one extends the ray to the principal plane 1, and then continues parallel to the axis. This determines

$$r_2^{(b)} = f_1 r_1'^{(b)} \qquad \text{and} \qquad r_2'^{(b)} = 0$$

for $r_1^{(b)} = -(d_1 - f_1)r_1'^{(b)}$ and arbitrary r_1'. From (b) one obtains two equations for B and D:

$$r_2^{(b)} = f_1 r_1'^{(b)} = -A(d_1 - f_1)r_1'^{(b)} + Br_1'^{(b)}$$

$$r_2'^{(b)} = 0 = -C(d_1 - f_1)r_1'^{(b)} + Dr_1'^{(b)}$$

Solving for B and D, using (5D.1) and (5D.2), one finds

$$B = \left(1 - \frac{d_2}{f_2}\right) d_1 + \frac{f_1}{f_2}\,d_2 \tag{5D.3}$$

$$D = -\frac{1}{f_2}\,d_1 + \frac{f_1}{f_2} \tag{5D.4}$$

The system of Fig. 5D.1 has four independent parameters d_1, d_2, f_1, and f_2. By choosing f_2 and d_2 one fixes A and C according to (5D.1) and (5D.2). The two equations (5D.3) and (5D.4) yield B and D for given f_1 and d_1. Note that the determinant

$$\det \begin{bmatrix} A & B \\ C & D \end{bmatrix} = \frac{f_1}{f_2}$$

All systems that have unity determinant have $f_1 = f_2$.

PROBLEMS

5.1. The electric field of a Gaussian beam (5.26) has an x component and a z component. The z component can be separated into a part that is in phase with u_{00} and one that is in quadrature with u_{00}. The in-phase component is responsible for the curvature of the field lines, which is equal to the curvature of the phase fronts. Prove this statement by evaluating Re $[E_x/E_z]$ and noting that (see Fig. P5.1)

$$\tan \theta \simeq \theta \simeq \frac{x}{R} = -\text{Re}\left[\frac{E_z}{E_x}\right]$$

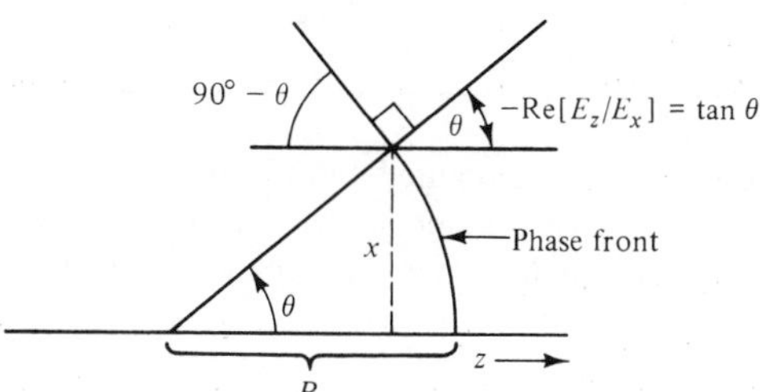

Figure P5.1

5.2. A Gaussian beam of wavelength $1/\lambda = k/2\pi$ is incident upon a Michelson interferometer. Its waist is w_0 at the half-silvered mirror position.

(a) At first, treating the Gaussian beam as if it were a plane wave, find the intensity distribution in the observation plane, assuming perfect alignment, as a function of l_a, l_b, and l (Fig. P5.2).

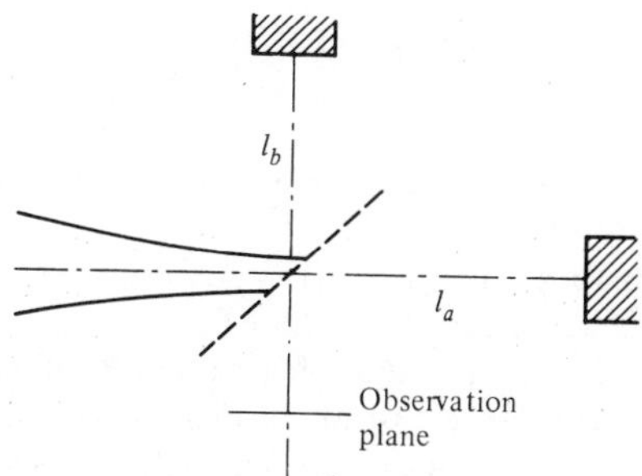

Figure P5.2

(b) Now take the finite waist size into account and repeat part (a). In particular, evaluate for $l_a = l_b$. Are there any visible fringes (zeros of intensity)? If $l_a \neq l_b$, will there be fringes?

5.3. The beam emerging from a laser diode (wavelength λ) may be approximated by the elliptical Gaussian cross section

$$u(x, y) = A \exp\left(-\frac{x^2}{w_{0x}^2}\right) \exp\left(-\frac{y^2}{w_{0y}^2}\right)$$

and a planar surface (Fig. P5.3).

(a) Prove that the well-known formulas for beam waist and radius of curvature of phase front may be applied separately to the x dependence and y dependence of elliptic Gaussian beam propagation.

(b) If $\lambda = 0.81\ \mu\text{m}$, $w_{0x} = 10\ \mu\text{m}$, and $w_{0y} = 1\ \mu\text{m}$, what are the angles of expansion of the two beam radii (in the x and y planes)?

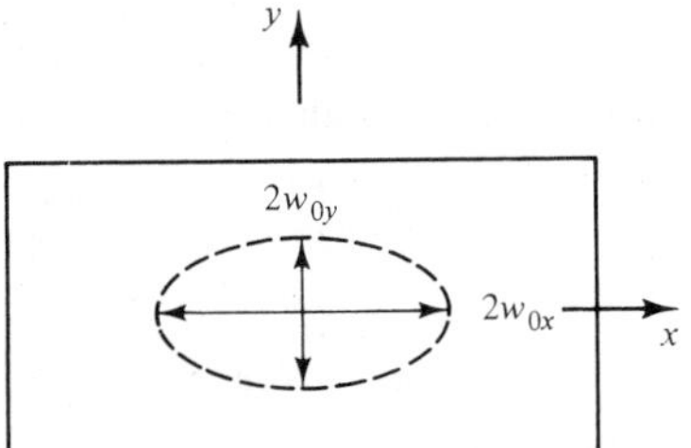

Figure P5.3

5.4. A Fabry–Perot resonator is illuminated by a Gaussian beam, with a minimum beam radius w_0 at the reference plane shown, through a lens of focal distance f. Find f and w_0 that will match the outside beam to the resonator fundamental mode.

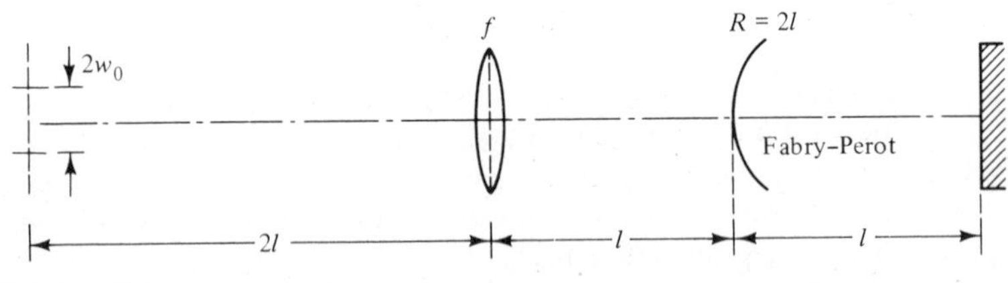

Figure P5.4

5.5. With reference to Fig. P5.5:

(a) Find the ray-optical $ABCD$ matrix for a planar interface between two media of index n_1 and n_2, respectively.

(b) Determine the $ABCD$ matrix from the q parameter transformation (i.e., from its effect upon the radius of curvature of the phase front).

(c) Assume now that the interface has radius of curvature R_0. Repeat part (a).

(d) Construct the $ABCD$ matrix for a cascade of two curved surfaces, close together (i.e., a thin lens). Compare with the thin lens $ABCD$ matrix.

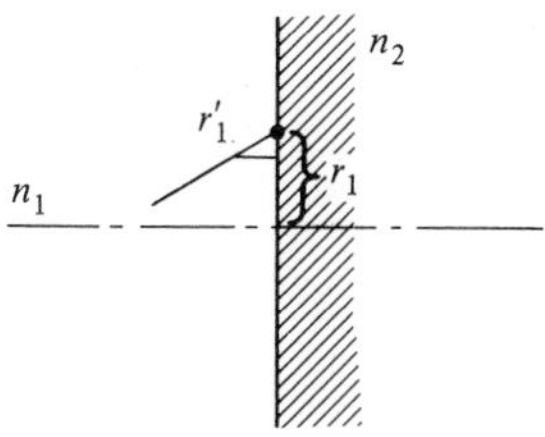

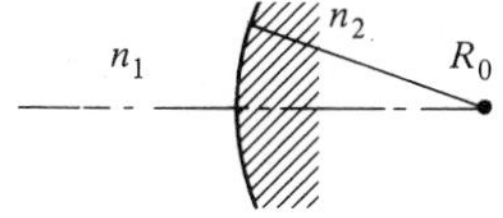

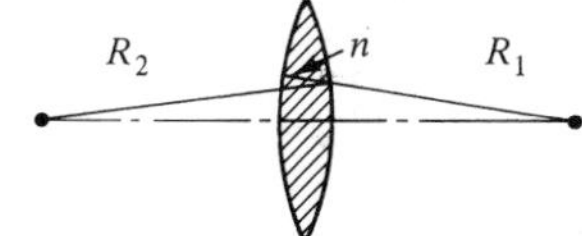

Figure P5.5

5.6. When we found a Gaussian beam solution in a curved-mirror resonator by inspection we assumed that the phase front matches the mirror surface. The purpose of this problem is to prove that this is always the case if we require that q repeat itself after one transit.

(a) Using the formalism of Section 5.6, show that Re $(1/q) = (A - D)/2C$ at the input reference plane of Fig. 5.21 is equal to $1/R_1$.

(b) Another question may be asked. Why do we require that the q parameter repeat itself after one transit? Suppose that it needs to repeat only after n transits. To disprove this hypothesis, consider the requirement that the q parameter repeat itself after two transits. Construct

$$\begin{bmatrix} A & B \\ C & D \end{bmatrix}^2$$

Find the q parameter corresponding to this new matrix. What do you find? What does this say about n transits, where n is an even number? *Hint:* Use the formula

$$q = \frac{A - D}{2C} \pm \sqrt{\left(\frac{A - D}{2C}\right)^2 + \frac{B}{C}}$$

5.7. Show that a Gaussian beam with a waist w_1 located at a distance d_1 from a lens with focal length f is transformed to a beam with a waist w_2 located at a distance d_2 on the other side of the lens (Fig. P5.7) given by

$$d_2 = \frac{f^2(d_1 - f)}{(d_1 - f)^2 + (\pi w_1^2/\lambda)^2} + f \tag{1}$$

and

$$\left(\frac{\pi w_2}{\lambda f}\right)^2 = \left(\frac{\pi w_1}{\lambda f}\right)^2 \frac{1}{[(d_1/f) - 1]^2 + (\pi w_1^2/\lambda f)^2} \tag{2}$$

or

$$w_2 = \left[\frac{1}{w_1^2}\left(1 - \frac{d_1}{f}\right)^2 + \frac{1}{f^2}\left(\frac{\pi w_1}{\lambda}\right)^2\right]^{-1/2}$$

Hint: Use the *ABCD* formalism and require that a pure imaginary q_1 transform into a pure imaginary q_2. Apply (1) to the case $w_1 = 0$ and compare with lens formula (5.85) for positions of object and image. Assume that $(\pi w_1^2/\lambda)^2 \ll (d_1 - f)^2$ and explore the correction to $1/d_2$ in the lens formula due to the fact that $w_1 \neq 0$. When is the correction most important?

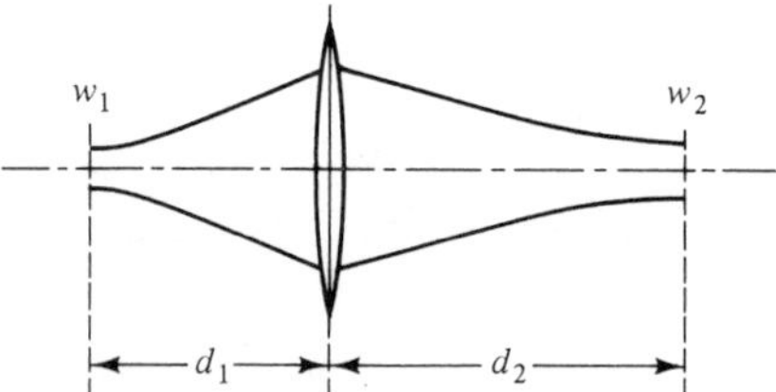

Figure P5.7

5.8. Determine d_1/f and d_2/f for a double convex lens to mode match the $m = 0$, $n = 0$, p resonant mode from a plane-spherical laser cavity into a confocal Fabry–Perot (Fig. P5.8). You may pick a realistic value of f, and determine d_1 and d_2 using the results of Problem 5.7. For $R_1 = 100$ cm, $l_1 = 99$ cm, $f = 20$ cm, and $l_2 = 10$ cm, determine d_1 and d_2.

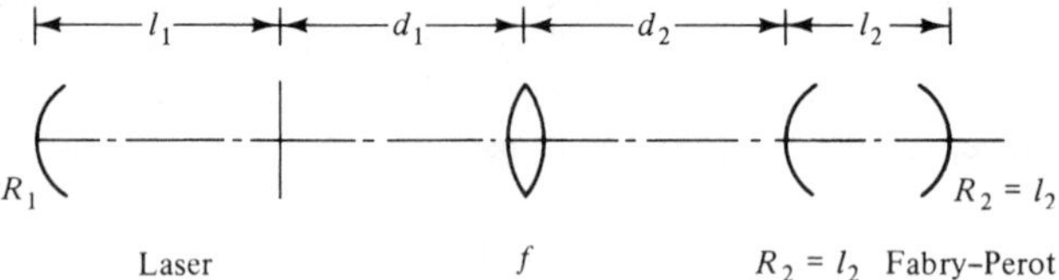

Figure P5.8

5.9. Using the results of Problem 4.7 derive the lowest order (Gaussian) solution for the two-dimensional wave equation in a way analogous to that of Section 5.1.

REFERENCES

[1] H. Kogelnik, Invited talk at International Quantum Electronics Conference, Kyoto, 1970.

[2] H. Kogelnik and T. Li, "Laser beams and resonators," *Appl. Opt.*, *5*, 1550–1567, 1966.

[3] H. Kogelnik, "Modes in optical resonators," in *Lasers*, A. K. Levine, ed., Marcel Dekker, New York, 1966, Chap. 6.

[4] A. E. Siegman and R. Arrathoon, "Modes in unstable optical resonators and lens waveguides," *IEEE J. Quantum Electron.*, *QE-3*, 156, 1967.

[5] A. E. Siegman, "A canonical formulation for analyzing multielement unstable resonators," *IEEE J. Quantum Electron.*, *QE-12*, 35, 1976.

[6] A. E. Siegman, *An Introduction to Lasers and Masers*, McGraw-Hill, New York, 1971.

[7] J. A. Arnaud, "Hamiltonian theory of beam mode propagation," in *Progress in Optics*, E. Wolf, ed., North-Holland, Amsterdam, 1973.

[8] A Yariv, *Introduction to Optical Electronics*, Holt, Rinehart and Winston, New York, 1976.

[9] S. Ramo, J. R. Whinnery, and T. van Duzer, *Fields and Waves in Communication Electronics*, Wiley, New York, 1965.

[10] K. Murata, *Prog. Opt.*, *5*, 199–245, 1966.

[11] M. Françon, *Optical Interferometry*, Academic Press, New York, 1966.

[12] M. Born and E. Wolf, *Principles of Optics*, Macmillan, New York, 1964.

[13] L. I. Schiff, *Quantum Mechanics*, McGraw-Hill, New York, 1968.

[14] C. Cohen-Tannoudji, B. Diu, and F. Laloë, *Quantum Mechanics*, Wiley, New York, 1977.

[15] H. Bateman, *Tables of Integral Transforms*, vol. 2, McGraw-Hill, New York, 1954, p. 290.

6

OPTICAL FIBERS AND GUIDING LAYERS

Optical fibers can guide optical waves by means of total internal reflection that occurs within a medium of stratified index. One model of an optical fiber is a medium with a parabolic index profile. [1] Such an index profile is, strictly speaking, unrealizable in that the index decreases parabolically from a given positive value on the axis of the medium; eventually, such an index has to become negative. The advantage of this model is that it yields simple solutions to the paraxial wave equation. They are the Hermite–Gaussian solutions encountered in the preceding chapter. A number of interesting features are already contained in the simplified model. One finds that different modes propagate with different phase velocities, one of the limitations of distortion-free propagation along long fiber guides.

A more realistic model of an optical fiber is one with a positive index extending to a specified radius, the outer radius of the fiber. [2] The full analysis is encumbered by the need for Bessel function solutions. Much of the physics may be gleaned from a simplified slab model of an optical guide. [3, 4] Dispersion of the modes, and the dependence of the number of guided modes upon frequency, are illustrated quite realistically by the slab model. For this reason a detailed study of the model is presented.

Next we take up the analysis of a dielectric slab between two media of different index, not necessarily symmetric. [5] This is a case of importance for integrated optics applications. Often guidance of the optical wave is provided by a "guiding layer" on top of a substrate. The guiding layer by itself does not limit the width of the mode parallel to the layer. Additional confinement may be provided by other means for which the reader is referred to the literature. [6]

We consider propagation along a single-mode fiber and the question of group velocity dispersion as it affects pulse propagation. The case of a Gaussian pulse propagating along a dispersive fiber is taken up in detail. It is shown how the spreading of such a pulse can be compensated by a grating pair. Next we evaluate the change of propagation constant caused by an index change of the fiber. This problem is of importance to sensor design that utilizes the index change of the fiber caused by the physical phenomenon to be measured. The orthogonality of modes is proven, followed by the related topic of power, energy, and energy velocity of a fiber mode.

6.1 WAVE EQUATION IN NONUNIFORM DIELECTRIC

The vector potential obeys the following equation in a source-free, nonuniform, dielectric medium [compare (1.27)]:

$$\nabla^2 \mathbf{A} + \omega^2 \mu_0 \varepsilon \mathbf{A} = -j\omega\mu_0 \nabla\varepsilon\Phi \tag{6.1}$$

Further the relation between $\mathbf{A}$ and Φ is (1.26)

$$\nabla \cdot \mathbf{A} + j\omega\mu_0 \varepsilon\Phi = 0 \tag{6.2}$$

The electric field is given by (1.23)

$$\mathbf{E} = -j\omega\mathbf{A} - \nabla\Phi \tag{6.3}$$

The magnetic field is given by (1.22)

$$\mu_0 \mathbf{H} = \nabla \times \mathbf{A} \tag{6.4}$$

The nonuniform dielectric couples the vector potential to the scalar potential, and vice versa. The coupling is weak when the spatial variation of ε is small over distances of the order of a wavelength. If we neglect this coupling entirely, we arrive at the wave equation for $\mathbf{A}$:

$$\nabla^2 \mathbf{A} + \omega^2 \mu_0 \varepsilon \mathbf{A} = 0 \tag{6.5}$$

We shall adhere to this approximate form of the wave equation for the analysis of propagation in optical fibers. We assume an axially uniform medium with radial variation and a particular polarization for $\mathbf{A}$, say parallel to y. We write

$$\mathbf{A} = \hat{y} u(x, y) e^{-j\beta z} \tag{6.6}$$

where β is the, as yet undetermined, propagation constant. The exponential factor $\exp(-j\beta z)$ contains the entire z dependence, so that the remaining factor $u(x, y)$ can be assumed z independent. We obtain from (6.5) the differential equation

$$\nabla_T^2 u + [\omega^2 \mu \varepsilon(\boldsymbol{\rho}) - \beta^2] u = 0 \tag{6.7}$$

where

$$\boldsymbol{\rho} = \hat{x}x + \hat{y}y$$

Equation (6.7) is the Schrödinger equation of a particle in a two-dimensional potential well $-\omega^2\mu\varepsilon(\boldsymbol{\rho})$. Its solutions are bounded, if and only if, there are local negative values of the "well function" $\beta^2 - \omega^2\mu\varepsilon(\boldsymbol{\rho})$.

Indeed, consider the identity obtained by multiplying (6.7) by u^* and integrating over the entire $x - y$ plane

$$\int_{-\infty}^{\infty}\int_{-\infty}^{\infty} u^* \nabla_T^2 u \, dx \, dy + \int_{-\infty}^{\infty}\int_{-\infty}^{\infty} [\omega^2\mu_0 \varepsilon(\boldsymbol{\rho}) - \beta^2] |u|^2 \, dx \, dy = 0$$

We may transform the first integral through integration by parts:

$$\int_{-\infty}^{\infty}\int_{-\infty}^{\infty} u^* \nabla_T^2 u \, dx \, dy = \int_{-\infty}^{\infty}\int_{-\infty}^{\infty} \nabla_T \cdot [u^* \nabla_T u] \, dx \, dy$$

$$- \int_{-\infty}^{\infty}\int_{-\infty}^{\infty} \nabla_T u^* \cdot \nabla_T u \, dx \, dy$$

The first integral on the right can be transformed into an integral over the contour C at infinity by Gauss' theorem applied to the (two-dimensional) volume of unit height in the z-direction

$$\iint \nabla_T \cdot [u^* \nabla_T u] \, dx \, dy = \oint_C ds \, \hat{\boldsymbol{n}} \cdot (u^* \nabla_T u)$$

where $\hat{\boldsymbol{n}}$ is the unit vector normal to the contour. This integral vanishes because u vanishes at infinity. Thus

$$\int_{-\infty}^{\infty}\int_{-\infty}^{\infty} \nabla_T u^* \cdot \nabla_T u \, dx \, dy + \int_{-\infty}^{\infty}\int_{-\infty}^{\infty} [\beta^2 - \omega^2\mu_0 \varepsilon(\boldsymbol{\rho})] |u|^2 \, dx \, dy = 0.$$

The first integral is positive. Therefore, the second integral must be negative, i.e.,

$$\beta^2 - \omega^2\mu_0 \varepsilon(\boldsymbol{\rho}) < 0$$

over at least part of the range of integration. Bounded solutions correspond to guided waves and are found only for specific values of the eigenvalue β^2. In the next three sections we look at three special cases.

6.2 PARABOLIC PROFILE OF DIELECTRIC CONSTANT [1]

We assume a dielectric constant profile that depends quadratically on the distance from the axis (Fig. 6.1):

$$\omega^2\mu_0 \varepsilon(\boldsymbol{\rho}) = \omega^2\mu_0 \varepsilon(0)\left(1 - \frac{x^2 + y^2}{h^2}\right) \tag{6.8}$$

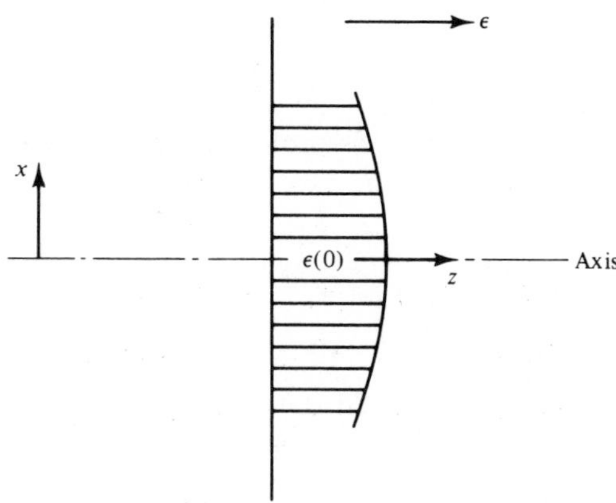

Figure 6.1 Axially uniform medium with ε a function of distance from the axis.

Denote by k_0 the propagation constant of an infinite parallel plane wave propagating in a medium of uniform dielectric constant ε(0), i.e., the value of ε on the axis

$$k_0 \equiv \omega\sqrt{\mu_0\,\varepsilon(0)}$$

When we introduce (6.8) into (6.5) and use the normalized variables

$$\xi \equiv \sqrt{\frac{k_0}{h}}\,x, \qquad \eta \equiv \sqrt{\frac{k_0}{h}}\,y \tag{6.9}$$

equation (6.8) reduces to the standard form

$$\frac{\partial^2 u}{\partial \xi^2} + \frac{\partial^2 u}{\partial \eta^2} + [\lambda - (\xi^2 + \eta^2)]u = 0 \tag{6.10}$$

where

$$\lambda = \frac{k_0^2 - \beta^2}{k_0}\,h \tag{6.11}$$

The solutions to this equation can be obtained by separation of variables, $u = X(\xi)Y(\eta)$. The equation for $X(\xi)$ becomes the total differential equation

$$\frac{d^2X}{d\xi^2} + (\lambda_x - \xi^2)X = 0 \tag{6.12}$$

and a corresponding equation for $Y(\eta)$. The eigenvalue λ is the sum of λ_x and λ_y. Equation (6.12) is of the form (5A.1), with the solutions $H_m(\xi)\exp(-\xi^2/2)$, and $\lambda_x = 2m + 1$. Equation (6.10) thus has the solution

$$u_{mn}(\xi, \eta) = H_m(\xi)H_n(\eta)e^{-(\xi^2+\eta^2)/2} \tag{6.13}$$

with

$$\lambda = 2(m + n) + 2, \qquad m, n = 0, 1, 2, \ldots \tag{6.14}$$

$u_{mn}(\xi, \eta)$ is the mode pattern, or mode profile of the mn mode. The fundamental mode, with $\lambda_{00} = 2$, has the pattern

$$u_{00}(x, y) = \sqrt{\frac{2}{\pi}}\frac{1}{w}\exp\left(-\frac{x^2 + y^2}{w^2}\right) \tag{6.15}$$

where

$$w^2 = \frac{2h}{k_0} = \frac{h\lambda}{\pi}\sqrt{\frac{\varepsilon_0}{\varepsilon(0)}} \tag{6.16}$$

This beam radius w should be compared with the q parameter in Section 5.6, obtained for a parabolic index profile:

$$n = n_0\left(1 - \frac{x^2 + y^2}{2h^2}\right) \simeq \sqrt{\frac{\varepsilon(0)}{\varepsilon_0}}\left(1 - \frac{x^2 + y^2}{2h^2}\right)$$

For small variations of index, the expression for the index profile is in one-to-one correspondence with (6.8) for the dielectric constant profile. The q parameter (5.107) of the GRIN lens is found to be identical with the one derived from (6.16).

The propagation constant of the mn mode is obtained from (6.11):

$$\beta_{mn}^2 = k_0^2\left(1 - \frac{\lambda_{mn}}{k_0 h}\right) \tag{6.17}$$

The propagation constant is smaller than k_0, thus indicating that the phase velocity of the mode is greater than the speed of an infinite parallel plane wave in a medium with a uniform dielectric constant $\varepsilon = \varepsilon(0)$, the value of ε at the center. This is the consequence of the finite transverse extent of the field of the mode, which samples the lower ε values farther away from the axis. Modes of higher order, with greater variation in the radial direction, have greater phase velocities. Such modes are composed of waves with $\boldsymbol{k}$ vectors of greater inclination with respect to the z axis and their fields reach farther out into regions of lower ε. The phase velocity is infinite when $\lambda_{mn} \sim k_0 h$. The radial extent of such a mode is roughly $\eta^2 \simeq \xi^2 \simeq \lambda_{mn}$. But this means that the mode has appreciable amplitude when

$$\xi^2 \simeq k_0 h$$

and, according to (6.9),

$$x^2 \simeq h^2, \qquad y^2 \simeq h^2$$

From (6.8) we see that $\varepsilon = 0$ at this radius and thus the vanishing of β_{mn} is not surprising. Modes with such high eigenvalues cease to be physically meaningful because $\varepsilon = 0$ is unrealistic. Of course, physical fibers do not have an infinitely large cross section. The radius of the physical fiber is much smaller than h and thus the modes found in the present analysis have to be picked among Hermite Gaussians of much lower order; only those modes can be accepted as appropriate solutions that have negligible intensities at the physical boundary of the fiber.

When $\lambda_{mn}/(k_0 h) \ll 1$, one may approximate (6.11) by

$$\beta_{mn} \simeq k_0\left(1 - \frac{m+n+1}{k_0 h}\right) \tag{6.18}$$

The inverse group velocity is then

$$\frac{d\beta_{mn}}{d\omega} = \sqrt{\mu_0\, \varepsilon(0)}$$

independent of mode number m and n. This result is peculiar to the parabolic index profile, which is, strictly speaking, unphysical. A physical profile leads, in general, to differences between group velocities of different modes (Sections 6.3 and 6.4).

The modes in a parabolic index profile are an approximate representation of modes in multimode fibers that are currently employed in optical-fiber communication technology. The advantage of multimode fibers (as opposed to single-mode fibers discussed in Section 6.5) is that incident waves with $\boldsymbol{k}$ vectors at a relatively large angle subtended with respect to the fiber axis are critically reflected and "captured" by the fiber. In this way, power is coupled more efficiently from a light-emitting diode (LED) or laser into the multimode fiber than into a single-mode fiber, which accepts only a narrow range of angles of $\boldsymbol{k}$, a small solid angle. The disadvantage of a multimode fiber is that the modes possess different phase velocities. Two modes of different phase velocities interfere at the fiber-output end. The interference varies with time due to thermal effects and mechanical vibrations that affect the phase factor $\beta_{mn} l$ produced by propagation of the mn mode over a length l of fiber. If many modes interfere, this effect averages out statistically over the total number of interfering modes and can be overcome. A multimode fiber can handle several-megabit communication rates over several kilometers. [7] The distance is not limited by fiber loss, but by dispersion. Li quotes [7] a pulse spreading of 10 ps/km for a multimode fiber with pulses from an LED of rms bandwidth of 35 nm operating at 1.3 μm wavelength. (Compare the discussion in Section 6.5.) For long-distance propagation and high-speed communications, single-mode fibers have to be employed.

6.3 THE DIELECTRIC SLAB GUIDE

A simple example of a dielectric waveguide is a slab of dielectric material of dielectric constant ε_i and thickness $2d$ (Fig. 6.2). This problem can be solved exactly, with no approximation in the wave equation for the vector potential.

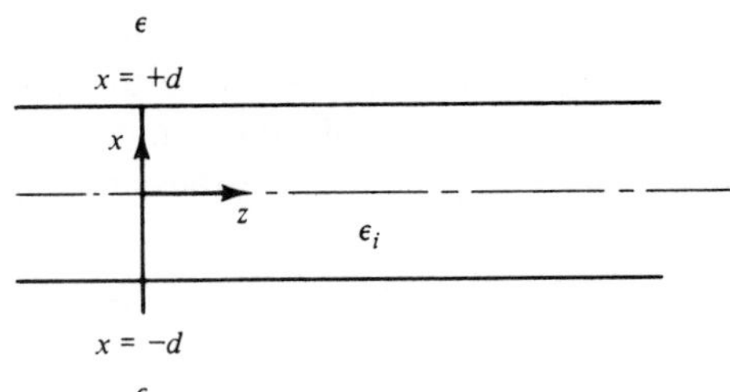

Figure 6.2 Dielectric slab geometry.

One uses solutions to the vector wave equations of the electric field and magnetic field inside and outside the slab and matches the boundary conditions on the slab surface.

Transverse Electric Modes

We start with the analysis of a transverse electric (TE) wave with the vector potential and the electric field polarized along $\hat{y}$, transverse to the direction of propagation, the z direction. Because the structure is symmetric, one picks the origin of the Cartesian coordinates at its center, and one looks for solutions for the electric field that are either symmetric, or antisymmetric with respect to the symmetry plane of the waveguide, the y-z plane. The symmetric solution may be written

$$\mathrm{E}_y = A \cos k_x x \, e^{-j\beta z}, \qquad |x| < d \tag{6.19}$$

$$\mathrm{E}_y = Be^{-j\beta z}e^{-\alpha_x x}, \qquad x > d \tag{6.20a}$$

$$= Be^{-j\beta z}e^{\alpha_x x}, \qquad x < -d \tag{6.20b}$$

The magnetic field follows from Faraday's law (1.41). Its component in the z direction is

$$\mathrm{H}_z = -\frac{jk_x}{\omega\mu_0} A \sin k_x x \, e^{-j\beta z}, \qquad |x| < d \tag{6.21}$$

$$\mathrm{H}_z = -\frac{j\alpha_x}{\omega\mu_0} Be^{-\alpha_x x}e^{-j\beta z}, \qquad x > d \tag{6.22a}$$

$$= \frac{j\alpha_x}{\omega\mu_0} Be^{\alpha_x x}e^{-j\beta z}, \qquad x < -d \tag{6.22b}$$

Taking advantage of symmetry, we need to match the boundary conditions only at $x = d$. Continuity of $\mathrm{E}_y/\mathrm{H}_z$ at $x = d$ gives

$$\tan k_x d = \frac{\alpha_x}{k_x} \tag{6.23}$$

We further have from the wave equation

$$\beta^2 - \alpha_x^2 = \omega^2 \mu_0 \varepsilon \tag{6.24}$$

$$\beta^2 + k_x^2 = \omega^2 \mu_0 \varepsilon_i \tag{6.25}$$

Combining these two equations, we find the determinantal equation

$$\frac{\alpha_x}{k_x} = \sqrt{\frac{\omega^2 \mu_0 (\varepsilon_i - \varepsilon)}{k_x^2} - 1} \tag{6.26}$$

The dispersion diagram, the dependence of the propagation constant β on frequency, is found by constructing an intersection in Fig. 6.3 of the functions $\tan k_x d$ and α_x/k_x for each frequency and mode and then evaluating β from (6.25). For decreasing ω, α_x/k_x moves toward the origin and intersections are lost, except for the intersection with the first branch of the tangent function. This corresponds to the dominant mode, $m = 0$, with no cutoff. All higher-order modes ($m > 0$) have a "cutoff"; they are not guided below a certain critical frequency, the cutoff-frequency.

The dispersion curves of Fig. 6.4 are obtained by the use of (6.25), with the value of k_x for every frequency evaluated from Fig. 6.3. At low frequencies,

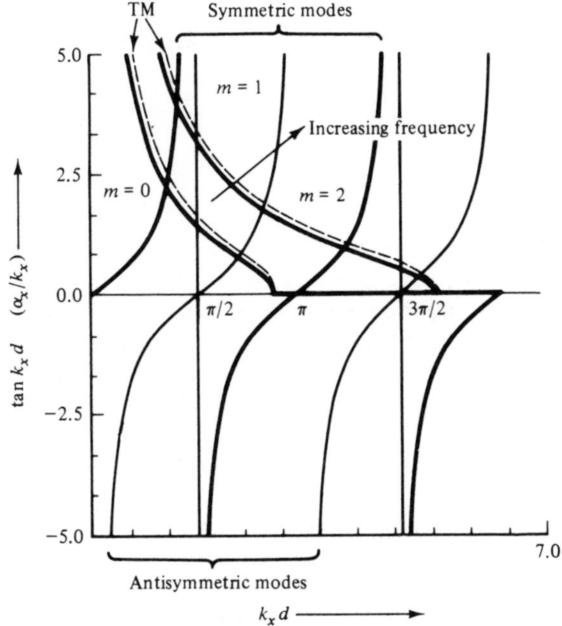

Figure 6.3 Graphical solution of (6.26) and (6.23). Construction for TM modes shown dashed. $\varepsilon_i/\varepsilon = 1.1$ is assumed.

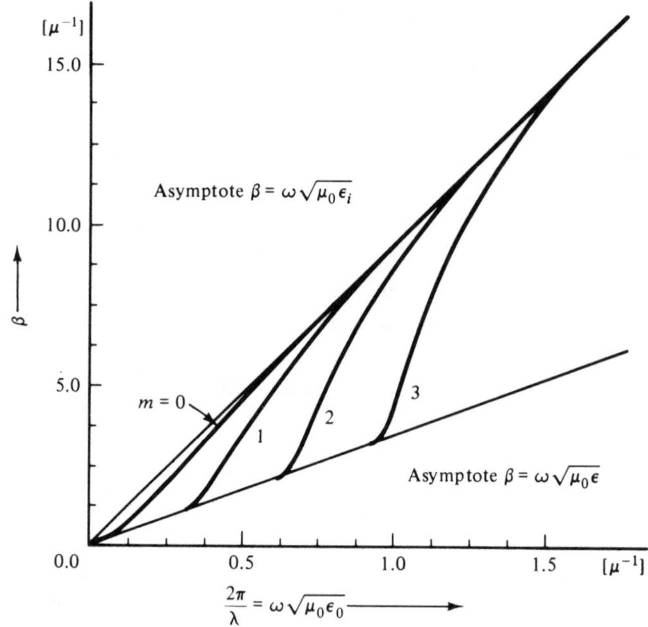

Figure 6.4 Dispersion diagram for TE waves in dielectric slab guide. $n_i = 9$, $n = 3.5$, $d = 0.6$ μm. The interior index is picked exaggeratedly large in order to separate the dispersion curves.

the fundamental mode acquires a small k_x. Hence $\tan k_x d \simeq k_x d$, and (6.23) and (6.26) yield

$$\omega^2\mu_0(\varepsilon_i - \varepsilon)d^2 - k_x^2 d^2 \simeq k_x^4 d^4$$

Neglecting $k_x^4 d^4$ compared with $k_x^2 d^2$, we have $k_x^2 = \omega^2\mu_0\ (\varepsilon_i - \varepsilon)$ and from (6.25), $\beta \simeq \omega\sqrt{\mu\varepsilon}$. The wave propagates at the speed characteristic of the external region. This may be explained by the fact that most of the field is outside the slab at low frequencies. When $\omega \to \infty$, $k_x d$ approaches $\pi/2$ and we find from (6.25) that $\beta \simeq \omega\sqrt{\mu_0\,\varepsilon_i}$. Now the propagation is at the speed characteristic of the interior of the slab. The field is confined to the interior (Fig. 6.5).

To complete the field expressions (6.19)–(6.22b), we note that

$$\mathrm{H}_x = -\frac{j}{\omega\mu_0}\frac{\partial E_y}{\partial z} = \begin{cases} \dfrac{-\beta}{\omega\mu_0} A \cos k_x x\, e^{-j\beta z}, & |x| < d \\[2ex] \dfrac{-\beta}{\omega\mu_0} B e^{-\alpha_x x} e^{-j\beta z}, & x > d \\[2ex] \dfrac{-\beta}{\omega\mu_0} B e^{\alpha_x x} e^{-j\beta z}, & x < d \end{cases} \tag{6.27}$$

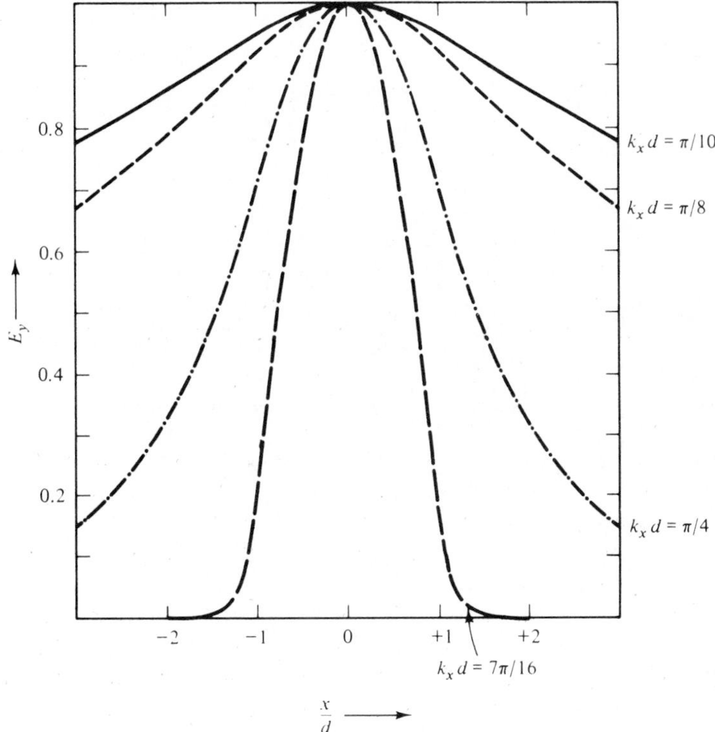

Figure 6.5 Field profile of dominant mode for three different frequencies.

Figure 6.6 shows a sketch of $\boldsymbol{E}$ and $\boldsymbol{H}$ as a function of z at one instant in time. For the construction of the sketch we note that H_x and E_y are in phase, according to (6.27). Both fields reach their maxima at the same time (i.e., the same position z along the axis) in the present case of a traveling wave. The z component of the magnetic field is 90° out of phase in space, and time. Because $\nabla \cdot \boldsymbol{H} = 0$, the magnetic field lines must close. To develop a picture of the magnetic field lines, it is helpful to note that $\mu_0 \boldsymbol{H}$ lines are the lines of equal height of the real part of the potential:

$$\psi = \begin{cases} j\dfrac{A}{\omega}\cos k_x x\, e^{-j\beta z}, & |x| < d \\[2ex] j\dfrac{A}{\omega}\cos k_x d\, e^{\mp\alpha_x x} e^{-j\beta z}, & |x| > d \end{cases}$$

multiplied by $\exp(j\omega t)$. Indeed,

$$-\hat{y} \times \left[\hat{x}\frac{\partial}{\partial x} + \hat{z}\frac{\partial}{\partial z}\right]\psi(x, z) = \mu_0 \mathbf{H} \tag{6.28}$$

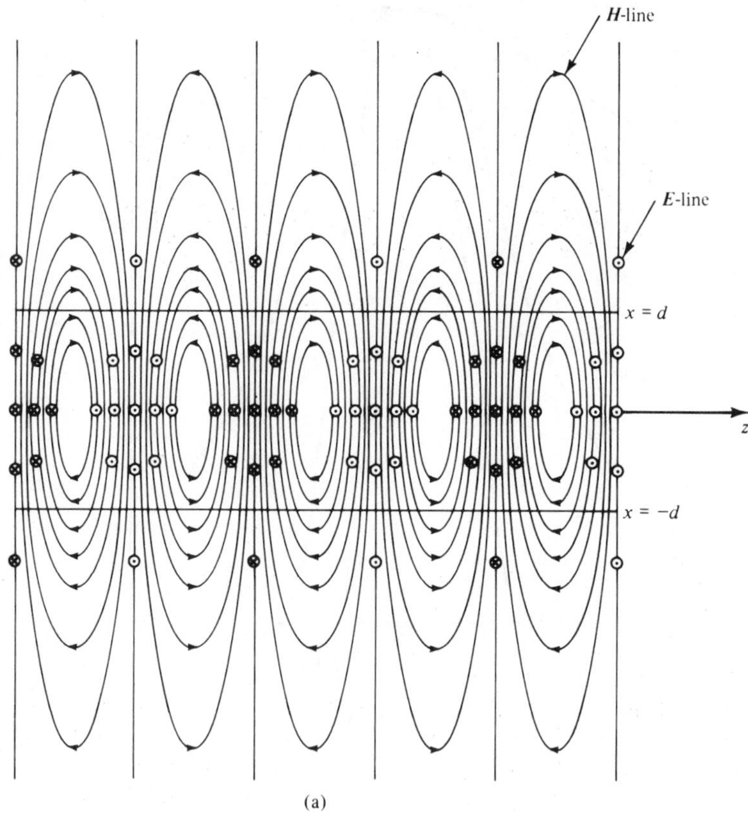

(a)

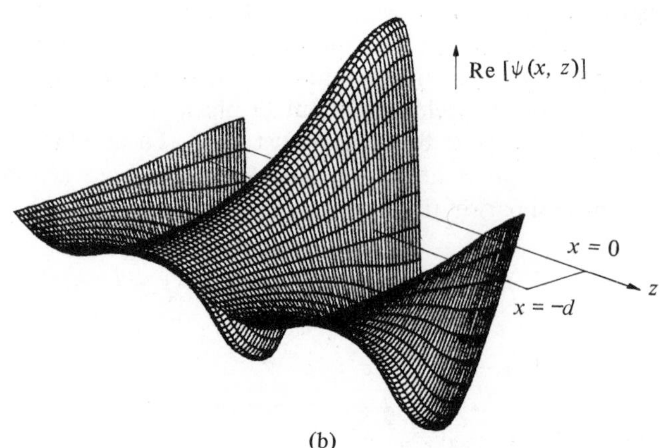

(b)

Figure 6.6 (a) Dominant TE mode ($m = 0$) of dielectric slab. With $\varepsilon = \varepsilon_0 n^2$, the wavelength λ for which graph is constructed is given by $nd/\lambda = 0.37$. (b) "Three-dimensional" plot of potential Re $[\psi(x, z)]$ for dominant TE mode.

as one may confirm by direct evaluation. The gradient produces the lines of steepest descent, the cross product is directed normal to them, along the lines of equal height.

It is easy to see that $\hat{y}\psi = \mathbf{A}$, the vector potential, whose curl is equal to $\mu_0 \mathbf{H}$, because

$$-\hat{y} \times \left[\hat{x}\frac{\partial}{\partial x} + \hat{z}\frac{\partial}{\partial z}\right]\psi(x, z) = \nabla \times [\hat{y}\psi(x, z)]$$

for the present example of a two-dimensional field. Figure 6.6b is a "three-dimensional" view of the potential Re $[\psi e^{j\omega t}]$ at one instant of time, $t = 0$.

Antisymmetric modes, with a nodal plane of the electric field at the center of the slab, can be shown to lead to the determinantal equation

$$\cot k_x d = -\frac{\alpha_x}{k_x} \tag{6.29}$$

The graphical solution of this equation is shown in Fig. 6.3. The field sketch of the lowest-order antisymmetric mode, $m = 1$, is shown in Fig. 6.7a.

Transverse Magnetic Modes

Transverse magnetic (TM) modes in the slab of Fig. 6.2 have a magnetic field along y, which is coupled by Maxwell's equations to an electric field in the x-z plane having both x and z components. The vector potential A is now not divergence-free; a scalar potential Φ is present, and no advantage occurs from the use of the vector potential for the solution of the field problem which can be solved exactly, with no approximations. We use the solutions to the wave equation for H_y, inside and outside the slab, the associated electric field, and match the boundary conditions at the interfaces.

The transverse magnetic (TM) modes with a symmetric x component of the electric field have a transverse $\boldsymbol{H}$ field

$$\mathrm{H}_y = A\cos k_x x\, e^{-j\beta z}, \qquad |x| < d \tag{6.30}$$

$$\mathrm{H}_y = Be^{-\alpha_x x}e^{-j\beta z}, \qquad x > d \tag{6.31}$$

$$= Be^{\alpha_x x}e^{-j\beta z}, \qquad x < -d$$

The electric field follows from Ampère's law (1.42):

$$\nabla \times \mathbf{H} = j\omega\varepsilon\mathbf{E} \tag{6.32}$$

so that its z component is

$$\mathrm{E}_z = \frac{1}{j\omega\varepsilon}\frac{\partial \mathrm{H}_y}{\partial x}\begin{cases} = -\dfrac{k_x}{j\omega\varepsilon_i}A\sin k_x x\, e^{-j\beta z}, & |x| < d \\[2ex] = -\dfrac{\alpha_x}{j\omega\varepsilon}Be^{\mp\alpha_x x}e^{-j\beta z}, & |x| > d \end{cases} \tag{6.33}$$

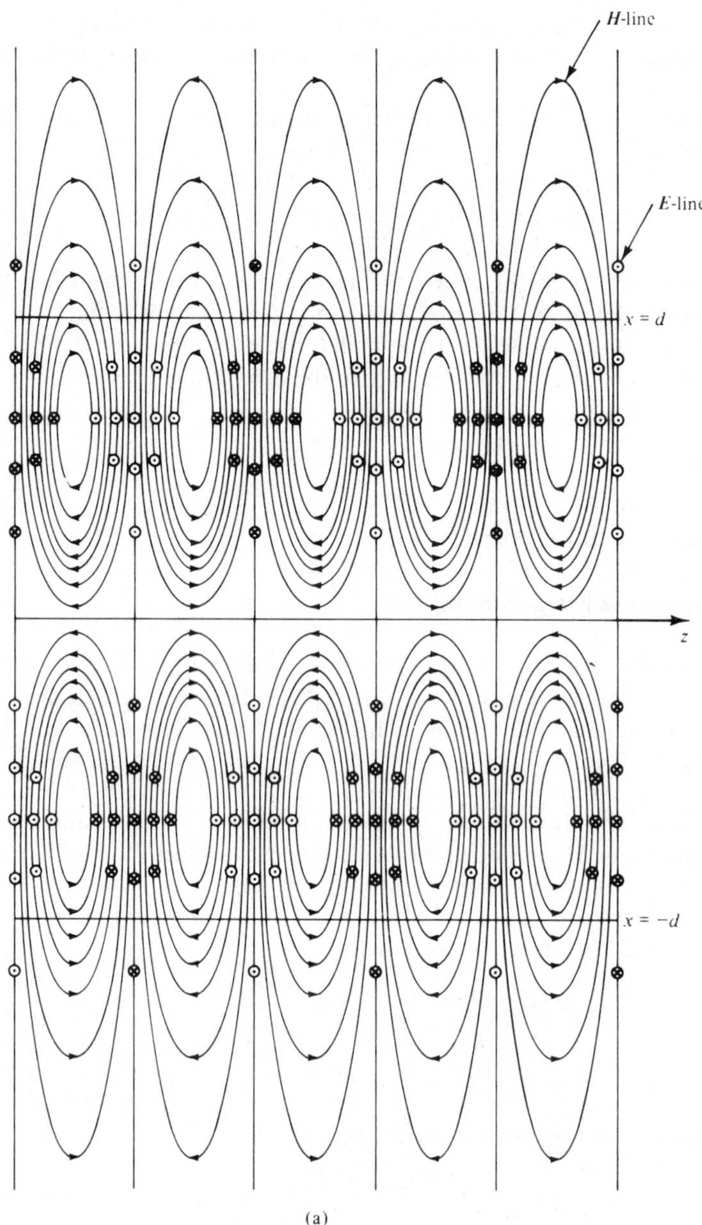

Figure 6.7 (a) First antisymmetric TE mode ($m = 1$) of dielectric slab. With $\varepsilon = \varepsilon_0 n^2$; the wavelength λ for which graph is constructed is given by $nd/\lambda = 1.4$. (b) "Three-dimensional" plot of potential Re $[\psi(x, z)]$ for lowest-order antisymmetric TE mode.

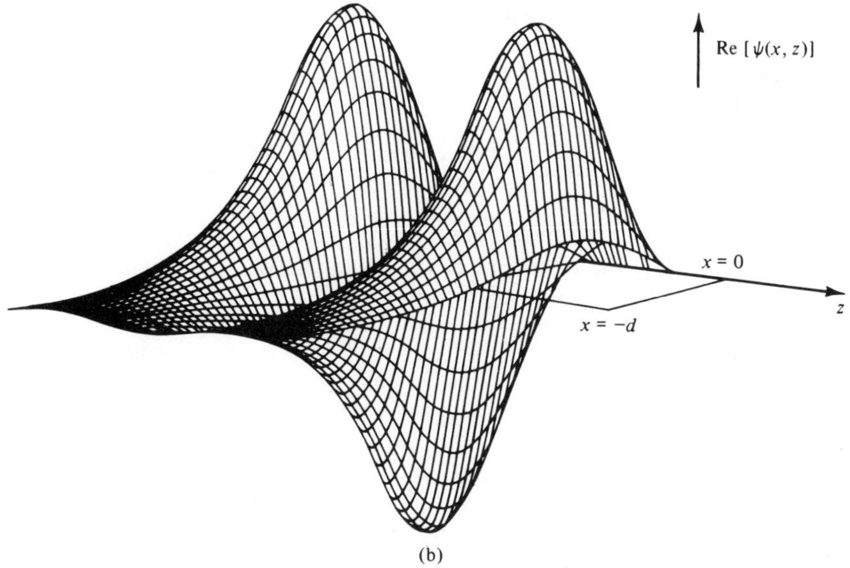

(b)

Figure 6.7 *(continued)*

Matching of the wave impedances gives the determinantal equation:

$$-\frac{E_z}{H_y} = \frac{k_x}{j\omega\varepsilon_i}\tan k_x d = \frac{\alpha_x}{j\omega\varepsilon} \tag{6.34}$$

Again we may construct a graphical solution to the transcendental equation:

$$\tan k_x d = \frac{\varepsilon_i}{\varepsilon}\frac{\alpha_x}{k_x} = \frac{\varepsilon_i}{\varepsilon}\sqrt{\frac{\omega^2\mu_0(\varepsilon_i - \varepsilon) - k_x^2}{k_x}} \tag{6.35}$$

as shown dashed in Fig. 6.3.

The plot is almost identical with the one for TE modes, except that the monotonically decreasing intersection curves lie higher by the factor $\varepsilon_i/\varepsilon$. This indicates a slightly larger k_x value and a lower β value for the TM modes than for the TE modes. The electric field at any instant of time can be represented by lines of equal height of the real part of the potential function

$$\Xi = \begin{cases} -\dfrac{j}{\omega\varepsilon_i} A \cos k_x x \, e^{-j\beta z}, & |x| < d \\[2ex] -\dfrac{j}{\omega\varepsilon} A \cos k_x d \, e^{\mp\alpha_x x} e^{-j\beta z}, & |x| > d \end{cases} \tag{6.36}$$

multiplied by exp $(j\omega t)$. This potential function is shown in Fig. 6.8b at one instant of time. Note that the potential is discontinuous at $|x| = d$. The electric and magnetic fields are depicted in Fig. 6.8a. The TM modes with an antisymmetric x component of the electric field lead similarly to the determinantal equation

$$\cot k_x d = -\frac{\varepsilon_i}{\varepsilon} \frac{\sqrt{\omega^2 \mu_0(\varepsilon_i - \varepsilon) - k_x^2}}{k_x} \tag{6.37}$$

The values of k_x are found by the graphical construction of Fig. 6.3. The mode has a cutoff because the argument of the cotangent must be greater than $\pi/2$ and k_x cannot go to zero as the frequency goes to zero [compare (6.37)].

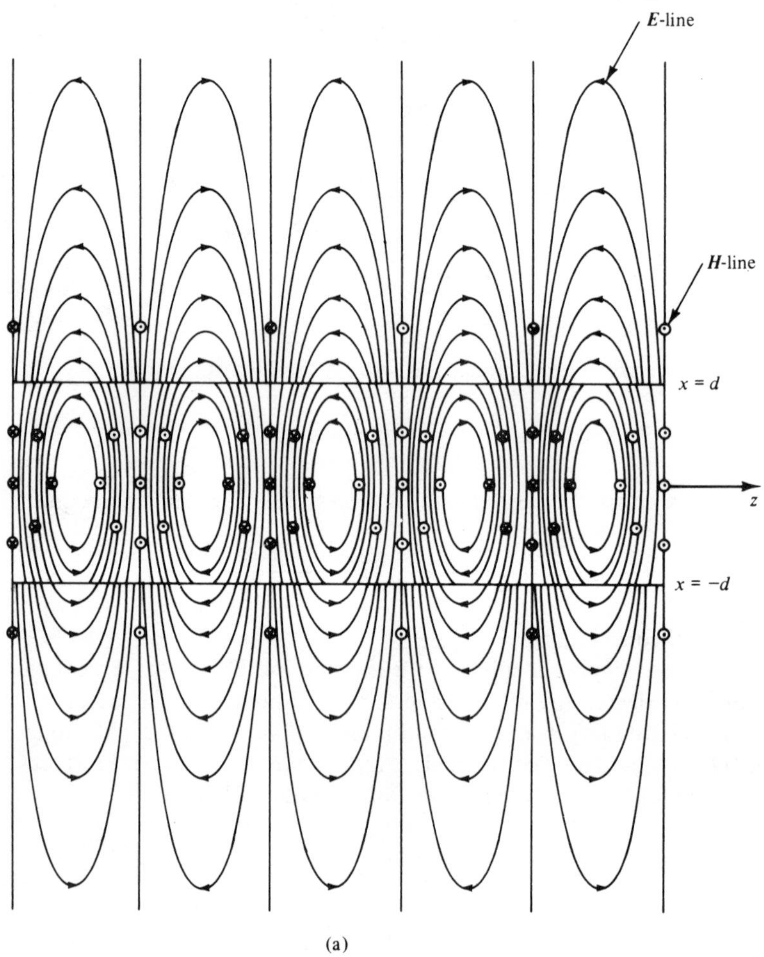

Figure 6.8 (a) Lowest-order TM mode of dielectric slab ($m = 0$). With $\varepsilon = \varepsilon_0 n^2$, the wavelength λ for which graph is constructed is given by $nd/\lambda = 0.35$. The transverse propagation constant k_x is the same as that of Fig. 6.6a. (b) "Three-dimensional" plot of potential Re $[\Xi(x, t)]$ for lowest-order TM mode.

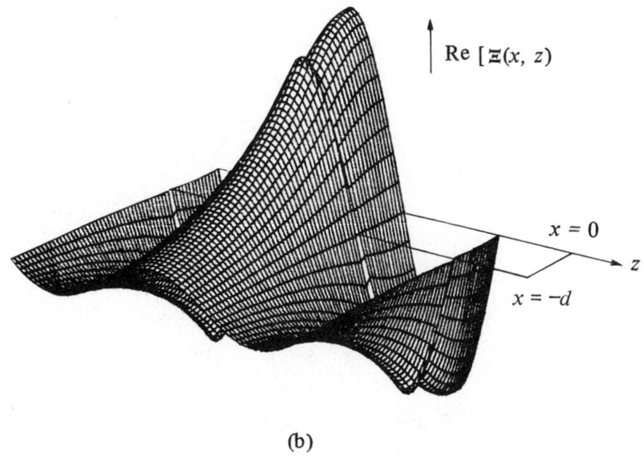

(b)

Figure 6.8 *(continued)*

Figure 6.9a shows a sketch of ***E*** and ***H*** for the first antisymmetric TM mode. Again in drawing the sketch we take advantage of the fact that the ***E*** lines are the lines of equal height of the real part of the potential (6.36) multiplied by $\exp(j\omega t)$.

In contrast with a metallic guide, [3, 4] we have found only a finite number of mode solutions for a dielectric guide. In a metallic guide, a mode below the cutoff frequency acquires an imaginary propagation constant. A dielectric waveguide mode below cutoff ceases to be guided; it becomes "leaky" and its field extends to infinity in the transverse directions. [8, 9] In general, matching of a given field distribution in the x-y plane requires a superposition of guided modes, leaky waves, and radiation fields.

6.4 TE MODES IN A GENERAL GUIDING LAYER [5]

We shall now develop the dispersion relation for TE modes in a guiding layer that is not symmetric, as assumed in the preceding section. We use the notation of Kogelnik and Ramaswamy [5] and develop normalized graphs for the dispersion characteristic. The geometry is shown in Fig. 6.10. The solution in the substrate is of the form (6.20b) and (6.22b). To distinguish it from the solution in the cover, we use the symbol α_s for the spatial decay constant. The impedance at the lower boundary of the film is

$$Z_s \equiv \frac{\mathrm{E}_y}{\mathrm{H}_z} = \frac{j\omega\mu_0}{\alpha_s} = j\sqrt{\frac{\mu_0}{\varepsilon_0}}\,\frac{k}{\alpha_s} \tag{6.38}$$

where

$$\alpha_s = \sqrt{\beta^2 - \omega^2\mu_0\,\varepsilon_s} \tag{6.39}$$

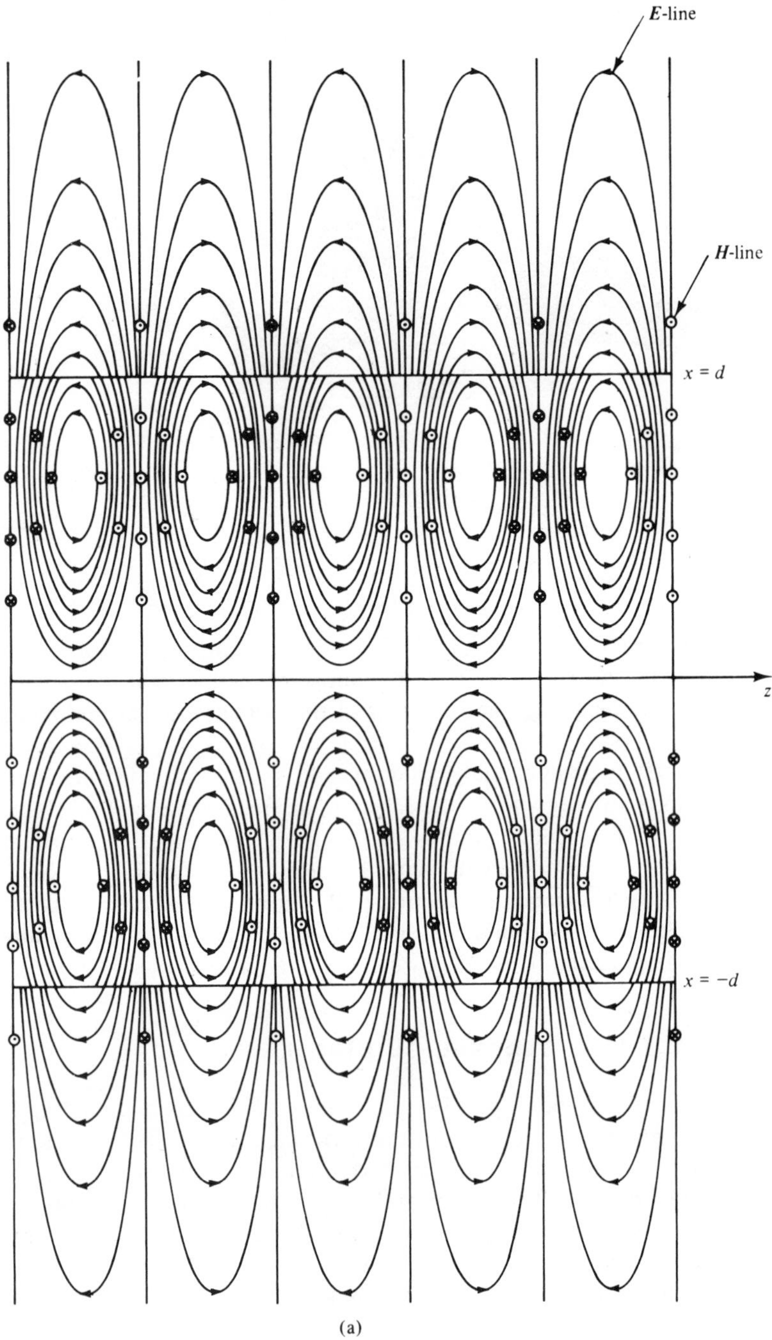

(a)

Figure 6.9 (a) First antisymmetric TM modes of dielectric slab. With $\varepsilon = \varepsilon_0 n^2$, the wavelength λ for which graph is constructed is given by $nd/\lambda = 1.18$. The transverse propagation constant k_x is the same as that of Fig. 6.7a. (b) "Three-dimensional" plot of potential Re $[\Xi(x, z)]$ for first antisymmetric TM mode.

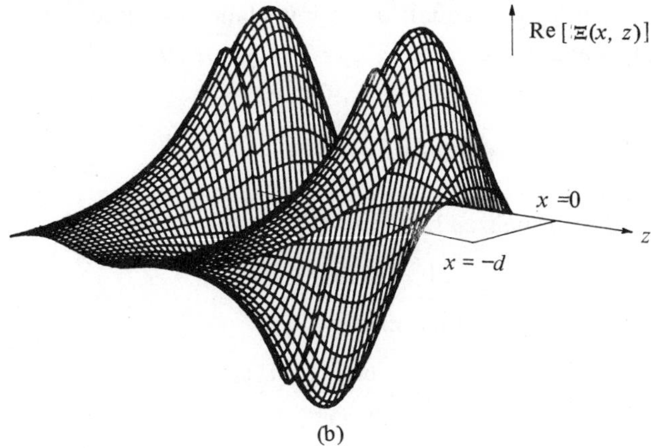

Figure 6.9 (*continued*)

and

$$k \equiv \omega\sqrt{\mu_0 \varepsilon_0} \tag{6.40}$$

This impedance is transformed by the film of characteristic impedance $Z_0^{(f)}$ [compare (2.6)]:

$$Z_0^{(f)} = \frac{1}{\cos \theta_f}\sqrt{\frac{\mu_0}{\varepsilon_f}} = \sqrt{\frac{\mu_0}{\varepsilon_0}}\,\frac{k}{k_x^{(f)}} \tag{6.41}$$

into an impedance Z according to (2.33):

$$\begin{aligned}\frac{Z}{Z_0^{(f)}} &= \frac{Z_s + jZ_0^{(f)} \tan k_x^{(f)} f}{Z_0^{(f)} + jZ_s \tan k_x^{(f)} f} \\ &= \frac{(j/\alpha_s) + (j/k_x^{(f)}) \tan k_x^{(f)} f}{(1/k_x^{(f)}) - (1/\alpha_s) \tan k_x^{(f)} f}\end{aligned} \tag{6.42}$$

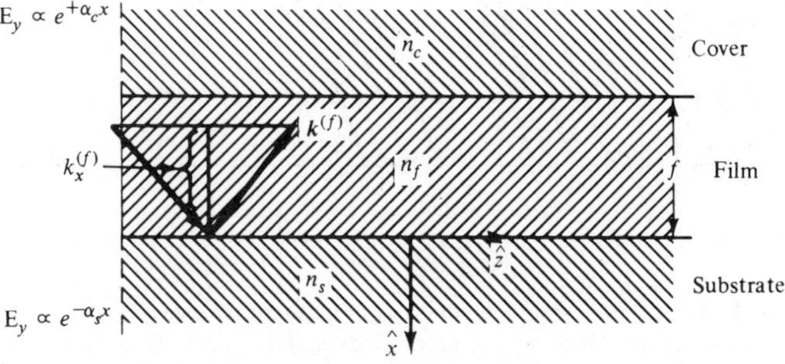

Figure 6.10 Guiding layer for TE waves.

This impedance has to be equal to the impedance presented by the exponentially decaying wave in the "cover" $Z_c = -j\sqrt{\mu_0/\varepsilon_0}\,(k/\alpha_c)$. Thus we obtain the determinantal equation

$$\tan k_x^{(f)} f = \frac{(\alpha_s/k_x^{(f)}) + (\alpha_c/k_x^{(f)})}{1 - (\alpha_s \alpha_c/k_x^{(f)2})} \tag{6.43}$$

We introduce the normalized frequency

$$V \equiv kf(n_f^2 - n_s^2)^{1/2} \tag{6.44}$$

The normalized propagation parameter

$$b = \frac{\beta^2 - \omega^2 \mu_0 \varepsilon_s}{\omega^2 \mu_0 (\varepsilon_f - \varepsilon_s)} \tag{6.45}$$

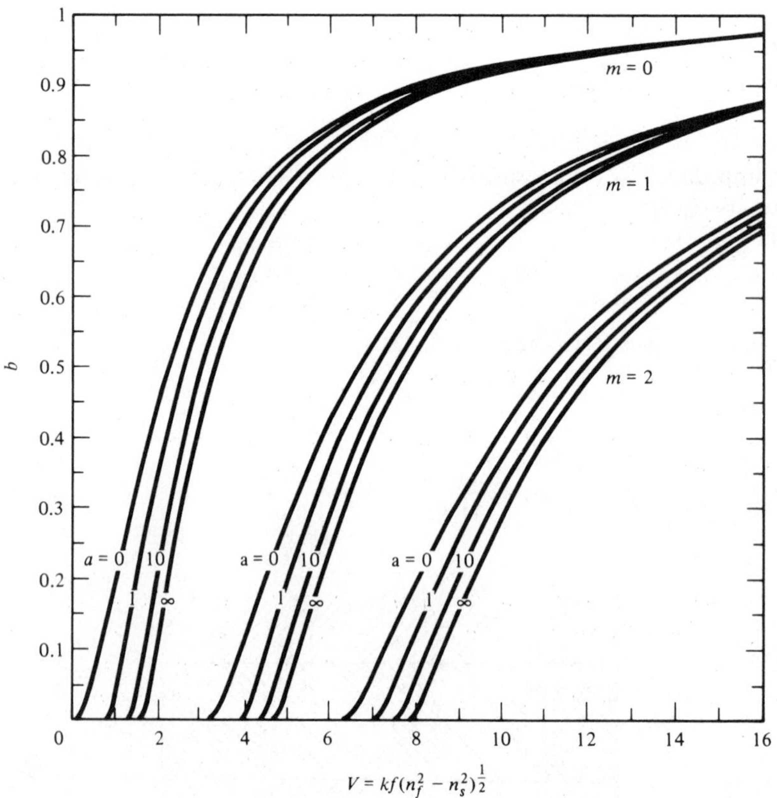

Figure 6.11 Guide index b as a function of normalized frequency V for the lowest three TE mode orders for various degrees of asymmetry. (From H. Kogelnik and V. Ramaswamy, "Scaling rules for thin-fiber optical waveguides," *Appl. Opt.*, *8*, 1857, 1974.)

and the asymmetry parameter

$$a \equiv \frac{n_s^2 - n_c^2}{n_f^2 - n_s^2} \tag{6.46}$$

Equation (6.43) may be written in the form

$$\tan V(1-b)^{1/2} = \frac{\sqrt{\dfrac{b}{1-b}} + \sqrt{\dfrac{b+a}{1-b}}}{1 - \dfrac{\sqrt{b(b+a)}}{1-b}} \tag{6.47}$$

A plot of V versus b with a as a parameter provides a universal plot for any guide; this is shown in Fig. 6.11. The figure shows that the dominant mode, $m = 0$, propagates at all frequencies only in the symmetric case, $a = 0$. This case has been analyzed in the preceding section. The curves $a = 0$, $m = 0, 1, 2$, correspond to the dispersion diagram, Fig. 6.4. The two asymptotes in that figure have been transformed by the change of variables (6.44) and (6.45) into the line $b = 0$ and $b = 1$, respectively.

Optical waveguides which provide two-dimensional confinement of the field are constructed from slab waveguides by a slight decrease of the index of the slab or decrease of thickness of the slab, in the region outside the "guiding region" as shown in Fig. 6.12. The field of a mode is concentrated in the region of highest index and/or greater thickness. It decays into the substrate and into the regions of decreased index or thickness of the film. The mode pattern is still roughly that of the slab guide except for its finite extent in the y direction. The modes are still called TE and TM, even though they do not possess strictly transverse electric or magnetic fields.

The same terminology is applied to waveguides that are produced by an

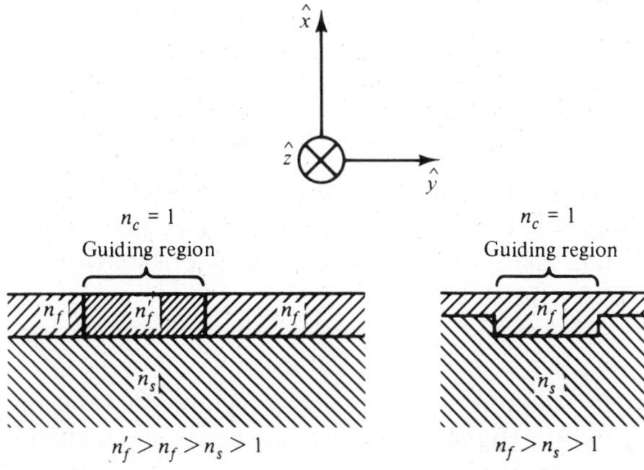

Figure 6.12 Optical waveguide cross sections.

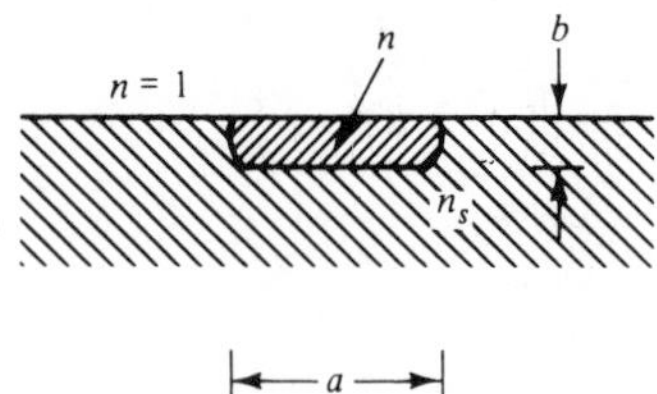

Figure 6.13 Waveguide formed of a local change of index. This waveguide is discussed in Section 7.7.

index change over a limited cross section (Fig. 6.13). The aspect ratio a/b is usually larger than unity so that the mode patterns and field polarizations are modifications of the modes of the infinitely wide slab.

6.5 PROPAGATION ALONG A SINGLE-MODE FIBER

We have looked at wave propagation in a medium with a parabolic index profile and in a dielectric slab guide. The former case was a model for a multimode fiber that propagates many modes. Propagation in a single mode offers advantages for optical communications. The beating between different modes that occurs in multimode propagation is eliminated. The propagation distance is now limited solely by the group velocity dispersion of one single mode. This is why the future of high-bit-rate optical communications is with single-mode fibers.

The (approximate) wave equation for the vector potential to be solved for a single-mode fiber with a uniform circular cylindrical dielectric core surrounded by a medium of lower index is (6.5) (Fig. 6.14a). An approximate solution is obtained by introducing (6.6) into the vector wave equation, which reduces to the scalar wave equation (6.7).

We have found in Section 6.1 that $\beta^2 < \omega^2\mu_0\,\varepsilon(\rho)$, at least in part of the range $-\infty < x < \infty$, $-\infty < y < \infty$. Thus we must have $\beta^2 < \omega^2\mu_0\,\varepsilon_i$ where ε_i is the dielectric constant of the core. Outside the cylindrical core $\beta^2 > \omega^2\mu_0\,\varepsilon$, as we shall now show. This follows from the fact that the solution outside the core must obey the equation

$$\nabla_T^2 u - (\beta^2 - \omega^2\mu\varepsilon)u = 0 \tag{6.48}$$

where ε is constant. If the solution is to decay exponentially to infinity, rather than be oscillatory, the constant $\beta^2 - \omega^2\mu\varepsilon$ in this equation must be negative or $\beta^2 > \omega^2\mu\varepsilon$. Thus, the propagation constant is bracketed between the limits (compare Fig. 6.4 for the slab)

$$\omega^2\mu_0\,\varepsilon < \beta^2 < \omega^2\mu_0\,\varepsilon_i$$

The lowest-order solution is cylindrically symmetric and has the profile sketched in Fig. 6.14b. Higher-order solutions have more than one extremum and thus have greater curvature , larger values of $\nabla_T^2 u/u$. By making $(\varepsilon_i - \varepsilon)/\varepsilon$ small enough, one can prevent solutions from occurring with greater curva-

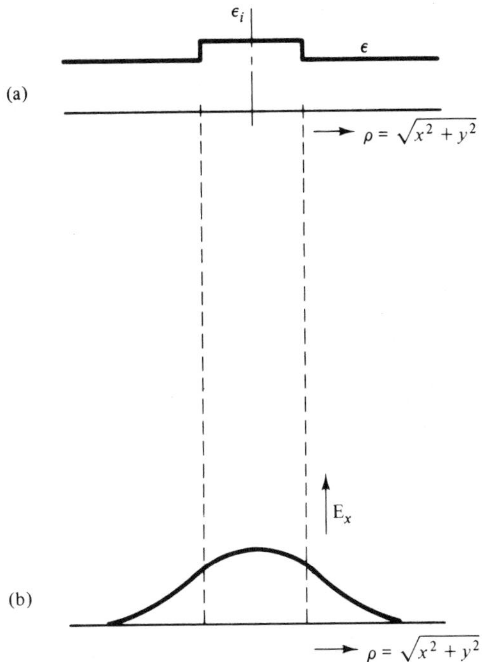

Figure 6.14 (a) Index profile; (b) amplitude distribution of the lowest-order solution.

tures at any given frequency and thus prevent the existence of any higher-order modes. This is the principle of the single-mode fiber. Of course, there are still two different polarizations possible in a circular cylindrical fiber (e.g., along $\hat{x}$ and along $\hat{y}$). The input excitation determines the polarization.

The actual solution of (6.48) involves Bessel functions and will not be pursued further. The reader is referred to the extensive literature on the subject. [2] We do not need to evaluate the dispersion curve $\beta(\omega)$ from the eigenvalue equation (6.48) in order to determine the group velocity dispersion, because, in practice, it is overshadowed by the material dispersion, the frequency dependence of the dielectric constant, $\varepsilon(\omega)$. Figure 6.15 shows a plot of dispersion versus wavelength in a single-mode optical fiber. In order to explain the meaning of the dispersion parameter we look more closely at pulse propagation along a single-mode fiber.

Suppose that we have solved the eigenvalue problem and have determined β of (6.48) as a function of frequency. The spatial dependence of the vector potential A_y is then

$$A_y = a(z)u(x, y) \tag{6.49}$$

where $a(z)$ is the amplitude with the z dependence $\exp[-j\beta(\omega)z]$. We may

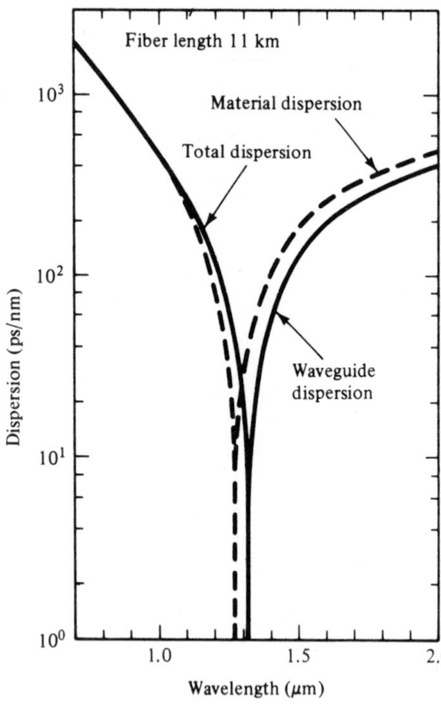

Figure 6.15 11-km-long single-mode fiber dispersion. Zero dispersion wavelength is 1.32 μm. (From J. I. Yamada, M. Saruwatari, K. Asatani, H. Tsuchiya, A. Kawana, K. Sugiyama, and T. Kimura, "High speed optical pulse transmission at 1.29 μm wavelength using low-loss single-mode fibers," *IEEE J. Quantum Electron.*, *QE-14*, 791–800, Nov. 1978 © 1978 IEEE.)

write for a(z) the differential equation

$$\frac{\partial \mathrm{a}}{\partial z} = -j\beta \mathrm{a} \tag{6.50}$$

Now, consider an excitation with a spectrum that occupies a narrow range of frequencies around ω_0 so that $\beta(\omega)$ can be expanded into a few terms:

$$\frac{\partial \mathrm{a}(z, \omega)}{\partial z} = -j\left(\beta(\omega_0) + \frac{d\beta}{d\omega}(\omega - \omega_0) + \frac{1}{2}\frac{d^2\beta}{d\omega^2}(\omega - \omega_0)^2\right)\mathrm{a} \tag{6.51}$$

The space–time dependence of a(z, ω) is obtained by inverse Fourier transformation

$$\mathrm{a}(z, t) = \int_0^\infty d\omega\, e^{j\omega t}\mathrm{a}(z, \omega) \tag{6.52}$$

where, consistent with the definitions introduced in Section 1.5, we use only positive frequency components; a(z, t) is complex and the real-time-dependent function is

$$\frac{1}{\sqrt{2}}[\mathrm{a}(z, t) + \mathrm{a}^*(z, t)]$$

If the spectrum $a(z, \omega)$ is narrow, and centered around ω_0, it is convenient to express it as a function of $(\omega - \omega_0)$. Further, it is convenient to factor the fast spatial dependence $\exp[-j\beta(\omega_0)z]$ by writing

$$a(z, \omega) = A(z, \omega - \omega_0) \exp[-j\beta(\omega_0)z] \tag{6.53}$$

This defines the complex "envelope" $A(z, \omega - \omega_0)$ of the wave, which varies slowly with respect to z. The Fourier transform of (6.53) introduces the space–time-dependent envelope $A(z, t)$:

$$\begin{aligned} a(z, t) &\equiv \int_{-\infty}^{\infty} d\omega\, e^{j\omega t} a(z, \omega) \\ &= \int_{-\infty}^{\infty} d(\omega - \omega_0) e^{j(\omega - \omega_0)t} A(z, \omega - \omega_0) \exp\{-j[\beta(\omega_0)z - \omega_0 t]\} \\ &\equiv A(z, t) \exp\{-j[\beta(\omega_0)z - \omega_0 t]\} \end{aligned} \tag{6.54}$$

With the definition of the envelope $A(z, \omega - \omega_0)$ we obtain from (6.51)

$$\frac{\partial}{\partial z} A(z, \omega - \omega_0) = -j\left[\frac{d\beta}{d\omega}(\omega - \omega_0) + \frac{1}{2}\frac{d^2\beta}{d\omega^2}(\omega - \omega_0)^2\right] A(z, \omega - \omega_0) \tag{6.55}$$

The inverse Fourier transform of (6.55) is

$$\left(\frac{\partial}{\partial z} + \frac{1}{v_g}\frac{\partial}{\partial t}\right) A(z, t) = \frac{j}{2}\frac{d^2\beta}{d\omega^2}\frac{\partial^2 A(z, t)}{\partial t^2} \tag{6.56}$$

where we have used the Fourier transform relationship

$$[j(\omega - \omega_0)]^n A(z, \omega - \omega_0) = \text{F.T.}\left[\frac{\partial^n}{\partial t^n} A(z, t)\right] \tag{6.57}$$

and where

$$\frac{1}{v_g} = \frac{d\beta}{d\omega}$$

is the inverse group velocity. The envelope $A(z, t)$ would propagate unchanged at the group velocity were it not for the term on the right-hand side, which produces "distortion." Note that it is imaginary (i.e., it contributes to phase).

There is a simple interpretation of the term $d^2\beta/d\omega^2$ that applies to pulses which are not Fourier transform limited, that is, whose spectral width is much wider than the inverse pulse width. In this limit, one may treat the pulse of width $1/\tau_p$ as made up of many different spectral components occupying the total bandwidth $\Delta\omega \gg 1/\tau_p$. Each of the pulses has its own inverse group velocity $1/v_g$, which, over the width of the spectrum, $\Delta\omega$, varies over the range of values

$$\Delta\left(\frac{1}{v_g}\right) = \frac{d^2\beta}{d\omega^2}\,\Delta\omega$$

Over a travel distance l, the pulses associated with the extreme edges of the spectrum get shifted by

$$\Delta\tau = \Delta\left(\frac{1}{v_g}\right)l = \frac{d^2\beta}{d\omega^2}\,\Delta\omega l$$

This is the quantity plotted in Fig. 6.15. The units of $\partial^2\beta/\partial\omega^2$ are ns/nm km assuming that $\Delta\tau$ is measured in nanoseconds, the bandwidth in nanometers (i.e., wavelength range $\Delta\lambda$ where $|\Delta\lambda/\lambda_0| = |\Delta\omega/\omega_0|$), and l in kilometers. Note that there is a minimum at the wavelength of 1.3 μm. This accounts for the present emphasis on the development of optical components operating in this frequency regime (e.g., quaternary compound laser diodes in InGaAsP). Also note that the dispersion is dominated by the material dispersion.

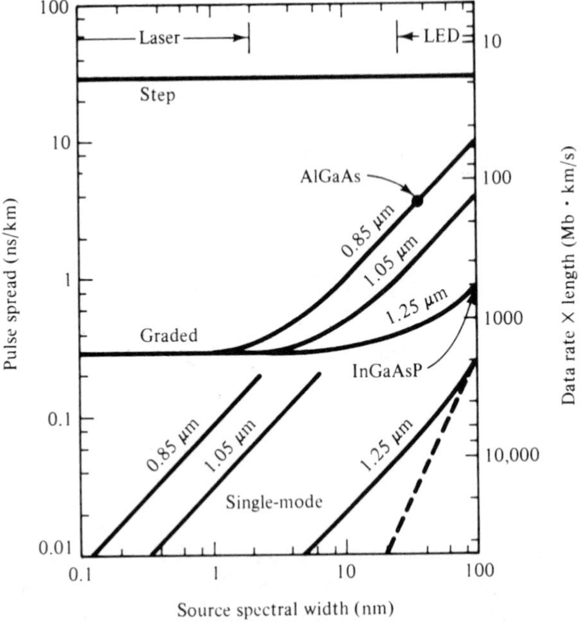

Figure 6.16 Pulse delay spread ($2\sigma_m$) and bandwidth (data rate × fiber length) versus source spectral width ($2\sigma_s$) for (1) multimode fiber of step-index profile, $GeO_2 \cdot B_2O_3 \cdot SiO_2$ glass, $\Delta = 0.01$, (2) multimode fiber of graded-index profile, $GeO_2 \cdot B_2O_3 \cdot SiO_2$ glass, $\Delta = 0.01$, bandwidth improvement factor = 100, and (3) single-mode fiber, $B_2O_3 \cdot SiO_2$ glass, $\Delta = 0.001$. The dashed line represents the lower limit of pulse spreading in silica fibers. (From T. Li, "Optical fiber communication—the state of the art," *IEEE Trans. Commun.*, 946–955, July 1978 © 1978 IEEE.)

In Fig. 6.16, the pulse spreading is plotted versus the spectral width of the source and together with this information the data rate times fiber length. Both cases of multimode propagation (top curves) and single-mode propagation (bottom curves) are shown.

6.6 PROPAGATION OF GAUSSIAN PULSES THROUGH DISPERSIVE SYSTEMS

In the preceding section, we noted that pulses with a given spectral width experience spreading in a dispersive fiber by virtue of the fact that different frequency components travel at different speeds. This somewhat qualitative argument is satisfactory for the understanding of propagation of pulses with relatively broad spectra, such as pulses produced by modulation of LEDs (light-emitting diodes). For the understanding of narrow-band pulses, particularly for so-called Fourier-transform-limited pulses for which the product of bandwidth $\Delta\omega/2\pi$ and pulse width τ_p is of order of 0.5 (different values pertain to different pulse shapes), we look more closely at the equation of pulse propagation, (6.56).

The equation for pulse propagation in a dispersive fiber (6.56) can be made to bear an uncanny resemblance to the paraxial wave equation (4.45) in two dimensions (say z and x, so that $\nabla_T^2 = \partial^2/\partial x^2$). We introduce the new variables

$$\tau = (t - t_0) - \frac{z - z_0}{v_g} \tag{6.58}$$

$$\xi = z - z_0 \tag{6.59}$$

with t_0 and z_0 as arbitrary constants. Note that $d\tau = 0$ when $dt = dz/v_g$ (i.e., when the position z is changed at the rate v_g). If we fix our attention on one such moving point, the variable ξ increases linearly with t, $\Delta\xi = v_g\, \Delta t$. Thus ξ measures the time (in distance units) elapsed in the observation of such a moving point.

With these new variables we can write (6.56) in the form

$$j\frac{\partial \text{A}}{\partial \xi} + \frac{1}{2}\frac{d^2\beta}{d\omega^2}\frac{\partial^2 \text{A}}{\partial \tau^2} = 0 \tag{6.60}$$

This equation is of the same form as the paraxial wave equation (4.45) in two dimensions with $\nabla_T^2 = \partial^2/\partial x^2$. Hence the solutions must be identical with those of the two-dimensional paraxial wave equation (Problem 5.9). The Gaussian beam in two dimensions is described by

$$\frac{1}{\sqrt{z + jb}} \exp\left[-\frac{jkx^2}{2(z + jb)}\right]$$

Comparison of the paraxial wave equation (4.45) with the propagation equa-

TABLE 6.1

Paraxial Wave Equation	Dispersion Problem
$u(x, z)$	$A(\tau, \xi)$
$\dfrac{-z}{k}$	$\xi \dfrac{d^2\beta}{d\omega^2}$
x	τ
$\dfrac{b}{k}$	α

tion (6.60) describing dispersion leads to the correspondences shown in Table 6.1. One must note that $d^2\beta/d\omega^2$ can be either positive (normal dispersion) or negative (anomalous dispersion). See Problem 6.5. In Table 6.1 this is taken into account by the sign of k assigned to the paraxial wave solution. Using the paraxial wave equation solution and the correspondence of Table 6.1, we obtain the solution to Gaussian pulse propagation along a dispersive fiber.

$$A(z, t) = \frac{A_0}{\sqrt{j\alpha + \delta_0 - z(d^2\beta/d\omega^2)}} \exp\left\{\frac{j(t - z/v_g)^2}{2[z(d^2\beta/d\omega^2) - j\alpha - \delta_0]}\right\} \tag{6.61}$$

where we have set $t_0 = z_0/v_g$, and $\delta_0 \equiv z_0(d^2\beta/d\omega^2)$ is the *chirp parameter*. Consider this solution at $z = 0$.

$$A(z = 0, t) = \frac{A_0}{\sqrt{j\alpha + \delta_0}} \exp\left[-\frac{\alpha t^2}{2(\alpha^2 + \delta_0^2)}\right] \exp\left[-\frac{j\delta_0 t^2}{2(\alpha^2 + \delta_0^2)}\right] \tag{6.62}$$

The pulse is "chirped"; its frequency, the derivative of phase with respect to time, is a function of time

$$\omega = \frac{d\phi}{dt} = \omega_0 - \frac{\delta_0 t}{\alpha^2 + \delta_0^2} \tag{6.63}$$

For a positive *chirp parameter* δ_0, high frequencies occur in the front part of the pulse ($t < 0$), low frequencies in the back part ($t > 0$). The reverse is true for a negative chirp parameter. The Fourier transform of $A(z = 0, t)e^{j\omega_0 t}$ is

$$a(z = 0, \omega) = \frac{A_0}{\sqrt{j2\pi}} \exp\left[-\frac{1}{2}(\alpha - j\delta_0)(\omega - \omega_0)^2\right]$$

Note that the full width of the spectrum $|a(\omega)|^2$ at half-maximum (measured in hertz) is independent of the chirp parameter δ_0:

$$\text{spectral FWHM} = \left[\frac{\ln 2}{\pi^2 \alpha}\right]^{1/2} \tag{6.64}$$

The same is not true for the temporal full width of the pulse intensity at half-maximum, which is:

$$\text{temporal FWHM} = 2\sqrt{(\ln 2)\alpha\left(1 + \frac{\delta_0^2}{\alpha^2}\right)} \tag{6.65}$$

When $\delta_0 \neq 0$, the pulse has a greater pulse width than is warranted from the spectrum; for $\delta_0 = 0$, the pulse is a Fourier-transform-limited Gaussian pulse, and the product of the half-widths is

$$\frac{2 \ln 2}{\pi} = 0.441$$

Suppose now that the pulse propagates along a lossless fiber of length l. According to (6.61), the amplitude at $z = l$ is

$$A(l, t) = \frac{A_0}{\sqrt{j\alpha + \delta}} \exp\left[-\frac{\alpha(t-\tau)^2}{2(\alpha^2 + \delta^2)}\right] \exp\left[-\frac{j\delta(t-\tau)^2}{2(\alpha^2 + \delta^2)}\right] \tag{6.66}$$

where τ is the time delay

$$\tau = \frac{l}{v_g} = \frac{d\beta}{d\omega}\, l \tag{6.67}$$

and

$$\delta = \delta_0 - \frac{d^2\beta}{d\omega^2}\, l \tag{6.68}$$

is the modified chirp parameter. The pulse is delayed by the group delay over the propagation distance l. Also, the pulse width is modified by modification of the chirp parameter. In particular, if $\delta_0 \neq 0$ and $\delta = 0$, the pulse has been *shortened*. This can be understood as follows: If the fiber has "normal dispersion" $d^2\beta/d\omega^2 > 0$, the group velocity of Fourier components of higher frequency is smaller than that of components of lower frequency. For $\delta_0 > 0$, the high-frequency components occur in the front portion of the pulse. These components travel more slowly than the low-frequency components and are overtaken by them. If the fiber length l is too long, the effect may be too large, δ becomes negative, according to (6.68), and the pulse broadens again.

Fibers exhibit normal dispersion at frequencies higher than a critical frequency and analomous dispersion at frequencies below the critical frequency (compare Fig. 6.15). Pulses propagating through fibers above the critical frequency acquire, eventually, a $\delta < 0$, as can be seen from (6.68). To recompress them one needs a system with $(d^2\beta/d\omega^2)l < 0$, a system of anomalous dispersion. Such a system is more conveniently realized artifically (i.e., not with an anomalously dispersive medium) by a grating pair. [10]

We shall now illustrate how this is done. Note that the propagation through the fiber is characterized by a frequency-dependent phase delay $\phi(\omega) = \beta(\omega)l$. In analogy, study the phase delay of a plane wave traveling

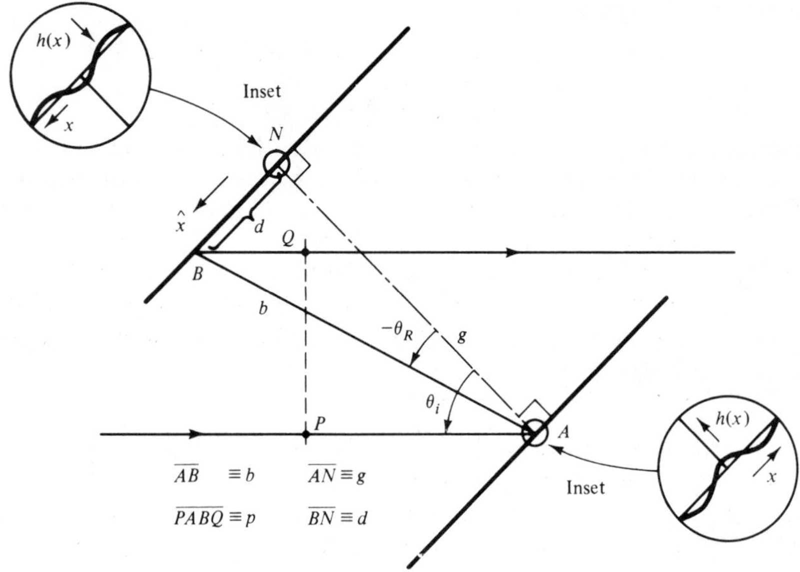

Figure 6.17 Grating pair for pulse compression.

through a pair of parallel, identical gratings as illustrated in Fig. 6.17 starting at the point P and emerging at the point Q. Without loss of generality, we assume that the peaks of each of the fundamental Fourier components of the grating corrugations occur at the intersections of the indicated normal. With the length $\overline{AB}$ defined as $b = g/(\cos\theta_R)$, where g is the spacing of the grating pair, we have for the travel distance

$$\overline{PABQ} \equiv p = b[1 + \cos(\theta_i + \theta_R)] = \frac{g}{\cos\theta_R}[1 + \cos(\theta_i + \theta_R)] \qquad (6.69)$$

where we note that θ_R is negative in Fig. 6.17, assuming that the grating reflection is of negative order, $m = -1$ in (2.50) and $\lambda/\Lambda > \sin\theta_i$, and the grating periodicity is small enough,

$$\sin\theta_R = \sin\theta_i - \frac{\lambda}{\Lambda} \qquad (6.70)$$

The phase shift ϕ of the wave over the path $\overline{PABQ}$ is

$$\phi = \frac{\omega}{c} p + C(\omega) \qquad (6.71)$$

where $C(\omega)$ is a phase correction due to reflection from the grating. It can be evaluated by referring the phase to the point N in Fig. 6.17. At that point, the reflected wave of order -1 experiences a phase delay of $-\pi/2$ according to (2.55). The phase of the incident wave is advanced by $k_{ix}d$ with respect to its phase at the point B. The reflected wave is delayed by $[k_{ix} - (2\pi/\Lambda)]d$, where d is the distance

$$\overline{BN} = -g \tan \theta_R \tag{6.72}$$

Therefore,

$$C(\omega) = -\pi - k_{ix} d + \left(k_{ix} - \frac{2\pi}{\Lambda}\right) d = -\pi + \frac{2\pi}{\Lambda} g \tan \theta_R \tag{6.73}$$

The phase $-\pi$ in (6.73) is due to the two reflections at A and B. The group delay is

$$\frac{d\phi}{d\omega} = \frac{p}{c} + \left(\frac{\omega}{c}\frac{dp}{d\omega} + \frac{dC}{d\omega}\right)$$

Use of (6.69), (6.70), and (6.73) shows that the term in parentheses vanishes and thus

$$\frac{d\phi}{d\omega} = \frac{p}{c}$$

The dispersion produced by the grating is characterized by

$$\frac{d^2\phi}{d\omega^2} = \frac{1}{c}\frac{dp}{d\omega} = -\frac{(1/\omega^2)(\lambda/\Lambda)(2\pi g/\Lambda)}{[1 - (\sin \theta_i - \lambda/\Lambda)^2]^{3/2}}$$

The second derivative is negative; the grating pair acts like a medium with anomalous dispersion of propagation constant β and thickness l, with

$$\frac{l\, d^2\beta}{d\omega^2} = \frac{d^2\phi}{d\omega^2}$$

6.7 INDEX CHANGE IN A FIBER

Single-mode fibers can be used as sensors of temperature, of strain, and of rotation rate (gyroscope). These applications monitor the change of index produced by the physical effect to be measured by measuring the change of the phase of a wave propagating through the fiber. Also, unwanted phase shifts are produced in fibers used for optical communications by index changes caused by temperature changes. In this section we evaluate the effect on the propagation constant of an index change in the fiber.

We start with equation (6.48) as applied to the unperturbed fiber. Suppose that this equation has been solved, the field profile $u(x, y)$ has been found, and β has been determined. Suppose next that $\varepsilon(x, y)$ is now changed from $\varepsilon(x, y)$ to $\varepsilon(x, y) + \delta\varepsilon(x, y)$. The new propagation constant is $\beta + \delta\beta$, the new field profile is $u + \delta u$. An equation can be found for the *perturbations* $\delta\beta$ and

δu by perturbing (6.48) to first order in these quantities:

$$\nabla_T^2 \, \delta u + \omega^2 \mu_0 \, \varepsilon \, \delta u + \omega^2 \mu_0 \, \delta\varepsilon u = 2\beta \, \delta\beta u + \beta^2 \, \delta u \tag{6.74}$$

We multiply the above by u^* and integrate over the cross section of the fiber:

$$2\beta \, \delta\beta \int da \, |u|^2 = \int da \, \omega^2 \mu_0 \, \delta\varepsilon \, |u|^2 + \int da \, [u^* \nabla_T^2 \, \delta u + \omega^2 \mu_0 \, \varepsilon u^* \, \delta u - \beta^2 u^* \, \delta u] \tag{6.75}$$

where $\int da$ indicates integration over the cross section. We show next that the second integral on the right-hand side vanishes. For this purpose, we take advantage of the identity that holds for any two functions $\phi(x, y)$, $\psi(x, y)$ that are twice differentiable:

$$\psi \nabla_T^2 \phi - \phi \nabla_T^2 \psi = \nabla_T \cdot [\psi \nabla_T \phi - \phi \nabla_T \psi] \tag{6.76}$$

We apply this identity with the identification $\phi = \delta u$ and $\psi = u^*$:

$$\int da \, [u^* \nabla_T^2 \, \delta u] = \int da \, [\delta u \, \nabla_T^2 u^*] + \int da \, \nabla_T \cdot [u^* \nabla_T \, \delta u - \delta u \, \nabla_T u^*] \tag{6.77}$$

The integral of the "two-dimensional" divergence of a vector can be transformed into a surface integral of the flux over the cylinder surface cutting the cross section in the contour C (see Fig. 6.18)

$$\int da \, \nabla_T \cdot [u^* \nabla_T \, \delta u - \delta u \, \nabla_T u^*] = \oint_C ds \, \boldsymbol{n} \cdot [u^* \nabla_T \, \delta u - \delta u \, \nabla_T u^*] \tag{6.78}$$

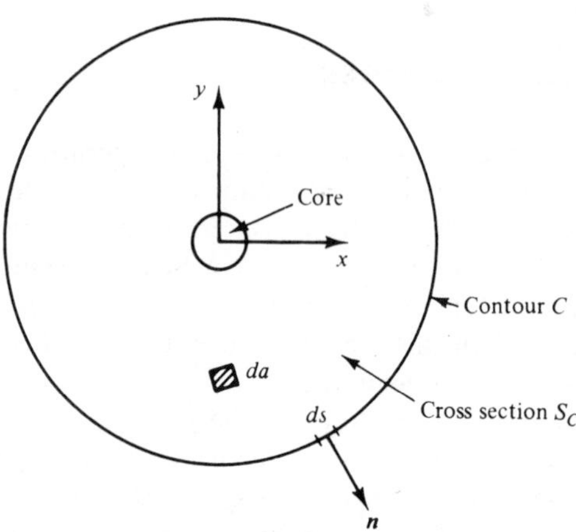

Figure 6.18 Cylinder cross section and contour C.

Because the integral $\int da$ extends over the infinite cross section, the contour C is at infinity. For guided modes, u and $u + \delta u$ vanish at infinity. Thus from (6.77) and (6.78)

$$\int da\ [u^* \nabla_T^2\ \delta u] = \int da\ [\delta u\ \nabla_T^2 u^*] \tag{6.79}$$

Next, we note that u^*, according to (6.48), obeys the equation

$$\nabla_T^2 u^* + \omega^2 \mu_0\, \varepsilon(x, y) u^* = \beta^2 u^* \tag{6.80}$$

where we used the fact that β is real. Combining (6.79) and (6.80) with (6.75) we obtain the perturbation formula for $\delta\beta$:

$$\delta\beta = \frac{\displaystyle\int da\ \omega^2 \mu_0\ \delta\varepsilon\ |u|^2}{\displaystyle 2\beta \int da\ |u|^2} \tag{6.81}$$

The perturbed field profile δu has dropped out! This is why the expression above is so useful; one need not know the perturbed field to evaluate the perturbed propagation constant to first order in $\delta\varepsilon$.

The formula can be used to show the sensitivity to index changes of the phase of a wave traveling through a long fiber of length l (many meters long). For a wavelength of 1μm, there are 10^6 wavelengths in 1 m. A change of $\delta\varepsilon/\varepsilon$ by 1 part in 10^9 changes $\delta\beta/\beta$ by the same order of magnitude. A wave propagating over a fiber of 1 km length experiences a change of phase of the order of 2π. Because fibers are so thin, it is possible to wind lengths of the order of 1 km around cylinders of size small enough to fit into a measurement instrument that responds to minute changes of optical index.

6.8 ORTHOGONALITY OF GUIDED WAVES [12]

Power conservation, which is obeyed by waves propagating along a guiding structure, implies "orthogonality" of the solutions, as we now proceed to show. This is a generalization of the orthogonality of Hermite–Gaussian solutions of the scalar wave equation for parabolic index fibers.

Power conservation requires that the integral of the real part of the complex Poynting vector over the cross section of the guiding structure be independent of distance along the structure, the z direction,

$$\frac{d}{dz} \int_{S_C} d\mathbf{a} \cdot (\mathbf{E} \times \mathbf{H}^* + \mathbf{E}^* \times \mathbf{H}) = 0 \tag{6.82}$$

where the notation is defined in Fig. 6.18 and $d\mathbf{a} = \hat{z}\, da$. Suppose that we define mode patterns by $\mathbf{e}_\mu$ and $\mathbf{h}_\mu$ for the electric and magnetic field, respectively, so that a superposition of waves is written

$$\mathbf{E} = \sum_{\mu} a_{\mu} e^{-j\beta_{\mu} z} \mathbf{e}_{\mu} \tag{6.83}$$

$$\mathbf{H} = \sum_{\mu} a_{\mu} e^{-j\beta_{\mu} z} \mathbf{h}_{\mu} \tag{6.84}$$

We use Greek subscripts for the mode number; one Greek letter subscript may stand for two Roman letter indices, $\mu \leftrightarrow mn$ if two-dimensional mode patterns are being considered. Introducing (6.83) and (6.84) into (6.82) gives

$$-\sum_{\mu, \nu} j(\beta_{\mu} - \beta_{\nu}^{*}) a_{\mu} a_{\nu}^{*} \int_{S_C} d\boldsymbol{a} \cdot (\mathbf{e}_{\mu} \times \mathbf{h}_{\nu}^{*} + \mathbf{e}_{\nu}^{*} \times \mathbf{h}_{\mu}) = 0 \tag{6.85}$$

Since a_{μ} and a_{ν} are arbitrary, we find that

$$\int_{S_C} d\boldsymbol{a} \cdot (\mathbf{e}_{\mu} \times \mathbf{h}_{\nu}^{*} + \mathbf{e}_{\nu}^{*} \times \mathbf{h}_{\mu}) = 0 \tag{6.86}$$

if $\beta_{\mu} - \beta_{\nu}^{*} \neq 0$. This is one form of the orthogonality condition.

A simpler form of the orthogonality relation is obtained by imposing on (6.86) the reciprocity constraint (3.5) combined with time reversal of Section 3.3. Consider first the operation of time reversal applied to a waveguide mode. Replacement of the electric field of mode ν, $\mathbf{e}_{\nu}$, by $\mathbf{e}_{\nu}^{*}$, of the magnetic field $\mathbf{h}_{\nu}$ by $-\mathbf{h}_{\nu}^{*}$ constructs a new, time-reversed, solution of Maxwell's equations with the spatial dependence $\exp(+j\beta_{\nu}^{*} z)$. This solution is a "backward wave" traveling in the $-z$ direction if $\exp(-j\beta_{\nu} z)$ was originally a forward wave.

Next, apply the reciprocity principle (3.5) to the solutions (a) and (b) with the interpretations $\mathbf{E}^{(a)} \rightarrow \mathbf{e}_{\mu}$, and $\mathbf{E}^{(b)} \rightarrow \mathbf{e}_{\nu}^{*}$. The surface S is taken as a surface with two parallel end planes S_C perpendicular to z at $z = 0$ and $z = z$, and bounded by a cylindrical surface of infinite radius. The dependence of the product $\mathbf{E}^{(a)} \times \mathbf{H}^{(b)}$ on z is of the form $\exp[-j(\beta_{\mu} - \beta_{\nu}^{*})z]$. We find from (3.5) that

$$\exp[-j(\beta_{\mu} - \beta_{\nu}^{*})z - 1] \int_{S_C} d\boldsymbol{a} \cdot (\mathbf{e}_{\mu} \times \mathbf{h}_{\nu}^{*} - \mathbf{e}_{\nu}^{*} \times \mathbf{h}_{\mu}) = 0 \tag{6.87}$$

This equality has to hold for arbitrary z. If $\beta_{\mu} - \beta_{\nu}^{*} \neq 0$, the integral over the cross section has to vanish. But this integral is of the same form as (6.86) except for the $-$ sign. Adding the two integrals, we have the simpler expression for orthogonality:

$$\int_{S_C} d\boldsymbol{a} \cdot \mathbf{e}_{\mu} \times \mathbf{h}_{\nu}^{*} = 0 \tag{6.88}$$

In the field expansions (6.83) and (6.84) we have not distinguished explicitly between forward and backward waves. Forward waves are waves with propagation constants $\beta_{\mu} > 0$, backward waves with $\beta_{\mu} < 0$. (Strictly, forward and backward waves are distinguished by the sign of the group velocity or its inverse, $d\beta_{\mu}/d\omega$. In all cases encountered in this book the more rigorous criterion coincides with the criterion $\beta_\mu \gtrless 0$.) Time reversal as applied above shows that every forward wave solution is paired with a backward wave solution.

Since it is usually advantageous to distinguish between forward and backward waves, we rewrite (6.83) and (6.84) in the form

$$\mathbf{E} = \sum_\mu \mathbf{e}_\mu(\mathrm{a}_\mu e^{-j\beta_\mu z} + \mathrm{b}_\mu e^{j\beta_\mu z}) \tag{6.89}$$

$$\mathbf{H} = \sum_\mu \mathbf{h}_\mu(\mathrm{a}_\mu e^{-j\beta_\mu z} - \mathrm{b}_\mu e^{j\beta_\mu z}) \tag{6.90}$$

where the summation is now over modes (i.e., wave pairs). We shall use the designation "*μth mode*" for a pair of forward and backward waves, of field patterns $\mathbf{e}_\mu$ and $\mathbf{h}_\mu$, and propagation constant $\pm\beta_\mu$, respectively.

The power passing a cross section is given by

$$\frac{1}{2}\,\mathrm{Re}\int_{S_C} \mathbf{E}\times\mathbf{H}^*\cdot d\mathbf{a} = \sum_\mu (|\mathrm{a}_\mu|^2 - |\mathrm{b}_\mu|^2)\,\frac{1}{2}\,\mathrm{Re}\int_{S_C} \mathbf{e}_\mu\times\mathbf{h}_\mu^*\cdot d\mathbf{a}$$

$$= \sum_\mu (|\mathrm{a}_\mu|^2 - |\mathrm{b}_\mu|^2). \tag{6.91}$$

The last expression applies for an appropriate normalization of the field patterns $\mathbf{e}_\mu$, $\mathbf{h}_\mu$. We have used the fact that the power can be written as the difference of the squares of the amplitudes of the counter-propagating waves in Section 3.2 in the definition of the scattering matrix of a twoport.

We have mentioned in Section 6.3 that an expansion of a field in a dielectric waveguide structure open to infinity requires, in general, guided modes and radiation modes. Radiation modes do not have the simple z-dependence $\exp(-j\beta_\nu z)$ with β_ν-real. The proof of orthogonality presented here has assumed such a simple dependence and, as formulated, applies only among guided modes. Because a good design of an optical guiding system must avoid coupling to radiation modes, it is usually adequate to represent the field as a superposition of guided modes as in (6.89) and (6.90).

6.9 POWER, ENERGY, AND GROUP VELOCITY

We have shown in Section 1.3 that the time-averaged power flow density in a plane wave, traveling in a dispersion-free medium, can be written as the product of the energy density times the velocity of energy transport $1/\sqrt{\mu\varepsilon}$. In a medium with dispersion, and/or in a nonplanar wave like in a wave along a fiber, this relation does not hold. Yet it is still legitimate to interpret the time-averaged power flow

$$\langle P\rangle = \tfrac{1}{2}\,\mathrm{Re}\int_{S_C} d\mathbf{a}\cdot\mathbf{E}\times\mathbf{H}^*$$

in a single forward traveling guided wave of narrow bandwidth as a product of the energy per unit length and a velocity of energy transport:

$$\langle P \rangle = v_e \int_{SC} \langle w \rangle \, da \tag{6.92}$$

Here the integration is over the cross section and v_e is the velocity of energy transport. The time average $\langle \cdot \rangle$ is taken over one cycle of the quasi-sinusoidal electromagnetic field. We shall now determine v_e.

Consider a fiber mode μ, with given field profiles $\mathbf{e}_\mu$ and $\mathbf{h}_\mu$, propagating with the spatial dependence $\exp(-j\beta_\mu z)$. We can construct by Fourier superposition a "packet" extending over a large, but limited, range in z and lasting over a long, but limited period of time. The spectrum of such a Fourier integral occupies a narrow band of frequencies centered around the "carrier frequency ω_0" so that $\beta(\omega)$ can be approximated by the truncated Taylor expansion

$$\beta_\mu(\omega) = \beta_\mu(\omega_0) + \frac{d\beta_\mu}{d\omega} \Delta\omega$$

where $\Delta\omega = \omega - \omega_0$. We shall omit henceforth the subscript μ referring to the mode μ under consideration in order to simplify the notation. One single Fourier component of the electric field is of the form

$$\mathbf{E}(\omega) = \mathrm{a}(\omega) e^{-j\beta(\omega)z} \mathbf{e}(x, y) \tag{6.93}$$

The inverse Fourier transform yields

$$\begin{aligned} \boldsymbol{E}(z, t) = \exp \{ j[\omega_0 t - \beta(\omega_0) z] \} \mathbf{e}(x, y) \int_{\text{band}} d\,\Delta\omega \; \mathrm{a}(\Delta\omega) \\ \cdot \exp\left[j\,\Delta\omega \left(t - \frac{d\beta}{d\omega} z \right) \right] + \text{c.c.} \end{aligned} \tag{6.94}$$

A similar expression can be written for $\boldsymbol{H}(z, t)$ in terms of the field pattern $\mathbf{h}(x, y)$ and the same amplitude $\mathrm{a}(\Delta\omega)$ [compare (6.89) and (6.90)]. The energy densities and the power flow density time-averaged over one cycle are proportional to the absolute values squared of the integral:

$$\int_{\text{band}} d\,\Delta\omega \; \mathrm{a}(\Delta\omega) \, \mathrm{e} \left[j\,\Delta\omega \left(t - \frac{d\beta}{d\omega} z \right) \right] \tag{6.95}$$

By choice of the spectrum $\mathrm{a}(\Delta\omega)$, the shape of the packet of

$$\int_{SC} \langle w \rangle \, da$$

as a function of z can be varied arbitrarily (except that the length of the packet must be consistent with the requirement that the bandwidth be maintained narrow). The packet travels undistorted at the *group velocity* $v_g = d\omega/d\beta$ (see Fig. 6.19). The local time-averaged energy and power flow can be varied at

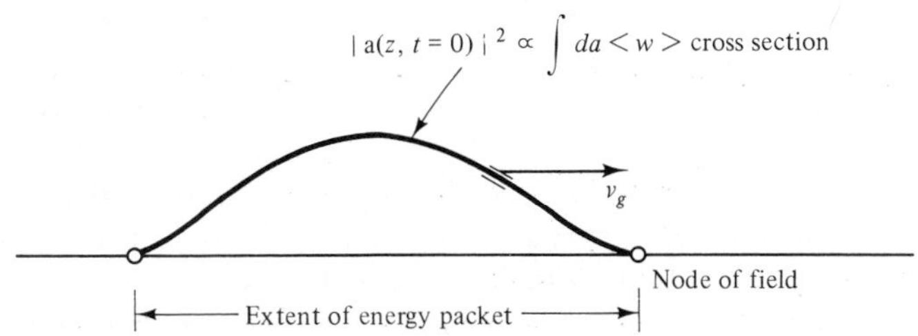

Figure 6.19 Energy "trapped" in wave packet.

will. The only way in which one may satisfy (6.92) is by setting $v_e = v_g$. Thus the energy velocity is equal to the group velocity.

Note that we treated $\mathbf{e}(x, y)$ and $\mathbf{h}(x, y)$ as frequency independent. The justification for the approximation is that the frequency dependence of the propagation constant appears in the exponent and thus is of much greater importance than the frequency dependence of $\mathbf{e}(x, y)$ and $\mathbf{h}(x, y)$; the latter may be ignored in comparison with the former.

If two counter-traveling waves of the same mode are present, then the powers $\langle P_+ \rangle$ and $\langle P_- \rangle$ of the two waves subtract as shown in (6.91). The net power is the difference of the powers associated with the forward and backward waves. The energy per unit length $\int_{SC} \langle w \rangle \, da$, is the sum of the energies associated with each

$$\int_{SC} \langle w \rangle \, da = \frac{1}{v_g} \{ |\mathrm{a}_\mu|^2 + |\mathrm{b}_\mu|^2 \} = \frac{1}{v_g} \{ \langle P_+ \rangle + \langle P_- \rangle \}. \tag{6.96}$$

6.10 SUMMARY

We studied the wave equation in a nonuniform dielectric and in particular with a parabolic profile to arrive at the modes in a multimode optical fiber. The fundamental mode was intimately related to the mode in a GRIN configuration as derived in the preceding chapter. The modes were infinite in number, but only a finite number had real propagation constants. The changeover from real to imaginary propagation constants occurred when the mode profile extended into regions of negative dielectric constant, an unphysical situation.

The dielectric slab guide was analyzed exactly, with no approximations. We found the TE and TM modes in the symmetric slab and the TE modes in the asymmetric guiding layer. The dispersion relation shows that the mode profile extends into the region exterior to the slab at low frequencies; at high frequencies the profile is confined to the interior of the slab.

The single-mode fiber avoids the mode interference effects generally associated with multimode propagation. The dispersion of single-mode fibers is much smaller than for multimode fibers, so that higher bit rates can be handled by single-mode fibers. We studied pulse propagation along a single-mode fiber, pulse spreading, and ways of reversing it through the use of a grating pair that is equivalent to a medium with anomalous dispersion. We studied the change of propagation constant produced by a change of index as may be utilized in fiber sensors. We proved orthogonality of modes using power conservation and time reversibility. Finally, we defined a velocity of energy propagation and showed that it is equal to the group velocity when applied to an optical guide or fiber.

PROBLEMS

6.1. Suppose that a GaAlAs laser operating at 8300 Å is pulsed on, producing pulses 10 ns long. The laser under pulsed operation usually oscillates in several "axial" modes that are separated in frequency by $c/2nl$, where $n = 3.5$, $l = 200\ \mu$m. Suppose that three modes are oscillating. Estimate the pulse spreading (difference between pulse envelope speeds) of highest and lowest frequency for an 11-km fiber from Fig. 6.15. If the spreading is to be no more than 20%, what is the fiber length?

6.2. Find the minimum thickness of a guiding layer of GaAs (n_f) on AlGaAs (n_s) required so that the lowest-order mode just starts propagating at $\lambda = 1\ \mu$m. Assume the values $n_c = 1$, $n_s = 3.5$, and $n_f = n_s \times 1.03$. Use Fig. 6.11.

6.3. A cylindrical rod with a parabolic index profile images an illumination in the object plane into the image plane with no magnification (Fig. P6.3). You may consider the input $u(x_0, y_0)$ expanded into an infinite set of Hermite–Gaussians. At a certain distance, all Hermite–Gaussians are in phase with multiples of 2π. What is the shortest distance over which this happens, in terms of k_0 and h? If the radius is 1 mm, the wavelength $\lambda = 6328$ Å (He–Ne laser) and n changes from 1.7 to 1.6 within that radial distance, what is l? What is the radius w of the fundamental Gaussian mode?

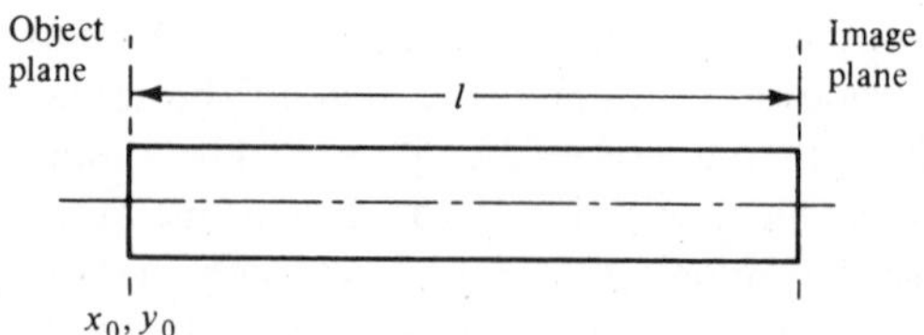

Figure P6.3

6.4. A slab waveguide is to be so designed that the first higher-order TE mode is not propagating at a given frequency (Fig. P6.4). Develop a criterion for the index profile for this to be the case. With $\Delta n = 0.02n$ and $n = 1.7$, find the limit on d for $\lambda = 6328$ Å.

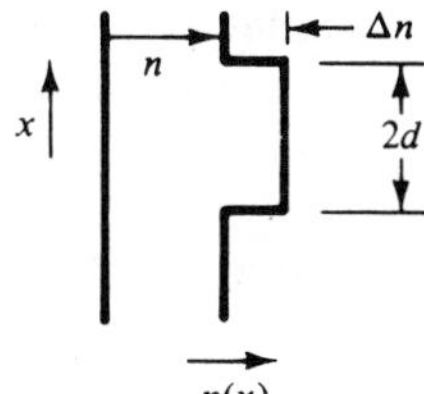

Figure P6.4

6.5. Determine the dispersion parameter $d^2\beta/d\omega^2$ for the model of dielectric developed in Problems 1.4 and 1.5. Set $\sigma = 0$, and sketch the plot of $d^2\beta/d\omega^2$ versus ω.

6.6. By analogy with the paraxial wave equation, show that dispersion produces the Fourier transform of the input pulse after long-distance propagation, except for a phase factor.

6.7. Using the Gaussian modes derived from the scalar wave equation for a fiber with a parabolic index profile:

(a) Derive the field pattern $\mathbf{e}_{mn}(x, y)$ and $\mathbf{h}_{mn}(x, y)$ in analogy with (5.55) and (5.56).

(b) Show that the orthogonality relation (6.88) holds, using the orthogonality properties of the Hermite–Gaussians.

6.8. Show that the potentials (6.36), multiplied by ε_i and ε, respectively, are related to a vector potential-like field whose curl yields the displacement flux density.

REFERENCES

[1] H. Kogelnik, "On the propagation of Gaussian beams of light through lenslike media including those with a loss and gain variation," *Appl. Opt.*, *4*, 1562, 1965.

[2] D. Marcuse, *Light Transmission Optics*, Van Nostrand, Princeton, N.J., 1972.

[3] R. B. Adler, L. J. Chu, and R. M. Fano, *Electromagnetic Energy Transmission and Radiation*, Wiley, New York, 1960.

[4] S. Ramo, J. R. Whinnery, and T. Van Duzer, *Fields and Waves in Communication Electronics*, Wiley, New York, 1965.

[5] H. Kogelnik and V. Ramaswamy, "Scaling rules for thin-fiber optical waveguides," *Appl. Opt.*, *8*, 1857, 1974.

[6] P. K. Tien, "Integrated optics and new wave phenomena in optical waveguides," *Rev. Mod. Phys.*, *49*, 361–420, 1977.

[7] T. Li, "Optical fiber communication—the state of the art," *IEEE Trans. Commun.*, 946–955, July 1978.

[8] T. Tamir and A. A. Oliner, "Guided complex waves, Part I," *Proc. IEE*, *110*, 310, 1963.

[9] T. Tamir and A. A. Oliner, "Guided complex waves, Part II," *Proc. IEE*, *110*, 325, 1963.

[10] E. B. Treacy, "Optical pulse compression with diffraction gratings," *IEEE J. Quantum Electron.*, *QE-5*, 454–459, Sept. 1969.

[11] J. I. Yamada, M. Saruwatari, K. Asatani, H. Tsuchiya, A. Kawana, K. Sugiyama, and T. Kimura, "High speed optical pulse transmission at 1.29 μm wavelength using low-loss single-mode fibers," *IEEE J. Quantum Electron.*, *QE-14*, 791–800, Nov. 1978.

[12] R. B. Adler, "Waves in inhomogeneous cylindrical structures," *Proc. IRE*, *40*, 339–348, 1952.

7

COUPLING OF MODES— RESONATORS AND COUPLERS

The phenomenon of resonance is a general one that can take many forms. When we speak of resonance we usually visualize a spring mass model or an *LC* circuit. The basic equation is, in both cases, a second-order differential equation. If an isolated resonator were the only physical system of interest, not much would be gained by studying it by the formalism developed in this chapter. Often one is interested in the coupling of a resonator to one or more input waveguides, or to another resonator. Here is where the analysis developed in this chapter is particularly useful. It is based on very few physical concepts and employs a minimum of algebra.

In Section 7.1 we develop the basic equation for the "positive frequency amplitude" of a resonance mode. Section 7.2 treats the case when the mode is coupled to an input wave or waveguide. The power fed into the mode as a function of frequency is found, and critical coupling is discussed. Secton 7.3 applies the formalism to the computation of laser threshold gain and power output. Section 7.4 treats a transmission resonator and compares the result with the Fabry–Perot transmission resonator (or filter) of Chapter 3. Section 7.5 develops the formalism of coupling between two resonators and determines the time dependence of the solution for the coupled system. In the remainder of the chapter we convert the coupling-in-time formalism to coupling in space. We discuss the optical waveguide coupler which can be used as a tunable filter or an optical switch.

The coupling of modes formalism is, basically, a perturbation analysis that finds a wide range of application. It is particularly convenient for the treatment of optical nonlinearities and the interaction of optical waves with acoustic waves, Chapters 9 and 13.

7.1 THE POSITIVE FREQUENCY AMPLITUDE OF A MODE

The coupling-of-modes formalism developed in this chapter is general and applies to numerous physical systems possessing resonant modes or propagating modes. It is helpful, however, to focus on a simple example to illustrate the meaning of the physical parameters used. We choose a simple LC circuit for the purpose of illustration. Its equations are (see Fig. 7.1)

$$v = L\frac{di}{dt} \tag{7.1}$$

$$i = -C\frac{dv}{dt} \tag{7.2}$$

The two coupled first-order differential equations lead to the second-order differential equation for the voltage

$$\frac{d^2v}{dt^2} + \omega_0^2 v = 0 \tag{7.3}$$

where

$$\omega_0^2 = \frac{1}{LC} \tag{7.4}$$

Instead of the coupled first-order differential equations we may derive two uncoupled first-order differential equations, by defining the complex variables

$$\mathrm{a}_\pm = \sqrt{\frac{C}{2}}\left(v \pm j\sqrt{\frac{L}{C}}\,i\right) \tag{7.5}$$

Through addition and subtraction of (7.1) and (7.2) with appropriate multipliers, we obtain

$$\frac{d\mathrm{a}_+}{dt} = j\omega_0\,\mathrm{a}_+ \tag{7.6}$$

$$\frac{d\mathrm{a}_-}{dt} = -j\omega_0\,\mathrm{a}_- \tag{7.7}$$

To understand better the meaning of the amplitudes a_+ and a_-, consider a_+

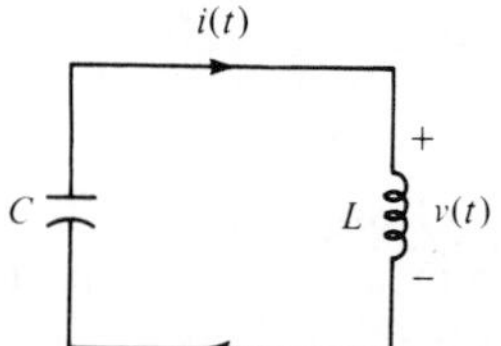

Figure 7.1 An LC circuit.

and its relation to the voltage and/or current in the resonant circuit. Solutions to (7.1) and (7.2) are

$$v(t) = |\mathrm{V}| \cos(\omega_0 t + \phi) \tag{7.8}$$

$$i(t) = \sqrt{\frac{C}{L}}\, |\mathrm{V}| \sin(\omega_0 t + \phi) \tag{7.9}$$

where $|\mathrm{V}|$ is the peak amplitude of the voltage in the LC circuit and ϕ is the phase, $\phi = \arg \mathrm{V}$. Therefore,

$$\mathrm{a}_+ = \sqrt{\frac{C}{2}}\,[|\mathrm{V}| \cos(\omega_0 t + \phi) + j|\mathrm{V}| \sin(\omega_0 t + \phi)] = \sqrt{\frac{C}{2}}\, \mathrm{V} e^{j\omega_0 t} \tag{7.10}$$

$\mathrm{a}_+(t)$ has the dependence $\exp(j\omega_0 t)$ [consistent with (7.6)] and is so normalized that

$$|\mathrm{a}_+|^2 = \frac{C}{2} |\mathrm{V}|^2 = W \tag{7.11}$$

where W is the energy in the circuit. a_+ is the *positive-frequency component* of the mode amplitude. The resonant mode is fully described by (7.6) alone, since (7.7) is simply the complex conjugate of (7.6). The obvious advantage of the new formalism is the reduction of a set of coupled differential equations to two uncoupled equations (i.e., the system of equations has been separated). Note that we have already emphasized the usefulness of working only with positive frequencies in connection with the response of the Michelson interferometer (Section 2.7). We shall drop the subscript + henceforth.

If the circuit is lossy, the loss may be represented by a conductance G in parallel with L and C (see Fig. 7.2). The introduction of loss into the mode amplitude equation (7.6) is simple, if the loss is small. Then

$$\frac{d\mathrm{a}}{dt} = j\omega_0 \mathrm{a} - \frac{1}{\tau_0} \mathrm{a} \tag{7.12}$$

where $1/\tau_0$ is the decay rate due to the loss. This decay rate can be computed without resorting to the modified circuit equations. Because $|\mathrm{a}|^2$ decays as $\exp(-2t/\tau_0)$ and the time rate of energy decrease goes into the dissipation power P_d:

$$\frac{dW}{dt} = -\frac{2}{\tau_0} W = -P_d \tag{7.13}$$

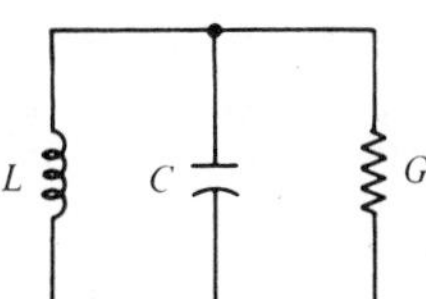

Figure 7.2 An LC circuit with loss.

It is a simple matter to compute P_d in the limit when the loss is small and can be treated as a perturbation. Indeed, the time-averaged loss is

$$P_d = \frac{1}{2} G|\mathrm{V}|^2 = \frac{G}{C}|\mathrm{a}|^2 \tag{7.14}$$

where we have used (7.11). Thus

$$\frac{P_d}{\omega_0 W} = \frac{G}{\omega_0 C} = \frac{2}{\omega_0 \tau_0} = \frac{1}{Q_0} \tag{7.15}$$

where the dimensionless quantity $2/\omega_0 \tau_0$ is the inverse unloaded Q or quality factor (of the "unloaded" resonator, that is, not connected to an external "load").

If we had started from the equations of the circuit (Fig. 7.2) and had computed the complex frequency $s = -(1/\tau) + j\omega$ exactly, we would have obtained

$$s = -\frac{1}{\tau} + j\omega = -\frac{G}{2C} + j\sqrt{\frac{1}{LC} - \frac{G^2}{4C^2}} \tag{7.16}$$

Note that the decay rate has been properly evaluated by the perturbation approach. There is, however, a correction to the frequency $\omega_0 = 1/\sqrt{LC}$ that is of second order in G/C. This correction is not recovered by the approximate treatment. Recall that the reduction of the second-order differential equation of the lossless resonator to two uncoupled first-order differential equations was a rigorous mathematical step. However, the introduction of loss by modification of the *separated* equations is only approximately correct. The introduction of loss actually "couples" the two equations for a_+ and a_-, a fact expressed by the change of the real part of the resonator frequency *to second order* in the normalized loss parameter $(G/\omega_0 C)$.

Other perturbations of the resonant circuit or resonant mode can be performed similarly to the introduction of loss. An example of a perturbation is the connection of a resonance circuit to an external transmission line, or coupling of the Fabry–Perot resonator to waves incident upon it from the "outside" by making the mirror partially transmitting.

7.2 THE SINGLE RESONATOR WITH INPUT WAVE OR WAVEGUIDE

The equation for the positive-frequency component of the mode amplitude, a, states that a varies with time as $\exp(j\omega_0 t)\exp(-t/\tau_0)$. If the resonator is coupled to an external waveguide or to the outside space by a partially transmitting mirror as shown in Fig. 7.3, two effects must be accounted for:

(α) Modification of decay rate

(β) The excitation of mode amplitude a by incident wave

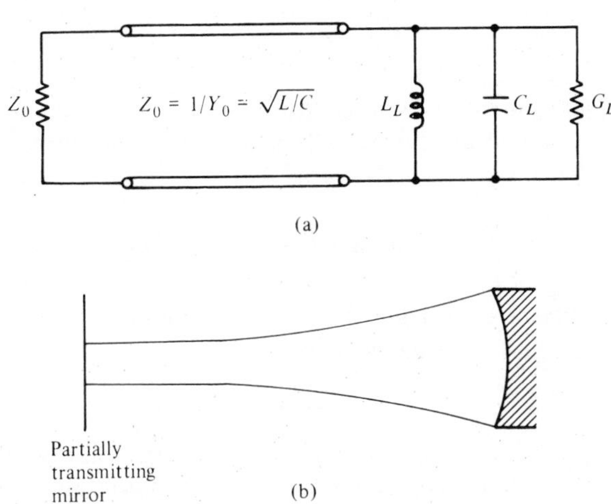

Figure 7.3 (a) *GLC* circuit terminated in matched transmission line; (b) optical resonator with partially transmitting mirror.

The *decay rate* of the mode *is modified* because now power is not only dissipated internally but also escapes into the waveguide or outside space. Equation (7.12) must be modified:

$$(\alpha) \quad \frac{d\mathrm{a}}{dt} = j\omega_0 \mathrm{a} - \left(\frac{1}{\tau_0} + \frac{1}{\tau_e}\right)\mathrm{a} \tag{7.17}$$

where $1/\tau_e$ expresses the additional rate of decay due to escaping power (the subscript e stands for "external"). We form the time rate of change of energy, from (7.17):

$$\frac{dW}{dt} = \mathrm{a}^* \frac{d\mathrm{a}}{dt} + \mathrm{a}\frac{d\mathrm{a}^*}{dt} = -2\left(\frac{1}{\tau_0} + \frac{1}{\tau_e}\right)W \tag{7.18}$$

In this way the net power lost by the resonance to dissipation and escape from the resonator is treated as a Taylor expansion to first order in the parameters determining the dissipated and escaping power. The escaping power P_e is given by

$$P_e = \frac{2}{\tau_e} W$$

and the resulting "external" Q of the resonator is

$$\frac{P_e}{\omega_0 W} = \frac{2}{\omega_0 \tau_e} = \frac{1}{Q_{\text{ext}}} \tag{7.19}$$

For the configuration of Fig. 7.3a in which the line is terminated in a matched load, we obtain

$$\frac{1}{Q_{\text{ext}}} = \frac{Y_0}{\omega_0 C_L} \tag{7.20}$$

The *waveguide may carry a wave traveling toward the resonator* of amplitude s_+ due to a source as shown in Fig. 7.4. Then the impinging wave s_+ serves as a drive for a, or

$$(\beta) \quad \frac{da}{dt} = j\omega_0 a - \left(\frac{1}{\tau_0} + \frac{1}{\tau_e}\right)a + \kappa s_+ \tag{7.21}$$

where κ is a coefficient expressing the degree of coupling between the resonator and the wave s_+. We normalize s_+ so that

$$|s_+|^2 = \text{power carried by incident wave}$$

in contrast to $|a|^2$, which is normalized to the energy. The synthesis of two formalisms, incidence of a waveguide wave on a resonator, and the time evolution of the amplitude of excitation in the resonator, calls for renewed attention to the symbols used for the various amplitudes. We have chosen a for the amplitude in the resonator. Before, in Chapter 6, we introduced wave amplitudes a_μ and b_μ for the forward and backward waves of the μ-th mode and in Chapter 3 we used a and b for incident and reflected waves. Since the alphabet is limited, and esoteric letters to denote wave amplitudes are undesirable, we chose here s_+ to designate a (source) wave incident upon the resonator. The reflected wave will be denoted by s_-. In the notation of Chapter 2 these would be designated by a and b respectively and in the notation of Chapter 6 by a_μ and b_μ denoting the μ-th mode.

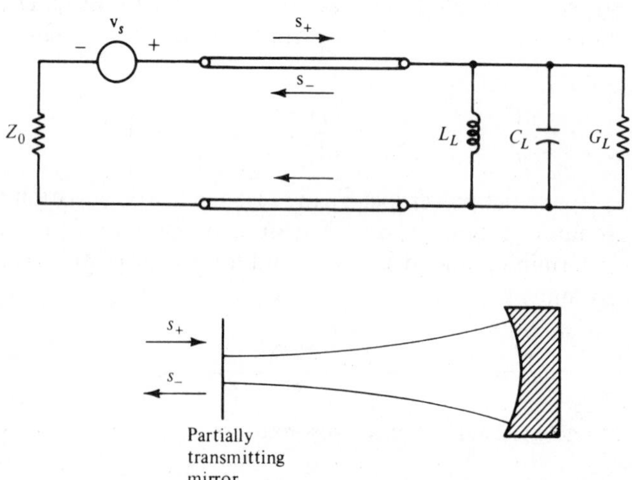

Figure 7.4 Resonators with external excitation.

Since only one mode will be considered at a time, the subscript μ is not really required. (The generalization to more modes will be obvious to the reader after reading Section 7.4 referring to two guides or modes.)

If the source is at frequency ω, $s_+ \propto \exp(j\omega t)$, then the response is at the same frequency and we find, from (7.21), that

$$a = \frac{\kappa s_+}{j(\omega - \omega_0) + [(1/\tau_0) + (1/\tau_e)]} \tag{7.22}$$

We shall now show that κ and τ_e are related. This could be done by setting up the equations for the *GLC* circuit of Fig. 7.4 as coupled to the source via the transmission line. Aside from the fact that such a procedure would be more cumbersome, the derivation would be model dependent and thus would lack generality. Instead, one may use a general property of Maxwell's equations for lossless media that is shared by other physical equations, like the equations of the propagation of sound (i.e., the property of time reversibility). Time reversibility implies the following: Given a set of mathematical functions that are solutions of the system; then these can be "run backward" in time, t is changed into $-t$, the power flow is reversed (e.g., in the case of an electromagnetic wave, if $\boldsymbol{E}$ is retained, $\boldsymbol{H}$ is changed into $-\boldsymbol{H}$), and the resulting mathematical functions remain solutions of the system. This property has been used in Section 3.3 to derive relations for the scattering matrix, and in Section 6.8 for a convenient form of the mode-orthogonality relations. We apply time reversibility to the present case with $1/\tau_0 = 0$ (i.e., with no internal loss) to derive an expression for κ. If there is no source, $s_+ = 0$, the mode decays at the rate $1/\tau_e$, because the wave s_- traveling away from the resonator in the connecting guide carries away power. From (7.17) and energy conservation

$$\frac{d}{dt}|a|^2 = -\frac{2}{\tau_e}|a|^2 = -|s_-|^2 \tag{7.23}$$

Next consider the time-reversed solution. The wave traveling away from the resonator s_- is converted into an incident wave s_+. Furthermore, the energy in the resonator builds up, rather than decays, with the time dependence $\exp[+(2/\tau_e)t]$. The time-reversed solution of the positive frequency amplitude $a_+(t)$ becomes a negative frequency amplitude $a_-(t)$, because the time dependence $\exp(j\omega t)$ has become $\exp(-j\omega t)$. In keeping with our convention, which deals with "positive-frequency" amplitudes only, we switch from $a_+(t)$ to $a_-(t)$ before reversing time, so that the equation obeyed by the time-reversed solution is given by (7.21) still containing $j\omega_0 a$ on the right-hand side, that is, is written for the positive-frequency amplitude of the time-reversed solution.

Denote the positive frequency amplitude of the time-reversed solution by $\tilde{a}$. It is growing, instead of decaying, and from (7.23),

$$\frac{d}{dt}|\tilde{a}|^2 = \frac{2}{\tau_e}|\tilde{a}|^2 \tag{7.24}$$

The time-reversed solution is *driven* by the incident wave $\tilde{s}_+$ at frequency ω_0, and grows at the rate $1/\tau_e$. Thus the frequency of the drive is

$$\omega = \omega_0 - \frac{j}{\tau_e}$$

Introducing this frequency into (7.22) with $1/\tau_0 = 0$, we find that

$$\tilde{a} = \frac{\kappa \tilde{s}_+}{2/\tau_e} \tag{7.25}$$

Further, since $|\tilde{s}_+|^2$ of the time-reversed solution is equal to $|s_-|^2$ of the original solution, and $|\tilde{a}| = |a|$ at $t = 0$, one has, from (7.23),

$$|\tilde{s}_+|^2 = \frac{2}{\tau_e}|a|^2 = \frac{2}{\tau_e}|\tilde{a}|^2 \tag{7.26}$$

Combining (7.25) and (7.26) we find that

$$|\kappa| = \sqrt{\frac{2}{\tau_e}} \tag{7.27}$$

The phase of κ can be disposed of by noting that the phase of a relative to s_+ can be defined arbitrarily. Thus (7.21) becomes

$$\frac{da}{dt} = j\omega_0 a - \left(\frac{1}{\tau_0} + \frac{1}{\tau_e}\right)a + \sqrt{\frac{2}{\tau_e}}\, s_+ \tag{7.28}$$

This is the equation describing excitation of the resonator mode by an incident wave. The resonator mode is described by three parameters: its resonance frequency ω_0, the decay rate of the amplitude caused by internal losses $1/\tau_0$, and the decay rate of the amplitude due to the power leaking from the resonator, $1/\tau_e$.

The Reflection Coefficient

It is of interest to develop an equation for s_-, so that one may evaluate the reflection coefficient $\Gamma = s_-/s_+$ for an arbitrary excitation. The system is linear, so that s_- is the sum of a term proportional to s_+ and a term proportional to a:

$$s_- = c_s s_+ + c_a a \tag{7.29}$$

We have already gone through the thought experiment of evaluating s_- for the case when $s_+ = 0$. Then, according to (7.23),

$$s_- = \sqrt{\frac{2}{\tau_e}}\, a = c_a a \tag{7.30}$$

so that

$$c_a = \sqrt{\frac{2}{\tau_e}} \tag{7.31}$$

where we have fixed the phase factor of c_a to be unity by a choice of reference plane at which s_- is to be evaluated. The second coefficient c_s can be evaluated by energy conservation. Indeed, the net power flowing into the resonator must be equal to the rate of buildup of energy within the resonator added to the rate of energy dissipation:

$$|s_+|^2 - |s_-|^2 = \frac{d}{dt}|a|^2 + 2\left(\frac{1}{\tau_0}\right)|a|^2 \tag{7.32}$$

On the other hand, from (7.28) we have

$$\frac{d}{dt}|a|^2 = -2\left(\frac{1}{\tau_0} + \frac{1}{\tau_e}\right)|a|^2 + \sqrt{\frac{2}{\tau_e}}(a^* s_+ + a s_+^*) \tag{7.33}$$

Comparison of (7.32) and (7.33) gives

$$-\frac{2}{\tau_e}|a|^2 + \sqrt{\frac{2}{\tau_e}}(a^* s_+ + a s_+^*) = |s_+|^2 - |s_-|^2 \tag{7.34}$$

Eliminating a from (7.34) by means of (7.29) and (7.31), we obtain

$$c_s = -1 \tag{7.35}$$

and thus

$$s_- = -s_+ + \sqrt{\frac{2}{\tau_e}}\, a \tag{7.36}$$

Equations (7.28) and (7.36) are the fundamental equations of resonators coupled to one input "port" (waveguide or optical mode). As an example of their application we may compute the reflection coefficient as a function of drive frequency ω. The reflection coefficient of the resonator in the steady state [1] is obtained from (7.36) and (7.22):

$$\Gamma = \frac{s_-}{s_+} = \frac{(1/\tau_e) - (1/\tau_0) - j(\omega - \omega_0)}{(1/\tau_e) + (1/\tau_0) + j(\omega - \omega_0)} \tag{7.37}$$

This is a circle in the Γ plane (Fig. 7.5). At resonance

$$\Gamma_{\text{res}} = \frac{(1/\tau_e) - (1/\tau_0)}{(1/\tau_e) + (1/\tau_0)}$$

No reflection is experienced when $\tau_e = \tau_0$; the external Q, Q_{ext}, is equal to the unloaded Q, Q_0. The power entering the cavity for a *given incident power* $|s_+|^2$ is maximized. This is the condition of critical match. When $|\omega - \omega_0| \gg (1/\tau_0) + (1/\tau_e)$, the excitation is far from resonance, $\Gamma = -1$, the condition of perfect

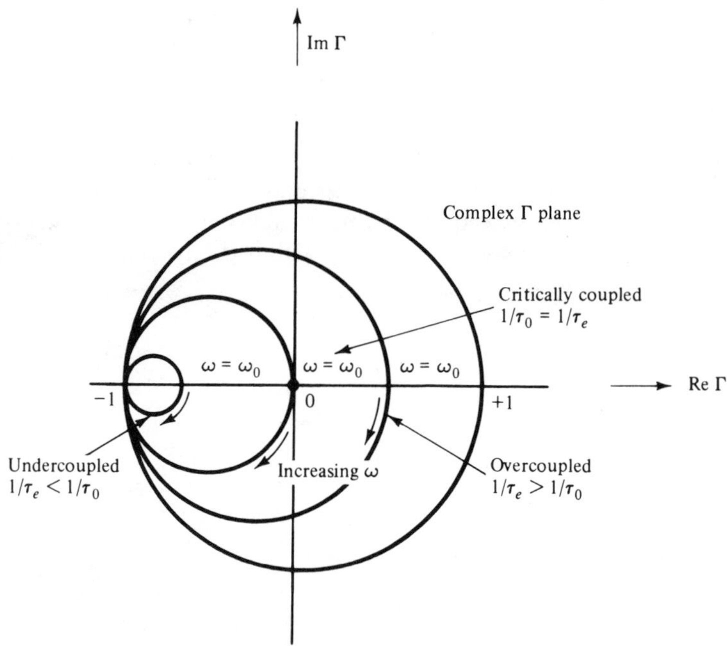

Figure 7.5 Loci of $\Gamma(\omega)$ in Γ plane for various values of τ_e/τ_0.

reflection (see Fig. 7.5). When $1/\tau_e > 1/\tau_0$, the rate of power escape from an initially excited resonator is greater than the rate of internal dissipation. This is the so-called overcoupled case. When $1/\tau_e < 1/\tau_0$ the resonator is said to be undercoupled.

The Admittance

A reflection coefficient can be related to a normalized impedance (or admittance) according to the definition (2.38)

$$\frac{Y}{Y_0} = \frac{1-\Gamma}{1+\Gamma} = \frac{\tau_e}{\tau_0}[1 + j(\omega - \omega_0)\tau_0] = \frac{Q_{\text{ext}}}{Q_0}\left(1 + 2j\,\frac{\omega - \omega_0}{\omega_0}\,Q_0\right) \qquad (7.38)$$

This is the admittance as a function of frequency of a parallel GLC, circuit (see Fig. 7.4) normalized to the admittance Y_0 of the feed-transmission line and expanded to first order in $\omega - \omega_0$ around the resonance frequency ω_0:

$$\frac{Y}{Y_0} \simeq \frac{G_L}{Y_0}\left(1 + 2j\,\frac{\omega - \omega_0}{\omega_0}\,\frac{\omega_0 C_L}{G_L}\right) = \frac{Q_{\text{ext}}}{Q_0}\left(1 + 2j\,\frac{\omega - \omega_0}{\omega_0}\,Q_0\right) \qquad (7.39)$$

Here we have used (7.15) and (7.20) for the Q's of the circuit. Thus the formalism developed for the mode of a general resonator at resonance frequency ω_0 leads to the admittance as a function of frequency of a parallel

GLC circuit. A choice of reference plane at a distance of a quarter wavelength from the present one would have resulted in the impedance of a series *RLC* circuit.

Equation (7.28) is convenient for the study of transient excitation of a resonator. Suppose that the incident wave is a step excitation at $t = 0$ at frequency ω.

$$s_+(t) = \begin{cases} S_+ e^{j\omega t}, & t > 0 \\ 0, & t < 0 \end{cases}$$

The response is the superposition of the steady-state response and the homogeneous solution to give $a = 0$ at $t = 0$:

$$a(t) = \frac{\sqrt{2/\tau_e}\, S_+}{j(\omega - \omega_0) + 1/\tau} [e^{j\omega t} - e^{(j\omega_0 - 1/\tau)t}] \tag{7.40}$$

where

$$\frac{1}{\tau} = \frac{1}{\tau_0} + \frac{1}{\tau_e}$$

The steady state is established after the transient buildup that lasts for a time of the order of τ.

7.3 COMPUTATION OF QUALITY FACTORS, LASER THRESHOLD, AND OUTPUT POWER

The advantage of the perturbation approach presented in the preceding section is that one may evaluate quite easily the different Q factors for a Fabry–Perot resonator for a given mirror transmissivity and internal loss. We turn first to the evaluation of the external Q, treating the resonator as loss-free, and then treat the loss separately. We concentrate on the resonator of Fig. 7.6. At resonance, a partially standing wave exists inside the resonator, which would be a perfect standing wave if the mirrors were totally reflecting. Then the two counter-propagating traveling waves, of which the standing wave is composed, would be of equal amplitudes. The energy W in the resonator is, as defined,

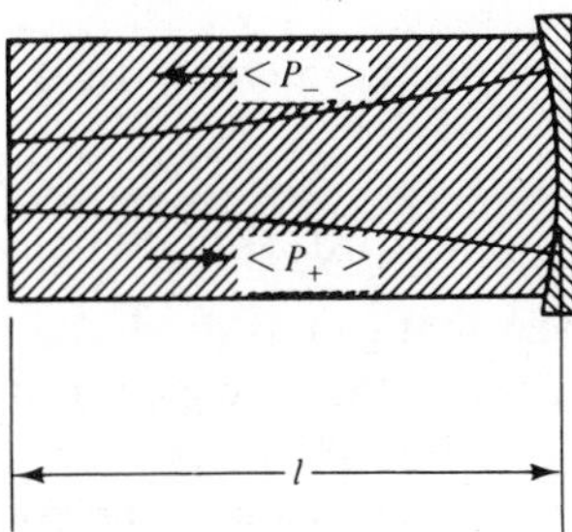

Figure 7.6 Resonator filled with medium of index n.

$W = |a|^2$. In the limit of high reflectivity $|a|^2/2$ is the energy associated with each of the oppositely directed traveling waves. We shall treat the general case of a resonator filled with a medium of index $n(\omega)$. The group velocity $v_g = 1/(d\beta/d\omega)$, where $\beta = \omega n/c$ gives the ratio of power to energy storage per unit length as derived in Section 6.9. Thus, the powers in the two counter-traveling waves, in the limit when the reflectivity of the mirrors is high, is approximately (compare 6.96)

$$\langle P_{\pm} \rangle = \frac{1}{2} \frac{|a|^2}{l} v_g \tag{7.41}$$

where l is the length of the resonator. The power P_e escaping through the partially transmitting mirror of transmissivity $t^2 \equiv T$ is

$$P_e = T\langle P_- \rangle = \frac{T}{2} \frac{|a|^2}{l} v_g \tag{7.42}$$

The external Q is thus

$$\frac{1}{Q_{\text{ext}}} = \frac{P_e}{\omega_0 W} = \frac{2}{\omega_0 \tau_e} = \frac{T v_g}{2\omega_0 l} \tag{7.43}$$

An interesting result can be obtained from (7.43) and (7.42); one may determine the ratio of power $\langle P_+ \rangle$ in one of the two counter-traveling waves inside the resonator at resonance to the power in the incident wave $|s_+|^2$ exciting the resonator. From (7.26) and using (7.43) and (7.41) with zero internal resonator loss,

$$|s_+|^2 = \frac{2}{\tau_e} |a|^2 = \frac{T v_g}{2l} |a|^2 = T\langle P_+ \rangle \tag{7.44}$$

The internal power is $1/T$ times the power of the incident wave.

Next we determine the unloaded Q. Suppose that the medium filling the resonator has a spatial decay rate α for the field, 2α for the power. The power lost per unit length is $2\alpha\langle P_{\pm} \rangle$ for each of the counter-traveling waves. If the loss is small as assumed, the powers $\langle P_{\pm} \rangle$ in the two counter-traveling waves are approximately constant along the length of the resonator, and the total loss, or power dissipated P_d, is equal to

$$P_d = 4\alpha l \langle P_{\pm} \rangle = 2\alpha |a|^2 v_g$$

Thus the inverse Q due to internal loss is

$$\frac{1}{Q_0} \equiv \frac{2}{\omega_0 \tau_0} = \frac{P_d}{\omega_0 W} = \frac{2\alpha v_g}{\omega_0} \tag{7.46}$$

Suppose next that gain is produced by some form of "pumping" [2] over a length l_g of the resonator. The gain medium, if acting by itself, would cause a rate of growth of the mode amplitude, $1/\tau_g$, due to the generated power P_g that can be derived by the same argument as the one used to obtain the rate of

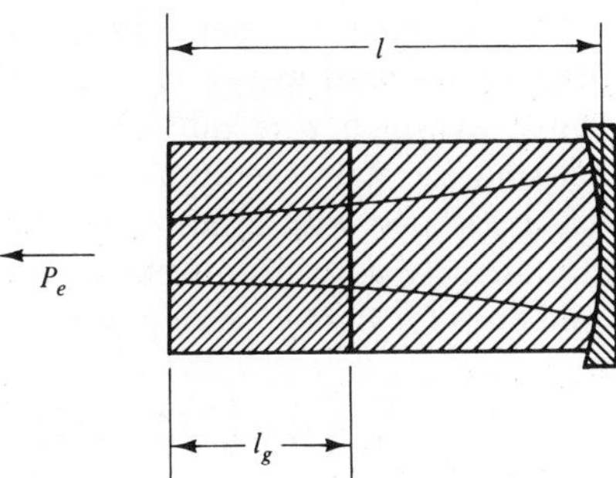

Figure 7.7 Resonator containing gain medium.

decay of the amplitude due to the dissipated power P_d. Suppose the spatial growth rate of the amplitude of a traveling wave is α_g. Then (Fig. 7.7)

$$P_g = 4\alpha_g l_g \langle P_{\pm} \rangle \tag{7.47}$$

The associated rate of growth $1/\tau_g$ is given by (compare (7.46))

$$\frac{2}{\omega_0 \tau_g} = \frac{P_g}{\omega_0 W} = \frac{2\alpha_g l_g v_g}{\omega_0 l} \tag{7.48}$$

The equation for the mode amplitude in the laser is now

$$\frac{da}{dt} = \left(j\omega_0 - \frac{1}{\tau_0} - \frac{1}{\tau_e} + \frac{1}{\tau_g}\right)a + \sqrt{\frac{2}{\tau_e}}\, s_+ \tag{7.49}$$

The solution of this equation with no drive, $s_+ = 0$, is

$$a = A \exp\left(j\omega_0 - \frac{1}{\tau_0} - \frac{1}{\tau_e} + \frac{1}{\tau_g}\right)t$$

If the laser is to oscillate in the steady state with no drive, $s_+ = 0$, one must have

$$\frac{1}{\tau_g} = \frac{1}{\tau_0} + \frac{1}{\tau_e}$$

or

$$\alpha_g = \frac{l}{l_g}\alpha + \frac{T}{4l_g}. \tag{7.50}$$

This is the gain coefficient which must be achieved to reach *threshold*, the gain level for self-starting of the oscillator. As threshold is exceeded, gain saturation (decrease of gain with increasing intensity in the medium) will assure that the value (7.50) of α_g is maintained at the operating power level of the laser. A common form of gain saturation is [2]

$$\alpha_g(I) = \frac{\alpha_g^0}{1 + I/I_s} \tag{7.51}$$

where α_g^0 is the gain at small signal levels ($I \to 0$), I is the intensity (in the forward and backward waves in a resonator, and I_s is the saturation intensity, the intensity at which α_g is half of its value at zero intensity. Suppose that the beam passing the gain medium has cross-sectional area A_g. We ignore the subtleties associated with a nonuniform intensity across the beam profile and treat the intensity as constant across the area A_g. Since the gain must obey (7.50), one finds for the two-way power inside the resonator,

$$IA_g = I_s A_g \left(2\alpha_g^0 - \frac{l}{l_g} 2\alpha - \frac{T}{2l_g}\right) \Big/ \left(\frac{l}{l_g} 2\alpha + \frac{T}{2l_g}\right)$$

The power P_t transmitted through the mirror and escaping the resonator is $\frac{1}{2}T$ times this amount. In the limit when $2\alpha l \ll T$ and $2\alpha_g^0 l_g \gg T/2$, we have for the power P_t from the laser:

$$P_t = I_s A_g 2\alpha_g^0 l_g$$

It is proportional to the small-signal gain and to the saturation intensity. The preceding analysis shows how the parameters of a resonator can be identified from the parameters characterizing loss, gain, and mirror transmission. It also showed how laser operation, a phenomenon that is nonlinear since the gain depends on the intensity in the resonator, can be described by the very same set of equations. An example of the use of the same formalism to describe another nonlinear phenomenon, injection locking of an oscillator, is relegated to Appendix 7A.

7.4 THE TRANSMISSION RESONATOR

If a resonator is connected to two guides, or power is coupled in and out at two mirrors as in the Fabry–Perot transmission resonator, then (7.21) is modified to

$$\frac{da}{dt} = j\omega_0 \mathrm{a} - \left(\frac{1}{\tau_0} + \frac{1}{\tau_{e1}} + \frac{1}{\tau_{e2}}\right)\mathrm{a} + \kappa_1 \mathrm{s}_{+1} + \kappa_2 \mathrm{s}_{+2} \tag{7.52}$$

Here $1/\tau_{e1}$ and $1/\tau_{e2}$ express the contribution to the mode decay of the free-running resonator due to the power escaping into each of the two guides. The same argument as before applied to one guide at the time, with the other guide disconnected gives

$$\kappa_1 = \sqrt{\frac{2}{\tau_{e1}}}, \qquad \kappa_2 = \sqrt{\frac{2}{\tau_{e2}}} \tag{7.53}$$

The power transmitted to guide 2 from guide 1 is

$$|s_{-2}|^2 = \frac{2|\mathrm{a}|^2}{\tau_{e2}} = \frac{(4\tau^2/\tau_{e1}\tau_{e2})|\mathrm{s}_{+1}|^2}{(\omega - \omega_0)^2\tau^2 + 1} \tag{7.54}$$

where $1/\tau = (1/\tau_{e1}) + (1/\tau_{e2}) + (1/\tau_0)$. When there is no loss, $1/\tau = (1/\tau_{e1}) + (1/\tau_{e2})$, and the transmitted power is at resonance, $\omega = \omega_0$:

$$|s_{-2}|^2 = \frac{4/\tau_{e1}\tau_{e2}}{(1/\tau_{e1}) + (1/\tau_{e2})]^2}|s_{+1}|^2 \tag{7.55}$$

Total transmission occurs for a symmetric resonator $1/\tau_{e1} = 1/\tau_{e2}$. When the couplings to the two-ports are not the same, the transmission is less. The present derivation checks with the high-finesse approximation for the transmission of a Fabry–Perot interferometer at, and near, resonance of one of its modes, (3.49).

The reflected waves s_{-1} and s_{-2} are evaluated using a generalization of (7.36):

$$s_{-1} = -s_{+1} + \sqrt{\frac{2}{\tau_{e1}}}\,a \tag{7.56}$$

and similarly for s_{-2}.

7.5 COUPLING OF TWO RESONATOR MODES

The simple formalism for the time evolution of a resonant mode is particularly well suited for the description of coupling between two resonant modes. Examples of such coupled modes abound. Figure 7.8 shows two coupled *LC* circuits and two optical resonators coupled by the partially transmitting mirror that is common to each of the two separate resonators. The coupling of quantum mechanical energy levels by a perturbation of the wave functions is a further example; the coupling of two penduli suspended from a string is another one.

Consider the equation of motion of the amplitudes a_1 and a_2 of the modes of two uncoupled lossless resonators of natural frequencies ω_1 and ω_2:

$$\frac{da_1}{dt} = j\omega_1 a_1 \tag{7.57}$$

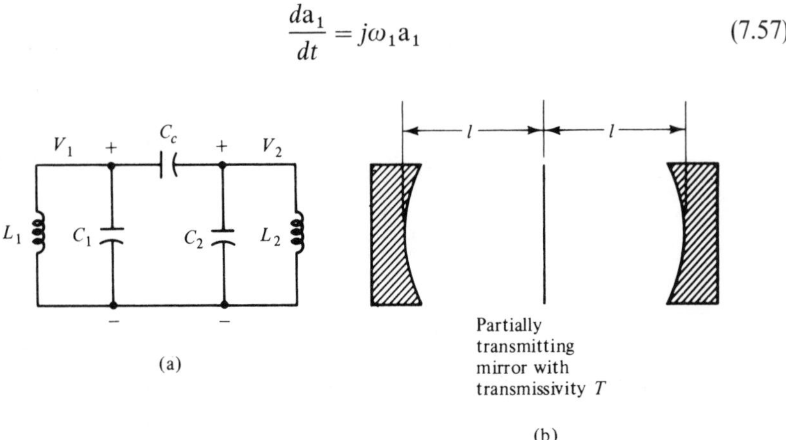

Figure 7.8 Examples of coupled resonators.

$$\frac{da_2}{dt} = j\omega_2 a_2 \tag{7.58}$$

Suppose that the modes are coupled through some small perturbation of the system, such as the small connecting capacitor in Fig. 7.8a, or by a change of the totally reflecting mirror between the two resonators in Fig. 7.8b to a partially transmitting mirror. This change can be described by

$$\frac{da_1}{dt} = j\omega_1 a_1 + \kappa_{12} a_2 \tag{7.59}$$

$$\frac{da_2}{dt} = j\omega_2 a_2 + \kappa_{21} a_1 \tag{7.60}$$

where κ_{12} and κ_{21} are the coupling coefficients. Weak coupling means that $|\kappa_{12}| \ll \omega_1$ and $|\kappa_{21}| \ll \omega_2$. The simple format of the coupled equations (7.59) and (7.60) is not immediately self-evident. Indeed, offhand one might expect that κ_{12} and κ_{21} would be linear integrodifferential operators. However, if the coupling is weak, $|\kappa_{12} a_2|$ is, on the average, small compared with $|\omega_1 a_1|$; then the coupling will affect the time evolution of a_1 and a_2 only when ω_1 is close to ω_2. The time dependence will thus be roughly like $\exp\{j[(\omega_1 + \omega_2)/2]t\}$. Any differentiation of such a time dependence produces the factor $j[(\omega_1 + \omega_2)/2]$ and additional, much smaller, terms. Therefore, any differentiation involved in the coupling term can be replaced by $j(\omega_1 + \omega_2)/2$, an integration by $-2j/(\omega_1 + \omega_2)$; κ_{12} and κ_{21} can be treated as complex numbers rather than operators.

Energy conservation imposes a restriction on κ_{12} and κ_{21}. The time rate of change of energy, which must vanish, is derived from (7.59) and (7.60):

$$\begin{aligned}\frac{d}{dt}(|a_1|^2 + |a_2|^2) &= a_1 \frac{da_1^*}{dt} + a_1^* \frac{da_1}{dt} + a_2 \frac{da_2^*}{dt} + a_2^* \frac{da_2}{dt} \\ &= a_1^* \kappa_{12} a_2 + a_1 \kappa_{12}^* a_2^* + a_2^* \kappa_{21} a_1 + a_2 \kappa_{21}^* a_1^* \\ &= 0\end{aligned}$$

Because the initial amplitudes and phases of a_1 and a_2 can be set arbitrarily, the coupling coefficients must be related by

$$\kappa_{12} + \kappa_{21}^* = 0 \tag{7.61}$$

Transient Response

We solve for the natural frequencies of the coupled systems. One assumes a dependence $\exp(j\omega t)$, and then obtains two homogeneous equations in the amplitudes a_1 and a_2 from (7.59) and (7.60). The determinant has to vanish in order to obtain a nontrivial solution, thereby yielding the roots

$$\omega = \frac{\omega_1 + \omega_2}{2} \pm \sqrt{\left(\frac{\omega_1 - \omega_2}{2}\right)^2 + |\kappa_{12}|^2} \equiv \frac{\omega_1 + \omega_2}{2} \pm \Omega_0$$

The two frequencies of the coupled system are "forced apart" by the coupling. In particular, when $\omega_1 = \omega_2$, the difference between the two natural frequencies of the coupled modes is $2\Omega_0 = 2|\kappa_{12}|$. Suppose, initially, that at $t = 0$, $a_1(0)$ and $a_2(0)$ are specified. Then, using the two solutions with the time dependences

$$\exp\left(j\,\frac{\omega_1 + \omega_2}{2}\,t\right) \exp\,(\pm j\Omega_0\, t)$$

one finds

$$\mathrm{a}_1(t) = \left[\mathrm{a}_1(0)\left(\cos\,\Omega_0\, t - j\,\frac{\omega_2 - \omega_1}{2\Omega_0}\,\sin\,\Omega_0\, t\right) + \frac{\kappa_{12}}{\Omega_0}\,\mathrm{a}_2(0)\,\sin\,\Omega_0\, t\right] \cdot e^{j[(\omega_1 + \omega_2)/2]t} \tag{7.62}$$

$$\mathrm{a}_2(t) = \left[\frac{\kappa_{21}}{\Omega_0}\,\mathrm{a}_1(0)\,\sin\,\Omega_0\, t + \mathrm{a}_2(0)\left(\cos\,\Omega_0\, t - j\,\frac{\omega_1 - \omega_2}{2\Omega_0}\,\sin\,\Omega_0\, t\right)\right] \cdot e^{j[(\omega_1 + \omega_2)/2]t} \tag{7.63}$$

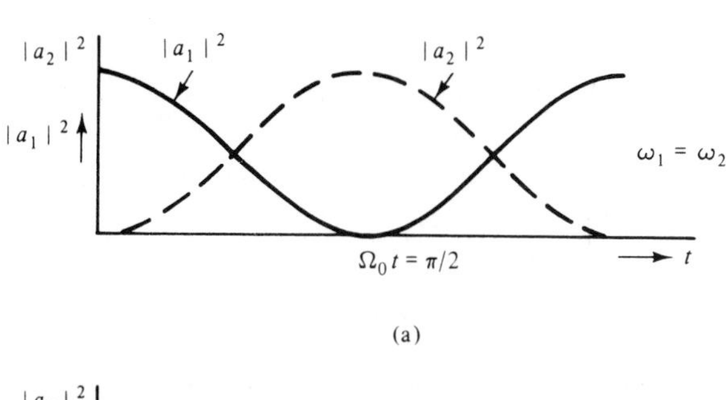

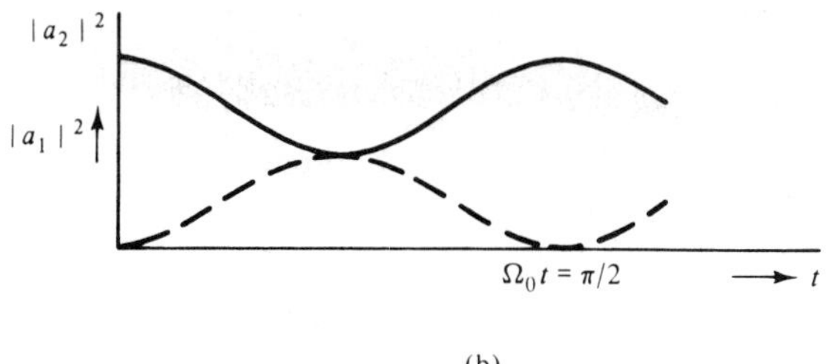

Figure 7.9 Energy in modes 1 and 2 as a function of time: (a) symmetric case; (b) general case.

Consider the case when $a_2(0) = 0$ and $\omega_1 = \omega_2$. Mode 1 is fully excited at $t = 0$, but at $\Omega_0 t = \pi/2$ all the excitation is in mode 2. At $\Omega_0 t = \pi$, the excitation has returned to mode 1 and mode 2 is unexcited. The process repeats itself. The excitation is transferred back and forth with the frequency $2\Omega_0 = 2|\kappa_{12}|$, which is determined by the coupling. Figure 7.9 shows the energy in modes 1 and 2 as a function of time. When $\omega_1 \neq \omega_2$, the transfer from mode 1 to mode 2 is not complete, as shown in Fig. 7.9.

Evaluation of the Coupling Coefficients

The coupling coefficient κ_{12} can be evaluated for any physical system using energy considerations. Consider first the two LC circuits, in the absence of the coupling capacitor C_c. Then

$$\omega_1 = \frac{1}{\sqrt{L_1 C_1}}, \qquad \omega_2 = \frac{1}{\sqrt{L_2 C_2}}$$

Through the coupling capacitor, power is supplied from circuit 1 to circuit 2. We have, from (7.60),

$$\frac{d|\mathrm{a}_2|^2}{dt} = \kappa_{21}\mathrm{a}_1\mathrm{a}_2^* + \kappa_{21}^*\mathrm{a}_1^*\mathrm{a}_2 \tag{7.64}$$

From the physical circuit, we conclude that the power P_{21} flowing from circuit 1 to circuit 2 is

$$P_{21} = \left\langle v_2 C_c \frac{d}{dt}(v_1 - v_2)\right\rangle \tag{7.65}$$

where the angular brackets indicate a time average over a few cycles of the resonance at ω_2. If one introduces the complex voltage envelope quantities $\mathrm{V}_1(t)$ and $\mathrm{V}_2(t)$ so that $v_1(t)$ can be written

$$v_1(t) = \tfrac{1}{2}[\mathrm{V}_1(t)e^{j\omega_1 t} + \mathrm{V}_1^*(t)e^{-j\omega_1 t}] \tag{7.66}$$

and a similar expression for $v_2(t)$, one obtains for P_{21}:

$$P_{21} = \tfrac{1}{4}(j\omega_1 C_c \mathrm{V}_1\mathrm{V}_2^* e^{j(\omega_1-\omega_2)t} - j\omega_1 C_c \mathrm{V}_1^*\mathrm{V}_2 e^{-j(\omega_1-\omega_2)t}) \tag{7.67}$$

Here we ignore the contribution of $d\mathrm{V}_1/dt$ compared with $j\omega_1\mathrm{V}_1$. Introduction of the definition (7.10) for a_1 and a_2 gives

$$P_{21} = \frac{1}{2}\left[\left(\frac{j\omega_1 C_c}{\sqrt{C_1 C_2}}\mathrm{a}_1\mathrm{a}_2^* - \frac{j\omega_1 C_c}{\sqrt{C_1 C_2}}\mathrm{a}_1^*\mathrm{a}_2\right)\right. \tag{7.68}$$

Comparison with (7.64) yields

$$\kappa_{21} = \frac{j\omega_1 C_c}{2\sqrt{C_1 C_2}}$$

A similar derivation leads to

$$\kappa_{12} = \frac{j\omega_2 C_c}{2\sqrt{C_1 C_2}}$$

Because $\kappa_{12} + \kappa_{21}^* = 0$, we must interpret ω_1 and ω_2 in the equations above as the arithmetic average $(\omega_1 + \omega_2)/2$, or the geometric average $\sqrt{\omega_1 \omega_2}$, a correction that is of the same order as the terms that have been neglected in the analysis (e.g., $d\mathrm{V}_1/dt$ compared with $j\omega_1 \mathrm{V}_1$).

We can check the analysis against the special symmetric case $L_1 = L_2 = L$, $C_1 = C_2 = C$, and $\omega_1 = \omega_2 = \omega_0$. The symmetric mode, $\mathrm{V}_1 = \mathrm{V}_2$, leaves the coupling capacitor unexcited and thus has the frequency

$$\omega_s = \frac{1}{\sqrt{LC}}$$

The antisymmetric mode $\mathrm{V}_1 = -\mathrm{V}_2$ is equivalent to the circuit shown in Fig. 7.10 and has the frequency

$$\omega_a = \frac{1}{\sqrt{L(C + 2C_c)}}$$

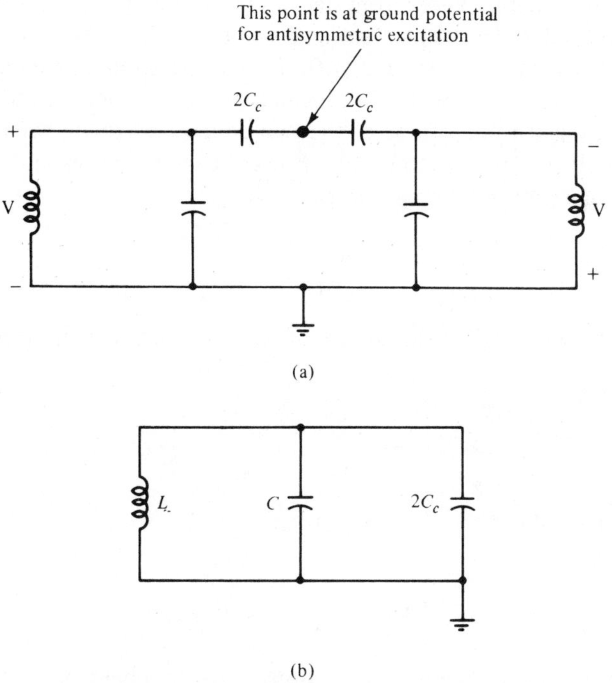

Figure 7.10 Circuit for evaluation of resonance of antisymmetric eigenmode of coupled system: (a) separation of coupling capacitor; (b) equivalent circuit of antisymmetrically excited system.

The split of the resonance frequencies is, for $C_c \ll C$

$$\omega_s - \omega_a = \omega_s \frac{C_c}{C}$$

From the coupling of modes analysis we find the split to be equal to $2|\kappa_{12}| = \omega_0(C_c/C)$, which checks. There is, however, a discrepancy between the exact analysis and the coupling of modes analysis, if one identifies the uncoupled systems 1 and 2 as those remaining after the coupling capacitor C_c has been removed. The coupling of modes analysis yields

$$\omega_{s,a} = \frac{1}{\sqrt{LC}}\left(1 \pm \frac{1}{2}\frac{C_c}{C}\right)$$

The average frequency $(\omega_s + \omega_a)/2$ is predicted incorrectly. This is a common difficulty of coupling of modes that predicts "splitting" of the frequencies easily and correctly to first order in coupling, but from which the average frequency after coupling must be extracted with care and sophistication. In the present example we should have used $C + C_c$ for the capacitance of the uncoupled circuits. In most cases, however, we are interested in the splitting due to coupling and not in the average frequency, because the splitting determines the rate at which energy passes from one circuit to the other, and vice versa.

Next, consider the coupled resonators of Fig. 7.8b. Assume them to be symmetric. The equation of motion (7.60) for mode 2 should be compared with (7.28) for the resonator excited by an incident wave s_+ from the outside. In the present case s_+ is formed of one of the two counter-traveling waves in resonator 1 that make up the standing wave. Since the energy in resonator 1 is $|a_1|^2$, the power in one wave is (compare section 7.3)

$$\frac{\frac{1}{2}|a_1|^2}{l}\,c$$

which must be set equal to the square of the wave incident upon resonator 2, $|s_+|^2$. Thus

$$|\kappa_{21}|a_1 = \sqrt{\frac{2}{\tau_e}}\,s_+ = \sqrt{\frac{2}{\tau_e}}\sqrt{\frac{c}{2l}}\,a_1$$

and thus, using (7.43) with $v_g = c$, we obtain

$$|\kappa_{21}| = \sqrt{\frac{c}{\tau_e l}} = \sqrt{T}\,\frac{c}{2l} \tag{7.69}$$

The phase of κ_{21} depends upon the definition of the phase of a_1. If a definite phase is picked by relating a_1 to the electric field in the resonator, the phase κ_{21} is fixed. In most applications, however, the phase is irrelevant, and κ_{21} can be picked real and positive.

7.6 COUPLING OF MODES IN SPACE [3–6]

The preceding section treated coupling between two modes in time. The initial conditions were set up at $t = 0$, and the modes developed from these initial conditions in a "transient" manner.

Another problem of great practical interest is coupling of modes* in space in the steady state. For example, two optical waveguides are coupled to each other via their fringing fields (see Fig. 7.11). A wave set up initially in one guide is transferred to the other guide. Because the transfer can be controlled electrically, this mechanism can be used for switching of guided optical radiation. Another coupling is effected between forward and backward waves by periodic perturbations on an optical waveguide. These perturbations can be produced in optical waveguides by integrated-optics fabrication methods and can be used to build the equivalent of a mirror into an optical waveguide without interrupting the waveguide physically. Nonlinear optical phenomena couple waves at different frequencies. This process is also akin to the coupling of modes analysis presented here.

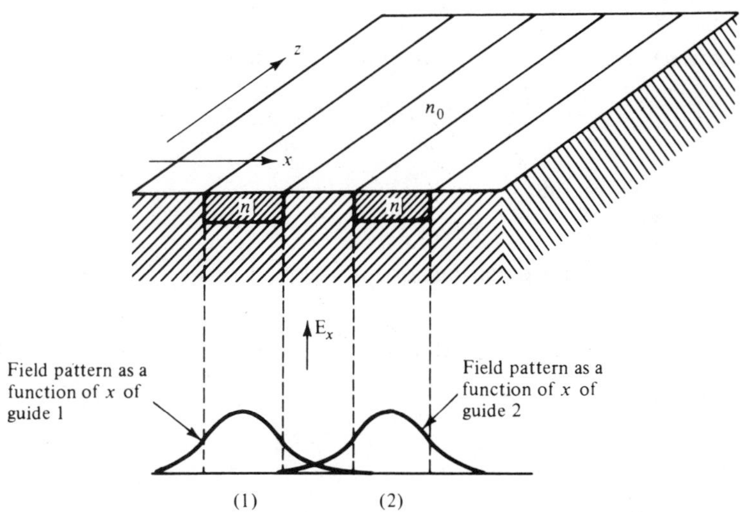

Figure 7.11 Cross-sectional view of index profile forming two optical guides.

Consider two waves a_1 and a_2, of modes 1 and 2, which, in the absence of coupling, have propagation constants β_1 and β_2. They obey the equations

$$\frac{da_1}{dz} = -j\beta_1 a_1 \tag{7.70}$$

*The term "coupling of modes" is deeply ingrained in the literature and can be traced back to J. R. Pierce's original work on the subject [3]. Actually, in coupling of modes in space, waves are coupled, each mode of a waveguide consisting of a forward and a backward wave.

$$\frac{da_2}{dz} = -j\beta_2 a_2 \tag{7.71}$$

Suppose next that the two waves are weakly coupled by some means, so that a_1 is affected by a_2 and a_2 is affected by a_1. Then the equations become

$$\frac{da_1}{dz} = -j\beta_1 a_1 + \kappa_{12} a_2 \tag{7.72}$$

$$\frac{da_2}{dz} = -j\beta_2 a_2 + \kappa_{21} a_1 \tag{7.73}$$

If power is to be conserved, there are restrictions imposed on κ_{12} and κ_{21}. Weak coupling implies that we may evaluate the power in the two waves disregarding the coupling. We normalize a_1 and a_2 so that the power in the modes is $|a_1|^2$ and $|a_2|^2$. Because the waves may carry power in opposite directions, we must distinguish the directions of power flow by a sign. We define $p_{1,2} = \pm 1$ depending upon whether the power flow is in the plus or minus z direction. The net power P is

$$P = p_1 |a_1|^2 + p_2 |a_2|^2 \tag{7.74}$$

Power conservation requires that the power be independent of distance z.

$$\frac{dP}{dz} = p_1 \frac{d|a_1|^2}{dz} + p_2 \frac{d|a_2|^2}{dz} = 0 \tag{7.75}$$

from which it follows that

$$p_1 \kappa_{12} + p_2 \kappa_{21}^* = 0 \tag{7.76}$$

The determinantal equation for an assumed exp $(-j\beta z)$ dependence is, from (7.72) and (7.73),

$$(\beta - \beta_1)(\beta - \beta_2) + \kappa_{12}\kappa_{21} = 0 \tag{7.77}$$

with the solution

$$\beta = \frac{\beta_1 + \beta_2}{2} \pm \sqrt{\left(\frac{\beta_1 - \beta_2}{2}\right)^2 - \kappa_{12}\kappa_{21}} \tag{7.78}$$

For waves carrying power in the same direction, $p_1 p_2 = +1$, $\kappa_{12}\kappa_{21} = -|\kappa_{12}|^2$, and β is always real. For $p_1 p_2 = -1$ (i.e., waves carrying power in opposite directions), $\kappa_{12}\kappa_{21} = |\kappa_{12}|^2$ and β is complex for

$$\left|\frac{\beta_1 - \beta_2}{2}\right| < |\kappa_{12}|$$

Note that appreciable coupling can occur only if $|\beta_1 - \beta_2|$ is of order $|\kappa_{12}|$, which is small compared with $|\beta_1|$ and $|\beta_2|$ (weak-coupling assumption). Thus $\beta_1 \simeq \beta_2$ and the phase velocities of the two waves must be of the same sign. Nonetheless, the power flow can be in opposite directions ($p_1 p_2 = -1$) if the group velocities are in opposite directions. This is the case of coupling of forward and backward waves by a periodic structure as analyzed in the next chapter. The power flows can be in opposite directions also for codirectional group velocities if the energy assigned to one of the waves is negative [4, 7, 8]. The concept of negative energy is a common one in plasma physics, but is also

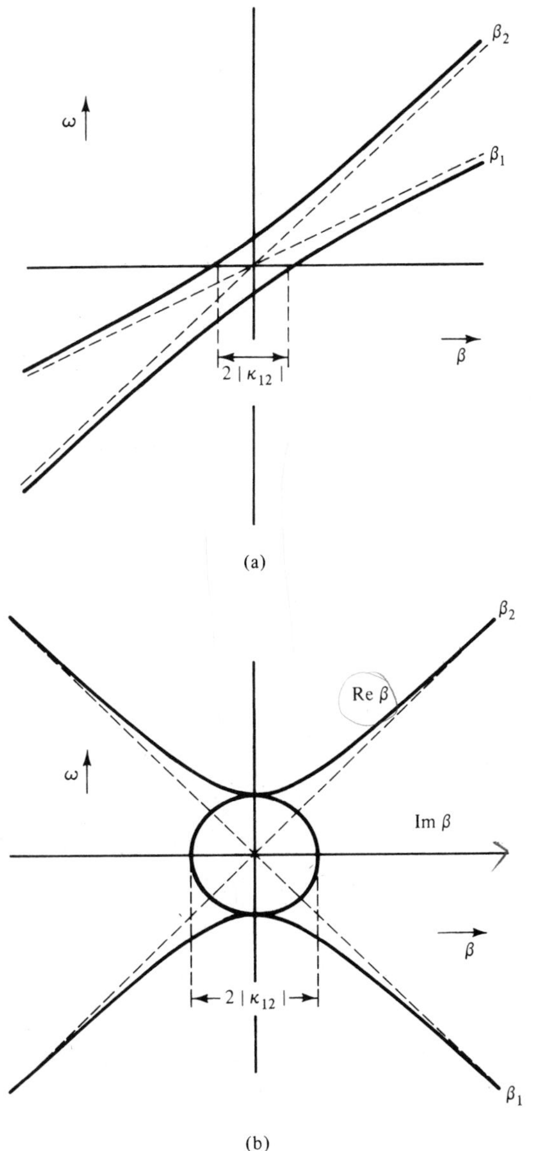

Figure 7.12 Dispersion diagram for coupling of modes (linear dependence of β_1 and β_2 upon ω is assumed).

useful in nonlinear optics with some modification. We shall bring it up in Chapter 13.

Suppose that both β_1 and β_2 are frequency dependent, with different group velocities. For each ω there is a pair of values β_1 and β_2 which yield the propagation constant β of the coupled modes via (7.78). If both waves have the same direction of group velocity, the dispersion curves β_1 and β_2 may cross as shown in Fig. 7.12a and $\beta = (\beta_1 + \beta_2)/2 \pm |\kappa_{12}|$. The propagation constant β approaches asymptotically the unperturbed propagation constants far from the crossover.

Another case is the one of opposite group velocities as shown in Fig. 7.12b. We shall see in Chapter 8 how the two dispersion curves β_1 and β_2 of the two waves can be made to cross. Suffice it to point out here that a crossing of two such curves leads to exponentially growing and decaying solutions.

Consider the case of codirectional, positive, group velocities, $p_1 = p_2 = +1$. Suppose that waves $a_1(0)$ and $a_2(0)$ are launched at $z = 0$. Then the solution is analogous to the coupling-of-modes-in-time solution, (7.62) and (7.63):

$$a_1(z) = \left[a_1(0)\left(\cos \beta_0 z + j\,\frac{\beta_2 - \beta_1}{2\beta_0} \sin \beta_0 z \right) + \frac{\kappa_{12}}{\beta_0}\, a_2(0) \sin \beta_0 z \right] \cdot e^{-j[(\beta_1 + \beta_2)/2]z} \tag{7.79}$$

$$a_2(z) = \left[\frac{\kappa_{21}}{\beta_0}\, a_1(0) \sin \beta_0 z + a_2(0)\left(\cos \beta_0 z + j\,\frac{\beta_1 - \beta_2}{2\beta_0} \sin \beta_0 z \right) \right] \cdot e^{-j[(\beta_1 + \beta_2)/2]z} \tag{7.80}$$

where

$$\beta_0 = \sqrt{\left(\frac{\beta_1 - \beta_2}{2}\right)^2 + |\kappa_{12}|^2} \tag{7.81}$$

The dependence upon z of the waves excited at $z = 0$ is of the same form as the transient dependence upon t of the resonator mode amplitudes. If $a_2(0) = 0$, then Fig. 7.9 gives the spatial dependence of $|a_1(z)|^2$ and $|a_2(z)|^2$, with $\Omega_0 t$ replaced by $\beta_0 z$. The solutions (7.79) and (7.80) underlie the operation of optical couplers, switches, and tunable filters, as discussed in the next section.

7.7 APPLICATIONS TO WAVEGUIDE COUPLERS

Several important applications of waveguide couplers have been published [9–13]. We shall discuss two: a tunable filter and the optical waveguide switch. Before we describe them analytically, a few words as to how coupled waveguide structures are made. Laboratory versions are made usually on $LiNbO_3$ because of its large linear electrooptic effect (Chapter 11) and the ease with which waveguides can be fabricated. The linear electrooptic effect enables

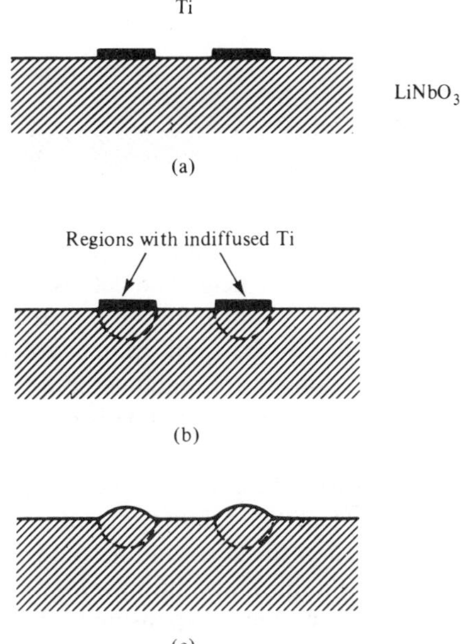

Figure 7.13 Ti indiffusion into $LiNbO_3$ for formation of waveguide.

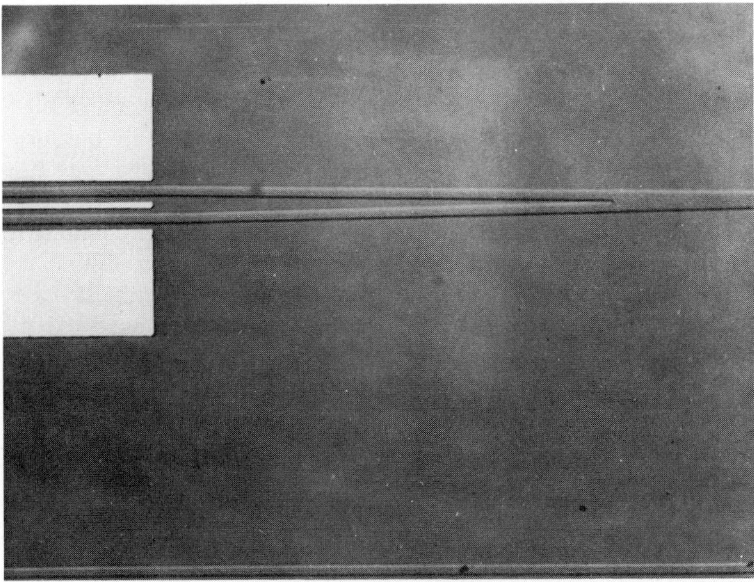

Figure 7.14 Waveguide "*Y*" formed by Ti indiffusion in $LiNbO_3$. Electrodes visible at right. (Photograph courtesy of Dr. F. J. Leonberger, M. I. T. Lincoln Laboratory.)

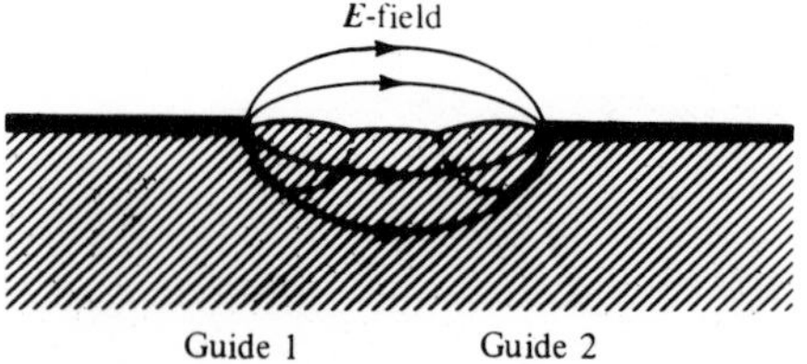

Figure 7.15 E field produced with voltage applied to electrodes.

one to change the index, and thus β_1 and/or β_2 of the two coupled waveguides, by an applied electric field. The details of this effect will be discussed later; all we need to know at this juncture is that the propagation constants β_1 and β_2 can be controlled by the application of voltages.

The waveguides are manufactured by Ti indiffusion. Where a waveguide is to be formed, a thin layer of Ti is deposited by photolithographic means on the $LiNbO_3$ crystal surface (Fig. 7.13a). The structure is heated up, which causes the Ti to diffuse into the $LiNbO_3$, producing cylindrical volumes with Ti interspersed in the $LiNbO_3$ structure (Fig. 7.13b). These regions have an index higher than the surrounding pure $LiNbO_3$ and form guides completely analogous to the fiber and slab guides discussed in the preceding section. The indiffusion causes local expansions which cause the surface to bulge (Fig. 7.13c). These can be observed under a microscope (Fig. 7.14). Finally, electrodes are deposited on top of the crystal surface. With a voltage applied to the electrodes the electric fields change β_1 and β_2 in the two guides (Fig. 7.15).

The Tunable Filter

The tunable filter, first proposed by H. F. Taylor [10] and constructed by Alferness and Cross [13], is based on the dispersion diagram of Fig. 7.12a. Two waveguides with different dispersion curves β_1 and β_2 are put into proximity over a length l (Fig. 7.16). With zero applied voltage they have the dispersion curves shown in Fig.7.17.

Suppose that guide 1 is excited at $z = 0$, the start of the coupling region; guide 2 is unexcited. The power transferred to guide 2 is, from (7.80),

$$|a_2(l)|^2 = \left|\frac{\kappa_{21}}{\beta_0}\right|^2 \sin^2 \beta_0 l \, |a_1(0)|^2 \tag{7.82}$$

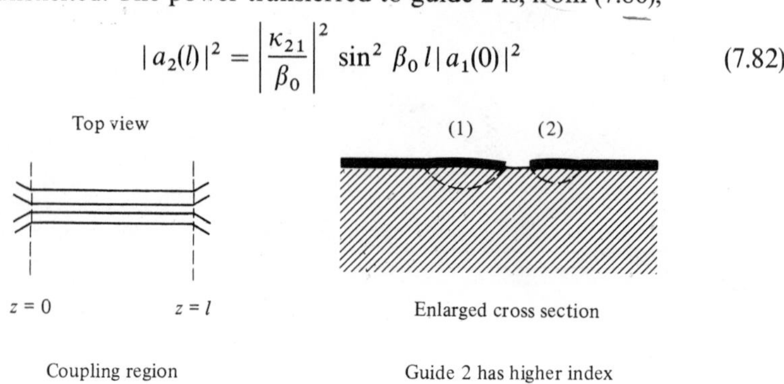

Figure 7.16 Two unequal coupled waveguides.

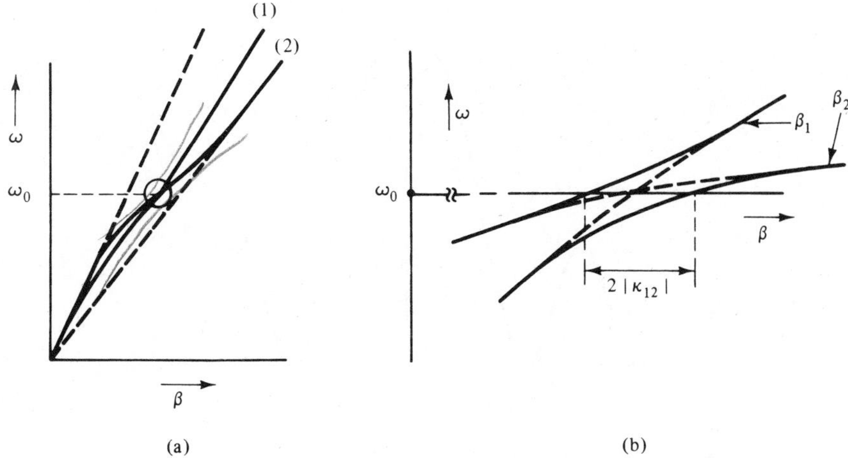

Figure 7.17 Dispersion diagram of the two unequal guides of Fig. 7.16: (a) exaggerated dispersion diagram; (b) dispersion diagram at crossover (encircled region of part a).

Suppose that $|\kappa_{21}|l$ is chosen so that full transfer occurs when $\beta_1 = \beta_2$ at frequency ω_0, $|\kappa_{21}|l = 3\pi/2$. Figure 7.18 shows how the power transfer efficiency changes as a function of frequency due to the fact that

$$\beta_1 - \beta_2 = \left(\frac{\partial \beta_1}{\partial \omega} - \frac{\partial \beta_2}{\partial \omega}\right)(\omega - \omega_0)$$

is a function of frequency.

The operation of the tunable filter is based on the fact that the propagation constants can be changed by an applied voltage. The crossover frequency ω_0 can be shifted and thus the transfer function is tunable. The coupling constant κ_{21} is independent of the applied voltage for all practical purposes. If $|\kappa_{21}|l$ is chosen to be $\pi/2$, the resulting transfer has lower sidelobes, but the width of the main lobe as a function of frequency is wider.

The Optical Waveguide Switch

A coupler of length l such that $|\kappa_{12}|l = \pi/2$ transfers the entire power from waveguide 1 to waveguide 2 when $\beta_1 = \beta_2$. If the waveguides are symmetric, the transfer is only weakly frequency dependent, because $\beta_1 = \beta_2$ at all frequencies, and $|\kappa_{12}|$ is generally not a strong function of frequency. With a voltage applied onto electrodes such as those shown in Fig. 7.15, the β's can be shifted in opposite directions, so that $\beta_1 \neq \beta_2$. When

$$\frac{\beta_1 - \beta_2}{2} = \sqrt{3}\,|\kappa_{12}| \tag{7.83}$$

no transfer occurs between the guides. This is the principle of the optical switch.

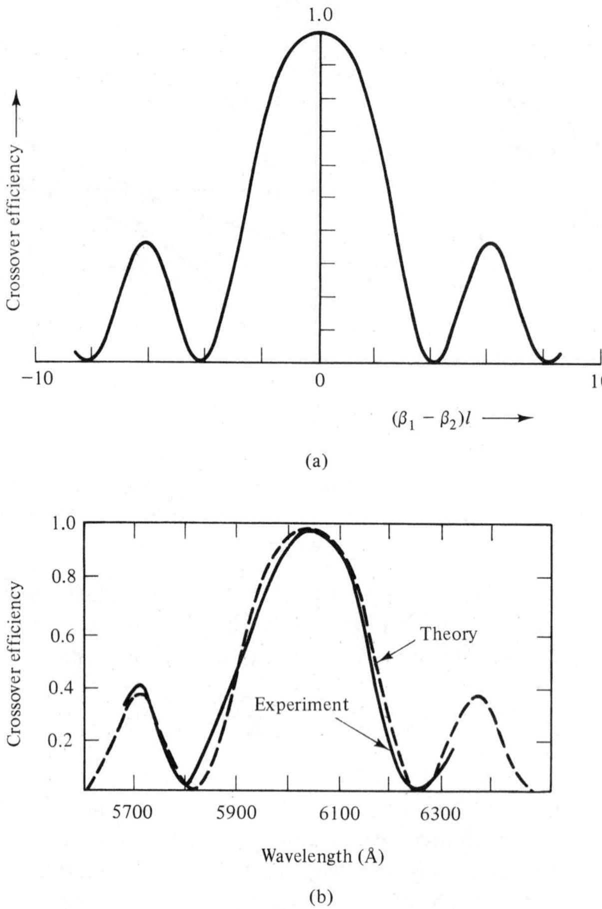

Figure 7.18 Power transfer between guides of Fig. 7.16 as a function of frequency: (a) theory; (b) experiment. (From R. C. Alferness and R. V. Schmidt, "Tunable optical waveguide directional coupler filter," *Appl. Phys. Lett.*, *33*, no. 2, 161–163, July 1978.)

A practical switch must achieve full transfer and extinction ratios better than 25 dB. Full transfer occurs only when $|\kappa_{12}|l = \pi/2$. Because the microstructure tolerances are difficult to control and $|\kappa_{12}|$ is a strong function of guide spacing (see Section 7.8), it is desirable to build into the structure flexibility so that adjustments can be made after fabrication. The $\Delta\beta$ coupler proposed by Kogelnik and Schmidt [11] is such a structure. It consists of two couplers back to back, with electrodes driven "push-pull" as shown in Fig. 7.19. Even when $|\kappa_{12}|l \neq \pi/2$, but slightly larger, full transfer can be effected by an applied voltage of appropriate amplitude. To prove this we use the solutions (7.79) and (7.80), and apply them to two couplers in cascade, of length $l/2$ each, with $\beta_1 - \beta_2$ reversed in the second section. The output at

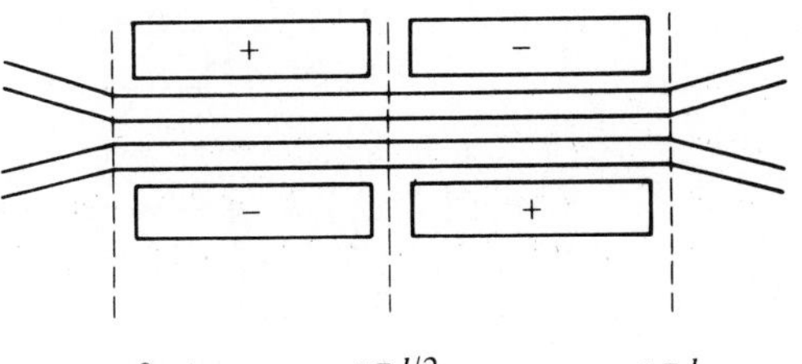

Figure 7.19 The $\Delta\beta$ coupler.

$z = l$ is found to be

$$a_1(l) = a_1(0)\left[\cos^2\left(\beta_0 \frac{l}{2}\right) - \left(\frac{2|\kappa_{12}|^2}{\beta_0^2} - 1\right)\sin^2\left(\beta_0 \frac{l}{2}\right)\right] \cdot \exp\left(-j\,\frac{\beta_1+\beta_2}{2}\,l\right) \tag{7.84}$$

$$a_2(l) = a_1(0)\, 2\,\frac{\kappa_{21}}{\beta_0}\sin\beta_0\frac{l}{2}\left[\cos\beta_0\frac{l}{2} - j\,\frac{\beta_1-\beta_2}{2\beta_0}\sin\beta_0\frac{l}{2}\right] \cdot \exp\left(-j\,\frac{\beta_1+\beta_2}{2}\,l\right) \tag{7.85}$$

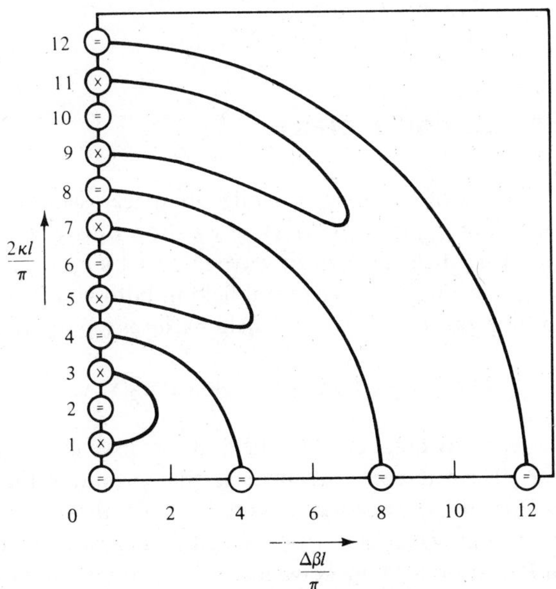

Figure 7.20 Switching diagram for a switched coupler with two sections of alternating $\Delta\beta$. The ⊗ sign marks the cross-state conditions and the ⊜ sign marks the parallel-state conditions. (From H. Kogelnik and R. V. Schmidt, "Switched directional couplers with alternating $\Delta\beta$," *IEEE J. Quantum Electron.*, *QE-12*, no. 7, 396–401, July 1976 © 1976 IEEE.)

Full power transfer is effected when $a_1(l) = 0$, that is, when

$$\cot\left(\beta_0 \frac{l}{2}\right) = \sqrt{\frac{\kappa^2 - \left(\dfrac{\beta_1 - \beta_2}{2}\right)^2}{\kappa^2 + \left(\dfrac{\beta_1 - \beta_2}{2}\right)^2}}$$

where

$$\beta_0 = \sqrt{\kappa^2 + \left(\frac{\beta_1 - \beta_2}{2}\right)^2}$$

and $|\kappa_{12}| \equiv \kappa$. No transfer occurs when

$$\beta_0 \frac{l}{2} = m\pi$$

These conditions are illustrated in Fig. 7.20. Kogelnik and Schmidt have constructed a diagram which shows the transfer (cross) states and the no-transfer (parallel) states in the plane of $2\kappa l/\pi$, $(\beta_1 - \beta_2)l/\pi$ (Fig. 7.20). The curves are loci of the full-transfer state (⊗) or no-transfer state (⊜). The regions between the curves correspond to mixed states.

7.8 THE COUPLING COEFFICIENT

The preceding analysis of coupling of modes in space assumed knowledge of the coupling coefficient κ_{12} $(= \pm\kappa_{21}^*)$. Here we show one way of evaluating it. Denote the normalized field pattern of waveguide 1 by $\mathbf{e}_1(x, y)$, that of waveguide 2 by $\mathbf{e}_2(x, y)$ (see Fig. 7.21). The total field in both waveguides is then, by assumption, the superposition of the two field patterns (compare Fig. 7.11):

$$\mathbf{E}(x, y, z) = \mathrm{a}_1(z)\mathbf{e}_1(x, y) + \mathrm{a}_2(z)\mathbf{e}_2(x, y) \tag{7.86}$$

The field $\mathrm{a}_1\mathbf{e}_1$ is, by definition, the field in the absence of waveguide 2; in other words, the dielectric constant increase that produces the waveguide is "thought away." Figure 7.21 shows a sketch of a typical field pattern. The power transferred from waveguide 1 to waveguide 2 is caused by the polarization current $j\omega\mathbf{P}_{21}$ produced in waveguide 2 by the field of waveguide 1, within waveguide 2:

$$j\omega\mathbf{P}_{21} = j\omega(\varepsilon_i - \varepsilon)\mathrm{a}_1\mathbf{e}_1(x, y) \tag{7.87}$$

Note that only $(\varepsilon_i - \varepsilon)$ appears because the polarization current $j\omega\varepsilon\mathrm{a}_1\mathbf{e}_1$ flows

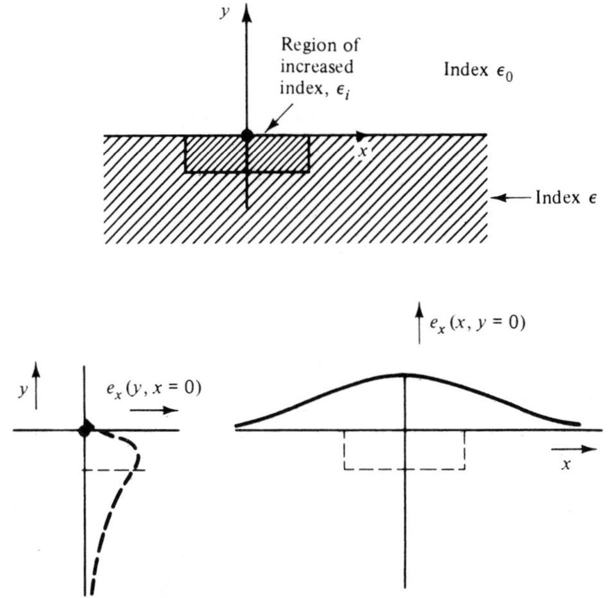

Figure 7.21 Typical field pattern of TE mode.

in the absence of guide 2 and must be subtracted. The power transferred is

$$-\frac{1}{4}\left[\int_{\substack{\text{cross section}\\ \text{of guide 2}}} \mathbf{E}_2^* \cdot (j\omega \mathbf{P}_{21})\, da + \text{c.c.}\right]$$

$$= -\frac{1}{4}\left[j\omega a_1 a_2^* \int_{\substack{\text{cross section}\\ \text{of guide 2}}} da(\varepsilon_i - \varepsilon)\mathbf{e}_1 \cdot \mathbf{e}_2^* + \text{c.c.}\right] \tag{7.88}$$

From coupling of modes we know that the power transfer is

$$\frac{d|a_2|^2}{dz} = \kappa_{21} a_1 a_2^* + \kappa_{21}^* a_1^* a_2 \tag{7.89}$$

Comparison of (7.88) and (7.89) gives

$$\kappa_{21} = -\frac{j\omega}{4} \int_{\substack{\text{cross section}\\ \text{of guide 2}}} da\, (\varepsilon_i - \varepsilon)\mathbf{e}_1 \cdot \mathbf{e}_2^* \tag{7.90}$$

The same approach gives

$$\kappa_{12} = -\frac{j\omega}{4} \int_{\substack{\text{cross section}\\ \text{of guide 1}}} da\, (\varepsilon_i - \varepsilon)\mathbf{e}_2 \cdot \mathbf{e}_1^* \tag{7.91}$$

One can show that $\kappa_{12} = -\kappa_{21}^*$ even when symmetry is not maintained. This confirms the general proof that led to (7.76).

Consider a coupled-waveguide structure consisting of two slab waveguides of Section 6.3 as an example. For TE waves in the two slabs the overlap integral evaluates to

$$\kappa_{12} = -j \frac{\alpha_x k_x^2}{[k_x^2 + \alpha_x^2](\beta_0 d + \beta_0/\alpha_x)} e^{-\alpha_x(D-2d)}$$

In the derivation we have used the determinantal equation (6.26) and the expressions (6.24) and (6.25). Note the exponential dependence of κ_{12} upon the separation distance D between the centers of the two slabs.

It can be shown that the expressions for the coupling coefficients (7.90) and (7.91) follow from a *variational principle* in which (7.86) is used as a trial field solution. Because a variational principle is insensitive to the approximations made in the trial field, this alternative way of deriving the equations puts them on a firmer foundation. [4, 15]

7.9 SUMMARY

This chapter covers material of utmost importance for the analysis of optical waveguide devices—coupling of modes in time and space. In both these cases, the coupling is assumed weak and the time rate of change, or spatial rate of change, is expressed by a first-order derivative; higher-order derivatives are ignored. This accounts for the simplicity of the resulting equations. The approximation is an excellent one for optical devices because the rates of change per cycle, or per wavelength, are small in practically all optical devices.

The equation of a resonant mode coupled to an input waveguide, or wave, contains only the essential parameters describing the internal loss $(1/\tau_0)$ and the coupling to the "outside" $(1/\tau_e)$. Time reversibility was used to derive the coupling to the wave, or waveguide mode, exciting the resonance. This general approach yields model independent relations which, when applied to the electrical case, result in the well-known admittance, a function of frequency. Coupling of modes in time between two resonant modes is described by two coupled first-order differential equations. Energy conservation relates the two coupling coefficients. The solution of the determinantal equation leads to sinusoidal time dependence of the coupled modes. The coupling of modes in space proceeds analogously. The constraint on the coupling constants now results from power conservation. Spatially periodic solutions are found for coupling between two modes (waves) of codirected power flow directions, spatially growing and decaying solutions for coupled waves of counterdirected power flow directions.

The equations for the resonator were applied to threshold computation of a laser oscillator. Whereas we are not studying the physics of laser gain, the simple equation for gain saturation contains all the information necessary for the analysis of steady-state output power of a laser oscillator. The coupling of modes in space was used for the analysis of a tunable filter and an optical switch, both components that are fundamental to integrated optical waveguide systems.

APPENDIX 7A

Injection Locking of an Oscillator

One type of modulation of lasers for communication purposes is phase or frequency-modulation. Frequency modulated signals are amplified conveniently by injection locking an oscillator to a much weaker injection signal. The injection signal is the input signal and the power emitted by the injection-locked laser is the output power. The dynamics of injection locking follow from (7.49). Suppose that the injection signal is of frequency ω

$$\mathrm{s}_+ = \mathrm{S}_+ \mathrm{e}^{j\omega t} \tag{7A-1}$$

Separating a into a product of an amplitude and time dependent phase factor

$$\mathrm{a} = A\mathrm{e}^{j[\omega t + \phi(t)]} \tag{7A-2}$$

where A is real. Introducing (7A-1) into (7.49) and separating the equation into real and imaginary parts, gives

$$\frac{dA}{dt} = \left(\frac{1}{\tau_g} - \frac{1}{\tau_e} - \frac{1}{\tau_0}\right)A + \sqrt{2/\tau_e}\, S_+ \cos \phi(t) \tag{7A-3}$$

and

$$\frac{d\phi}{dt} = (\omega_0 - \omega) - \sqrt{2/\tau_e}\, \frac{S_+}{\mathrm{A}} \sin \phi(t) \tag{7A-4}$$

The first of these two equations serves to determine the time dependence of the amplitude A in the resonator, subject to the dependence of the gain parameter $1/\tau_g$ on the energy A^2 due to saturation; the second equation determines the time dependent phase. If the injection signal is small, one may ignore the dependence of A on S_+ in the phase equation, treating A as a constant and identifying A^2 with the energy of the free-running oscillator. Then (7A-4) is the locking equation of Adler [16]. If the injection locking is successful, a oscillates at the frequency ω of the injection signal. Then $d\phi/dt = 0$. This is possible, according to (7A-4), only if

$$(\omega_0 - \omega)^2 \tau_e^2 < 2S_+^2\, \tau_e/A^2 \tag{7A-5}$$

The square of the deviation of the frequency of the locking signal normalized to the energy escape rate $2/\tau_e$ must be smaller than the ratio of the injected power S_+^2 to the rate of energy escape $2A^2/\tau_e$ from the resonator. Power amplification of the locking signal can be obtained as long as the frequency deviation $(\omega - \omega_0)$ of the locking signal is less than $2/\tau_e$. Of course, our starting assumption was that the amplification was large and thus we have assumed that $|\omega - \omega_0| \ll 2/\tau_e$. We denote by ω_L the value of $\omega - \omega_0$ which is given by the equality sign in (7A-5). This is the locking range, ω_L

$$\omega_L \equiv \sqrt{2/\tau_e}\, S_+/A \tag{7A-6}$$

In terms of $\Delta\omega = \omega - \omega_0$, and the locking range ω_L, one may write the locking equation (7A-4)

$$\frac{d\phi}{dt} = -\Delta\omega - \omega_L \sin\phi \tag{7A-7}$$

The solution of the equation is [17]

$$\phi = -2\tan^{-1}\left[\sqrt{1 - (\omega_L/\Delta\omega)^2}\,\tan\left(\tfrac{1}{2}\sqrt{\Delta\omega^2 - \omega_L^2}\,t\right) + (\omega_L/\Delta\omega)\right] \tag{7A-8}$$

The solution is periodic in time for $|\Delta\omega| > \omega_L$, when the oscillator is unlocked. This represents the unsuccessful attempt at "capture" of the phase of the oscillator by the phase of the injection signal every time the two phases approach each other. When $|\Delta\omega| < \omega_L$, the square roots become imaginary and the internal tangent becomes tanh. The phase goes through a transient adjustment to end up at

$$\phi = -2\tan^{-1}\left[(\omega_L/\Delta\omega) - \sqrt{(\Delta\omega/\omega_L)^2 - j}\right] \tag{7A-9}$$

which is equivalent to

$$\sin\phi \equiv -\frac{\Delta\omega}{\omega_L}$$

as determined by the steady state solution of (7A-7).

PROBLEMS

7.1. In the coupling-of-modes formalism, the analysis of the excitation of a resonant circuit is reduced to a first-order differential equation. Here you are asked to start with a simple circuit and arrive at the first-order differential equation in a series of steps reminiscent of the paraxial approximation (slow envelope variation approximation).

Consider a series *RLC* circuit terminating a transmission line of characteristic impedance Z_0. For a given voltage of the incident wave $v_+(t)$, the equivalent circuit is shown in the figure below. [17]

$$v_+(t) = \tfrac{1}{2}[v_+ e^{j\omega t} + v_+^* e^{-j\omega t}]$$

(a) Write down the two differential equations relating v_L to i, and v_C to i, using Kirchhoff's voltage law and the definition

$$a_{\pm} = \sqrt{C/2}\,(v_c \pm j\sqrt{L/C}\, i)$$

Obtain equation for a_+ in terms of V_+ and a_-. In the "high Q-approximation"

$$\frac{Z_0 + R}{\omega L} \simeq \frac{Z_0 + R}{\omega_0 L} \ll 1$$

one may ignore the coupling of a_+ to a_- and to $V_+^* e^{-j\omega t}$.

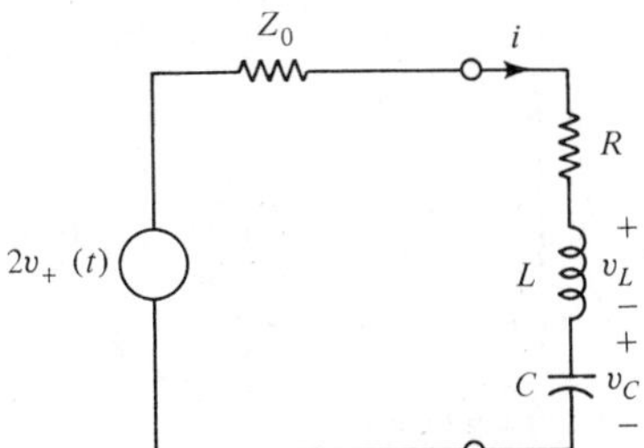

Figure P7.1

(b) Express $1/\tau_0$, $1/\tau_e$ and κ in terms of the circuit parameters. Remember

$$\frac{1}{2}\frac{|V_+|^2}{Z_0} \equiv |s_+|^2$$

where s_+ is the normalized amplitude of the incident wave.

7.2. With reference to Fig. P7.2:

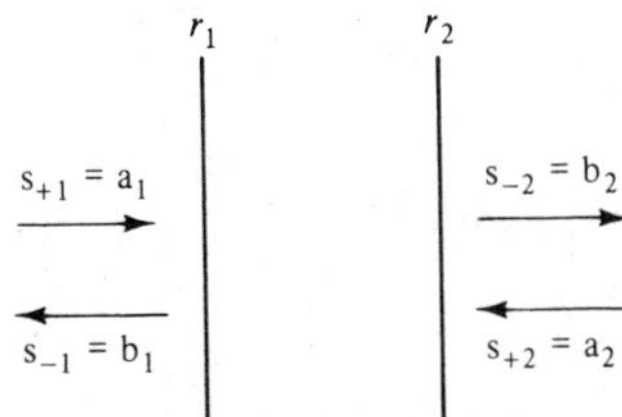

Figure P7.2

(a) Construct the scattering matrix of a resonator with two ports of access.

(b) Compare the resulting expression for the case of a lossless resonator ($1/\tau_0 = 0$) with the result for the asymmetric Fabry–Perot ($\cos\theta = 1$). Express $1/\tau_{e1}$ and $1/\tau_{e2}$ in terms of r_1 and r_2.

(c) Derive the full width at half maximum $\delta f_{1/2}$ for the symmetric case from the coupling-of-modes equation and compare with (3.54) for $\cos\theta = 1$.

7.3. Apply the transient solution (7.40) to a lossless symmetric Fabry–Perot transmission resonator described by the coupling-of-modes formalism ($\tau_0 = \infty$). Use (7.54), which gives

$$|s_{-2}|^2 = \frac{2|a(t)|^2}{\tau_{e2}}$$

in the transient case as well. How long does it take to obtain 90% transmission? Express your answer in terms of $\tau_{e1} = \tau_{e2} = \tau_e$ (set $\omega = \omega_0$).

7.4. In a waveguide switch, the propagation constants β_1 and β_2 are changed by an applied voltage (linear electrooptic effect). In a coupler, in which $|\kappa| l = \pi/2$, full transfer occurs from guide 1 to guide 2. What is the value of $|\beta_1 - \beta_2|$ so that no transfer occurs?

7.5. Confirm (7.69) by direct evaluation of the resonance frequencies of the system shown in Fig. P7.5. Take advantage of the fact that

$$a_{1,2} = -b_{1,2} e^{-2j(\omega/c)l}$$

and write a determinantal equation. Expand $\exp[-j(\omega/c)l]$ around $(\omega_0/c)l = n\pi$ and find the frequency separation of the coupled modes $\Delta\omega$.

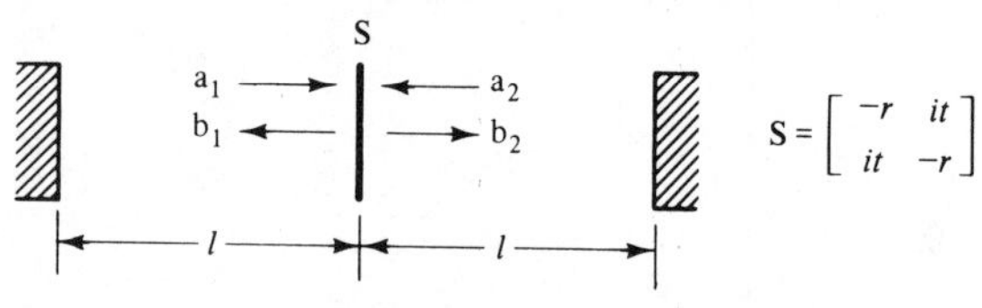

Figure P7.5

7.6. With reference to Fig. P7.6:

(a) Write down the general coupling of modes solution $a_1(z)$ and $a_2(z)$ for a synchronous ($\beta_1 = \beta_2$) waveguide coupler, and solve them with specified $a_1(0)$ and $a_2(0)$.

(b) Specialize on a length $z = l$ so that $|\kappa| l = \pi/4$. Show that you can control the output power $|a_1(l)|^2$ in guide 1 by varying the relative phase between $a_1(0)$ and $a_2(0) = e^{j\phi} a_1(0)$.

(c) Compare the response of part (b) with the half-silvered mirror excited as shown in Fig. P7.6.

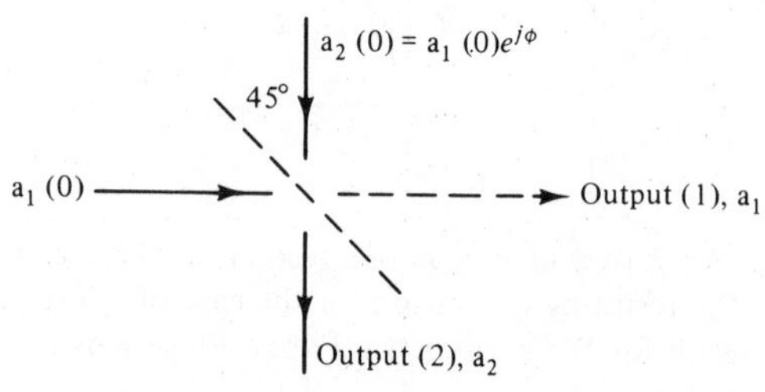

Figure P7.6

7.7. Consider the optical waveguide device shown in Fig. P7.7. This problem is concerned with the action of the coupler, given the fact that

$$a_1(0) = \frac{1}{\sqrt{2}} a_i e^{-j\phi}, \qquad a_2(0) = \frac{1}{\sqrt{2}} a_i e^{+j\phi}$$

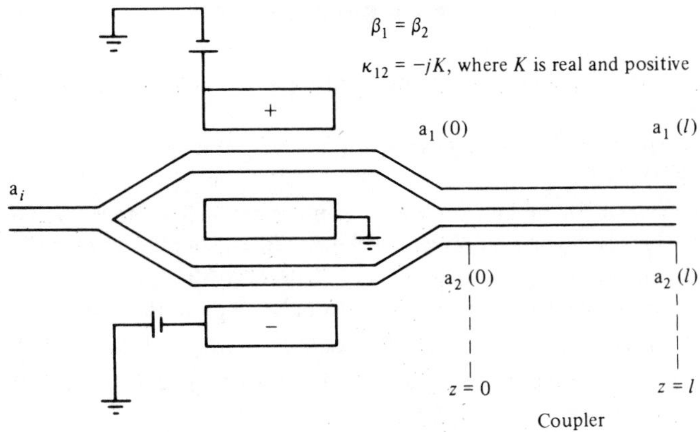

Figure P7.7

where ϕ is the phase shift produced in the electrode structure to the left of $z = 0$ (see Section 12.4)

(a) Find $a_1(l)$ and $a_2(l)$.

(b) With $Kl = \pi/2$ and $Kl = \pi/4$, obtain $|a_1(l)|^2$ and $|a_2(l)|^2$.

(c) Suppose that $\phi = (\pi/2)\sin(\omega t) - (\pi/4)$. Give a graphical construction of $|a_2(l)|^2$ for $Kl = \pi/4$.

7.8. The equations of mode coupling between two identical optical waveguides are

$$\frac{da_1}{dz} = -j\beta a_1 + j\kappa a_2$$

$$\frac{da_2}{dz} = -j\beta a_2 + j\kappa a_1$$

where κ is real and positive and $k = \omega/v$. A similar set of equations holds for the backward waves, b (see Fig. P7.9).

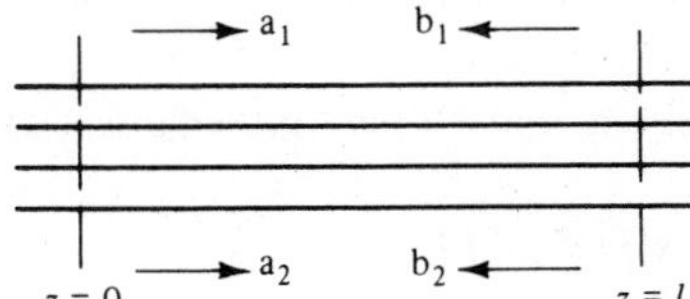

Figure P7.9

(a) Find the two solutions $a_1(z)$ and $a_2(z)$ of this coupled set of equations.

(b) Suppose that perfect reflection is provided at two cross sections a distance l apart $b_1 = -a_1$; $b_2 = -a_2$ at $z = 0$ and l. What are the resonance frequencies of this structure?

REFERENCES

[1] Compare J. C. Slater, *Microwave Electronics*, Van Nostrand, Princeton, N.J., 1950.

[2] A. Yariv, *Introduction to Optical Electronics*, Holt, Rinehart and Winston, New York, 1976.

[3] J. R. Pierce, "Coupling of modes of propagation," *J. Appl. Phys.*, *25*, 179–183, 1954. This is the original work which introduced coupling of modes.

[4] H. A. Haus, "Electron beam waves in microwave tubes," *Proc. of Symposium on Electronics*, Polytechnic Institute of Brooklyn, Apr. 8–10, 1958.

[5] W. Louisell, *Coupled Mode and Parametric Electronics*, Wiley, New York, 1960.

[6] A. Yariv, "Coupled mode theory of guided wave optics," *IEEE J. Quantum Electron.*, *QE-9*, 919, 1973.

[7] P. A. Sturrock, "Kinematics of growing waves," *Phys. Rev.*, *112*, no. 5, 1488–1503, Dec. 1958.

[8] H. A. Haus, "Power-flow relations in lossless nonlinear media," *Trans. IRE, PG-MTT*, 317–324, July 1958.

[9] E. A. J. Marcatili, "Dielectric rectangular waveguide and directional coupler for integrated optics," *Bell Syst. Tech. J.*, *48*, 2071–2102, Sept. 1969.

[10] H. F. Taylor, "Optical switching and modulation in parallel dielectric waveguides," *J. Appl. Phys.*, *44*, 3257–3262, July 1973.

[11] H. Kogelnik and R. V. Schmidt, "Switched directional couplers with alternating $\Delta\beta$," *IEEE J. Quantum Electron.*, *QE-12*, no. 7, 396–401, July 1976.

[12] J. Noda, M. Fukuma, and O. Mihami, "Design calculations for directional couplers fabricated by Ti-diffused $LiNbO_3$ waveguides," *Appl. Opt.*, *20*, no. 13, 2284–2298, July 1981.

[13] R. C. Alferness and P. S. Cross, "Filter characteristics of codirectionally coupled waveguides with weighted coupling," *IEEE J. Quantum Electron.*, *QE-14*, no. 11, 843–847, Nov. 1978.

[14] R. C. Alferness and R. V. Schmidt, "Tunable optical waveguide directional coupler filter," *Appl. Phys. Lett.*, *33*, no. 2, 161–163, July 1978.

[15] H. A. Haus and M. N. Islam, "Application of variational principle to systems with radiation loss," to be published, 1982.

[16] R. Adler, "A study of locking phenomena in oscillators," Proc. I.R.E., *34*, 351–357, June 1946.

[17] H. B. Dwight, "Tables of Integrals and Other Mathematical Data," MacMillan, 1961, p. 99.

[18] R. B. Adler, L. J. Chu, and R. M. Fano, *Electromagnetic Energy Transmission and Radiation*, Wiley, New York, 1960, p. 155.

8

DISTRIBUTED FEEDBACK STRUCTURES

Distributed feedback structures have been proposed by Kogelnik and Shank [1] for integrated optics applications. Distributed feedback (DFB) grating structures may act as mirrors or filters [2].

DFB structures couple counter-traveling waves, that is waves of opposite group velocities. Such coupling leads to exponentially growing and decaying coupled wave solutions, which were obtained in the preceding chapter. In this chapter we develop the equations for distributed feedback gratings. We analyze a reflection filter and derive the reflection as a function of frequency. In Chapter 2, multiple reflecting layers, one quarter wavelength thick, were shown to give very high reflectivity. The analysis presented here applies to these layers if the reflection coefficient of one double layer is much less than unity. The frequency dependence of the reflectivity is obtained quite simply and from it the dependence of the reflection bandwidth on the layer parameters.

We show that a DFB grating can also act in the manner of a transmission resonator, although with rather broad transmission bands. Two gratings, shifted by an odd integer number of quarter wavelengths produce a very narrow, high-Q, transmission band at and near the frequency of peak reflection of the individual gratings. We evaluate the quality factors so that the analysis of a transmission resonator of Chapter 7 can be applied to it. Finally, we show how the coupling coefficient between the counter-traveling waves can be evaluated for a given set of parameters describing the structure.

8.1 THE EQUATIONS OF DISTRIBUTED FEEDBACK STRUCTURES

In the preceding chapter we mentioned briefly the phenomenon of coupling between two waves of opposite group velocity. To produce appreciable coupling, synchronism was necessary; that is, the two propagation constants could not differ by much more than the magnitude of the coupling coefficient $|\kappa_{12}|$.

Generally, waves with opposite group velocities have widely different propagation constants. How, then, can one bring the wave with a positive group velocity to be synchronous with the wave with negative group velocity? In this section we show how this can be done.

Consider a structure of the type shown in Fig. 8.1a. Denote the wave with positive group velocity, the "forward" wave in the guiding structure, by a, the wave with negative group velocity the "backward" wave, by b. If there is no periodicity, then the differential equations obeyed by a and b are

$$\frac{d\mathrm{a}}{dz} = -j\beta\mathrm{a} \tag{8.1}$$

$$\frac{d\mathrm{b}}{dz} = j\beta\mathrm{b} \tag{8.2}$$

Suppose, however, that a periodic perturbation of the guiding structure is introduced (see Fig. 8.1b). The polarization current associated with the electromagnetic field of the wave a $\propto \exp(-j\beta z)$ traveling along the periodic

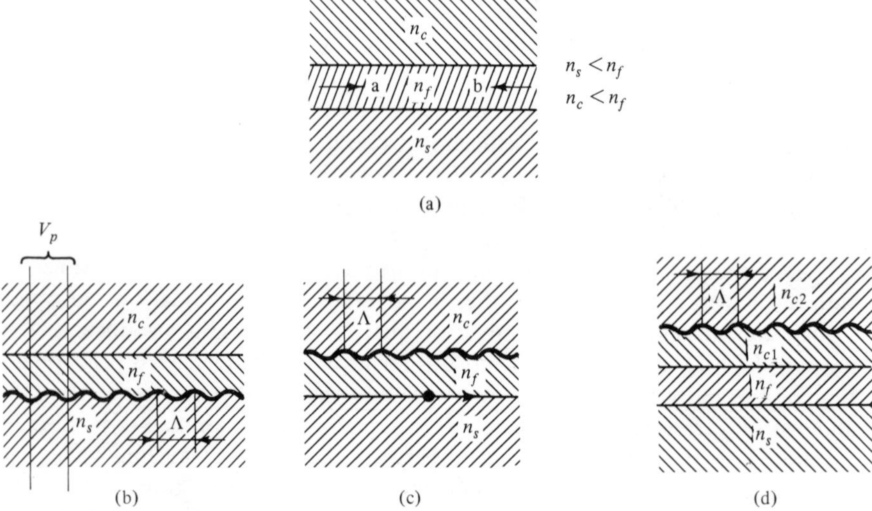

Figure 8.1 (a) Guiding structure with no periodic coupling; (b), (c), and (d) guiding structures with coupling; (*d*) $n_{c2} < n_{c1} < n_f$.

structure acquires modulation components (spatial sidebands) because the fields get modulated by the perturbation with the spatial dependence cos $2(\pi/\Lambda)z$. These sidebands are called *space harmonics*. They are part of the field structure of the waves and are capable of coupling to another wave with a propagation constant close to the propagation constant of the space harmonic. In the present case of a cosinusoidal perturbation and a spatial dependence of a of the form exp $(-j\beta z)$, the sidebands follow from

$$\exp(-j\beta z)\cos\left(2\frac{\pi}{\Lambda}\right)z = \frac{1}{2}\left\{\exp\left[-j\left(\beta - \frac{2\pi}{\Lambda}\right)z\right] + \exp\left[-j\left(\beta + \frac{2\pi}{\Lambda}\right)z\right]\right\}$$

If $\beta - (2\pi/\Lambda)$ is close to $-\beta$, these polarization currents radiate in the backward direction and the radiation superimposes coherently over extended distances l, obeying the inequality

$$l\left(\beta - \frac{2\pi}{\Lambda}\right) \leq -\beta l \pm \pi$$

or

$$\left|\beta - \frac{\pi}{\Lambda}\right| \leq \frac{\pi}{2l}$$

The exponential with the argument $[\beta + (2\pi/\Lambda)]z$ does not produce backward radiation because its spatial dependence differs greatly from that of exp $(+j\beta z)$. The effect of coupling of a to b can be included in (8.2) by introducing a coupling term produced by the backward-radiating polarization currents

$$\frac{d\mathrm{b}}{dz} = j\beta\mathrm{b} + \kappa_{ba}\mathrm{a}e^{+j(2\pi/\Lambda)z} \tag{8.3}$$

A similar effect is produced by the interaction of the backward wave with the periodic perturbation

$$\frac{d\mathrm{a}}{dz} = -j\beta\mathrm{a} + \kappa_{ab}\mathrm{b}e^{-j(2\pi/\Lambda)z} \tag{8.4}$$

The equations above can be reduced to coupling-of-modes equations with space-independent coefficients by introducing the new variables

$$\mathrm{a} = \mathrm{A}(z)e^{-j(\pi/\Lambda)z}, \qquad \mathrm{b} = \mathrm{B}(z)e^{j(\pi/\Lambda)z}$$

with the result

$$\frac{d\mathrm{A}}{dz} = -j\left(\beta - \frac{\pi}{\Lambda}\right)\mathrm{A} + \kappa_{ab}\mathrm{B} \tag{8.5}$$

$$\frac{d\mathrm{B}}{dz} = j\left(\beta - \frac{\pi}{\Lambda}\right)\mathrm{B} + \kappa_{ba}\mathrm{A} \tag{8.6}$$

These equations look like the coupling-of-modes equations (7.72) and (7.73). The roles of β_1 and β_2 are played by $\pm[\beta - (\pi/\Lambda)]$. The inverse group velocities $d\beta_1/d\omega$ and $d\beta_2/d\omega$ are of opposite sign, and synchronism occurs when

$$\frac{\beta_1 - \beta_2}{2} \leftrightarrow \beta - \frac{\pi}{\Lambda} = 0$$

Because the waves have oppositely directed group velocities, (7.76) holds with $p_1 \leftrightarrow p_a = +1$, $p_2 \leftrightarrow p_b = -1$, and $\kappa_{ab} = \kappa_{ba}^*$.

Equations (8.5) and (8.6) can be simplified in appearance by the introduction of the detuning parameter

$$\delta \equiv \beta - \frac{\pi}{\Lambda} \tag{8.7}$$

which measures the deviation of the propagation constant from π/Λ. In the neighborhood of the frequency ω_0 for which $\beta(\omega_0) = \pi/\Lambda$, we have

$$\beta = \beta(\omega_0) + \frac{d\beta}{d\omega}(\omega - \omega_0)$$

and thus

$$\delta = \frac{\omega - \omega_0}{v_g} \tag{8.8}$$

where v_g is the group velocity $d\omega/d\beta$. With the detuning parameter δ and

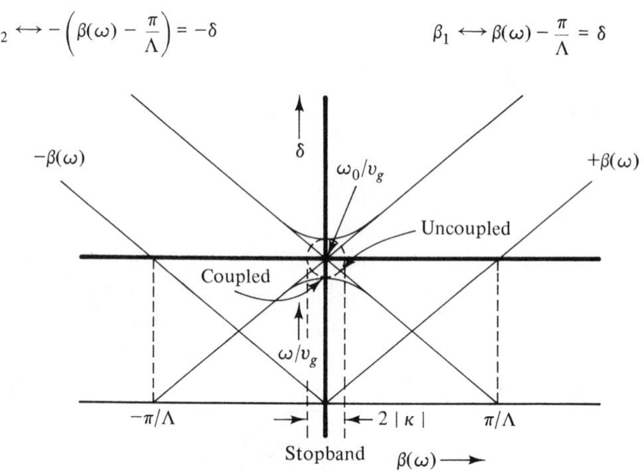

Figure 8.2 Dispersion diagram from (8.5) and (8.6) with the unperturbed propagation constant $\beta(\omega)$ proportional to ω.

$\kappa_{ab} = \kappa_{ba}^* \equiv \kappa$, (8.5) and (8.6) assume the simple form

$$\frac{dA}{dz} = -j\delta A + \kappa B \tag{8.9}$$

$$\frac{dB}{dz} = j\delta B + \kappa^* A \tag{8.10}$$

Figure 8.2 shows the dispersion diagram corresponding to (8.9) and (8.10) with the solutions (7.78) and $\kappa_{12}\kappa_{21} \leftrightarrow \kappa_{ab}\kappa_{ba} = |\kappa|^2 > 0$. The propagation constant β of the coupled system is

$$\beta = \pm\sqrt{\delta^2 - |\kappa|^2}$$

The unperturbed dispersion curves for $\pm\beta(\omega)$ are taken as straight lines. Note that the variables A(z) and B(z) have removed from them the spatial dependences exp $[\pm j(\pi/\Lambda)z]$. They are envelope quantities analogous to the envelope quantity A(z, $\omega - \omega_0$) defined in (6.54).

8.2 REFLECTION FILTER [1, 2]

The equations of "distributed feedback" (DFB) (8.9) and (8.10) describe periodic perturbations that extend over a given length of guiding structure. A DFB structure of length l acts as a filter. In this section we derive the filter response of such a structure.

Consider the reflection from a periodic structure of length l, with one end matched, B = 0 at $z = 0$ (see Fig. 8.3). For $|\delta| < |\kappa|$, the solutions of (8.5) and (8.6) are of the form exp $(\mp\gamma z)$, where

$$\pm\gamma = \pm\sqrt{|\kappa|^2 - \delta^2} \tag{8.11}$$

with $\kappa = \kappa_{ab} = \kappa_{ba}^*$ and $\delta \equiv \beta - (\pi/\Lambda)$. The solutions are growing and decaying exponentials, whereas they are periodic functions in the range $|\delta| > |\kappa|$.

The general solutions with arbitrary constants are

$$A = A_+ e^{-\gamma z} + A_- e^{+\gamma z} \tag{8.12}$$

$$B = B_+ e^{-\gamma z} + B_- e^{+\gamma z} \tag{8.13}$$

where only two of the four constants are independent. From (8.9) we find a

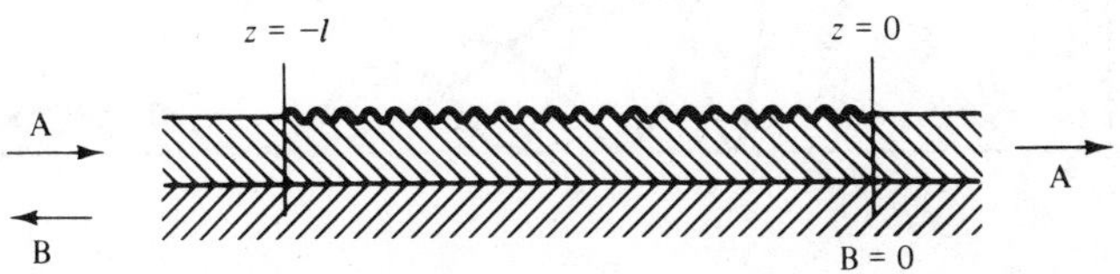

Figure 8.3 Reflection filter.

relation between $B_\pm$ and $A_\pm$:

$$B_\pm = \frac{\mp\gamma + j\delta}{\kappa} A_\pm \tag{8.14}$$

At $z = 0$ there is no reflected wave, $B_+ = -B_-$, and thus

$$B = -2B_+ \sinh \gamma z \tag{8.15}$$

Using (8.10), we obtain for A(z)

$$A = -2B_+\left(\frac{\gamma}{\kappa^*} \cosh \gamma z - \frac{j\delta}{\kappa^*} \sinh \gamma z\right) \tag{8.16}$$

The reflection coefficient $\Gamma = B/A$ at $z = -l$ is

$$\Gamma(-l) = -\frac{\sinh \gamma l}{(\gamma/\kappa^*) \cosh \gamma l + (j\delta/\kappa^*) \sinh \gamma l} \tag{8.17}$$

The analysis can be generalized to cover an arbitrary reflection at $z = 0$, $\Gamma(0)$. One then finds after some simple manipulation

$$\Gamma(z) = \frac{(\kappa^*/\kappa) + \Gamma(0)[(\gamma/\kappa) \coth \gamma z + j(\delta/\kappa)]}{\Gamma(0) + [(\gamma/\kappa) \coth \gamma z - j\delta/\kappa]} \tag{8.18}$$

Equation (8.17) is a special case of (8.18) with $\Gamma(0) = 0$ and $z = -l$. Figure 8.4 shows $\Gamma(-l)$ of (8.17) as a function of δ—the frequency parameter. The reflection coefficient is zero at a set of frequencies within the passband. At these frequencies the backward wave has a sinusoidal distribution within the struc-

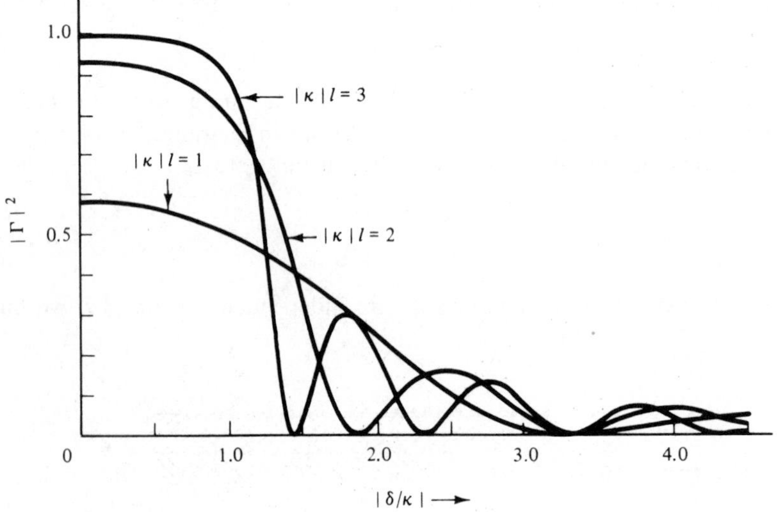

Figure 8.4 Reflection $|\Gamma|^2$ as a function of $|\delta/\kappa|$ for uniform distributed feedback structure of length l.

ture, according to (8.15), with γ pure imaginary, and vanishes at the two end planes of the structure. These frequencies are the resonance frequencies of the periodic structure acting as a "distributed" Fabry–Perot transmission resonator. Instead of a single pair of mirrors, the periodic structure has a large number of reflecting "mirrors," one each for each pair of periods of the structure lying symmetrically with respect to the center plane of the structure.

In Fig. 8.4 $|\kappa|$ is kept constant at its value for $\delta = 0$, in other words is treated as if it were frequency independent (δ independent). This approximation is consistent with the approximations inherent in coupled mode theory. The two modes a and b couple within a normalized frequency range $\pm\delta$ of the order of $\pm|\kappa|$. A frequency dependence of $|\kappa|$ can be ignored as it is of second order (i.e. equal to the product of δ and $d\kappa/d\delta$).

The preceding analysis applies to any periodic structure. In particular, it applies to multiple dielectric layers as described in Section 2.4, provided, of course, that the reflection from each layer is weak so that the difference equation governing the amplitudes at the discrete interfaces can be replaced by a differential equation. To relate the present analysis to the multiple-layer problem, we have to evaluate κ_{ab} for a multiple-layer structure. This is done with the aid of Fig. 8.5. The two layers forming one period of the periodic structure are defined as having indices $n + \Delta n$ and $n - \Delta n$ immersed in a medium of index n. We compute the reflection from these two layers, that are one quarter wavelength thick at the resonance wavelength ($\lambda/2n = \Lambda$).

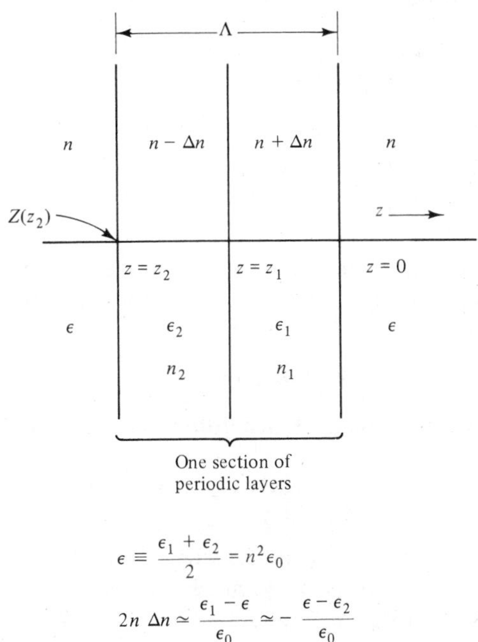

Figure 8.5 Two dielectric layers in a uniform medium of index n.

The analysis in Section 2.5 gives the wave impedance seen at the reference plane 2, under normal incidence (see Fig. 2.10):

$$Z(z_2) = \sqrt{\frac{\mu_0}{\varepsilon_0} \frac{n_1^2}{n n_2^2}} \tag{8.19}$$

and

$$\frac{n_1}{n_2} = \frac{n + \Delta n}{n - \Delta n} \tag{8.20}$$

Therefore,

$$Z(z_2) = \left(\frac{n + \Delta n}{n - \Delta n}\right)^2 \frac{1}{n} \sqrt{\frac{\mu_0}{\varepsilon_0}} \simeq \left(1 + \frac{4\ \Delta n}{n}\right) \frac{1}{n} \sqrt{\frac{\mu_0}{\varepsilon_0}} \tag{8.21}$$

The reflection of one section of the periodic layers is, by definition, equal to the reflection coefficient Γ produced by one double layer, one section of the periodic structure [see (2.39)]. From (8.10) with $dz = -\Lambda$, $\Delta \mathrm{B} = -\kappa^* \Lambda \mathrm{A} = \Gamma \mathrm{A}$, or

$$-\kappa^* \Lambda = \Gamma = \frac{(1 + 4\ \Delta n/n) - 1}{1 + 4\ \Delta n/n + 1} \simeq \frac{2\ \Delta n}{n} \tag{8.22}$$

The discrete approach would have given for m sections for the normalized impedance seen at the input, $Z/Z_0 = nZ/\sqrt{\mu_0/\varepsilon_0}$:

$$\frac{Z}{Z_0} = \left(\frac{n_1}{n_2}\right)^{2m} = \left(\frac{n_1}{n_2}\right)^{2(l/\Lambda)} \tag{8.23}$$

Introducing (8.20) for n_1/n_2 and (8.22) for $|\kappa|$,

$$\frac{Z}{Z_0} \simeq \left(1 + \frac{4\ \Delta n}{n}\right)^{l/\Lambda} = \left[\left(1 + \frac{4\ \Delta n}{n}\right)^{n/4\Delta n}\right]^{2|\kappa| l} \Rightarrow e^{2|\kappa| l} \quad \text{in limit } \frac{\Delta n}{n} \to 0$$

The term in brackets approaches e as $\Delta n/n$ is made very small. On the other hand, the coupling-of-modes approach gives for the reflection coefficient, [(8.17) with $\delta = 0$]

$$\Gamma(-l) = \tanh(|\kappa| l) \tag{8.24}$$

The impedance at the input plane is, according to (2.38) and (8.24),

$$\frac{Z}{Z_0} = \frac{1 + \Gamma}{1 - \Gamma} = e^{2|\kappa| l}$$

Thus both approaches give the same answer in the limit of small $\Delta n/n$. The coupling-of-modes approach is very much simpler and gives simple expressions over the entire frequency range.

Therefore, within the approximations of coupled mode theory, Fig. 8.4 gives the frequency dependence of the reflection from a multiple layer reflector for $|\kappa|$ evaluated at "center band", $\delta = 0$. As mentioned earlier this is a legitimate assumption. The reflection filter ceases to act as a reflector outside the stopband, for $|\delta| > |\kappa|$, because the reflections from individual layers cease to interfere constructively. A weak frequency dependence of $|\kappa|$ goes unnoticed.

The bandwidth over which the transmissivity $T = 1 - |\Gamma|^2$ of the filter has increased by a factor of 2 follows from (8.17). The transmissivity is

$$T = \frac{1 - \tanh^2 \gamma l}{1 + \dfrac{\delta^2}{\gamma^2} \tanh^2 \gamma l}$$

We concentrate on the small transmissivity, high reflectivity, limit $\gamma l \gg 1$. The function $\tanh \gamma l$ can be approximated by $1 - 2e^{-2\gamma l}$ and the transmissivity by

$$T \simeq 4 \frac{\gamma^2}{|\kappa|^2} \exp - 2\gamma l$$

The dependence on frequency of T is dominated by the exponential and the multiplier $\gamma^2/|\kappa|^2$ can be replaced by unity. Further the change of γ, $\Delta\gamma$, is $\Delta\gamma/|\kappa| \simeq -\frac{1}{2}\delta^2/|\kappa|^2$ in the limit of small $\delta/|\kappa|$. With these approximations, the change of $\delta_{1/2}$ which produces a doubling of the transmissivity is

$$\left|\frac{\delta_{1/2}}{\kappa}\right| = \sqrt{\frac{\ln 2}{|\kappa| l}}$$

If we introduce the specific expression for $|\kappa|\Lambda$ from (8.22) we find for the fractional fullwidth at half maximum

$$\frac{\Delta\omega_{1/2}}{\omega} = 2\frac{v_g}{\omega}|\delta_{1/2}| = \frac{2}{\pi}\sqrt{2 \ln 2}\sqrt{\frac{1}{m}\frac{\Delta n}{n}}$$

where we have set $v_g = c/n$, $\Lambda = \lambda/2n$ and $(l/\Lambda) = m$, the number of layer pairs. The minimum transmissivity is

$$T = 4 \exp\left(-4\frac{\Delta n}{n} m\right)$$

The minimum of the transmissivity is controlled by the product $m \times \Delta n/n$, the bandwidth by the ratio $(\Delta n/n)/m$.

8.3 HIGH-Q TRANSMISSION RESONATOR [3]

The periodic structure of the preceding section acts as a reflector in its stopband ($|\delta| < |\kappa|$), and has a set of transmission resonances in the passband ($|\delta| > |\kappa|$). It is possible to achieve transmission within the stopband, if one spaces two periodic structures by one (or an odd multiple of) quarter wave-

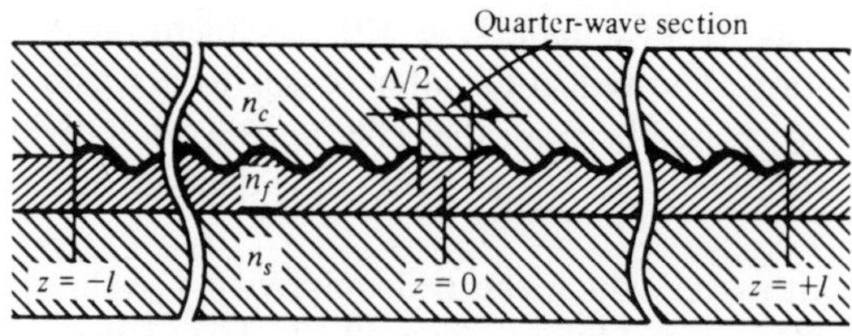

Figure 8.6 Transmission resonator.

length(s) (see Fig. 8.6). This transmission resonance is made narrower than the transmission resonances of the single structure, because the reflectivity of each of the gratings is exploited at the frequency of peak reflection.

The subsequent analysis is an approximate one in which we employ the fact that the Q of the resonance is very high; the energy storage in the structure is high compared with the power escaping the structure within one cycle. The formalism of Sections 7.2 and 7.4 can be adapted to the present case, thus obviating the need for a detailed analysis of the two structures in cascade. We start by considering two semi-infinite structures separated by $\lambda/4$, where λ corresponds to the wavelength at the center of the stopband. This is the resonance of the "closed" resonator. The power escaping from a resonator of finite length is obtained by a simple perturbation analysis. The parameters τ_{e1} and τ_{e2} (the structure is symmetric and thus $\tau_{e1} = \tau_{e2}$) can be found from the ratio of energy storage to escaping power and the formalism of Section 7.4 can be applied.

Consider a semiinfinite grating occupying the region $0 < z < \infty$. If excited at $z = 0$, the solutions for A and B are

$$\mathrm{A} = \mathrm{A}_+ e^{-|\kappa| z} \tag{8.25}$$

$$\mathrm{B} = -\frac{|\kappa|}{\kappa} \mathrm{A}_+ e^{-|\kappa| z} \tag{8.26}$$

A semi-infinite grating occupying the region $-\infty < z \leq 0$ has the solution

$$\mathrm{A} = \mathrm{A}_- e^{+|\kappa| z} \tag{8.27}$$

$$\mathrm{B} = \frac{|\kappa|}{\kappa} \mathrm{A}_- e^{+|\kappa| z} \tag{8.28}$$

At $z = 0$ we get (see Fig. 8.7)

$$\Gamma = \frac{\mathrm{B}}{\mathrm{A}} = -\frac{|\kappa|}{\kappa}, \qquad z = 0_+$$

$$\Gamma = \frac{\mathrm{B}}{\mathrm{A}} = \frac{|\kappa|}{\kappa}, \qquad z = 0_-$$

If we insert a quarter-wave section, we can match the reflection coefficient at $z = 0_-$, $\Gamma(z = 0_-)$, to the reflection coefficient at $z = 0_+$, $\Gamma(z = 0_+)$, because Γ

is transformed into $-\Gamma$ by such a section. To recognize this, we take note of the fact that the coupling-of-modes formalism for a periodic structure was written in terms of "spatial envelope" quantities A(z) and B(z). The actual wave amplitudes a(z) and b(z) are obtained from A(z) and B(z) through multiplication by exp $[-j(\pi/\Lambda)z]$ and exp $[+j(\pi/\Lambda)z]$, respectively. At the center of the stopband, the waveguide wavelength is 2Λ. Insertion of a section of quarter wavelength changes the exponential factors of A and B by $-j$ and $+j$, respectively. The amplitude distribution in the resonator is shown in Fig. 8.7.

If the two grating sections are not semi-infinite, the resonator is "open" to the outside world and possesses a noninfinite $Q_{\text{ext},1}$ and $Q_{\text{ext},2}$. Once these are evaluated together with the unloaded Q, one may obtain the power transmitted as a function of frequency from (7.54). We shall now compute $Q_{\text{ext},1}$ and $Q_{\text{ext},2}$ from a simple approximate analysis.

If the decay is pronounced, $|kl| \gg 1$, then a reflection at $z = \pm l$ does not appreciably affect the boundary condition at $z = 0$. We may start with assumed amplitudes A_+ and B_+ at $z = l$. Then, for $z > 0$,

$$A = A_+ e^{-|\kappa|z} + A_- e^{+|\kappa|z}$$

$$B = -\frac{|\kappa|}{\kappa}(A_+ e^{-|\kappa|z} - A_- e^{+|\kappa|z})$$

where A_- is to be evaluated from A_+ using the boundary condition $B = 0$ at $z = l$. One finds that

$$A_- = A_+ e^{-2|\kappa|l}$$

and thus

$$A(z = l) = 2A_+ e^{-|\kappa|l}$$

The power escaping at this port 2 is

$$P_2 = 4|A_+|^2 e^{-2|\kappa|l}$$

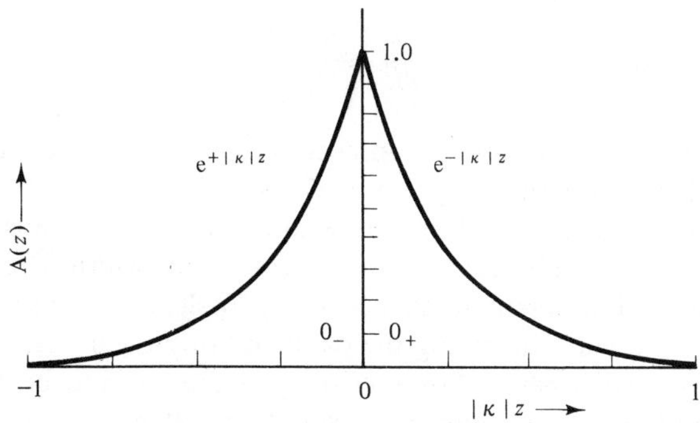

Figure 8.7 Amplitude $A(z)$ in two semi-infinite gratings separated by quarter-wave section. On the scale of the drawing, the quarter-wave section is of negligible length.

The energy storage is obtained from the energy per unit length, which in turn is related to the power by the group velocity, (6.96). The group velocities of the waves traveling in opposite directions are of opposite sign; so are their powers—the energies are positive, as they must be. The energy per unit length $\langle w \rangle$ is

$$\langle w \rangle = \frac{1}{v_g}(|\mathbf{A}|^2 + |\mathbf{B}|^2)$$

When the spatial decays are pronounced, we may use an infinitely long structure to evaluate W approximately:

$$W = \int_{-l}^{l} \langle w \rangle \, dz \simeq \int_{-\infty}^{\infty} \frac{1}{v_g}(|\mathbf{A}|^2 + |\mathbf{B}|^2)\, dz = \frac{2|\mathbf{A}_+|^2}{v_g|\kappa|}$$

The external Q is thus obtained from

$$\frac{1}{Q_{\text{ext},2}} = \frac{P_2}{\omega_0 W} = \frac{2v_g|\kappa|}{\omega_0} e^{-2|\kappa|l} = \frac{v_g}{\pi c}\frac{\lambda}{l}|\kappa| l e^{-2|\kappa|l}$$

The same external Q is found for port 1. The external Q is proportional to l/λ, where λ is the free-space wavelength, and $\exp(2|\kappa|l/|\kappa|l)$. A structure with a large value of $|\kappa|l$ can achieve a very high Q_{ext}. If there is a decay constant α inside the structure, the power p lost per unit length is

$$p = 2\alpha(|\mathbf{A}|^2 + |\mathbf{B}|^2)$$

and the net power lost is

$$P_0 = \int_{-l}^{l} p \, dz \simeq \int_{-\infty}^{\infty} 2\alpha(|\mathbf{A}|^2 + |\mathbf{B}|^2)\, dz = \frac{2\alpha}{|\kappa|}|\mathbf{A}_+|^2$$

Thus the unloaded Q is given by

$$\frac{1}{Q_0} = \frac{\alpha v_g}{\omega_0}$$

From these parameters follows the power transmission as a function of frequency around the resonance as given by (7.54).

It is a simple exercise in algebra to evaluate the power transmitted over a frequency band much wider than the single resonance width. Figure 8.8 shows a plot of transmission versus frequency for a lossless transmission resonator ($\alpha = 1$) over a bandwidth including the passband of the periodic structure. The structure is assumed lossfree. For the analysis the spacing between the two gratings is maintained at a quarter-waveguide wavelength. This is reasonable because the frequency range represented by a graph like Fig. 8.8 is generally a very narrow one for practical values of $|\kappa|$. The narrow central transmission

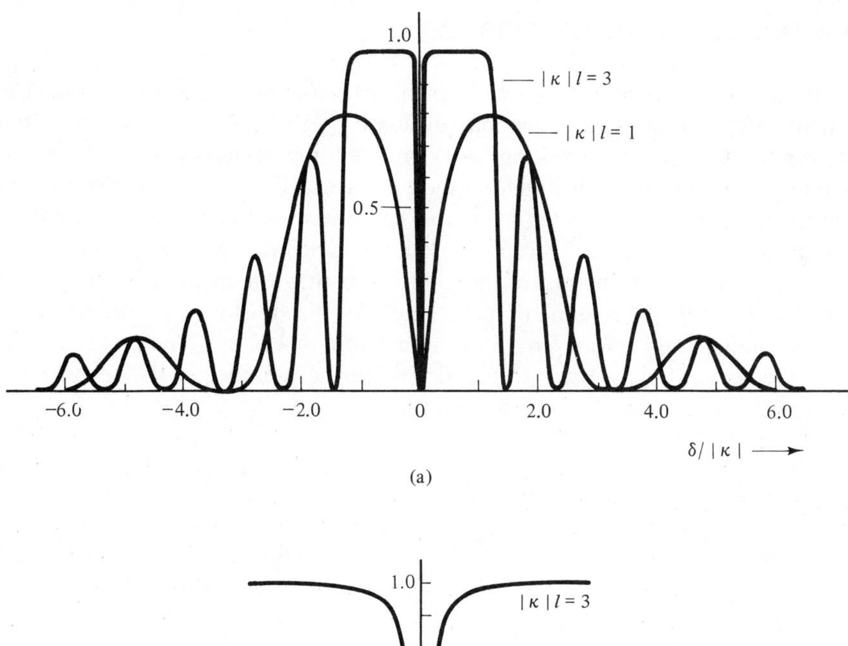

Figure 8.8 (a) Transmission characteristic of distributed feedback structure with quarter-wave section; (b) expanded central part of plot.

band is the resonance analyzed above and its full width at half maximum is

$$\Delta\omega_{1/2} = \frac{2}{\tau_{e1}} + \frac{2}{\tau_{e2}} = \frac{4}{\tau_e} = 4v_g |\kappa| e^{-2|\kappa|l}$$

or in normalized units

$$\Delta\delta_{1/2} = 4|\kappa| e^{-2|\kappa|l}$$

The bandwidth depends exponentially on $|\kappa|l$. Thus very narrow bandwidths are achievable by this split-grating resonator.

8.4 THE COUPLING COEFFICIENT

Thus far we assumed that κ_{ab} was known, either through measurement or from theory. Here we show how one may evaluate κ_{ab} theoretically for the structure shown in Fig. 8.1b. By taking away and adding dielectric material to the guiding structure in a periodic manner, we are perturbing the dielectric constant periodically by $\delta\varepsilon = (n_f^2 - n_s^2)\varepsilon_0$. The volume of the perturbation has the z dependence $\cos(2\pi/\Lambda)z$, that is the volume is positive or negative depending upon whether the excursion of the interface is into the medium with index n_f or away from it. We assume that the height of the sinusoidal perturbation h is much smaller than a wavelength in the medium, $h \ll \lambda/n_f$. The power P_{AB} fed into the forward wave per unit length due to the presence of the perturbation is, from the coupling-of-modes formalism,

$$P_{AB} = \mathrm{A}^* \frac{d\mathrm{A}}{dz} + \mathrm{A}\frac{d\mathrm{A}^*}{dz} = \kappa_{ab}\mathrm{BA}^* + \kappa_{ba}^*\mathrm{AB}^* \tag{8.29}$$

From the field-theoretical point of view, this is equal to the power generated in the volume V_p, occupied by one period of the grating, divided by Λ (see Fig. 8.1b):

$$P_{AB} = -\frac{1}{4\Lambda}\int_{V_p} \mathbf{E}_A^* \cdot (j\omega\mathbf{P}_B)\,dv + \text{c.c.} \tag{8.30}$$

Here $\mathbf{E}_A$ is the complex field associated with the wave A, and $j\omega\mathbf{P}_B$ is the polarization current density responsible for the coupling associated with the field B and caused by the perturbation. It is nonzero only within the volume δV of the dielectric perturbation. In a two dimensional structure, such as the one of Fig. 8.1b, the volume elements of δV are

$$d(\delta V) = h\,dz\,\cos\frac{2\pi}{\Lambda}z$$

and of unit width in the third dimension. Now, the A and B waves have field patterns $\mathbf{e}_A(\boldsymbol{\rho})$ and $\mathbf{e}_B(\boldsymbol{\rho})$, functions of the transverse coordinate vector $\boldsymbol{\rho} \equiv \hat{x}x + \hat{y}y$; for example,

$$\mathbf{E}_A = \mathrm{A}\mathbf{e}_A(\boldsymbol{\rho})e^{-j\beta z} = \mathrm{A}\mathbf{e}_A(\boldsymbol{\rho})e^{-(j\pi/\Lambda)z} \tag{8.31}$$

at synchronism. A similar expression holds for $\mathbf{E}_B$. The power per unit length P_{AB} of (8.30) is, therefore

$$P_{AB} = -\frac{\mathrm{A}^*\mathrm{B}}{4\Lambda}\int_{z=0}^{z=\Lambda} j\omega\varepsilon_0(n_f^2 - n_s^2)h\cos\frac{2\pi}{\Lambda}z \cdot e^{j(2\pi/\Lambda)z}\mathbf{e}_A^*(x=-d)\cdot\mathbf{e}_B(x=-d)\,dz + \text{c.c.} \tag{8.32}$$

where the coordinate of the interface is taken at $x = -d$. The integral yields

$$P_{AB} = -\frac{j\omega}{8}\varepsilon_0(n_f^2 - n_s^2)h\,\mathbf{e}_A^*(x = -d)\cdot\mathbf{e}_B(x = -d)\;\mathrm{A^*B} + \text{c.c.} \tag{8.33}$$

Comparison of (8.29) and (8.33) gives

$$\kappa_{ab} = -\frac{j\omega}{8}\varepsilon_0(n_f^2 - n_s^2)h\,\mathbf{e}_A^*(x = -d)\cdot\mathbf{e}_B(x = -d) \tag{8.34}$$

We use the field solution for the symmetric waveguide of Section 6.3 as an example, with the field profile shown in Fig. 6.5. Then we must interpret $n_f^2 - n_s^2$ as $n_i^2 - n^2$, where $\varepsilon_0 n_i^2 = \varepsilon_i$ is the dielectric constant within the slab and $\varepsilon_0 n^2 = \varepsilon$ is the dielectric constant outside. We obtain

$$\begin{aligned}\kappa_{ab} &= -\frac{j\omega^2\mu_0}{4\beta}k_x(\varepsilon_i - \varepsilon)h\frac{\cos k_x d}{k_x d + \frac{1}{2}\sin 2k_x d + \dfrac{k_x}{\alpha_x}\cos^2 k_x d}\\ &= -j\frac{\pi^2}{\beta\lambda^2}\left(\frac{\varepsilon_i - \varepsilon}{\varepsilon_0}\right)\frac{k_x h}{\cos k_x d}\frac{1}{\left[k_x d + \dfrac{k_x}{\alpha_x}\right]\left[1 + \dfrac{\alpha_x}{k_x}\right]}\end{aligned} \tag{8.35}$$

In the second expression we have used the dispersion relation (6.23).

8.5 SUMMARY

A distributed feedback structure is an example of a structure that couples forward and backward modes (waves). A periodic spatial variation of period Λ produces spatial sidebands $\beta \pm 2\pi/\Lambda$ on a wave of propagation constant β. The sideband $\beta - 2\pi/\Lambda$ can be made synchronous to a wave of propagation constant $-\beta$. Such coupling can lead to exponential solutions that decay with distance from their point (or cross section) of excitation.

The frequency range within which the coupling-of-modes equations give decaying solutions is the stopband. Filter structures can be made of gratings of finite length, yielding maximum reflection within the stopband, and full transmission at certain resonance frequencies within the passband. Two grating structures displaced with respect to each other by an odd-integer multiple of a quarter wavelength exhibit a transmission resonance at the center of the stopband of narrow width with a high external Q, because the gratings are utilized at the frequency of maximum reflection.

We derived a formula for the coupling coefficient between the forward and backward waves. We used the expression for the power fed into wave A by wave B in terms of the coupling coefficient κ_{AB} and equated it to the power produced by the polarization current density of wave B as caused by the periodic perturbation of the waveguide.

PROBLEMS

8.1. The coupling coefficient κ_{ab} is a function of the position of the reference plane. If the reference plane is shifted by a fraction X of Λ in the $+z$ direction, the phase of κ_{ab} changes.

(a) Determine the change of κ_{ab} as a function of X at resonance ($\delta = 0$).

(b) Do the same for κ_{ba}. Is the power conservation condition $\kappa_{ab} = \kappa_{ba}^*$ still satisfied?

8.2. Multiple-layer mirror coatings produce frequency-dependent reflectivities. Consider an infinite parallel plane wave incident upon a layered medium (Fig. P8.2a). Use the coupling of modes equations

$$\frac{dA}{dz} = -j\,\delta A + \kappa B$$

$$\frac{dB}{dz} = j\,\delta B + \kappa^* A$$

(a) For a semi-infinite layered medium extending from $z = 0$ to $z = \infty$, determine Γ for $|\delta| < |\kappa|$.

(b) Suppose that you change the angle of incidence, from 0, to a general angle θ_i (Fig. P8.2b). What happens to the stopband expressed in terms of $\omega - \omega_0$, which for $\theta_i = 0$ was defined by

$$|\delta| = \left|\frac{\omega - \omega_0}{v_g}\right| \le |\kappa|$$

Assume that $v = c/n$, where n is an average index, whereas the medium to the left of $z = 0$ has index n_0.

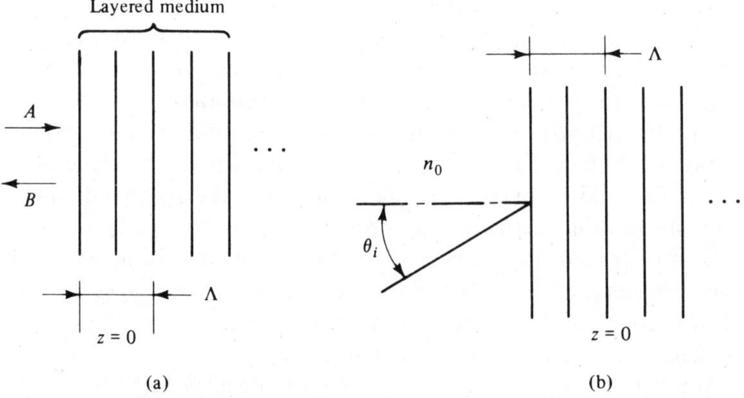

Figure P8.2

8.3. Complete transmission occurs in a "reflection filter" when

$$\sqrt{\delta^2 - |\kappa|^2}\; l = \pi$$

Then (8.15) and (8.16) show that $B = 0$ at $z = 0$ and $z = l$. This is a resonance of the reflection filter.

The external Q, $Q_{e1} = \omega_0 \tau_{e1}/2$, of the resonance can be found according to (7.43). Indeed, for $s_{+2} = 0$, at resonance for a lossless symmetric resonator $1/\tau_0 = 0$, $\tau_{e1} = \tau_{e2} = \tau_e$, one has

$$|a|^2 = \frac{\tau_e}{2} |s_{+1}|^2$$

where $|a|^2$ is the stored energy and $|s_{+1}|^2$ is the incident power. Using (8.15) and (8.16), evaluate the external Q. Compare with the external Q of Section 8.3. Why can the latter have higher Q?

Note: The $|a|$ of the resonator formalism and the a of (8.1) are different physical quantities; the former is the amplitude of the resonance mode; the latter is the amplitude of the forward wave and is a function of z.

8.4. (a) Show that a parallel resonant circuit (Fig. P8.4a) near its resonance frequency, connected across a transmission line at a cross section at which the reflection coefficient is Γ_L, transforms Γ_L into

$$\Gamma = \frac{[2\Gamma_L/(1 + \Gamma_L)] - j(\omega - \omega_0)\tau_e}{[2/(1 + \Gamma_L)] + j(\omega - \omega_0)\tau_e} \tag{1}$$

where $2/\omega_0 \tau_e = Y_0/\omega_0 C$ and $\omega_0 = 1/\sqrt{LC}$.

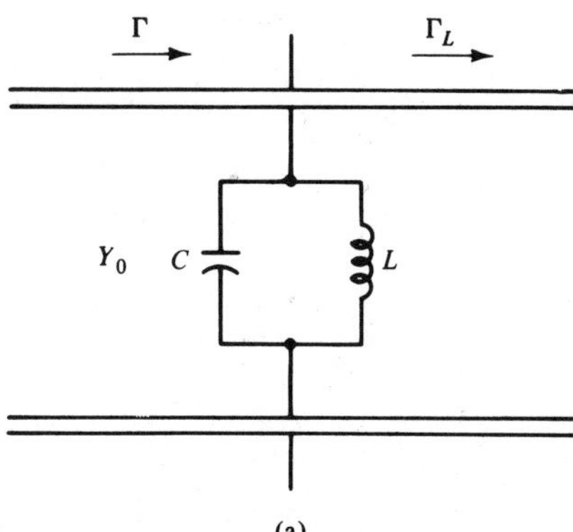

Figure P8.4a

(b) Two gratings in cascade spaced by $\lambda/4$ act as a transmission resonator and have the equivalent circuit of Fig. P8.4a in the neighborhood of the resonant frequency, small δ/κ. The reference planes are picked as shown in Fig. P8.4b. Prove that the transformation is of the form (1) by expanding to first order in δ/κ the numerator and denominator of Γ resulting from (8.18). Use that fact that $\kappa = -jK$ with K real and positive as shown in (8.35). Ignore $e^{-2\gamma l}$ and δ^2/κ^2 compared with unity and use the fact that

$$\coth \gamma l \simeq 1 + 2e^{-2\gamma l}$$

Show that the τ_e parameter found from this analysis agrees with that of p. 246.

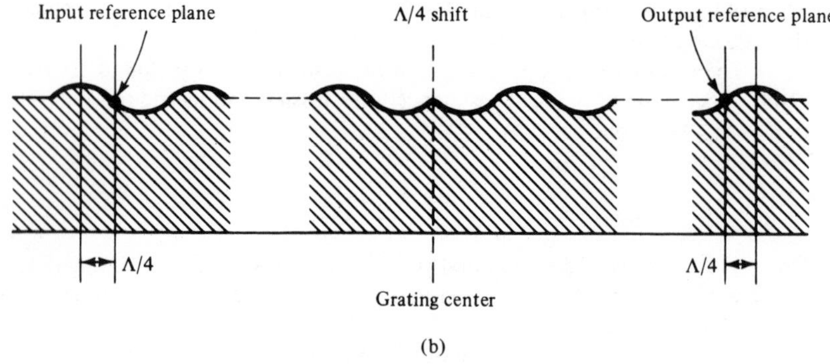

(b)

Figure P8.4b

8.5. Show that the grating arrangement shown in Fig. P8.5a is equivalent to the circuit of Fig. P8.5b. Find the values of $Y_0 L_a$, $Y_0 L_b$, $C_a Y_0$, and C_b/Y_0 in terms of the grating parameter. Use the results of P8.4.

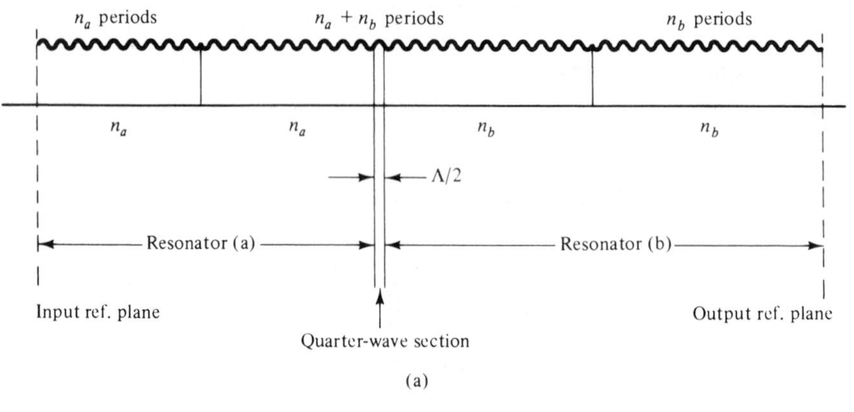

(a)

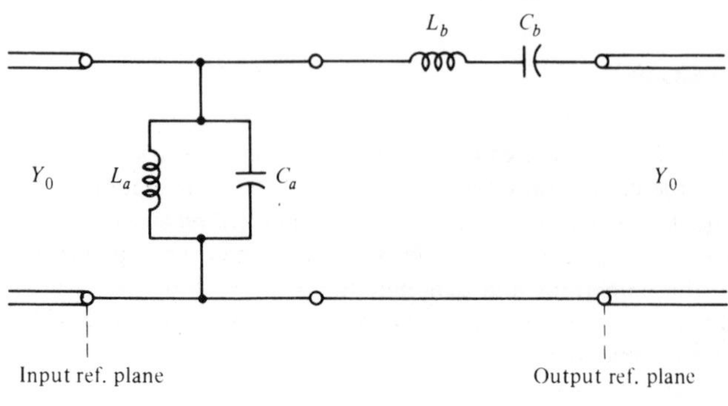

(b)

Figure P8.5

REFERENCES

[1] H. Kogelnik and C. V. Shank, "Coupled-wave theory of distributed feedback lasers," *J. Appl. Phys.*, *43*, no. 5, 2328–2335, May 1972.

[2] D. C. Flanders, H. Kogelnik, C. V. Shank, and R. D. Stanley, "Narrow-band grating filters for thin-film optical waveguides," *Appl. Phys. Lett.*, *25*, 651–652, Dec. 1974.

[3] H. A. Haus and C. V. Shank, "Antisymmetric taper of distributed feedback lasers," *IEEE J. Quantum Electron.*, *QE-12*, 532–539, Sept. 1976.

9

ACOUSTO-OPTIC MODULATORS

Thus far we have analyzed structures that were time independent. We studied reflection from periodic spatial perturbations and discussed its application to transmission resonator and filter design.

An optical wave propagating through a medium with periodic time and space variations acquires modulation sidebands. This interaction can be used to modulate and deflect optical waves. Because acoustic waves propagate at a velocity that is five to six orders of magnitude smaller than the speed of light, acoustic waves have wavelengths comparable to optical waves if the acoustic frequency is five to six orders of magnitude smaller than the optical frequency. Traveling or standing acoustic waves in an acousto-optic medium set up traveling or standing layers of spatial index inhomogeneities. Optical waves can scatter from the stratified index layers.

In this chapter we derive the so-called momentum and energy conservation relations, which impose conditions on the wave vectors and frequencies of the interacting acoustic and optical waves. We analyze the interaction between an incident optical wave and the scattering acoustic wave by the coupling-of-modes formalism. Under proper conditions, we find the existence of a diffracted optical wave which is frequency- and direction-shifted from the incident optical wave. The interaction is evaluated quantitatively in a simple geometry. Then we look at an acousto-optic amplitude modulator. The chapter continues with the discussion of active modelocking of lasers. Such "modelocking" via a modulator inserted in the laser resonator causes the laser to emit short pulses. Because of the very broad absolute bandwidths of the laser media, the pulses can be made short (typically 50 ps). Finally, we evaluate the power requirements of an acousto-optic modulator in terms of the photoelastic constant.

9.1 ACOUSTO-OPTIC WAVE COUPLER [1]

In an acousto-optic modulator an acoustic wave sets up a spatial modulation of the index of an acousto-optic medium; an optical wave diffracts from the index modulation. In a linear acousto-optic medium, the change of index n is proportional to the strain. We now study in greater detail the diffraction as predicted by Maxwell's equations.

The equation for the electric field may be derived from (1.1) and (1.2) by taking the curl of the former and introducing it into the latter (with $\boldsymbol{M} = \boldsymbol{J} = 0$):

$$\nabla \times (\nabla \times \boldsymbol{E}) = -\varepsilon_0 \mu_0 \frac{\partial^2 \boldsymbol{E}}{\partial t^2} - \mu_0 \frac{\partial^2 \boldsymbol{P}}{\partial t^2} \tag{9.1}$$

The displacement density $\varepsilon_0 \boldsymbol{E} + \boldsymbol{P}$ can be separated into a part due to the time-independent background index n, and a part due to the space–time-dependent index Δn produced by the acoustic wave,

$$\varepsilon_0 \boldsymbol{E} + \boldsymbol{P} = \varepsilon_0[n + \Delta n(\boldsymbol{r}, t)]^2 \boldsymbol{E} \simeq \varepsilon_0 n^2 \boldsymbol{E} + 2\varepsilon_0 n \, \Delta n(\boldsymbol{r}, t)\boldsymbol{E} \tag{9.2}$$

where we ignore terms quadratic in Δn, because $\Delta n \ll n$ is a valid assumption in most cases of interest. Combining (9.1) and (9.2), we obtain

$$\nabla \times (\nabla \times \boldsymbol{E}) + \mu_0 \varepsilon_0 n^2 \frac{\partial^2 \boldsymbol{E}}{\partial t^2} = -\mu_0 \varepsilon_0 \frac{\partial^2}{\partial t^2} [2n \, \Delta n(\boldsymbol{r}, t)\boldsymbol{E}] \tag{9.3}$$

We consider a plane acoustic wave with the propatation vector $\boldsymbol{k}_s$ in the x-z plane as shown in Fig. 9.1. The index change produced by the wave of fre-

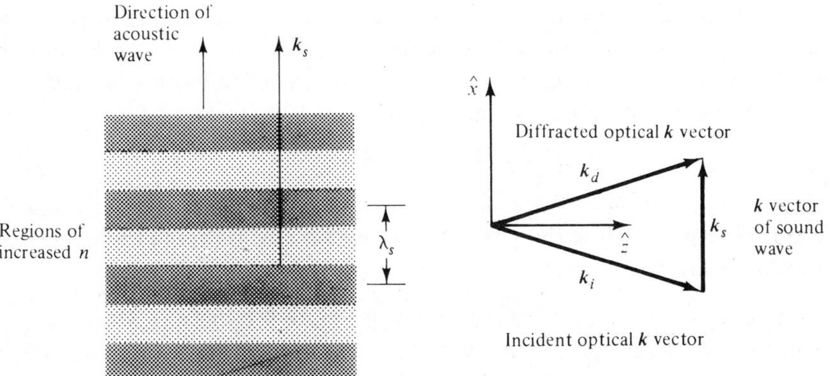

Figure 9.1 Bulk grating effect and k-vector diagram.

quency ω_s has the same spatial dependence as the plane wave

$$\Delta n(\boldsymbol{r}, t) = \Delta n \cos(\omega_s t - \boldsymbol{k}_s \cdot \boldsymbol{r})$$

$$= \frac{\Delta n}{2}[e^{j(\omega_s t - \boldsymbol{k}_s \cdot \boldsymbol{r})} + e^{-j(\omega_s t - \boldsymbol{k}_s \cdot \boldsymbol{r})}] \tag{9.4}$$

where we denote henceforth by Δn, without the argument $\boldsymbol{r}$, t, the peak amplitude of the index modulation.

Consider, now, Gauss's law for the electric field as affected by the index change $\Delta n(\boldsymbol{r}, t)$. The dielectric constant is a function of time and space:

$$\nabla \cdot \varepsilon \boldsymbol{E} = \rho = \varepsilon \nabla \cdot \boldsymbol{E} + \boldsymbol{E} \cdot \nabla \varepsilon \tag{9.5}$$

with

$$\varepsilon = \varepsilon_0[n + \Delta n(\boldsymbol{r}, t)]^2 \simeq \varepsilon_0[n^2 + 2n\,\Delta n(\boldsymbol{r}, t)]$$

For the assumed space–time dependence of Δn and thus ε, the gradient of ε is in the x-z plane. If $\boldsymbol{E}$ is polarized along y, as we shall assume, $\boldsymbol{E} \cdot \nabla \varepsilon = 0$ and, in the absence of free charge ($\rho = 0$), the divergence of $\boldsymbol{E}$ is zero according to (9.5). We obtain, from (9.3),

$$\nabla^2 \boldsymbol{E} - \mu_0 \varepsilon_0 n^2 \frac{\partial^2 \boldsymbol{E}}{\partial t^2} = \mu_0 \varepsilon_0 \frac{\partial^2}{\partial t^2}[2n\,\Delta n(\boldsymbol{r}, t)\boldsymbol{E}] \tag{9.6}$$

The time-dependent index multiplying the incident plane wave with the E field

$$\mathbf{E} \propto \exp[-j(\boldsymbol{k}_i \cdot \boldsymbol{r} - \omega_i t)]$$

produces a source for the diffracted wave on the right-hand side of (9.6) with the frequency

$$\omega_d = \pm\omega_s + \omega_i \tag{9.7}$$

and with the spatial dependence

$$\exp[-j(\boldsymbol{k}_i \pm \boldsymbol{k}_s) \cdot \boldsymbol{r}]$$

In order that the diffracted waves generated by the source not interfere destructively, the spatial dependence of the source $\boldsymbol{k}$ vector $\pm\boldsymbol{k}_s + \boldsymbol{k}_i$ must be close to the propagation constant $\boldsymbol{k}_d$ of the diffracted wave at the frequency ω_d, the frequency of the source. In general, if

$$\boldsymbol{k}_s + \boldsymbol{k}_i = \boldsymbol{k}_d \tag{9.8}$$

the vector equality $-\boldsymbol{k}_s + \boldsymbol{k}_i = \boldsymbol{k}_d$ will not hold. Only one diffracted wave is excited. Equation (9.7) is the *energy conservation law* and (9.8) the *momentum conservation law.**

Usually, $\omega_s \ll \omega_i$ and thus $|\boldsymbol{k}_d| \simeq |\boldsymbol{k}_i|$. Figure 9.1 shows a $\boldsymbol{k}$ diagram for the purpose of analysis. We analyze the excitation of the $\boldsymbol{k}_d$ wave as the $\boldsymbol{k}_i$

*In quantum mechanics, the momentum of a quasi-particle is $\hbar \boldsymbol{k}$, its energy $\hbar\omega$, where $\hbar$ is Planck's constant divided by 2π.

wave proceeds through a slab of an acousto-optic medium. We find an approximate solution of (9.6) by writing the incident and diffracted waves in terms of a product, one factor of which is the natural space–time dependence of the plane wave; the other factor accounts for the slow spatial variation of the field amplitudes as produced by the acoustic coupling between the two waves:

$$\mathbf{E}_i = \hat{y}\mathrm{A}_i(z)e^{-j(\boldsymbol{k}_i \cdot \boldsymbol{r} - \omega_i t)} \tag{9.9}$$

$$\mathbf{E}_d = \hat{y}\mathrm{A}_d(z)e^{-j(\boldsymbol{k}_d \cdot \boldsymbol{r} - \omega_d t)} \tag{9.10}$$

where the z dependences of $\mathrm{A}_i(z)$ and $\mathrm{A}_d(z)$ are the result of the coupling of the incident and diffracted waves by the index modulation. We apply (9.6) to the diffracted wave as excited by the source due to the incident wave and the acoustic index modulation, disregarding second-order derivatives of $\mathrm{A}_d(z)$:

$$-(\boldsymbol{k}_d^2 - \omega_d^2 \mu_0 \varepsilon_0 n^2)\mathrm{A}_d - 2j\boldsymbol{k}_d \cdot \nabla \mathrm{A}_d \simeq -\frac{\omega_d^2}{c^2} n \,\Delta n\, \mathrm{A}_i \tag{9.11}$$

We have retained only that *source term* on the right-hand side which has the correct frequency and spatial dependence. Because $k_d = \omega_d \sqrt{\mu_0 \varepsilon_0}\, n$ and $\boldsymbol{k}_d \cdot \nabla \mathrm{A}_d(z)$ is $\cos\theta\, k_d(d\mathrm{A}_d/dz)$, we obtain

$$\frac{d\mathrm{A}_d}{dz} = -j\frac{\omega_d}{2c}\frac{\Delta n}{\cos\theta}\mathrm{A}_i \tag{9.12}$$

One may write an equation analogous to (9.11) for the incident wave as affected by the diffracted wave. One finds again from (9.6):

$$\frac{d\mathrm{A}_i}{dz} = -j\frac{\omega_i}{2c}\frac{\Delta n}{\cos\theta}\mathrm{A}_d \tag{9.13}$$

These are coupling of modes equations in space. The coupling coefficients are

$$-j\frac{\omega_i}{2c}\frac{\Delta n}{\cos\theta} \quad \text{and} \quad -j\frac{\omega_d}{2c}\frac{\Delta n}{\cos\theta}$$

They are not exactly equal, in that $\omega_i \neq \omega_d$. There is no mistake, here. We shall learn later that the present system does not obey power conservation—after all, it is driven mechanically (acoustically) and the acoustic power can produce or extract optical power. However, ω_d differs from ω_i by 1 part in 10^5 and therefore we need not be particularly concerned about this slight discrepancy at this point. In all other respects, the system behaves just like a lossless coupled-wave system. The solutions are, for $\mathrm{A}_d = 0$ at $z = 0$,

$$\mathrm{A}_i(z) = \mathrm{A}_i(0)\cos|\kappa|z \tag{9.14}$$

$$\mathrm{A}_d(z) = -j\mathrm{A}_i(0)\sin|\kappa|z \tag{9.15}$$

where

$$|\kappa| = \frac{\sqrt{\omega_i \omega_d}}{2c} \frac{\Delta n}{\cos \theta} \simeq \frac{\omega_i}{c} \frac{\Delta n}{2 \cos \theta}$$

The incident wave is depleted by the diffracted wave as it travels along the medium supporting the acoustic wave. The diffracted wave is shifted in frequency. If the travel distance is long enough ($|\kappa| z > \pi/2$), the diffracted wave feeds back into the incident wave. This assumes, of course, that the two waves do not separate spatially—that $l \sin \theta$ is much smaller than the beam cross section. The obvious applications of the process of Fig. 9.2 are optical beam deflection and frequency shifting.

Equations (9.14) and (9.15) have another interpretation. Indeed, returning to (9.4), we note that we could set $\omega_s = 0$ in this equation. The resulting expression describes a time-independent "bulk grating," a cosinusoidal "frozen" index variation. The diffracted wave vector must still obey the "momentum conservation" law (9.8), but the diffracted wave has the same frequency as the incident wave, $\omega_d = \omega_i$. The same coupling-of-modes equations result, (9.9) and (9.10), except for the fact that $\omega_d = \omega_i$. The coupling coefficients are now strictly identical. The solutions (9.14) and (9.15) apply. They describe the interaction of the diffracted wave with the incident wave in a *bulk grating.*

9.2 AN ACOUSTO-OPTIC AMPLITUDE MODULATOR

In the preceding section we considered the interaction of a *traveling* acoustic wave with an optical wave. The incident wave is depleted as it travels through the acoustic medium and emerges with a time-independent intensity according to (9.14); the diffracted wave is shifted in frequency and direction. The intensity of an incident wave is modulated if it interacts with a *standing acoustic wave.* This principle is used, for example, in the acousto-optic Ge modulator of CO_2 radiation ($\lambda = 10.6\ \mu$m) (Fig. 9.2) or a quartz modulator for visible light.

A transducer is excited electrically and sets up acoustic standing waves in the crystal, $\exp(j\boldsymbol{k}_s \cdot \boldsymbol{r})$ and $\exp[-(j\boldsymbol{k}_s \cdot \boldsymbol{r})]$. The simplest case results when the incident wave vector is set so that the "momentum" conservation law is strictly obeyed (Fig. 9.3), for one of the acoustic traveling waves

$$\boldsymbol{k}_d = \boldsymbol{k}_s + \boldsymbol{k}_i \tag{9.16}$$

The momentum conservation law is not obeyed with the oppositely traveling acoustic wave. The index variation is now

$$\begin{aligned} n(\boldsymbol{r}, t) = \Delta n \sin \omega_s t \cos (\boldsymbol{k}_s \cdot \boldsymbol{r}) = \frac{\Delta n}{4j} \{&\exp [j(\omega_s t - \boldsymbol{k}_s \cdot \boldsymbol{r})] \\ &+ \exp [j(\omega_s t + \boldsymbol{k}_s \cdot \boldsymbol{r})] - \exp [-j(\omega_s t - \boldsymbol{k}_s \cdot \boldsymbol{r})] \\ &- \exp [-j(\omega_s t + \boldsymbol{k}_s \cdot \boldsymbol{r})]\} \end{aligned} \tag{9.17}$$

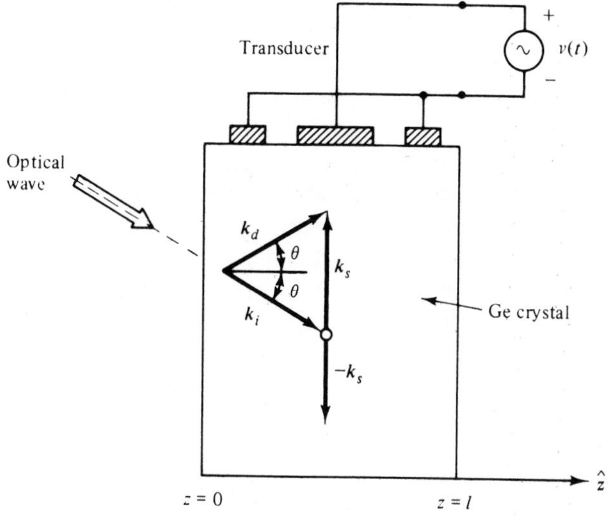

Figure 9.2 Typical $\boldsymbol{k}$-vector diagram for acousto-optic diffraction.

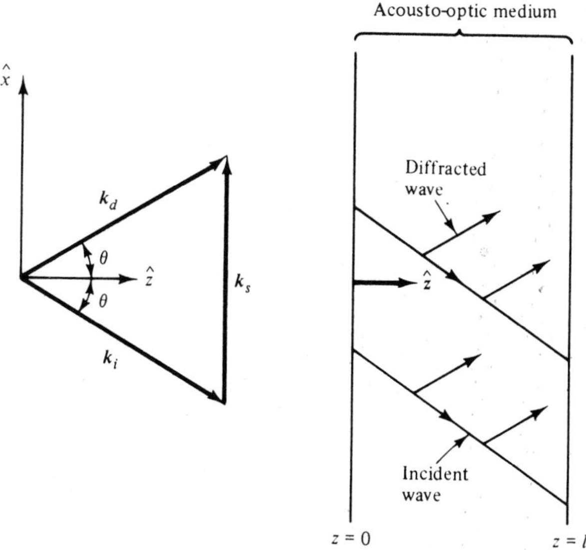

Figure 9.3 Acousto-optic amplitude modulator.

The problem in Section 9.1 was simplified by the fact that only one acoustic wave was present, with the space–time dependence exp $[\pm j(\omega_s t - \boldsymbol{k}_s \cdot \boldsymbol{r})]$ [compare (9.4) with (9.17)]. The incident wave at ω_i produced a diffracted wave at $\omega_d = \omega_i + \omega_s$ with the propagation vector $\boldsymbol{k}_i + \boldsymbol{k}_s = \boldsymbol{k}_d$ and the diffracted wave reacted back by producing $\omega_d - \omega_s = \omega_i$ with propagation vector $\boldsymbol{k}_i = \boldsymbol{k}_d - \boldsymbol{k}_s$. Now we have four dependences introduced by the index variation, exp $[\pm j(\omega_s t \pm \boldsymbol{k}_s \cdot \boldsymbol{r})]$. The incident wave at frequency ω_i can produce a diffracted wave at frequency $\omega_i \pm \omega_s$ with the propagation vector $\boldsymbol{k}_d = \boldsymbol{k}_i + \boldsymbol{k}_s$, which reacts back to produce $(\omega_i \pm \omega_s) \pm \omega_s$ with the propagation vector $\boldsymbol{k}_i = \boldsymbol{k}_d - \boldsymbol{k}_s$. Since the acoustic frequency ω_s is small compared with the optical frequency, waves produced at $\omega_i \pm 2\omega_s$ remain *phase-matched*; they satisfy the relation $\boldsymbol{k}_d = \boldsymbol{k}_i + \boldsymbol{k}_s$, for $\omega_d = \omega_i \pm m\omega_s$ for a very large range of m values. The sidebands *cascade*.

The simplest way of treating the problem is in terms of an *adiabatic approximation*; the (bulk) grating set up by the acoustic wave is treated as time independent; that is, the scattering of the electromagnetic wave is evaluated from a time-independent (bulk) grating. Then the depth of the spatial index modulation is varied slowly with time. The alternative, more rigorous approach is presented in Appendix 9A.

We utilize (9.14) and (9.15) as applied to the bulk grating analysis. At $z = l$,

$$\mathrm{A}_i(l) = \mathrm{A}_i(0) \cos\left(\frac{\omega_i}{c} \frac{\Delta n}{2 \cos \theta} l\right) \tag{9.18}$$

$$\mathrm{A}_d(l) = -j\mathrm{A}_i(0) \sin\left(\frac{\omega_i}{c} \frac{\Delta n}{2 \cos \theta} l\right) \tag{9.19}$$

Suppose now that Δn varies with time, so that $\Delta n(t) \to \Delta n \sin \omega_s t$. A graphical construction of the amplitudes $\mathrm{A}_i(l)$ and $\mathrm{A}_d(l)$ as functions of time is shown in Fig. 9.4.

A Fourier decomposition of $\mathrm{A}_i(l)$ and $\mathrm{A}_d(l)$ is accomplished by making use of the Bessel-function identity [2]

$$\exp(jx \sin \omega_s t) = \sum_{m=-\infty}^{\infty} J_m(x) e^{jm\omega_s t} \tag{9.20}$$

From the above one may construct

$$\cos(x \sin \omega_s t) = \sum_{m \text{ even}} J_m(x) e^{jm\omega_s t} \tag{9.21}$$

and

$$\sin(x \sin \omega_s t) = -j \sum_{m \text{ odd}} J_m(x) e^{jm\omega_s t} \tag{9.22}$$

where we use the fact that $J_m(-x) = (-1)^m J_m(x)$. The incident wave amplitude

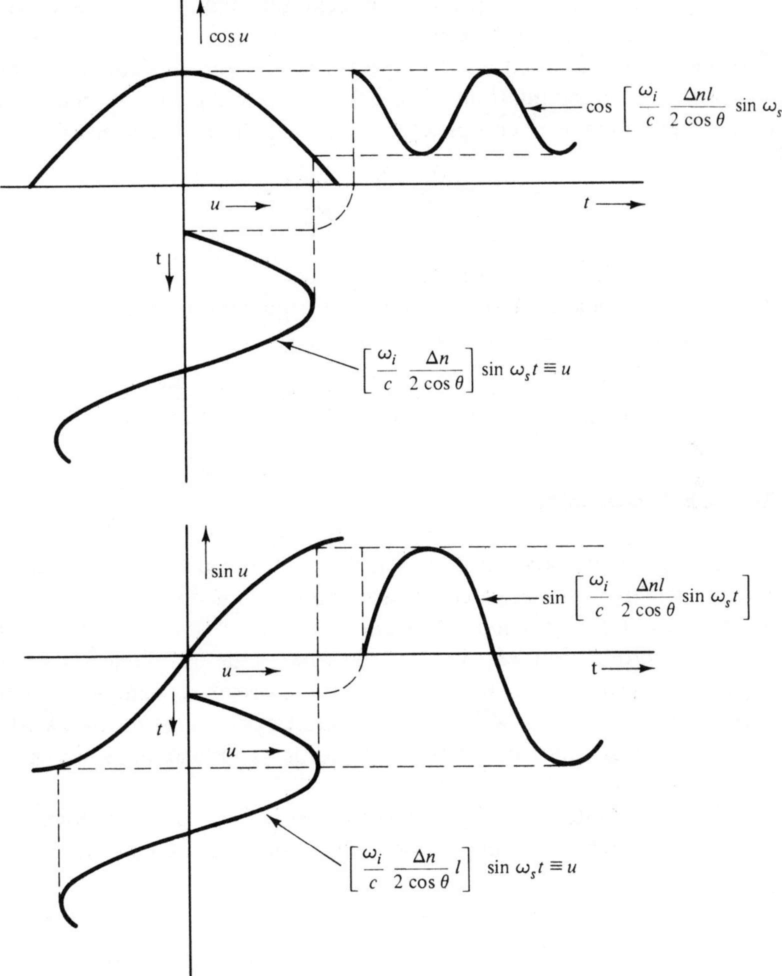

Figure 9.4 Time dependence of A_i and A_d.

decomposed into Fourier components is

$$A_i(l) = A_i(0) \sum_{m \text{ even}} e^{jm\omega_s t} J_m\left(\frac{\omega_i}{c}\frac{\Delta nl}{2\cos\theta}\right) \tag{9.23}$$

The diffracted wave is

$$A_d(l) = -A_i(0) \sum_{m \text{ odd}} e^{jm\omega_s t} J_m\left(\frac{\omega_i}{c}\frac{\Delta nl}{2\cos\theta}\right) \tag{9.24}$$

This is the same result as obtained in Appendix 9A through the coupling of the infinite set of sidebands. The modulation obeys the condition of adiabaticity if the transit time of the optical field through the crystal is much shorter than the period of the modulation $2\pi/\omega_s$.

Two facts may be noted. First, the sidebands of the incident wave are at even harmonics, the modulation of the diffracted wave at odd harmonics. Second, the fundamental is completely canceled if the modulation obeys the relation

$$\frac{\omega_i}{c}\frac{\Delta n l}{2\cos\theta} = 2.405$$

the first zero of the Bessel function J_0.

If the modulation depth is "shallow," the argument in (9.23) is small, and the two sidebands at $2\omega_s$ are ($\omega_i \to \omega$)

$$\frac{A_i(l, \omega + 2\omega_s)}{A_i(l, \omega)} = J_2\left(\frac{\omega}{c} l \frac{\Delta n}{2\cos\theta}\right) \simeq \frac{1}{8}\left(\frac{\omega}{c} l \frac{\Delta n}{2\cos\theta}\right)^2 \tag{9.25}$$

9.3 ACTIVE MODELOCKING [3, 4]

We have learned in the preceding section that an optical beam is amplitude-modulated when it passes through an acousto-optic medium under acoustic standing-wave excitation. Such a modulation may serve several purposes. One of these is the production of short pulses from a laser oscillator—by the process of *active modelocking.* In this section we present an analysis of active modelocking. The theory fits well into the overall mathematical framework of this chapter, which is concerned with propagation of optical waves through a time-dependent system.

Consider the system of Fig. 9.5 consisting of two mirrors forming a Fabry–Perot resonator containing a laser gain medium, a lossy medium, and a

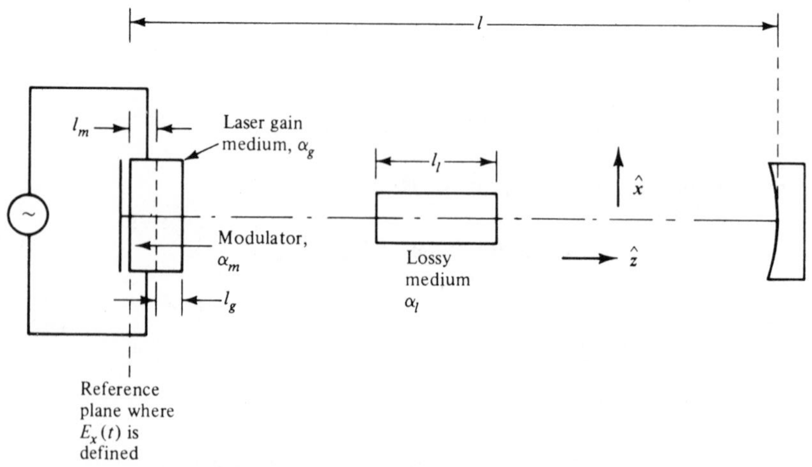

Figure 9.5 Schematic of modelocked laser.

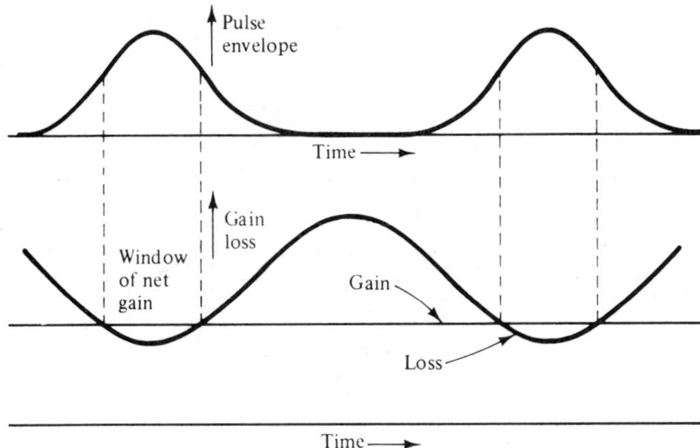

Figure 9.6 Pulse-envelope gain and loss as functions of time in modelocked laser resonator.

modulator. The modulator modulates the loss of the resonator at a frequency that is equal to, or near, the frequency of the mode separation of the Fabry–Perot modes of the resonator. This corresponds to a period of the loss variation equal or nearly equal to the round-trip time T_R of radiation in the resonator. (In an air-filled Fabry–Perot $\Delta f = c/2l = 1/T_R$.) The operation of modelocking can be understood schematically with the aid of Fig. 9.6. The radiation in the Fabry–Perot bouncing back and forth "sees" a time-independent gain and a time-dependent loss, which is equivalent to a time-dependent net gain. The part of the radiation that passes the modulator at an instant in time when it has minimum loss experiences maximum net gain. The radiation passing at other times experiences less gain, and even net loss. A pulse is formed which grows and narrows. In the steady state, the gain adjusts to equal the loss, and pulse narrowing in one pass is compensated by pulse broadening due to dispersion of the gain medium. Note that Fig. 9.6 and the above argument assume a specific time of passage of the pulse through the loss modulator. This is correct only when the transit time through the loss modulator is short compared with the width of the optical pulse and the modulator is adjacent to one of the mirrors.

The qualitative description can be put on a quantitative basis.

Effects of Gain, Loss, and Modulator

The (laser) gain medium of length l_g is described [1] by a frequency-dependent gain coefficient* $\alpha_g/\{1 + [(\omega - \omega_0)/\omega_g]^2\}$ where ω_0 is the center frequency of the gain medium, ω_g is its width, and $\alpha_g l_g$ is the integrated gain at line center.

*Strictly, the frequency-dependent gain coefficient requires a frequency-dependent imaginary part (i.e., phase shift of the light passing through the medium). This is required by the Kramers–Kronig relations [5]. The imaginary part does not affect the final result and will be ignored for simplicity.

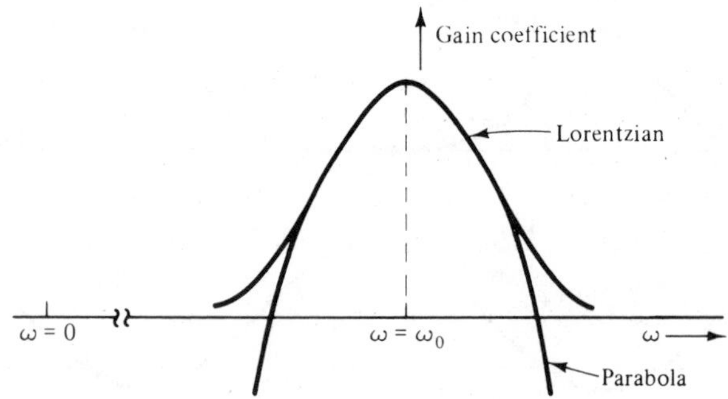

Figure 9.7 Lorentzian gain profile replaced by parabola.

A wave with the electric field polarized along x, $E_x(\omega) \propto a(\omega)$, traveling in the z direction and passing through the (laser) gain medium, emerges from it amplified by the exponential gain factor:

$$\exp\left\{\frac{\alpha_g l_g}{1 + [(\omega - \omega_0)/\omega_g]^2}\right\} a(\omega) \simeq \left[1 + \alpha_g l_g\left(1 - \left(\frac{\omega - \omega_0}{\omega_g}\right)^2\right)\right] a(\omega) \tag{9.26}$$

where we have assumed that $\alpha_g l_g \ll 1$ and $[(\omega - \omega_0)/\omega_g]^2 \ll 1$ so that expansions could be made in these parameters. Both assumptions are usually justified. The second assumption is legitimate, if the spectrum of the pulse train resulting from the modulation is narrow compared with the bandwidth of the gain. The *Lorentzian* gain profile $\alpha_g/[1 + (\omega - \omega_0)^2/\omega_g^2]$ is replaced by a parabola (Fig. 9.7). To justify the first assumption we note that the gain coefficient α_g is treated as time independent, the saturation of the gain due to the passage of one individual pulse is small, and the saturation is caused by the cumulative effect of the passage of many pulses. This assumption is valid if the response time of the gain medium is long compared with the round-trip time of the pulse.* The gain settles down to a time-averaged value which is just sufficient to overcome the assumedly small loss per pass.

In addition to the gain, there is time-independent loss due to unavoidable absorption, and power escape from the laser resonator. We lump together all these losses into a phenomenological loss coefficient α_l of a medium of length l_l, so that the net time-independent loss is represented by the loss factor

$$\exp(-\alpha_l l_l) a(\omega) \simeq (1 - \alpha_l l_l) a(\omega) \tag{9.27}$$

The wave traversing the modulation medium with the time-dependent modu-

*Instead of the saturation formula (7.51), we use its generalization to a time-dependent case

$$\frac{d\alpha_g}{dt} = -\frac{\alpha_g - \alpha_g^0}{\tau_g} - \frac{\alpha_g}{\tau_g}\frac{I}{I_s}$$

where τ_g is the relaxation time of the gain medium. In the steady state, $d/dt = 0$, this formula reduces to (7.51). When the time between pulses is much less than τ_g, the gain cannot follow the instantaneous intensity $I(t)$, but reacts only to its time average.

lated loss is best described in the time domain. If the medium of length l_m has the loss coefficient

$$\alpha_m l_m(1 - \cos \omega_M t)$$

the wave amplitude $A(t) \exp(j\omega_0 t)$ [where $A(t)$ is the envelope, ω_0 the "carrier frequency"] passing through it is modified by the factor

$$\begin{aligned} \text{modulation factor} &= \exp[-\alpha_m l_m(1 - \cos \omega_M t)] \\ &\simeq 1 - \alpha_m l_m(1 - \cos \omega_M t) \end{aligned} \tag{9.28}$$

We must describe the combined action of gain and loss on the amplitude of the wave in either the frequency domain or the time domain. We choose the latter. Multiplication of $a(\omega)$ by $j(\omega - \omega_0)$ in the frequency domain is equivalent to differentiation with respect to time of the envelope $A(t)$ [compare (6.57)].

The wave passes the three elements twice in one round trip in the resonator. Expanded to first order in net gain and loss the combined effect on $A(t)$ is

$$\left[1 + 2\alpha_g l_g\left(1 + \frac{1}{\omega_g^2}\frac{d^2}{dt^2}\right) - 2\alpha_l l_l - 2\alpha_m l_m(1 - \cos \omega_M t)\right]A(t) \equiv O\{A(t)\} \tag{9.29}$$

Equation (9.29) defines the operator O, which expresses the effect on the pulse of the combined effect of gain, loss, and modulation.

Equation of Steady-State Modelocking in the Time Domain

The function $O\{A(t)\}$ is delayed by the round-trip time T_R and returns as $A(t)$ so that:

$$O\{A(t - T_R)\} = A(t)$$

or

$$O\{A(t)\} = A(t + T_R) \tag{9.30}$$

In the steady state, $A(t)$ must be a periodic function with the period $2\pi/\omega_M$:

$$A\left(t - \frac{2\pi}{\omega_M}\right) = A(t) \tag{9.31}$$

If the modelocking drive period differs from the round-trip time T_R, the pulse must be pushed into synchronism by reshaping, assuming, of course, that steady-state modelocking has been achieved. If $2\pi/\omega_M < T_R$, the pulse is advanced by "shaving off" the trailing edge and amplifying the leading edge. The converse happens when $2\pi/\omega_M > T_R$(Fig. 9.8). Denote

$$\frac{2\pi}{\omega_M} = T_R + \delta T_R \tag{9.32}$$

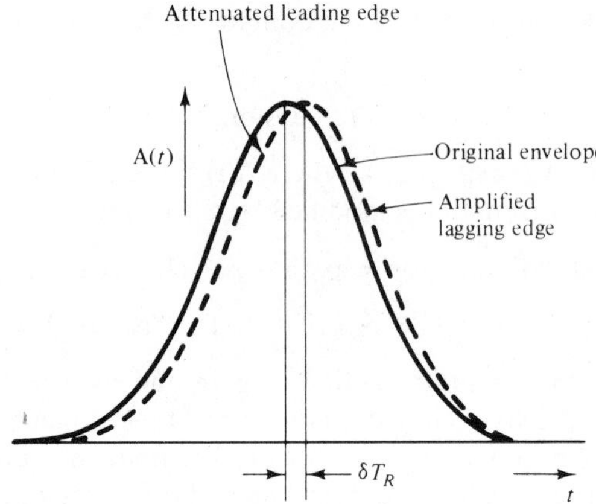

Figure 9.8 Retardation of pulse by amplification of lagging edge and attenuation of leading edge.

where δT_R is the change to the transit time wrought by pulse reshaping. Therefore, combining (9.30), (9.31), and (9.32), we have

$$\mathrm{A}(t) - \delta T_R \frac{d\mathrm{A}}{dt} = O\{\mathrm{A}(t)\} \tag{9.33}$$

where we assumed that δT_R is small and the Taylor expansion can be broken off with the first term. Combining (9.29) and (9.33), we obtain

$$\left[2\alpha_g l_g\left(1 + \frac{1}{\omega_g^2}\frac{d^2}{dt^2}\right) - 2\alpha_l l_l - 2\alpha_m l_m(1 - \cos\omega_M t) + \delta T_R \frac{d}{dt}\right]\mathrm{A}(t) = 0 \tag{9.34}$$

This is the equation of active modelocking. It is the Mathieu [2] equation, which yields periodic solutions with period $2\pi/\omega_M$. The equation can be simplified further under the assumption of strong modelocking—well-separated pulses occurring near the time instant of minimum loss $t = n(2\pi/\omega_M)$, with n an integer. Then, expanding $\cos\omega_M t$ around the minimum at $t = 0$,

$$\left[2\alpha_g l_g\left(1 + \frac{1}{\omega_g^2}\frac{d^2}{dt^2}\right) - 2\alpha_l l_l - \alpha_m l_m \omega_M^2 t^2 + \delta T_R \frac{d}{dt}\right]\mathrm{A}(t) = 0 \tag{9.35}$$

This equation yields approximate solutions for A(t) at and in the vicinity of $t = 0$. We shall study (9.35) in the case of "synchronism," $\delta T_R = 0$, which is achieved by proper adjustment of the modulation frequency ω_M. Then (9.35) is the Schrödinger equation for a particle in a parabolic potential with the solutions (see Appendix 5A)

$$\mathrm{A}(t) = H_\nu(\omega_p t)\exp\left(-\frac{\omega_p^2 t^2}{2}\right) \tag{9.36}$$

with

$$\omega_p = \sqrt[4]{\frac{\alpha_m l_m}{2\alpha_g l_g}}\sqrt{\omega_M \omega_g} \tag{9.37}$$

and

$$1 - \frac{\alpha_l l_l}{\alpha_g l_g} = \frac{\omega_p^2}{\omega_g^2}(2\nu + 1) \tag{9.38}$$

where H_ν is the Hermite polynomial of order ν. Equation (9.38) is a relation for the gain coefficient $\alpha_g l_g$ in terms of the depth of modulation $\alpha_m l_m$, the modulation frequency ω_M, and the line width ω_g.

Generally, $\omega_p^2 \ll |\omega_g^2|$, consistent with our assumption which allowed the expansion in (9.26), and the gain will not exceed the loss by much, $(\alpha_g l_g) \simeq (\alpha_l l_l)$. Equation (9.38) is then explicitly an expression for the excess gain $\alpha_g l_g/\alpha_l l_l - 1$. The equation defining the pulse shape (9.34) and its approximate form (9.35) are linear equations with time-dependent coefficients, they are satisfied by a pulse envelope function of arbitrary amplitude. The actual pulse amplitude is obtained from the expression for the excess gain, (9.38). The steady-state gain coefficient α_g is reduced from the small-signal gain α_g^0 by the intensity in the resonator:

$$\alpha_g = \frac{\alpha_g^0}{1 + (I/I_s)}$$

where I_s is the saturation intensity (compare footnote on p. 264). Thus (9.38) can be used to determine the intensity of the pulse train.

Gain saturation leads to stability of the fundamental Gaussian solution and instability of the higher-order Hermite–Gaussians, as we now proceed to show. When the fundamental solution is excited with its required gain, then the higher-order solutions have insufficient gain to be operating, they are unexcited. Conversely, if one of the higher-order solutions were set up, the lowest-order solution would experience a gain higher than that required for its steady-state operation and its amplitude would grow until the gain is depressed to the required steady-state value. All higher-order Hermite–Gaussians are unstable [4].

The discussion above leads to the following conclusions:

1. The steady-state pulse is Gaussian.
2. The pulse width is inversely proportional to the fourth root of the modulation depth $\alpha_m l_m$ and the square root of the gain bandwidth $\sqrt{\omega_g}$.

The full width at half maximum intensity of the pulse is

$$\text{FWHM} = \frac{2\sqrt{\ln 2}}{\omega_p}$$

For $\alpha_m l_m/\alpha_g l_g = 1$, $\omega_M/2\pi = 100$ MHz, and a gain bandwidth $\omega_g/\omega_0 = 0.2\%$ with the optical wavelength of a Nd : YAG laser, $\lambda = 1.06$ μm, we find that

$$\text{FWHM} = 42 \text{ ps}$$

A gain bandwidth of 0.2% is narrow compared with the bandwidth of the 1.06 μm optical transition in Nd YAG. It is found experimentally that shorter pulses are obtained when a bandwidth limiting element (for example a Fabry–Perot etalon) is introduced in the laser resonator. On the surface this contradicts the preceding analysis which predicts shorter pulses for greater ω_g. It turns out, however, that the deterioration of the modelocked pulses for an "excessive" bandwidth is accompanied by a broadening of the optical spectrum; different portions of the optical spectrum modelock incoherently with respect to each other. The individual pulses exhibit temporal substructures that can be observed in the autocorrelation of the pulse intensity obtained by second harmonic generation (Section 13.6). The break-up of the pulse spectrum is attributable to spontaneous emission noise which has been ignored in the analysis above [5, 6].

Equation of Steady-State Modelocking in Frequency Domain

For certain applications, it is more appropriate to represent the modelocking equation in the frequency domain. One of these applications is the case of a periodic gain or loss profile as a function of frequency as it occurs in a composite Fabry–Perot resonator [7]. Also, in the frequency domain the modelocking equation lends itself to an interpretation in terms of the locking of axial modes—the origin of the term "modelocking."

Consider the nth Fourier component A_n of the periodic function $A(t)$, with period $2\pi/\omega_M$. The time derivative d/dt multiplies A_n by $jn\omega_M$ and multiplication of $A(t)$ by $\cos \omega_M t$ couples A_{n+1} and A_{n-1} to A_n. We obtain from (9.34)

$$\left[2\alpha_g l_g\left[1 - \left(\frac{n\omega_M}{\omega_g}\right)^2\right] - 2(\alpha_l l_l + \alpha_m l_m) + jn\omega_M\, \delta T_R\right] A_n = -\alpha_m l_m (A_{n+1} + A_{n-1}) \tag{9.39}$$

This form of the equation can be cast into an equivalent-circuit representation (see Fig. 9.9) [4]. Interpret A_n as the complex "voltage" amplitude representing the field of the nth axial mode. The right-hand side is an injection current produced by A_{n+1} and A_{n-1}. The equivalent circuit of Fig. 9.9 shows the "modelocking" aspect of the process. The nth axial mode of the resonator is locked to the $(n - 1)$st and $(n + 1)$st mode through their modulation sidebands.

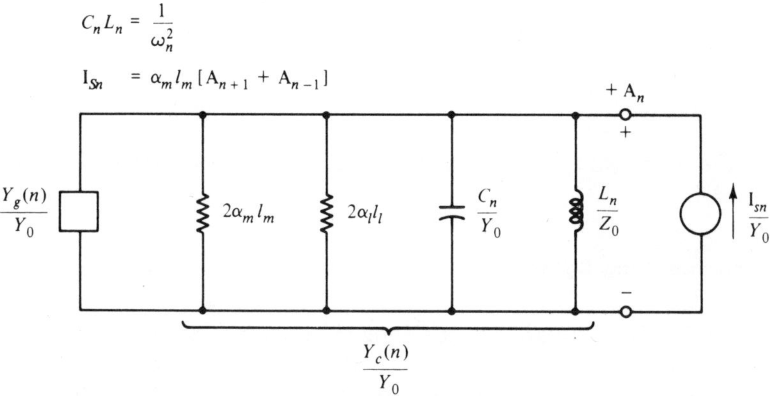

Figure 9.9 Equivalent circuit of (9.38).

The laser medium presents a (normalized) admittance to the nth mode:

$$\frac{Y_g(n)}{Y_0} = -2\alpha_g l_g\left[1 - \left(\frac{n\omega_M}{\omega_g}\right)^2\right] \tag{9.40}$$

The resonator (cavity) admittance of the nth mode is

$$\frac{Y_c(n)}{Y_0} = 2\alpha_l l_l\left(1 - \frac{jn\omega_M\,\delta T_R}{2\alpha_l l_l}\right) + 2\alpha_m l_m \tag{9.41}$$

Im $[Y_c(n)]$ increases with increasing n, the deviation from the center frequency ω_0. This is due to the increased detuning of the resonator modes of increasing n. Indeed, δT_R is the deviation of the round-trip time from "synchronism," from equality between the modulation frequency ω_M and the mode spacing $2\pi/T_R$ (which is $2\pi c/2l$ in a resonator with a medium of unity index); δT_R expresses the detuning. According to (9.32),

$$\omega_M - \frac{2\pi}{T_R} \simeq -\frac{2\pi}{T_R}\frac{\delta T_R}{T_R}$$

Thus the frequency deviation $\Delta\omega_n$ of the nth mode from its center frequency is given by

$$\Delta\omega_n = n\left(\omega_M - \frac{2\pi}{T_R}\right) \simeq -\frac{2\pi}{T_R}\,n\,\frac{\delta T_R}{T_R} \simeq -n\omega_M\frac{\delta T_R}{T_R} = -n\omega_M\frac{\delta T_R\,c}{2l} \tag{9.42}$$

With (9.42), we may write (9.41) in the form

$$\frac{Y_c(n)}{Y_0} = 2\alpha_l l_l\left(1 + 2j\frac{\Delta\omega_n}{\omega_n}Q_{0n}\right) + 2\alpha_m l_m$$

with

$$Q_{0n} = \frac{\omega_n l}{2\alpha_l l_l c}$$

This is the unloaded Q of the resonator due to the loss per pass, $2\alpha_l l_l$ [compare (7.46)].

Modelocking System

Modelocking with an acousto-optic modulator is illustrated in Fig. 9.10. A piezoelectric transducer, electrically driven at a preselected frequency, launches an acoustic wave into the quartz block. If the block depth D is equal to an integer number of acoustic half-wavelengths, a resonant standing acoustic wave is established. The acoustic wave produces a variation in refractive index periodic in time and space and this acts as a bulk diffraction grating, forming a loss element in the cavity by diffracting light out of the main beam. In quartz a driving frequency $f_s = 70$ MHz gives rise to a grating with a period of 4 μm. Furthermore, as this is a standing wave, the grating disappears and reestablishes itself periodically at twice the driving frequency so that the loss introduced into the cavity varies at twice the drive frequency $2f_s$ and its harmonics [see (9.23)].

The quartz block can be cut in the shape of a prism so that it also operates in the laser as the wavelength-selective element. Accurate temperature stabilization of the modulator is required because the dependence on temperature of the speed of sound in quartz. A temperature change of the order of 0.02°C detunes the acoustic resonance frequency sufficiently to reduce the modulation below the required level.

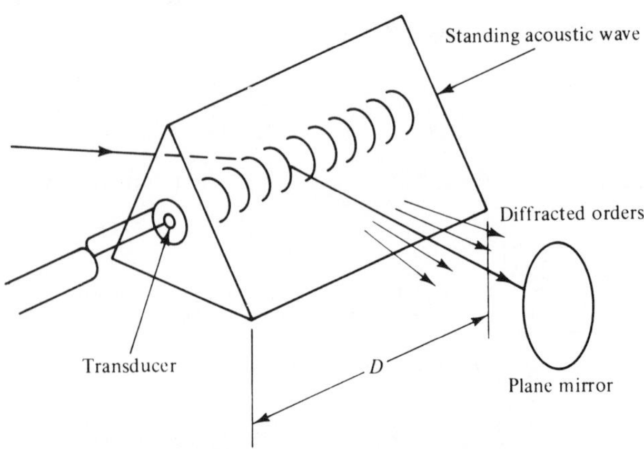

Figure 9.10 Configuration for laser modelocking by internal loss modulation.

9.4 ACOUSTIC DRIVE POWER FOR GIVEN MODULATION

The acousto-optic effect is described in terms of the photoelastic constant p, which gives the index change Δn in terms of the strain s:

$$\Delta n = -\frac{n^3 ps}{2} \tag{9.43}$$

Equation (9.43) stands, in fact, for a tensor equation, relating the stress tensor, a second-rank tensor, to the dielectric tensor, also of second rank (see Chapter 11). Thus p stands for a tensor of fourth rank. In an isotropic medium, a fourth-rank tensor of this nature is described by two constants, in an anisotropic medium by more. If the acoustic wave is chosen and the polarization of the electromagnetic field is selected, a relation of the type (9.43) can be written down, where the constant p is determined from the tensor components. If u_x is the displacement of the material in the x direction, and the wave travels along x, the strain is

$$s = \frac{\partial u_x}{\partial x} \tag{9.44}$$

The intensity I_s in the acoustic (sound) wave is equal to twice the kinetic energy density times the speed of sound (group velocity) v_s:

$$I_s = \tfrac{1}{2}\rho\omega_s^2 |u_x|^2 v_s \tag{9.45}$$

where ρ is the mass density. From (9.44) we find for s:

$$|s| = |k_s u_x| = \left|\frac{\omega_s}{v_s} u_x\right| \tag{9.46}$$

and thus the relation between the sound intensity and the strain s is

$$I_s = \tfrac{1}{2}\rho v_s^3 |s|^2 \tag{9.47}$$

Therefore,

$$|\Delta n| = \frac{n^3 p}{2}\sqrt{\frac{2I_s}{\rho v_s^3}} \tag{9.48}$$

The index change is proportional to the square root of the acoustic power. This equation holds for the traveling-wave modulator. The standing-wave modulator has a peak strain proportional to

$$|s| = k_s(|u_x^+| + |u_x^-|) = 2k_s|u_x^+|$$

where u_x^+ and u_x^- are the displacements associated with the forward and backward traveling wave. The intensities $I_\pm$ associated with each of the two countertraveling waves are still given by (9.47). Therefore, the peak of the index grating amplitude, n, is given by

$$\Delta n = n^3 p \sqrt{\frac{2I_s^+}{\rho v_s^3}} \tag{9.49}$$

We shall now evaluate the acoustic power required to achieve a given depth of modulation in the modelocked laser of the preceding section. From (9.25) we find the sidebands produced by passage of an optical amplitude A(t) through the modulator. According to (9.28), the amplitude A(t) passing twice through the modulating element is multiplied by $2\alpha_m l_m(1 - \cos \omega_M t)$, where $\omega_M = 2\omega_s$ with ω_s = frequency of the sound wave. The multiplier produces on a Fourier component $A_n \exp(j\omega_n t)$ sidebands $\alpha_m l_m A_n$ at frequencies $\omega_n \pm \omega_M$. Thus the ratio of each of the first two sidebands to the fundamental is $\alpha_m l_m$. Comparison with (9.25) shows that

$$\alpha_m l_m = J_2\left(\frac{\omega}{c} l_m \frac{\Delta n}{2 \cos \theta}\right) \simeq \frac{1}{8}\left(\frac{\omega}{c} l_m \frac{\Delta n}{2 \cos \theta}\right)^2 \tag{9.50}$$

Assuming that $\cos \theta \simeq 1$ and solving for Δn, we find that

$$\Delta n = \frac{4\sqrt{2\alpha_m l_m}}{(\omega/c) l_m} \tag{9.51}$$

If we introduce (9.49), we obtain an expression for the acoustic intensity I_s^+ in terms of the modulation $\alpha_m l_m$.

We assume that $l_m = 1$ cm and $\lambda = 1.06$ μm. We take from Table 12-1 of Ref. 1 the following values:

$$p = 0.2, \quad \rho = 2.2 \times 10^3 \text{ kg/m}^3, \quad v_s = 5.97 \times 10^3 \text{ m/s}, \quad n = 1.46$$

For a peak loss $\alpha_m l_m = 0.25$, we estimate $I_s^+ = 137$ W/cm^2. The power required to produce this internal intensity can be evaluated using (7.28) for the acoustic resonance. We find at the resonance frequency, assuming an ideal electroacoustic transducer:

$$|a|^2 = \frac{(2/\tau_e)|s_+|^2}{1/\tau^2}$$

where $|a|^2$ is the acoustic energy and $|s_+|^2$ is the incident electrical power. The acoustic intensity I_s^+ in a resonator of cross section $\mathscr{A}$ and length l_s is

$$I_s^+ = v_s \frac{|a|^2}{2l_s \mathscr{A}} = \frac{\tau^2}{\tau_e} \frac{v_s}{l_s \mathscr{A}} |s_+|^2 \tag{9.52}$$

If the acoustic resonator is critically matched, for maximum power delivery to the resonator, $\tau_e = 2\tau = \tau_0$. The decay rate $1/\tau_0$ due to internal losses is [compare (7.46)]

$$\tau_0 = \frac{1}{\alpha_s v_s}$$

where α_s is the spatial (amplitude) decay rate of the acoustic wave, and v_s is its group velocity. We have from (9.52) for the acoustic power inside the resonator

$$\mathscr{A}I_s^+ = \frac{|s_+|^2}{4\alpha_s l_s}$$

The traveling-wave acoustic power internal to the resonator is enhanced by the factor $1/4\alpha_s l_s$ over the incident electrical power. With a realistic transducer, of efficiency η, the above is reduced by η.

9.5 SUMMARY

An acoustic wave of frequency ω_s propagating through an acousto-optic material produces a traveling index grating which diffracts an incident optical wave of frequency ω_i. The diffracted wave is shifted in frequency, $\omega_d = \omega_i \pm \omega_s$, and has a propagation vector $\boldsymbol{k}_d$ that obeys the momentum conservation relation, $\boldsymbol{k}_d = \boldsymbol{k}_i \pm \boldsymbol{k}_s$. The diffracted wave couples back into the incident wave and the coupling-of-modes equation describes this mutual interaction. Acousto-optic amplitude modulators employ a standing acoustic wave. A simple approach treats the standing-wave grating as a time-independent bulk grating, finds the solutions for the incident and diffracted waves, and then introduces the time variation of the grating. Alternatively, the modulation can be analyzed by an infinite set of coupled mode equations. Both approaches give the same answer under the assumption that phase matching is maintained for all orders of diffraction.

The results of the acousto-optic amplitude modulator were used to characterize an actively modelocked laser system. We introduced the effect of the finite gain–bandwidth, which has the tendency of broadening a pulse in the process of amplifying it. The amplitude modulator counteracts this broadening effect so that a steady-state pulse train can be produced by an amplitude modulator driven at a frequency equal to the inverse of the round-trip time in the resonator. We evaluated the power necessary to achieve modelocking with quartz as the acousto-optic material.

APPENDIX 9A

Acousto-optic Amplitude Modulator; Alternative Analysis

We start with the system of Fig. 9.3 and single out the mth-order sideband of the incident wave of frequency $\omega_i + m\omega_s$ $(m \gtrless 0)$. Denote its amplitude by $A_i^{(m)}$. The source in the wave equation (9.6) driving $A_d^{(m)}$ is due to the incident wave at frequencies $\omega_i + (m \pm 1)\omega_s$. The equations are analogous to (9.12) and (9.13), except that each of the sidebands of the incident wave couples to a pair of diffracted waves, and vice versa. Also, instead of $\Delta n/2$, we have coupling proportional to $\Delta n/4j$ [compare (9.4) and (9.17)]:

$$\frac{dA_d^{(m)}}{dz} = -\frac{\omega_d^{(m)}}{4c}\frac{\Delta n}{\cos\theta}\left[A_i^{(m-1)} - A_i^{(m+1)}\right] \tag{9A.1}$$

and

$$\frac{dA_i^{(m)}}{dz} = -\frac{\omega_i^{(m)}}{4c}\frac{\Delta n}{\cos\theta}\left[A_d^{(m-1)} - A_d^{(m+1)}\right] \tag{9A.2}$$

This is an infinite set of equations for the infinite set of sidebands. A simple solution is found for the case when we approximate the multipliers $\omega_d^{(m)}/c \simeq \omega_i^{(m)}/c \simeq \omega_i^{(0)}/c$, or ω_i/c for short, which is a good approximation in view of the smallness of the acoustic frequency compared with the optical frequency. The solution is based on the recursion formula for Bessel functions [2]:

$$\frac{dJ_p}{dx} = \frac{1}{2}(J_{p-1} - J_{p+1}) \tag{9A.3}$$

with

$$x \equiv \frac{\omega_i}{2c}\frac{\Delta n}{\cos\theta} z$$

The solutions to (9A.1) and (9A.2) are of the form

$$A_i^{(m)} \propto J_m(x) \qquad \text{and} \qquad A_d^{(m)} \propto -J_m(x) \tag{9A.4}$$

If the incident wave starts out at frequency $\omega_i^{(0)}$ at $z = 0$ with amplitude $A_i(0)$, it must be of the dependence $J_0(x)$, and its sidebands are even-ordered Bessel functions, those of the diffracted wave are of odd order. At $z = l$ the amplitudes are

$$A_i^{(m)} = J_m\left(\frac{\omega_i}{2c}\frac{\Delta n}{\cos\theta} l\right) A_i(0), \qquad m \text{ even} \tag{9A.5}$$

$$A_d^{(m)} = -J_m\left(\frac{\omega_i}{2c}\frac{\Delta n}{\cos\theta} l\right) A_i(0), \qquad m \text{ odd} \tag{9A.6}$$

These solutions check with (9.23) and (9.24).

PROBLEMS

9.1. Suppose that an optical beam is to be deflected by a $LiNbO_3$ Bragg deflector. The highest acoustic frequency without excessive loss is 1 GHz. How much deflection is possible? The speed of sound in $LiNbO_3$ is 7.4 km/s. $\lambda = 6328$ Å, $n = 2.3$.

9.2. Consider a medium supporting a traveling sound wave of dependence $\exp[j(\omega_s t - \boldsymbol{k}_s \cdot \boldsymbol{r})]$ + c.c. producing an index variation

$$n(\boldsymbol{r}, t) = n_0 + \frac{\Delta n}{2}[\exp(j\omega_s t - j\boldsymbol{k}_s \cdot \boldsymbol{r}) + \text{c.c.}]$$

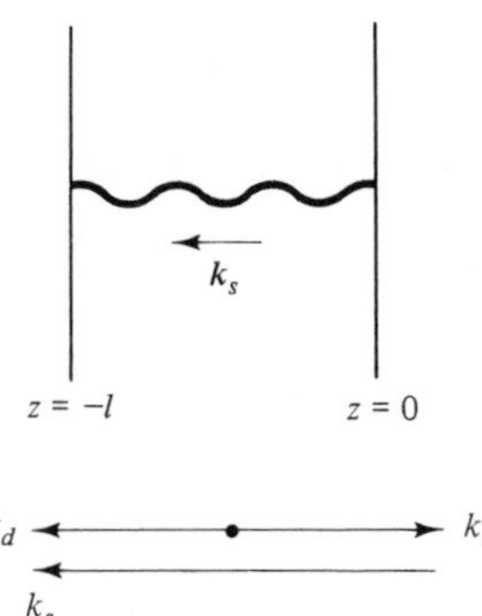

Figure P9.2

An incident (infinite parallel plane) optical beam is " scattered" by the index corrugation into a "diffracted beam" $\boldsymbol{k}_d$ (see Fig. P9.2 for the assumed phase-matched condition).

(a) What is the frequency of the scattered beam for the assumed diagram of $\boldsymbol{k}$ vectors?

(b) Set up coupling-of-modes equations for the slow envelopes A_i and A_d of the incident and "diffracted" or scattered beam.

(c) Subject to the boundary condition (note the directions of propagation of E_i and E_d), $A_i(0) = A$, $A_d(l) = 0$, solve the coupling-of-modes equations.

(d) For $\omega_s \rightarrow 0$ compare your results with the analysis of a grating reflector (Section 8.2).

9.3. Find the solutions of the modelocking equation (9.35) in the asynchronous case, $\delta T_R \neq 0$. For this purpose, use the substitution $A(t) = f(t) \exp(\alpha t)$ and remove the first derivative by proper choice of α.

Determine the excess gain, and the time shift of the pulse, relative to the time of minimum loss, as functions of δT_R. Consider only the fundamental solution.

9.4. Consider an infinite array of identical coupled waveguides. The equations are

$$\frac{da_n}{dz} = -j\beta a_n + \kappa(a_{n-1} - a_{n+1})$$

(a) Show that the equations obey power conservation, with κ picked real.

(b) Find the general solution of this equation with guide 0 excited at $z = 0$, all others unexcited. *Hint:* Use envelope quantities, $a_n(z) = A_n(z) \exp(-j\beta z)$ and the recursion formula for Bessel functions, (9A.3).

(c) From a graph of the Bessel functions [2] visualize the spreading of power.

REFERENCES

[1] A. Yariv, *Introduction to Optical Electronics*, Holt, Rinehart and Winston, New York, 1976.

[2] H. Abramowitz and I. A. Stegun, eds., *Handbook of Mathematical Functions*, Dover, New York, 1972.

[3] D. J. Kuizenga and A. E. Siegman, "FM and AM modelocking of the homogeneous laser—Part I: Theory," *IEEE J. Quantum Electron.*, *QE-6*, 694–708, Nov. 1970.

[4] H. A. Haus, "A theory of forced modelocking," *IEEE J. Quantum Electron., QE-11*, no. 7, 323–330, July 1975.

[5] L. D. Landau and E. M. Lifshitz, *Electrodynamics of Continuous Media*, Pergamon Press, Elmsford, 1960.

[6] H. A. Haus and P. T. Ho, "Effect of noise on active mode locking of a diode laser, *IEEE Journal of Quantum Electronics, QE-15*, no. 11, 1258–1265, November 1979.

[7] H. A. Haus, "Modelocking of semiconductor laser diodes," *Jap. J. Appl. Phys., 20*, no. 6, 1007–1020, June 1981.

10

SOME NONLINEAR SYSTEMS

Nonlinear optical phenomena are of great practical interest. The nonlinearity of optical fibers (dependence of index on intensity) can lead to undesirable phase modulations at relatively low powers, because the light is confined to a small cross section over long propagation distances (many kilometers), so that small powers (mW) can result in high intensities with large attendant nonlinear effects. The nonlinearity can also lead to picosecond pulse formation, as recently observed [1]. The phenomenon of pulse formation due to index nonlinearity is similar to (although not identical to) the process of pulse formation due to saturable (intensity-dependent) loss. The latter phenomenon has found wide application in the generation of picosecond light pulses and goes under the name "passive (saturable-absorber) modelocking." Self-focusing of optical beams is one of the major difficulties that had to be overcome by the laser fusion program. Because intense beams tend to self-focus in materials with an index that increases with increasing intensity, they can produce focused intensities that lead to optical damage.

Nonlinear optics is an extensive field initiated by the pioneering efforts of many researchers, in particular N. Bloembergen, who received the Nobel prize for his work. Some aspects of nonlinear optics are easily understood and need a minimum amount of background for their study. This is the case with the topics covered in this chapter. A somewhat more detailed treatment of nonlinear optics is given in Chapters 12 and 13 after coverage of linear propagation in anisotropic media.

In this chapter we study self-focusing, pulse formation, and passive modelocking. Finally, we present a nonlinear resonator model, which describes succinctly the phenomenon of bistability that is currently the topic of intensive research because of its potential for optical memory and optical logic operations.

10.1 SELF-FOCUSING

When a beam of finite transverse dimensions propagates through a nonlinear medium, with an index that depends on the optical intensity in the medium, the index within the beam is different from that outside the beam. If the index increases with increasing intensity, an "optical waveguide" is induced in the region where the beam is propagating; diffraction is counteracted by guidance of radiation by the region of increased index. This guidance need not be in the steady state, because the index profile depends on the intensity profile, and vice versa. If steady-state guidance occurs, or if the beam is progressively more and more confined and finally focused into a small spot, the process is called "self-focusing."

In a medium in which ε is a function of position, the vector potential $\mathbf{A}$ obeys approximately the wave equation (6.5)

$$\nabla^2 \mathbf{A} + \omega^2 \mu_0 \varepsilon \mathbf{A} = 0 \tag{10.1}$$

where

$$\omega^2 \mu_0 \varepsilon = \omega^2 \mu_0 \varepsilon_0 n^2 \tag{10.2}$$

with n the index. We separate ∇^2 into transverse and longitudinal derivatives. We assume, as before, that the vector potential is y-directed,

$$\mathbf{A} \propto \hat{y} u(x, y, z) e^{-jk_0 z} \tag{10.3}$$

where u varies slowly with z. We normalize $u(x, y, z)$ so that $|u|^2$ is the intensity. We further define

$$k_0 = \omega \sqrt{\mu_0 \varepsilon_0}\, n_0 \tag{10.4}$$

as the propagation constant at the chosen frequency ω_0, at the "unperturbed" value of index, n_0. We obtain from (10.1)

$$\nabla_T^2 u - 2jk_0 \frac{\partial u}{\partial z} = -\omega^2 \mu_0 \varepsilon_0 (n^2 - n_0^2) u \tag{10.5}$$

We specialize first to the two-dimensional case, $\nabla_T^2 = \partial^2/\partial x^2$. We consider the case when the index is intensity dependent and expand n to first order in intensity I:

$$n = n_0 + n_2 I = n_0 + n_2 |u|^2 \tag{10.6}$$

where n_2 is the coefficient of the index nonlinearity. Introducing (10.6) into (10.5), and ignoring terms quadratic in $n_2 |u|^2$, because they are presumably much smaller than the linear term, we obtain

$$-2jk_0 \frac{\partial u}{\partial z} + \frac{\partial^2}{\partial x^2} u + 2k_0^2 \frac{n_2}{n_0} |u|^2 u = 0 \tag{10.7}$$

If we introduce the variable

$$q \equiv \frac{z}{2k_0} \tag{10.8}$$

and the parameter

$$\kappa \equiv 2k_0^2 \frac{n_2}{n_0} \tag{10.9}$$

the equation above is put into the standard form of the *nonlinear Schrödinger* equation:

$$-j \frac{\partial u}{\partial q} + \frac{\partial^2 u}{\partial x^2} + \kappa |u|^2 u = 0 \tag{10.10}$$

For $\kappa > 0$ (i.e., $n_2 > 0$) the equation above has the solution [2]

$$u = \sqrt{\frac{2}{\kappa}}\, \eta \frac{\exp [j(\xi^2 - \eta^2)q + j\xi x - j\phi]}{\cosh [\eta(x - x_0) + 2\eta\xi q]} \tag{10.11}$$

as one may confirm by direct substitution. ξ, η, x_0, and ϕ are arbitrary parameters.

To understand the physical meaning of the solutions, we consider first the simple case of $\xi = 0$. Ignoring the uninteresting phase ϕ, we obtain for (10.11)

$$ue^{-jk_0 z} = \sqrt{\frac{2}{\kappa}}\, \eta \frac{\exp (-j\eta^2 q)}{\cosh \eta(x - x_0)} e^{-jk_0 z}$$

This is a beam with an x-dependent profile of width proportional to $1/\eta$ and a propagation constant $k = k_0[1 + (\eta^2/2k_0^2)]$. The propagation constant is increased from the value (10.4) by a term proportional to η^2 where η is the inverse beam width. The area integral of $\sqrt{\kappa/2} \int_{-\infty}^{\infty} |u|\, dx$ evaluates to* 2π independent of the beam parameters.

Next consider the case of $\xi \neq 0$. The center of the profile, which is at x_0 when $z = 0$, shifts with increasing z to $x_0 - \xi(z/k_0)$ (see Fig. 10.1). The beam is inclined with respect to the axis by the angle $\theta = \xi/k_0$. Together with this inclination goes an inclination of the phase fronts, which are now represented

*The integral is easily carried out using the following transformation:

$$\int_{-\infty}^{\infty} \frac{dx}{\cosh x} = \int_{-\infty}^{\infty} \frac{2e^x\, dx}{e^{2x} + 1} = \int_{-\infty}^{\infty} \frac{2\, dy}{y^2 + 1} = 2\pi$$

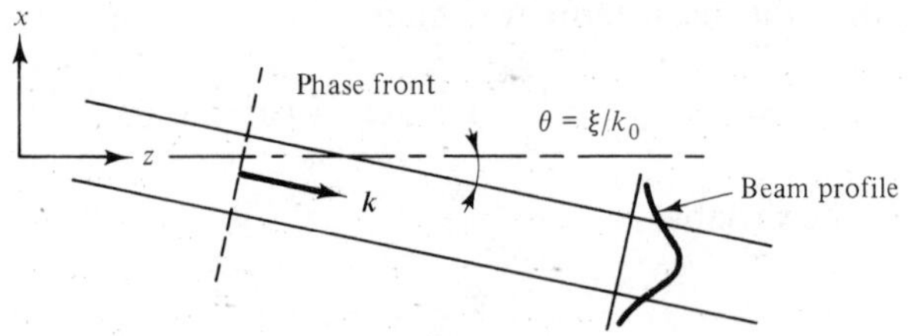

Figure 10.1 Single beam inclined with respect to z axis.

by the phase factor

$$\exp\left\{-j\left[k_0\left(1+\frac{\eta^2-\xi^2}{2k_0^2}\right)z+j\xi x\right]\right\}$$

The k vector of this expression also has the inclination angle ξ/k_0 with respect to the z axis. The correction to the z component of the k vector, $-\xi^2/2k_0$, is the paraxial approximation to $\cos\theta - 1$.

The physical interpretation of a steady-state self-focusing solution is simple if one recalls the mode analysis of the dielectric slab guide. A transverse variation of index, such that a maximum occurs at some value of x (say $x = 0$) and the index decreases in both the $+x$ and $-x$ directions, produces a dielectric slab guide capable of supporting one or more transverse modes. A nonlinear medium with an index that increases with intensity produces such an index profile. The only difference with the linear case is that the intensity profile must be consistent with the index profile—a consistency assured by the solution of the nonlinear Schrödinger equation.

The remarkable feature of the nonlinear Schrödinger equation is that it can be solved for more complicated boundary conditions $[u(x, z = 0)]$ and the nonlinear "transient" solution can be studied in general. The equation is one of a class that can be treated by the inverse scattering method [2–6]. It has been shown [7] that an initial $\sqrt{\kappa/2}\, u(x, 0)$, which integrates to an "area" greater than π and less than 3π, eventually settles down to a steady-state beam of area 2π. [5–7].

Focusing in three dimensions is a more difficult problem and no closed-form solutions have been found. The Laplacian operator for a function with cylindrical symmetry is

$$\frac{\partial^2}{\partial x^2}+\frac{\partial^2}{\partial y^2}=\frac{1}{\rho}\frac{\partial}{\partial \rho}\left(\rho\frac{\partial}{\partial \rho}\right)$$

for a problem of cylindrical symmetry. The beam is still guided by the increased dielectric constant of the intense center of the beam. The integral $\int_0^\infty |u(\rho)|^2\, \rho\, d\rho$ is now independent of the scale parameter of the solution (just as $\int_{-\infty}^{\infty} |u(x)|\, dx$ did not depend on η). Hence the self-focused beam possesses a typical net power, independent of the beam parameters [8, 9].

There is a very fundamental difference between the two- and three-dimensional self-focusing solutions. If the "area" exceeds 2π in two dimensions, the beam settles down to a steady-state profile after a transient adjustment. On the other hand, if the power of the three-dimensional beam exceeds the critical power, the paraxial equation predicts that the beam focuses to a point catastrophically [10]. This behavior is responsible for optical damage in high-powered laser systems.

10.2 SOLITON PROPAGATION IN FIBERS [1]

In the preceding section, we studied the propagation of beams whose tendency to spread in the transverse direction by diffraction is counterbalanced by the guiding tendency of an index that increases with intensity. A process that is described by mathematics identical to those of the two-dimensional self-focusing process is "soliton" formation in dispersive nonlinear fibers. A soliton is a pulse excitation of a nonlinear dispersive medium which propagates without distortion. The spreading of the pulse that would be caused by the dispersion acting alone is counteracted via the nonlinear phase modulation of the pulse by the nonlinearity of the medium.

We start with the expression (6.81) for the propagation constant of the lowest-order mode with profile $u(x, y)$ on a single-mode fiber, perturbed by a dielectric constant change $\delta\varepsilon$ which, expressed in terms of the index change δn, is $\delta\varepsilon = 2n\,\delta n \varepsilon_0$.

$$\delta\beta = \frac{\omega^2 \mu_0 \varepsilon_0}{\beta} \frac{\int da\, n\, \delta n |u|^2}{\int da\, |u|^2} \tag{10.12}$$

The propagation constant β is also a function of frequency. We use the same expansion as in Section 6.5, (6.51), and simply supplement it with the $\delta\beta$ contributed by δn:

$$\beta = \beta(\omega_0) + (\omega - \omega_0)\frac{d\beta}{d\omega} + \frac{1}{2}(\omega - \omega_0)^2 \frac{d^2\beta}{d\omega^2} + \frac{\omega_0^2 \mu_0 \varepsilon_0}{\beta(\omega_0)} \frac{\int da\, n\, \delta n |u|^2}{\int da\, |u|^2} \tag{10.13}$$

Here we substituted the carrier frequency ω_0 for ω and $\beta(\omega_0)$ for β in $\delta\beta$ of (10.12), treating the expansion as one to lowest significant powers of $(\omega - \omega_0)$ and δn. Introducing the envelope $\mathrm{A}(z, \omega - \omega_0)$ defined by $\mathrm{a}(z, \omega) = \mathrm{A}(z, \omega - \omega_0) \exp\{j[\omega_0 t - \beta(\omega_0)z]\}$ and carrying out the inverse Fourier transform as in (6.55), we arrive at the equation

$$\left(\frac{\partial}{\partial z} + \frac{1}{v_g}\frac{\partial}{\partial t}\right)\mathrm{A}(z, t) = \frac{j}{2}\frac{d^2\beta}{d\omega^2}\frac{\partial^2 \mathrm{A}(z, t)}{\partial t^2} - j\frac{\omega_0^2 \mu_0 \varepsilon_0}{\beta(\omega_0)} \frac{\int da\, n\, \delta n\, |u|^2}{\int da\, |u|^2}\mathrm{A}(z, t) \tag{10.14}$$

A transformation of the independent variables

$$\tau \equiv t - \frac{z}{v_g} \tag{10.15}$$

and

$$\zeta = z \tag{10.16}$$

casts this equation into the form (compare (6.60))

$$j\frac{\partial \mathrm{A}}{\partial \zeta} + \frac{1}{2}\frac{d^2\beta}{d\omega^2}\frac{\partial^2 \mathrm{A}}{\partial \tau^2} - \frac{\omega_0^2 \mu_0 \varepsilon_0}{\beta(\omega_0)} \frac{\int da\, n\, \delta n\, |u|^2}{\int da\, |u|^2}\mathrm{A} = 0 \tag{10.17}$$

Thus far, δn was an index perturbation imposed by some unspecified means. We specialize on a nonlinear index perturbation which is proportional to the intensity of the pulse traveling along the fiber

$$\delta n = n_2 I \tag{10.18}$$

where n_2 is the first order coefficient of expansion of n in powers of intensity, I, where

$$I = |\mathrm{A}(z, t)|^2\, |u|^2.$$

Thus (10.17) can be rewritten

$$j\frac{\partial \mathrm{A}}{\partial \zeta} + \frac{1}{2}\frac{d^2\beta}{d\omega^2}\frac{\partial^2 \mathrm{A}}{\partial \tau^2} - \kappa |\mathrm{A}|^2\, \mathrm{A} = 0 \tag{10.19}$$

where

$$\kappa \equiv \frac{\omega_0^2 \mu_0 \varepsilon_0}{\beta(\omega_0)} \frac{\int da\, n_0 n_2 |u|^4}{\int da |u|^2} \tag{10.20}$$

In the derivation of (10.20) we have used the concept of a time dependent propagation constant because $\kappa|\mathrm{A}|^2$, the contribution to β of the nonlinear medium, varies with time. Since β is normally considered to be a function of ω, this raises some questions. Remember that $\mathrm{A}(z, t)$ is the pulse envelope of a carrier with the time dependence $\exp(j\omega_0 t)$. A propagation constant can be assigned to the carrier frequency ω_0. If the spectrum of $\mathrm{A}(z, t)$ is narrow, its width is much smaller than ω_0, the assignment can be made at $\omega = \omega_0$ unambiguously. If the index changes with time, β varies with time.

Equation (10.19) is identical in form with the equation of self-focusing (10.10), and thus must have identical solutions, provided that $d^2\beta/d\omega^2 < 0$ (i.e., the fiber has anomalous dispersion). This is the case to the right (large wavelength) side of the zero dispersion point of the fiber in Fig. 6.15. Again, we may conclude that pulse-like solutions are allowed that are secant hyperbolics and are stable (see Fig. 10.2). The dispersion is counteracted by the nonlinear index change. From the existing solutions describing the formation of such pulses

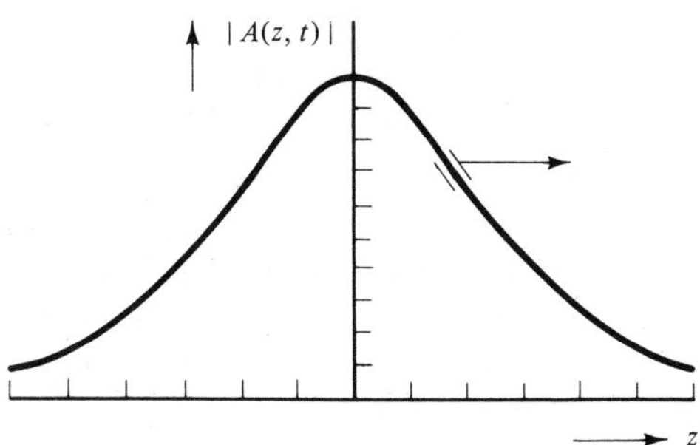

Figure 10.2 Pulse-like solution at a given time instant t. The speed of propagation of the envelope is

$$\frac{v_g}{1 - \sqrt{2}\,\xi v_g \sqrt{\dfrac{d^2\beta}{d\omega^2}}}$$

The parameter ξ measures the frequency deviation of the carrier frequency ω_c from ω_0:

$$\omega_c - \omega_0 = \sqrt{2}\,\xi\left[\left|\frac{d^2\beta}{d\omega^2}\right|\right]^{-1/2}$$

[5–7] we may conclude that an initial pulse whose area in normalized units lies between π and 3π forms into a steady-state pulse of area 2π. Some typical numbers are given in Ref. 1.

We learn an additional fact from the soliton solution. Anomalous dispersion ($d^2\beta/d\omega^2 < 0$) tends to produce a positive chirp parameter according to (6.68), that is, chirped pulses with the high frequencies in the front of the pulse, the low frequencies in the back. Because the nonlinear index counteracts anomalous dispersion, it must, if acting alone, produce pulses that have the low-frequency components in the front end, the high-frequency components in the back end of the pulse (see Problem 10.3). This phenomenon is utilized in pulse shortening as described in the next section.

The work on soliton formation and propagation along optical fibers is motivated by their unique propagation characteristic. A soliton, once formed, has a standard width–height product ("area") and propagates undisturbed along the fiber in spite of fiber dispersion. This is an attractive feature for transmission of a pulse-coded message in which the presence of a pulse in a time slot indicates a "one," absence indicates a "zero." Such a message could travel along the fiber without distortion. A soliton-forming structure can also be used as a pulse standardizer. A pulse of arbitrary shape, of area between π and 3π, emerges as a secant hyperbolic pulse.

10.3 SATURABLE ABSORBER MODELOCKING [11, 12]

In soliton formation, we witnessed a process in which pulse spreading by dispersion is balanced by a nonlinear index change. This process is "conservative," or "reactive," in that both the dispersion and the phase changes due to index variation do not change the power of the wave upon which they act.

Saturable absorber modelocking is a competition between a dissipative process and a process with gain. In this respect there is a fundamental difference between saturable absorber modelocking and the soliton process of the preceding section. Yet, in the steady state, they act very similarly. The gain dispersion (variation of gain with frequency due to the finite bandwidth of the gain medium) tends to narrow the spectrum and thus broaden a pulse traveling through the gain medium. A *saturable absorber* whose loss decreases with increasing intensity attenuates the low-intensity "wings" of the pulse more than the high-intensity "center," thus narrowing a pulse passing through it. In the steady state, the two processes keep each other in balance.

We shall study first propagation of a plane wave through a uniform medium which is a mixture of a laser gain medium and saturable absorber loss medium. Then we shall show that the same equations apply to the propagation of a pulse inside a Fabry–Perot resonator containing thin "slabs" of gain and absorber media.

An imaginary part of the propagation constant indicates loss, if negative, gain, if positive. The presence of nonlinear dispersive gain can be represented in an expansion of the propagation constant β around ω_0 by* [compare (9.26)]

$$\beta = \beta(\omega_0) + j\alpha_g\left[1 - \left(\frac{\omega - \omega_0}{\omega_g}\right)^2\right] \tag{10.21}$$

where α_g is the spatial growth constant at the center of the gain line, at $\omega = \omega_0$. The presence of saturable loss can be accounted for simply by adding to the expansion of β in terms of ω a decay constant that depends upon intensity:

$$\beta = \beta(\omega_0) + j\alpha_g\left[1 - \left(\frac{\omega - \omega_0}{\omega_g}\right)^2\right] - \frac{j\alpha_l}{1 + I/I_s} \tag{10.22}$$

where I_s is that value of intensity for which α_l decreases to half its value, the *loss-saturation intensity*.

We may now derive an equation analogous to (6.56), except that the dispersion is now associated with the gain and it is augmented by a saturable loss term [compare (10.14)]

$$\left(\frac{\partial}{\partial z} + \frac{1}{v_g}\frac{\partial}{\partial t}\right)\mathrm{A}(z, t) = \alpha_g\left(1 + \frac{1}{\omega_g^2}\frac{\partial^2}{\partial t^2}\right)\mathrm{A}(z, t) - \frac{\alpha_l}{1 + I/I_s}\mathrm{A}(z, t) \tag{10.23}$$

To solve this equation, we approximate $1/(1 + I/I_s)$ by $1 - I/I_s$ and normalize the amplitude A so that $|\mathrm{A}|^2$ is the intensity. The result is

$$\left(\frac{\partial}{\partial z} + \frac{1}{v_g}\frac{\partial}{\partial t}\right)\mathrm{A} = \alpha_g\left(1 + \frac{1}{\omega_g^2}\frac{\partial^2}{\partial t^2}\right)\mathrm{A} - \alpha_l\left(1 - \frac{|\mathrm{A}|^2}{I_s}\right)\mathrm{A}$$

With the coordinate transformations (10.15) and (10.16), the equation

*Strictly speaking, it is not legitimate to introduce a frequency-dependent imaginary part of a material index without an accompanying frequency-dependent real part (Kramers–Kronig [13]). The real part could be included, but does not change the essential physics of the problem solution presented here in approximate form.

above becomes

$$-\frac{\partial}{\partial \zeta} A + (\alpha_g - \alpha_l)A + \frac{\alpha_g}{\omega_g^2}\frac{\partial^2}{\partial \tau^2} A + \alpha_l \frac{|A|^2}{I_s} A = 0 \tag{10.24}$$

This equation resembles (10.19) but differs in two respects: the coefficient $(\alpha_g - \alpha_l)$ is present and the derivative $\partial/\partial\zeta$ appears, rather than $j(\partial/\partial\zeta)$. The latter difference is fundamental. It is due to the difference in the physical mechanisms producing the pulse shaping. In soliton formation, the group-velocity dispersion and index change act on the phase; in saturable absorber modelocking, the gain dispersion and loss saturation act on the amplitude. In the absence of the nonlinearity, A obeys the Schrödinger equation in the case of group velocity dispersion, the diffusion equation in the case of gain dispersion.

When a steady-state pulse is formed

$$\frac{\partial}{\partial \zeta} = \frac{\partial}{\partial z} + \frac{1}{v_g}\frac{\partial}{\partial t} = 0$$

and (10.24) reduces to

$$0 = (\alpha_g - \alpha_l)A + \frac{\alpha_g}{\omega_g^2}\frac{\partial^2 A}{\partial t^2} + \alpha_l \frac{|A|^2}{I_s} A \tag{10.25}$$

The finite bandwidth of the gain described by (10.21) causes spreading of the pulse in time as described by the "temporal" diffusion operator $(\alpha_g/\omega_g^2)(\partial^2/\partial t^2)$. This diffusion in time can be balanced by the saturable loss term proportional to $|A|^2$. Equation (10.25) has the same form, and thus the same solution, as (10.10) in the steady state, when $j(\partial u/\partial q)$ can be replaced by u times a real, constant multiplier (i.e., when the parameter ξ is set equal to zero). We have

$$A = \frac{A_0}{\cosh \gamma(z - v_g t)} \tag{10.26}$$

where

$$v_g^2 \gamma^2 = \frac{\alpha_l}{2\alpha_g}\omega_g^2 \frac{|A_0|^2}{I_s} \tag{10.27}$$

and

$$\alpha_l - \alpha_g = \frac{\alpha_g}{\omega_g^2} v_g^2 \gamma^2 \tag{10.28}$$

Equation (10.27) expresses the inverse of the width of the pulse, $1/v_g\gamma$, in terms of the normalized loss saturation $(\alpha_l/\alpha_g)|A_0|^2/I_s$ and the gain bandwidth ω_g. Equation (10.28) shows that, in the absence of the pulse, the system possesses net loss. Net gain exists only when the loss is saturated at and near the peak of the pulse. This finding suggests that the system is stable to perturbations. It certainly is so for perturbations that occur in the space–time interval lying outside the interval occupied by the pulse. The fact that these equations are stable against arbitrary perturbations has been demonstrated by Hagelstein [14].

Thus far, we have studied pulse propagation in a uniform nonlinear medium that could be represented by a succession of infinitesimally thin slabs of a dispersive gain medium and saturable loss medium. Pulse formation in a Fabry–Perot resonator may be represented by the same equations if the saturable absorber is near one of the mirrors and the pulse length is much longer than the absorber length and its distance from the mirror. In fact, consider (9.34), in which a modulated loss was represented by

$$2\alpha_l l_l + 2\alpha_m l_m(1 - \cos \omega_M t)$$

If this loss is self-modulated through saturation of the loss-medium, this term is replaced by

$$2\alpha_l l_l\left(1 - \frac{|A|^2}{I_s}\right)$$

The resulting equation is

$$\left[2\alpha_g l_g\left(1 + \frac{1}{\omega_g^2}\frac{d^2}{dt^2}\right) - 2\alpha_l l_l\left(1 - \frac{|A|^2}{I_s}\right) - \delta T_R \frac{d}{dt}\right]A(t) = 0 \qquad (10.29)$$

which is in form identical with (10.25), except for the operator $\delta T_R\, d/dt$. This operator expresses deviations of the pulse return-time from the natural round-trip time T_R. No analog to this term exists in the propagation model developed earlier in this section, because the pulse was not assumed to return by reflection. In the case of active modelocking, δT_R was determined by the choice of the modulation frequency. In saturable absorber modelocking, δT_R is fixed by the solution to the problem—the system itself picks the deviation from the natural round-trip time. Indeed, because a solution to (10.25) has been found in the absence of the d/dt term, we know that (10.26) is also a solution to (10.29) with $\delta T_R = 0$.

Figure 10.3 shows schematically the action of a pulse (upper trace) passing through the absorber. A "temporal hole" is bleached into the absorption. There is net gain at the center of the pulse, net loss in its wings. The time-dependent net gain sharpens the pulse. This sharpening is counterbalanced by the dispersion of the laser medium.

Let us estimate the pulse width for parameters comparable to those picked for active modelocking. With $\omega_g/\omega_0 = 0.2\%$, $\lambda = 1.06\ \mu$m, $\alpha_g/\alpha_l \simeq 1$, and $|A_0|^2/I_s = 0.1$ (10% saturation of the absorber), we find for the FWHM of the secant hyperbolic pulse:

$$\text{FWHM} = \frac{2 \ln (1 + \sqrt{2})}{v_g \gamma} = \frac{1.76}{v_g \gamma} = 2.2 \text{ ps}$$

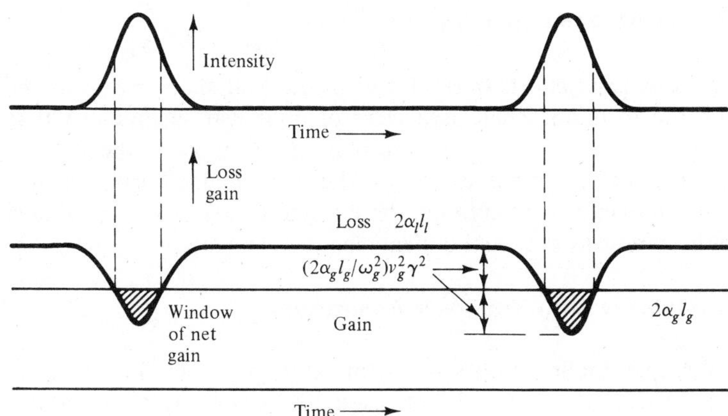

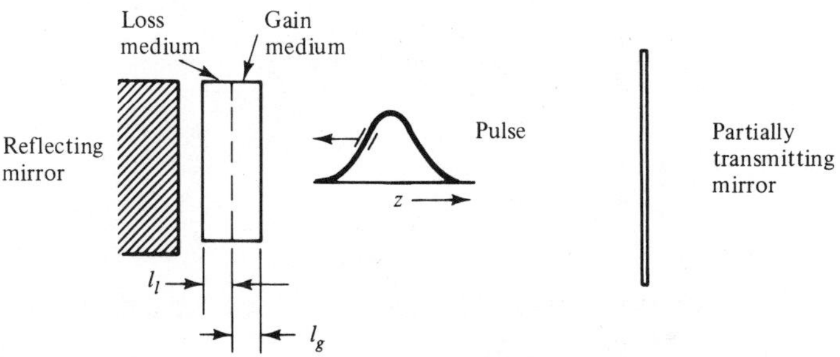

Figure 10.3 Passive modelocking.

Saturable absorber modelocking is capable of producing much shorter pulses than active modelocking, because it produces appreciable changes of loss on a time scale commensurate with the pulse width, whereas the changes of loss in active modelocking are on the time scale of the cavity round-trip time. The value above for the pulse width is shorter than the relaxation time of common absorbers. The present theory is not applicable to such a case, but a detailed theory has been worked out [15].

Pulses much shorter than the relaxation time of the saturable absorber have been produced [16] (90×10^{-15} s). Further pulse shortening has been achieved [17] by passing the pulse through an optical fiber. The spectrum of the pulse spreads due to the nonlinear index change ($n_2 > 0$). Such pulses are "chirped" with the front of the pulse containing the low frequencies, the back end the high frequencies; they acquire a negative chirp parameter [compare (6.63)]. With a grating pair like the one described in Section 6.6, pulses with a negative chirp parameter can be compressed. In this way, 30×10^{-15} s pulses have been generated at $\lambda = 6190$ Å. Such pulses contain only 14 optical periods within their width!

10.4 OPTICAL BISTABILITY [18–23]

We have discussed effects of nonlinear media that shape *pulses* as they travel through the medium. There is a class of nonlinear phenomena that exhibit hysteresis; that is, the *steady-state* response to an input of the system depends on the past history of the excitation. Because all memory elements involve hysteresis in one form or another, there is great interest in optical devices that exhibit hysteresis as potential optical memory elements.

Bistability with Nonlinear Absorption

A simple nonlinear optical system exhibiting bistability is shown in Fig. 10.4. It consists of a Fabry–Perot resonator containing a saturable-absorber medium. The equations for the system follow from (7.28). The equation for the amplitude a of the field in the resonator, excited by an incident wave s_+, is

$$\frac{da}{dt} = \left(j\omega_0 - \frac{1}{\tau}\right)a + \sqrt{\frac{2}{\tau_e}}\, s_+ \tag{10.30}$$

where

$$\frac{1}{\tau} = \frac{1}{\tau_0} + \frac{1}{\tau_e} \tag{10.31}$$

If the resonator contains a saturable absorber, then $1/\tau_0$ depends on the energy $|a|^2$, and one replaces $1/\tau_0$ with

$$\frac{1}{\tau_0} \to \frac{1}{\tau_0}\frac{1}{1 + |a|^2/a_0^2} \tag{10.32}$$

where a_0^2 is the energy in the resonator for which the loss decreases to one-half of its low-amplitude value. Substituting (10.32) into (10.30), we obtain

$$\frac{da}{dt} = j\omega_0 a - a\left(\frac{1}{\tau_0}\frac{1}{1 + |a|^2/a_0^2} + \frac{1}{\tau_e}\right) + \sqrt{\frac{2}{\tau_e}}\, s_+ \tag{10.33}$$

Consider an excitation at the resonance frequency of the resonator. Setting

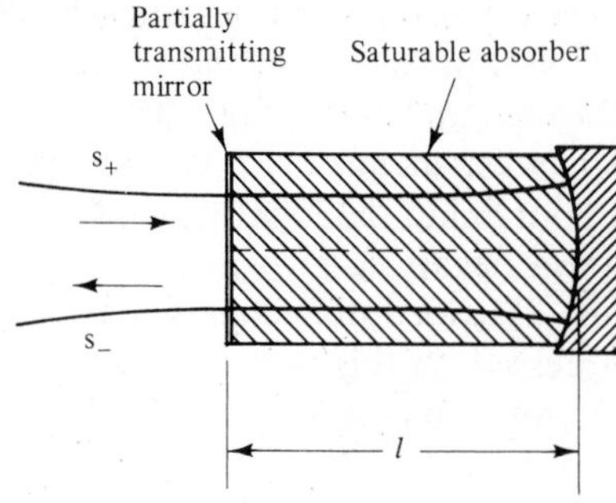

Figure 10.4 Fabry–Perot resonator containing saturable absorber.

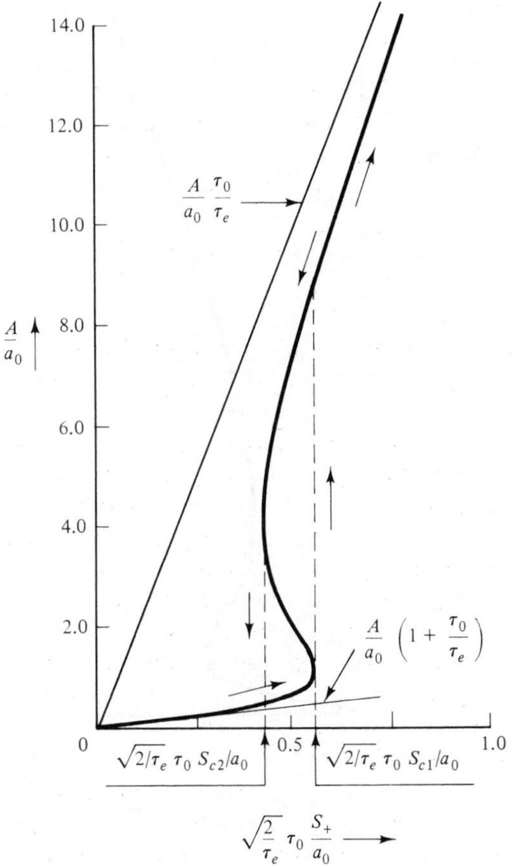

Figure 10.5 Plot of left-hand side of (10.34) versus A.

$s_+ = S_+ \exp(j\omega_0 t)$ and $a = A \exp(j\omega_0 t)$, in the steady state

$$\frac{A}{a_0}\left(\frac{1}{1 + A^2/a_0^2} + \frac{\tau_0}{\tau_e}\right) = \sqrt{\frac{2}{\tau_e}}\frac{\tau_0 S_+}{a_0} \tag{10.34}$$

Figure 10.5 shows the left-hand side of (10.34) as the abscissa versus the ordinate A/a_0. The intersection with a vertical line gives the operating point. Depending on the choice of parameters, we find either one or three intersection points. The latter is the case for $\tau_e/\tau_0 > 8$.

When there are three intersection points, the system follows the lower branch, as S_+ is increased to the critical value S_{c1}, and then jumps to the higher value. When S_+ is decreased from the high value, it follows the upper branch down to S_{c2} and then jumps down. The time evolution can be gleaned from (10.34). Indeed, the shaded regions in Fig. 10.6 show $d(A/a_0)/dt$ as a function of A/a_0. Without actually integrating (10.33) with respect to time, one

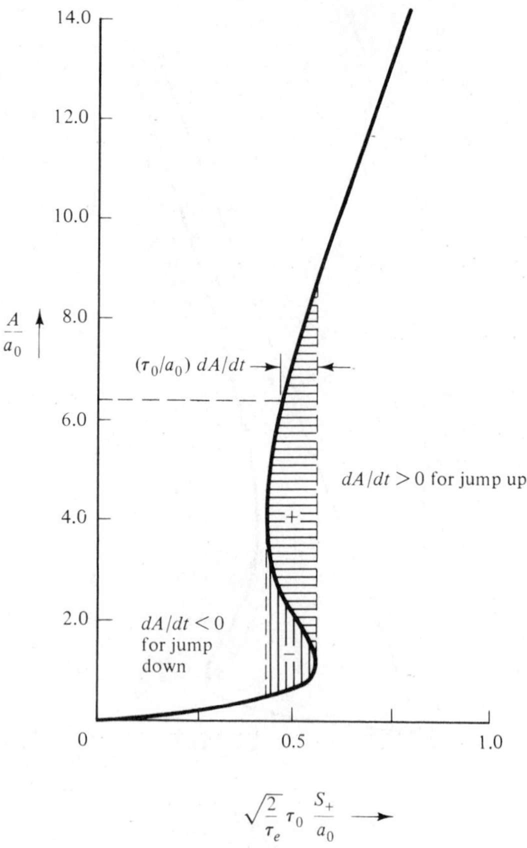

Figure 10.6 Graphical representation of dA/dt.

gathers from the graph how the transition occurs between the low and high branches when the system goes through the transition; the rate of change of A/a_0 is fastest when the horizontal width of the graph is greatest.

Bistability with Nonlinear Index

The bistable resonator discussed thus far used saturable absorption as the nonlinear process responsible for bistability. The first proposal for a bistable device suggested the use of saturable absorption [18]. The first clear-cut demonstration of bistability was made with a system with a "nonlinear" index, an index dependent on intensity [19]. The steady-state analysis of such a system is not greatly different from the one presented above.

Suppose that a cavity of length l is filled with a medium of index $n = n_0 + n_2 I$, where I is the intensity taken as the sum of the intensities of the counter traveling waves. The resonance frequency ω_0 of the resonator

$$\omega_0 = \frac{\pi c}{nl} \tag{10.35}$$

depends on the intensity $I \simeq c|\text{a}|^2/\mathscr{A}ln_0$, where $\mathscr{A}$ is the cross-sectional area of the optical beam assumed to be uniform over its cross section

$$\omega_0 = \frac{\pi c}{n_0 l(1 + |\text{a}|^2/a_0^2)} = \frac{\omega_{00}}{1 + |\text{a}|^2/a_0^2} \simeq \omega_{00}\left(1 - \frac{|\text{a}|^2}{a_0^2}\right) \tag{10.36}$$

where

$$a_0^2 = \frac{n_0^2}{n_2}\frac{\mathscr{A}l}{c} \tag{10.37}$$

The equation for the cavity is now

$$\frac{d\text{a}}{dt} = \left[j\omega_{00}\left(1 - \left|\frac{\text{a}}{a_0}\right|^2\right) - \frac{1}{\tau}\right]\text{a} + \sqrt{\frac{2}{\tau_e}}\,\text{s}_+ \tag{10.38}$$

The steady-state energy $|\text{a}|^2$ for a drive

$$\text{s}_+ = S_+ e^{j\omega t}$$

obeys the equation

$$\left|\frac{\text{a}}{a_0}\right|^2\left\{1 + \left[\omega - \omega_{00}\left(1 - \left|\frac{\text{a}}{a_0}\right|^2\right)\right]^2 \tau^2\right\} = \frac{2\tau^2}{\tau_e}\left|\frac{S_+}{a_0}\right|^2 \tag{10.39}$$

Figure 10.7 shows a plot of the left-hand side as a function of $|\text{a}/a_0|^2$. The plot has two extrema for positive values of $|\text{a}/a_0|^2$ (i.e., physical values) when the conditions are obeyed:

$$\omega < \omega_{00} \tag{10.40}$$

and

$$(\omega_{00} - \omega)\tau > \sqrt{3} \tag{10.41}$$

The first condition implies that the driving frequency must be smaller than the small-signal resonance frequency ω_{00} (this is under the assumption that $n_2 > 0$). A large intensity can lower the resonance frequency and bring the resonance closer to the driving frequency; two steady-state intensity values may then become possible. The second condition states that the frequency offset be sufficiently large compared with the resonator bandwidth. Figure 10.7 shows how the excitation of the system evolves under increasing input intensity $|S_+|^2$. At low intensities, we have one intersection. As one moves to higher input intensities, the normalized energy in the resonator rises to a critical value $|\text{a}_1/a_0|^2$. When the intensity increases further, the energy jumps to a higher value and then continues to rise. On the way down, with decreasing input intensity, the internal energy decreases to $|\text{a}_2/a_0|^2$ and then jumps to a lower

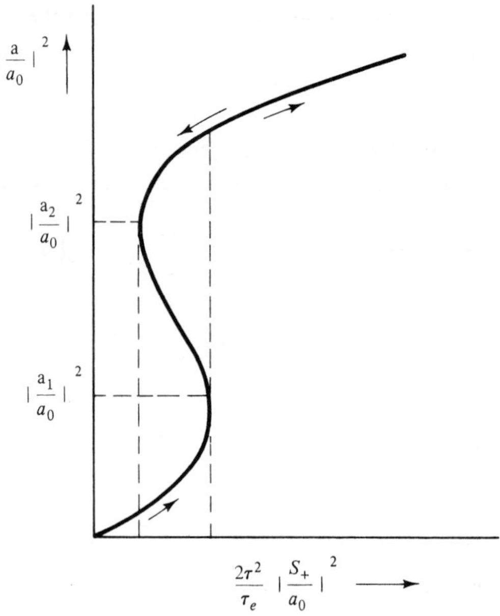

Figure 10.7 Bistable behavior of resonator with nonlinear index medium.

value. The transient is not analyzed as easily in this new case as in the case of saturable absorption. The amplitude a must be treated as a complex quantity, and hence (10.38) stands for two nonlinear coupled differential equations for Re [a] and Im [a]. We shall not pursue this problem any further.

An estimate of the powers and response times is of interest. Suppose the resonator is critically coupled for maximum power into the resonator at a given incident power $|s_+|^2$; then

$$\frac{1}{\tau_e} = \frac{1}{\tau_0} = \frac{1}{2\tau} \tag{10.42}$$

The frequency shift $\Delta\omega_0$ required in order to change the state is roughly equal to the inverse response time of the resonator, τ, according to (10.41). From (10.36)

$$|\Delta\omega_0| = \omega_{00}\frac{|a|^2}{a_0^2} \simeq \frac{1}{\tau} \tag{10.43}$$

The power into the resonator at resonance is according to (10.39)

$$|S_+|^2 = \frac{\tau_e}{2\tau^2}|a|^2 \simeq \frac{1}{\omega_{00}\tau}\frac{\tau_e}{2\tau^2}a_0^2 = \frac{n_0^2}{n_2}\frac{l\mathscr{A}}{c}\frac{4}{\omega_{00}\tau_e^2} \tag{10.44}$$

where we have introduced (10.42), (10.43) and (10.37).

If we specialize on an optical waveguide device, the cross-sectional area

$\mathscr{A}$ is of the order of $(\lambda/n_0)^2$. The decay rate $1/\tau_0$ is caused by the amplitude loss coefficient $\alpha[\mathrm{m}^{-1}]$ of the nonlinear medium filling the resonator. Thus [compare (7.46)]

$$\frac{2}{\tau_0} = \frac{2\alpha c}{n_0} \tag{10.45}$$

The decay rate $1/\tau_e$ for a "mirror" transmissivity T is given by (7.43). The mirror could be a DFB reflector; the result (7.43) still applies if the reflector is much shorter than l, or if one interprets l as an effective length of the total resonator. Combining (7.43), (10.42), (10.45) with (10.44) we obtain

$$|S_+|^2 = \frac{T}{2\pi} \frac{\lambda^3 \alpha}{n_2 n_0^2} \tag{10.46}$$

The quantity $(n_2 n_0^2/\lambda^3\alpha)$ in W^{-1} is a figure of merit of the nonlinear medium. The larger it is the smaller the power required to "activate" a bistable device. The transmissivity of the mirror $T(<1)$ can be used to reduce the power further at the cost of an increased response time of the resonator.

Consider GaAs as an example. The value of n_2 for $\lambda = 1.06\ \mu\mathrm{m}$ is [24]

$$n_2 = 1.57 \times 10^{-9}\ \mathrm{esu}.$$

To convert n_2 specified in esu into mks units, one expresses the electric field in V/m, the power in watts/m^2.

$$n = n_0 + n_2[\mathrm{mks}]\mathrm{I}[\mathrm{W/m^2}]$$

The conversion is

$$n_2[\mathrm{mks}] = \frac{4\pi}{3} \times 10^{-7} \frac{1}{n_0} n_2[\mathrm{esu}].$$

Further

$$n_0 = 3.2 \qquad \alpha = 6 \times 10^{-3}\ \mathrm{cm}^{-1}\ (\text{measured at } \lambda = 10\ \mu\mathrm{m})$$

The figure of merit is then

$$\frac{n_2 n_0^2}{\lambda^3 \alpha} = 2.95 \times 10^3\ \mathrm{W}^{-1}$$

The current interest in bistable devices lies in their hysteretic behavior. This allows information to be stored optically in a resonator, one bit per resonator. The disadvantage of the system is that it "stores" information only as long as there is an input s_+ to the resonator. The response time of the element is determined by either the response time of the medium, or the characteristic time of the resonator, whichever is longer.

10.5 SUMMARY

Some nonlinear optical phenomena can be analyzed and understood very simply by postulating a dependence of the index n upon the intensity I of the form $n = n_0 + n_2 I$. An intense optical beam increases the index $(n_2 > 0)$ at the peak of its profile and produces an optical waveguide. This phenomenon is the self-focusing phenomenon, which leads to a nonlinear Schrödinger equation. In two dimensions (slab geometry) the equation has a simple steady-state solution, a secant hyperbolic with a phase factor. The phase factor describes a beam with an axis inclined with respect to the z axis. The diffraction represented by the operator $\partial^2/\partial x^2$ is counteracted by the phase front bending caused by the index change.

The same two-dimensional nonlinear Schrödinger equation governs soliton propagation in an optical fiber. The process is not unlike the self-focusing phenomenon, Instead of the spatial diffraction operator $\partial^2/\partial x^2$, dispersion acts as "diffraction" in time represented by the operator $\frac{1}{2}(d^2\beta/d\omega^2)(\partial^2/\partial t^2)$. The nonlinear index produces phase changes that counteract the "diffraction." The attractive feature of solitons is their standard form—their normalized "area" is 2π. The hope is that such pulse standardization may find practical application.

The equation of passive modelocking has a similar steady-state solution as the nonlinear Schrödinger equation. The process is not a conservative one and the similarity does not extend to the transient analysis. Passive modelocking can lead to much shorter modelocked pulses than active modelocking because the shaping of the pulse is much more effective. This was borne out by a simple numerical example and is amply verified by experiment.

Bistable devices were analyzed using the resonator formalism developed in Section 7.2. Bistability can be produced by a saturable absorber. The resonator maintains a low internal energy when the loss is high until the input power reaches a critical switching threshold. Then the internal energy switches to a higher level and the absorber is saturated. On the reverse traversal of the characteristic, with input power decreasing, the high level of internal energy is maintained even when the input power is lowered below the original switching threshold. A nonlinear index material produces a very similar bistable behavior. Now, however, the two states correspond to different cavity resonance frequency regimes, one for high internal energy, the other for low internal energy.

PROBLEMS

10.1. A steady state, self focussed cylindrical beam possesses a characteristic power. From the discussion of Section 10.1 it follows that the equation for the profile can be put into the normalized form

$$\frac{1}{\xi}\frac{d}{d\xi}\left(\xi\frac{d}{d\xi}U\right) + |U|^2 U = \lambda U \tag{1}$$

where λ is an eigenvalue, $\xi = \rho/a$ with a a normalization distance, $U \equiv u/\sqrt{I_s}$ and I_s obeys the equation

$$2k_0^2 \frac{n_2}{n_0} I_s a^2 = 1.$$

The last equation defines a characteristic power. When (1) is solved [8, 9] a fundamental mode is obtained with the value of the integral

$$\int \xi \, d\xi \, |U|^2 = 1.86.$$

What is the steady state self focussing power in carbon disulfide, CS_2 ($n_2 = 1.2 \times 10^{-11}$ esu, $n_0 = 1.61$, $\lambda = 6328$ Å, [25].

10.2. Determine the energy in a soliton with a pulse-width parameter $\tau = 1$ ps propagating in a C2828 silicate glass single-mode fiber; $n_0 = 1.53$ and the n_2 parameter is 0.0208×10^{-11} in esu. [25] The dispersion is 10^2 ps/nm (compare Fig. 6.15). The mode-profile radius, w_0, which may be assumed Gaussian, is 1.5 μm. The wavelength of radiation is 1.4 μm.

10.3. Consider propagation of a Gaussian pulse

$$A\left(t - \frac{z}{v_g}\right) = A_0 \exp\left\{-\frac{[(t - z)/v_g]^2}{2\tau_p^2}\right\}$$

along a nondispersive fiber with nonlinear index $n_2 > 0$. Using (10.19), show that an initially unchirped pulse gets chirped. Evaluate the spectrum after a propagation distance l. Show that the spectrum has broadened.

10.4. In modelocking, the gain dispersion is counteracted by the gain time dependence. The pulsewidth for forced modelocking is proportional to

$$\frac{1}{\omega_p} = \sqrt[4]{2\alpha_g l_g / \alpha_m l_m} \frac{1}{\sqrt{\omega_M \omega_g}}$$

For saturable modelocking, with equal gain and loss lengths it is proportional to

$$\frac{1}{v_g \gamma} = \sqrt{2\alpha_g/\alpha_l} \frac{1}{\omega_g} \frac{\sqrt{I_s}}{|V_0|}$$

Show that in both cases, the pulsewidth is proportional to

$$\sqrt[4]{\alpha_g} \frac{1}{\sqrt{\omega_g} \text{ times } \sqrt{|\text{curvature of net gain vs. time}|}}$$

10.5. Determine the parameter range for which A/a_0 vs. S_+/a_0 in Fig. 10.5 is single valued.

10.6. In bistable devices it is desirable to make the ratio $|a_2|^2/|a_1|^2$ as large as possible [compare Fig. 10.7]. Find the largest achievable ratio.

REFERENCES

[1] L. F. Mollenauer, R. H. Stolen, and J. P. Gordon, "Experimental observation of picosecond pulse narrowing and solitons in optical fibers," *Phys. Rev. Lett.*, *45*, no. 13, 1095–1098, Sept. 1980.

[2] V. E. Zakharov and A. B. Shabat, "Exact theory of two-dimensional self-focussing and one-dimensional self-modulation of waves in nonlinear media," *Sov. Phys. JETP*, *34*, 118, 1972.

[3] H. A. Haus, "Physical interpretation of inverse scattering formalism applied to self-induced transparency," *Rev. Mod. Phys.*, *51*, no. 2, 331–339, Apr. 1979.

[4] G. B. Whitham, *Linear and Nonlinear Waves*, Wiley-Interscience, New York, 1974.

[5] D. J. Kaup, "Coherent pulse propagation: a comparison of the complete solution with the McCall–Hahn theory and others," *Phys. Rev. A*, *16*, no. 2, 704–719, Aug. 1977.

[6] D. J. Kaup, A. Reiman, and A. Bers, "Space–time evolution of nonlinear three-wave interactions, I. Interactions in an homogeneous medium," *Rev. Mod. Phys.*, *51*, no. 2, Apr. 1979.

[7] S. L. McCall and E. L. Hahn, "Self-induced transparency," *Phys. Rev.*, *183*, no. 2, 457–485, July 1969.

[8] R. Y. Chiao, E. Garmire, and C. H. Townes, "Self-trapping of optical beams," *Phys. Rev. Lett.*, *13*, no. 15, 479–482, Oct. 1964.

[9] H. A. Haus, "Higher order trapped light beam solutions," *Appl. Phys. Lett.*, *8*, 128–129, Mar. 1966.

[10] P. L. Kelley, "Self-focusing of optical beams," *Phys. Rev. Lett.*, *15*, no. 26, 1005–1008, Dec. 1965.

[11] H. A. Haus, "Theory of modelocking with a fast saturable absorber," *J. Appl. Phys.*, *46*, no. 7, 3049–3058, July 1975.

[12] H. A. Haus, "Modelocking of semiconductor laser diodes," *Jap. J. Appl. Phys.*, *20*, no. 6, 1007–1020, June 1981.

[13] L. D. Landau and E. M. Lifshitz, *Electrodynamics of Continuous Media*, Pergamon Press, Elmsford, N.Y., 1960.

[14] P. Hagelstein, "Periodic pulse solutions and stability in the fast absorber model," *IEEE J. Quantum Electron.*, *QE-14*, no. 6, 443–450, June 1978.

[15] H. A. Haus, "Theory of modelocking with a slow saturable absorber," *IEEE J. Quantum Electron.*, *QE-11*, 736–746, 1975.

[16] R. L. Fork, B. I. Greene, and C. V. Shank, "Generation of optical pulses shorter than 0.1 psec by colliding pulse modelocking," *Appl. Phys. Lett.*, *38*, 671–672, 1981.

[17] C. V. Shank, R. L. Fork, R. Yen, and R. H. Stolen, "Compression of femtosecond optical pulses," *Appl. Phys. Lett.*, *40*, 761–763, 1982.

[18] A. Szöke, V. Daneu, J. Goldhar, and N. A. Kurnit, "Bistable optical element and its applications," *Appl. Phys. Lett.*, *15*, 376–379, 1969.

[19] H. M. Gibbs, S. L. McCall, T. N. C. Venkatesan, A. C. Gossard, A. Passner, and W. Wiegmann, "Optical bistability in semiconductors," *Appl. Phys. Lett.*, *35*, 451–453, Sept. 1979.

[20] "Optical Bistability," *IEEE J. Quantum Electron.*, Mar. 1981 issue.

[21] H. M. Gibbs, S. L. McCall, A. C. Gossard, A. Passner, W. Wiegmann, and T. N. C. Venkatesan, "Controlling light with light: optical bistability and optical modulation," in *Laser Spectroscopy IV*, H. Walther and K. W. Rothe, eds., Springer-Verlag, Berlin, 1979.

[22] E. Garmire, S. D. Allen, and J. Marburger, "Multimode integrated optical bistability switch," *Opt. Lett.*, *3*, 69–71, Aug. 1979.

[23] P. W. Smith, "On the physical limits of digital optical switching and logic elements," Bell System Technical Journal, *61*, 1975–1994, Oct. 1982.

[24] Y. J. Chen and G. M. Carter, "Measurement of third order nonlinear susceptibilities by surface plasmons," *Appl. Phys. Lett.*, *41*, 307–309, 1982.

[25] T. Y. Chang, "Fast self-induced refractive index changes in optical media: A survey," *Opt. Eng.*, *20*, 220–232, Mar./Apr. 1981.

WAVE PROPAGATION IN ANISOTROPIC MEDIA

Nonlinear optical processes usually take place in anisotropic media. In particular, phase matching for second harmonic generation is accomplished by taking advantage of the anisotropy of the mixing crystal. Further, optical signals are modulated by means of anisotropies induced by applied electric fields. For this reason, we study linear wave propagation in anisotropic media and develop some basic relations for such media in anticipation of a treatment of nonlinear optics more detailed than the one given in Chapter 10. In Section 11.1 we review the Poynting theorem as applied to an anisotropic medium. From this relation and from reciprocity follows the fact that the $\bar{\bar{\varepsilon}}$ tensor must be a real symmetric tensor. Section 11.2 considers wave propagation in an anisotropic medium and constructs the index ellipsoid and the normal (or index) surface which gives the velocity of propagation of a wave along every spatial direction in the crystal. Section 11.4 studies wave-packet propagation in an anisotropic dispersive medium and derives the group velocity. Section 11.5 shows that the group velocity is also the velocity of energy propagation. Section 11.6 treats the propagation of Hermite–Gaussian beams in an anisotropic medium. For further detail the reader is referred to Refs. 1 and 2.

11.1 DIELECTRIC TENSOR FOR AN ANISOTROPIC MEDIUM AND POYNTING'S THEOREM

In a linear and anisotropic medium, general linear relations must exist among the three components $P_i (i = x, y, z)$ of the polarization density vector and the three components of the electric field, E_i. These may be written as follows:

$$P_i = \varepsilon_0 \sum_{j=x,y,z} \chi_{ij} E_j$$

where the coefficients define the susceptibility tensor χ_{ij}. We shall henceforth omit the summation sign and use the Einstein summation convention, which implies summation over repeated indices:

$$P_i = \varepsilon_0 \chi_{ij} E_j \tag{11.1}$$

It is often convenient to denote the three Cartesian coordinates by the three numbers 1, 2, and 3 instead of the letters x, y, and z. We shall use both notations. Another notation derived from vector notation is used:

$$\mathbf{P} = \varepsilon_0 \bar{\bar{\chi}} \cdot \mathbf{E} \tag{11.2}$$

which obviates reference to any particular coordinate system.

It is convenient to introduce the dielectric tensor with components ε_{ij} via the definition of the displacement density vector:

$$D_i \equiv \varepsilon_0 E_i + P_i = \varepsilon_0(I_{ij} + \chi_{ij})E_j = \varepsilon_{ij} E_j \tag{11.3}$$

Here I_{ij} is the identity tensor with unity diagonal components and zeros for its off-diagonal components. In vector notation

$$\mathbf{D} = \bar{\bar{\varepsilon}} \cdot \mathbf{E} \tag{11.4}$$

where $\bar{\bar{\varepsilon}}$ is the dielectric tensor defined by nine "components." Not all components are independent, as we shall show from the complex Poynting theorem derived for an anisotropic dielectric medium. Maxwell's equations, in the absence of magnetization, in the real, time-dependent form are [compare (1.7 through 1.10)]

$$\nabla \times \mathbf{E} = -\mu_0 \frac{\partial \mathbf{H}}{\partial t} \tag{11.5}$$

$$\nabla \times \mathbf{H} = \mathbf{J} + \frac{\partial}{\partial t} \mathbf{D} \tag{11.6}$$

$$\nabla \cdot \mathbf{D} = \rho \tag{11.7}$$

$$\nabla \cdot \mu_0 \mathbf{H} = 0 \tag{11.8}$$

where, in a linear anisotropic medium, (11.4) is the new constitutive law. Often one requires the relation inverse to (11.4):

$$\mathbf{E} = \bar{\bar{\kappa}} \cdot \mathbf{D} \tag{11.9}$$

The $\bar{\bar{\kappa}}$ tensor is the inverse of the $\bar{\bar{\varepsilon}}$ tensor:

$$\bar{\bar{\kappa}} = [\bar{\bar{\varepsilon}}]^{-1} \tag{11.10}$$

The constitutive laws (11.4) and (11.9) can be generalized to dispersive anisotropic media by writing

$$\mathbf{D} = \bar{\bar{\varepsilon}}(\omega) \cdot \mathbf{E} \tag{11.11}$$

and

$$\mathbf{E} = \bar{\bar{\kappa}}(\omega) \cdot \mathbf{D} \tag{11.12}$$

where $\bar{\bar{\varepsilon}}(\omega)$ and $\bar{\bar{\kappa}}(\omega)$ are, in general, complex tensors whose components are functions of frequency. Maxwell's equations in complex form with the generalized constitutive law (11.11) are

$$\nabla \times \mathbf{E} = -j\omega\mu_0 \mathbf{H} \tag{11.13}$$

$$\nabla \times \mathbf{H} = \mathbf{J} + j\omega\bar{\bar{\varepsilon}}(\omega) \cdot \mathbf{E} \tag{11.14}$$

$$\nabla \cdot [\bar{\bar{\varepsilon}}(\omega) \cdot \mathbf{E}] = \rho \tag{11.15}$$

$$\nabla \cdot \mu_0 \mathbf{H} = 0 \tag{11.16}$$

We shall now show by means of the complex Poynting theorem that the $\bar{\bar{\varepsilon}}$ tensor has special symmetry properties in a lossless medium. The complex Poynting theorem is derived in the same way as (1.48), with the result

$$\nabla \cdot (\mathbf{E} \times \mathbf{H}^*) - j\omega(\mathbf{E} \cdot \bar{\bar{\varepsilon}}^* \cdot \mathbf{E}^* - \mu_0 \mathbf{H} \cdot \mathbf{H}^*) + \mathbf{E} \cdot \mathbf{J}^* = 0 \tag{11.17}$$

In the sinusoidal steady state (ω real) in a lossless dielectric medium, the time-averaged power imparted to the dielectric must be zero:

$$\text{Re}\,[j\omega\mathbf{E} \cdot \bar{\bar{\varepsilon}}^* \cdot \mathbf{E}^*] = -\text{Im}\,[\omega\mathbf{E} \cdot \bar{\bar{\varepsilon}}^* \cdot \mathbf{E}^*] = 0$$

This must be true for an arbitrary choice of components of **E**. For example, if the only nonzero component of **E** is E_x, $\text{Im}\,[\varepsilon_{xx}^* |E_x|^2] = 0$ and therefore ε_{xx} must be real; similarly, ε_{yy} and ε_{zz} must be real. Further,

$$\text{Im}\,[\varepsilon_{xy}^* E_x E_y^* + \varepsilon_{yx}^* E_y E_x^*] = \text{Im}\,[(\varepsilon_{xy}^* - \varepsilon_{yx})E_x E_y^*] = 0$$

Because E_x and E_y^* are arbitrary, we must have

$$\varepsilon_{xy}^* = \varepsilon_{yx}$$

This proof is analogous to the one that led via energy conservation to a relationship between the coupling coefficients κ_{12} and κ_{21}, (7.61). The ε_{xy} corresponds to $j\kappa_{12}$, ε_{yx} to $j\kappa_{21}$.

Thus, in general,

$$\varepsilon_{ij} = \varepsilon_{ji}^* \tag{11.18}$$

for real ω; the $\bar{\bar{\varepsilon}}$ tensor is a *Hermitean* tensor. It is not symmetric in the sense that the transpose of the tensor is its complex conjugate. The proof of the reciprocity theorem of Section 3.1 fails for media with asymmetric ε tensors, as the reader may confirm easily by repeating the proof for an anisotropic dielectric medium with a tensor dielectric constant. We shall not study nonreciprocal media, all of which must have magnetic properties, and concentrate solely on lossless reciprocal media. These have *real symmetric dielectric tensors*. The inverse tensor $\bar{\bar{\kappa}} = \bar{\bar{\varepsilon}}^{-1}$ is also real and symmetric.

11.2 PROPAGATION OF WAVES IN ANISOTROPIC, LOSSLESS, RECIPROCAL MEDIA

Propagation of plane waves in anisotropic media is governed by Maxwell's equations on the one hand, and by the anisotropic dielectric constitutive law on the other hand. We study first the constraint imposed on the vectors **D** and **E** by Maxwell's equations and then introduce the constitutive relation between **D** and **E** to derive the propagation constants k for the waves in the anisotropic medium. We start by proving from Maxwell's equations that, under an assumed spatial dependence exp $(-j\boldsymbol{k} \cdot \boldsymbol{r})$, the following relation holds:

$$\frac{\mathbf{D}}{\varepsilon_0} = \frac{k^2}{\omega^2 \mu_0 \varepsilon_0} \mathbf{E}_\perp \tag{11.19}$$

where $\mathbf{E}_\perp$ is the component of **E** perpendicular to the propagation vector $\boldsymbol{k}$. Maxwell's equations in complex notation for a medium free of conduction current are [compare (11.14) and (11.13)]

$$\nabla \times \mathbf{H} = j\omega \mathbf{D} \tag{11.20}$$

$$\nabla \times \mathbf{E} = -j\omega\mu_0 \mathbf{H} \tag{11.21}$$

We take the curl of (11.21) and use (11.22) to replace **H**:

$$\nabla \times (\nabla \times \mathbf{E}) = \omega^2 \mu_0 \mathbf{D} \tag{11.22}$$

If the assumed spatial dependence of the fields is of the form $e^{-j\boldsymbol{k}\cdot\boldsymbol{r}}$, then

$$\nabla \times (\nabla \times \mathbf{E}) = k^2\left(\mathbf{E} - \frac{\boldsymbol{k} \cdot \mathbf{E}}{k^2} \boldsymbol{k}\right) = k^2 \mathbf{E}_\perp \tag{11.23}$$

Equation (11.23) shows that the operator $\nabla \times \nabla$ operating on a function of the form $\exp(-j\boldsymbol{k} \cdot \boldsymbol{r})$ produces a vector perpendicular to $\boldsymbol{k}$. According to (11.22) the vector **D** is parallel to $\nabla \times (\nabla \times \mathbf{E})$ and thus **D** itself must be perpendicular to $\boldsymbol{k}$. On the other hand, the fact that **D** is perpendicular to $\boldsymbol{k}$ follows from the divergence law in the absence of free charge, (11.7), which gives

$$-j\boldsymbol{k} \cdot \mathbf{D} = 0 \tag{11.24}$$

We have introduced an assumed spatial dependence into two of the four equations of Maxwell (Ampère's law and Faraday's law) and found that another Maxwell equation, Gauss' law for electric fields, was obeyed automatically.

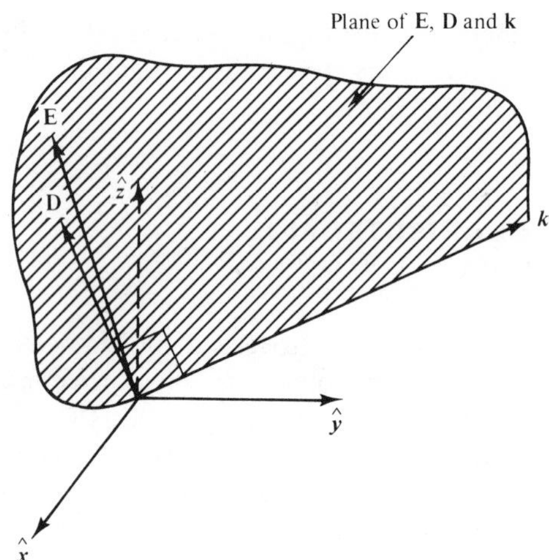

Figure 11.1 Orientations of **D**, **E**, and ***k***.

Returning to (11.23) and (11.22) we note that, when combined, they lead to (11.19) which thus has been proven. The relative orientations of **D**, **E** and ***k*** are shown in Fig. 11.1.

Next we consider the constitutive law expressed by (11.12), where $\bar{\bar{\kappa}}$ is a real symmetric tensor. By proper orientation of the coordinate system with respect to the medium, it is possible to diagonalize a symmetric tensor (Appendix 11A treats transformations of tensors and proves this assertion). Consider first the constitutive law in a coordinate system in which $\bar{\bar{\kappa}}$ is diagonal:

$$\bar{\bar{\kappa}} \equiv \begin{bmatrix} \kappa_{xx} & 0 & 0 \\ 0 & \kappa_{yy} & 0 \\ 0 & 0 & \kappa_{zz} \end{bmatrix} \tag{11.25}$$

The Cartesian coordinates x, y, and z have been chosen along the *principal axes* of the tensor.

Propagation along Principal Axes

If the ***k*** vector of the plane wave points along the z direction, then $D_z = 0$, according to (11.24). In this special case, E_z is zero as well because the $\bar{\bar{\kappa}}$ tensor is diagonal. The remaining x and y components of **D** and **E** are related by

$$E_x = \kappa_{xx} D_x \tag{11.26}$$

and

$$E_y = \kappa_{yy} D_y \tag{11.27}$$

When **D** is not linearly polarized along either $\hat{x}$ or $\hat{y}$, and $\kappa_{xx} \neq \kappa_{yy}$, then $\mathbf{E}_\perp$ is not parallel to **D** as required by (11.19). We conclude that a wave propagating along the principal axis in the z direction must be linearly polarized along one of the two remaining principal axes of the dielectric tensor. If the electric field is x-polarized, we find, from (11.19) and (11.26), that

$$k^2 = \frac{\omega^2 \mu_0}{\kappa_{xx}} \tag{11.28}$$

The phase velocity of the wave is $\sqrt{\kappa_{xx}/\mu_0}$. Similarly, we find from (11.27) and (11.19) for y-directed **D** and **E** vectors that

$$k^2 = \frac{\omega^2 \mu_0}{\kappa_{yy}} \tag{11.29}$$

The phase velocity of the wave is $\sqrt{\kappa_{yy}/\mu_0}$. The phase velocities are different for differently polarized waves traveling along the z direction. Crystals that propagate waves of different polarizations with different phase velocities are called *birefringent*. A similar argument carries through for a choice of the $\boldsymbol{k}$ vector along the $\hat{x}$ and $\hat{y}$ axes.

If a linearly polarized wave propagating in the z direction is incident normally upon a crystal interface, which is parallel to the x-y plane, the incident wave is partially reflected, and partially transmitted via the two wave solutions linearly polarized along $\hat{\boldsymbol{x}}$ and $\hat{\boldsymbol{y}}$ and propagating with their respective phase velocities. The matching of the boundary conditions is accomplished separately for each of the two polarizations (see Problem 11.4).

Propagation in a General Spatial Direction

Thus far we looked for propagation along the principal axes of the $\bar{\bar{\kappa}}$ tensor (or $\bar{\bar{\varepsilon}}$ tensor). We now show how one may determine the magnitude of the wave vector pointing in a general spatial direction and the polarization of the associated field. A wave with its $\boldsymbol{k}$ vector along a particular spatial direction possesses a **D** vector perpendicular to this direction; **D** lies in a plane perpendicular to $\boldsymbol{k}$. (See Fig. 11.1) The **E**-field component in this plane and parallel to **D** must be related to **D** by (11.19). In order to determine the allowed directions of **D** and **E**, we write the $\bar{\bar{\kappa}}$ tensor in a coordinate system whose $\hat{z}$ axis coincides with $\boldsymbol{k}$. The z component of **D** vanishes in this coordinate system, distinguished by primes, and thus the components of $\mathbf{E}_\perp$, perpendicular to $\boldsymbol{k}$, are

$$\mathrm{E}'_x = \kappa'_{xx} \mathrm{D}'_x + \kappa'_{xy} \mathrm{D}'_y \tag{11.30}$$

$$\mathrm{E}'_y = \kappa'_{yx} \mathrm{D}'_x + \kappa'_{yy} \mathrm{D}'_y \tag{11.31}$$

with $\kappa'_{yx} = \kappa'_{xy} \cdot \mathbf{E}_\perp$ will be parallel to **D** if, and only if,

$$\mathbf{E}_\perp = \lambda \mathbf{D} \tag{11.32}$$

where λ is a scalar. When this constraint is introduced into (11.30) and (11.31) we find a homogeneous equation for λ, an *eigenvalue equation.* Nontrivial

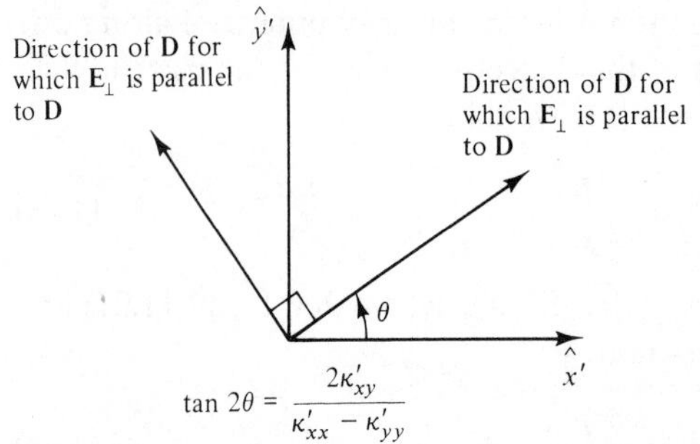

Figure 11.2 Directions of the eigenvectors.

solutions are found only if the determinant vanishes:

$$(\kappa'_{xx} - \lambda)(\kappa'_{yy} - \lambda) - \kappa'^2_{xy} = 0 \tag{11.33}$$

This quadratic equation has the two solutions for λ:

$$\lambda_\pm = \frac{\kappa'_{xx} + \kappa'_{yy}}{2} \pm \sqrt{\left(\frac{\kappa'_{xx} - \kappa'_{yy}}{2}\right)^2 + \kappa'^2_{xy}} \tag{11.34}$$

The directions of the *eigenvectors* pertaining to these two solutions are found by introducing these solutions into (11.30):

$$\frac{D'_y}{D'_x} = \frac{-\dfrac{\kappa'_{xx} - \kappa'_{yy}}{2} \pm \sqrt{\left(\dfrac{\kappa'_{xx} - \kappa'_{yy}}{2}\right)^2 + \kappa'^2_{xy}}}{\kappa'_{xy}} \tag{11.35}$$

The angle of inclination of **D** with respect to the x' axis, θ, has the tangent $\tan\theta = D'_y/D'_x$ (see Fig. 11.2). Through use of the identity

$$\tan 2\theta = \frac{2\tan\theta}{1 - \tan^2\theta}$$

and introducing (11.35), the above can be rewritten

$$\tan 2\theta = \frac{2\kappa'_{xy}}{\kappa'_{xx} - \kappa'_{yy}} \tag{11.36}$$

This equation yields two mutually perpendicular directions shown in Fig. 11.2. According to (11.21), the $\mathbf{E}_\perp$ field is related to **D** by the eigenvalues $\lambda_\pm$; the propagation constants associated with these two eigenvalues are, using (11.19),

$$k = \omega\sqrt{\frac{\mu_0}{\lambda_\pm}} = \omega\sqrt{\mu_0}\left[\frac{\kappa'_{xx} + \kappa'_{yy}}{2} \pm \sqrt{\left(\frac{\kappa'_{xx} - \kappa'_{yy}}{2}\right)^2 + \kappa'^2_{xy}}\right]^{-1/2} \tag{11.37}$$

The Index Ellipsoid

The operations involved in the procedure above are most easily visualized with the aid of the quadratic surface defined in terms of the $\bar{\bar{\kappa}}$-tensor components. In a coordinate system aligned with the principal axes, the equation of the surface is

$$\kappa_{xx} x^2 + \kappa_{yy} y^2 + \kappa_{zz} z^2 = \frac{1}{\varepsilon_0} \tag{11.38}$$

Since the κ_{ii}'s are all positive (see Appendix 11A), this is the equation of an ellipsoid. In the primed coordinates with $\hat{z}'$ parallel to the chosen direction of the $\boldsymbol{k}$ vector, the equation of the ellipsoid is

$$\kappa'_{xx} x'^2 + \kappa'_{yy} y'^2 + \kappa'_{zz} z'^2 + 2\kappa'_{yz} y'z' + 2\kappa'_{zx} z'x' + 2\kappa'_{xy} x'y' = \frac{1}{\varepsilon_0} \tag{11.39}$$

The plane perpendicular to the z' axis through the origin, $z' = 0$, cuts the ellipsoid in the ellipse (see Fig. 11.3a)

$$\kappa'_{xx} x'^2 + \kappa'_{yy} y'^2 + 2\kappa'_{xy} x'y' = \frac{1}{\varepsilon_0} \tag{11.40}$$

If we choose new x'' and y'' coordinates so that x'' is inclined with respect to x' by the angle θ (Fig. 11.3*b*),

$$\begin{aligned} x' &= x'' \cos\theta - y'' \sin\theta \\ y' &= x'' \sin\theta + y'' \cos\theta \end{aligned} \tag{11.41}$$

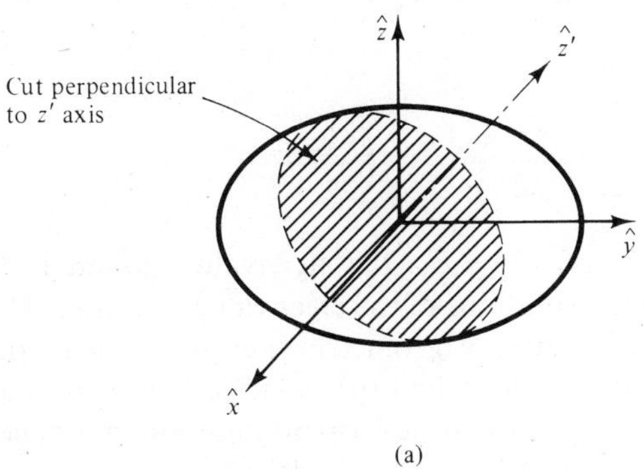

(a)

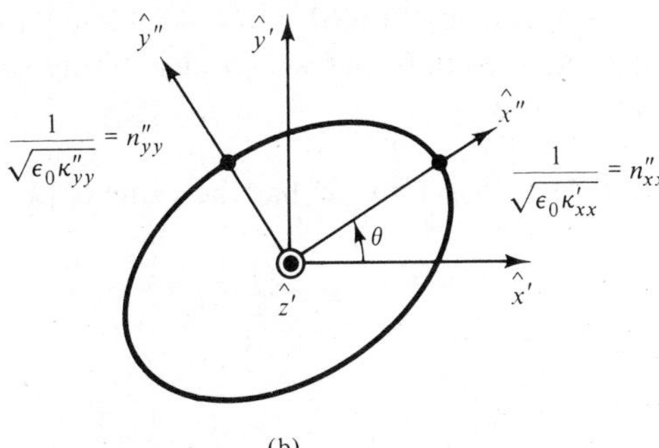

(b)

Figure 11.3 (a) Ellipsoid defined by (11.38). (*b*) Ellipse in plane perpendicular to z'.

we get a new equation for the ellipse, in terms of new components κ''_{ij}, $i, j = x, y$. The off-diagonal element is easily found to be

$$\kappa''_{xy} = \kappa'_{xy} \cos 2\theta + \tfrac{1}{2}(\kappa'_{yy} - \kappa'_{xx}) \sin 2\theta \tag{11.42}$$

With the choice

$$\tan 2\theta = \frac{2\kappa'_{xy}}{\kappa'_{xx} - \kappa'_{yy}}$$

the off-diagonal element κ''_{xy} is made equal to zero. This is the value of θ obtained previously in (11.36). If **D** points along either x'' or y'', the **E** field is parallel to **D**:

$$\mathrm{E}''_x = \kappa''_{xx} \mathrm{D}''_x \tag{11.43}$$

$$\mathrm{E}''_y = \kappa''_{yy} \mathrm{D}''_y \tag{11.44}$$

where

$$\kappa''_{xx} = \frac{\kappa'_{xx} - \kappa'_{yy}}{2} + \sqrt{\left(\frac{\kappa'_{xx} - \kappa'_{yy}}{2}\right)^2 + \kappa'^2_{xy}}$$

and

$$\kappa''_{yy} = \frac{\kappa'_{xx} - \kappa'_{yy}}{2} - \sqrt{\left(\frac{\kappa'_{xx} - \kappa'_{yy}}{2}\right)^2 + \kappa'^2_{xy}}$$

We see that $\kappa''_{xx} = \lambda_+$ and $\kappa''_{yy} = \lambda_-$, the eigenvalues found in (11.34). Thus (11.43) and (11.44) are the solutions of the eigenvalue equation (11.32).

The ellipsoid (11.38) gives a geometric interpretation to the eigenvalue formulation of (11.32). In order to find the polarization of the wave solutions, and their propagation constants for a $\boldsymbol{k}$ orientation along a general direction $\hat{z}'$, one proceeds as follows (see Figs. 11.3a and 11.3b):

1. One cuts the $\bar{\bar{\kappa}}$-tensor ellipsoid (11.38) by a plane perpendicular to $\hat{z}'$.
2. One constructs the coordinates x'' and y'' along the principal axes of the intersection ellipse.

The wave with its **D** vector along $\hat{x}''$ has the value of $|\boldsymbol{k}|$

$$k = \frac{\omega}{c} n''_{xx}$$

where

$$n''_{xx} = \frac{1}{\sqrt{\varepsilon_0 \kappa''_{xx}}}$$

and the wave with its **D** vector along $\hat{y}''$ has

$$k = \frac{\omega}{c} n''_{yy}$$

where

$$n''_{yy} = \frac{1}{\sqrt{\varepsilon_0 \kappa''_{yy}}}$$

The *indices* n''_{xx} and n''_{yy} are the principal axes of the intersection ellipse in Fig. 11.3. For this reason, the ellipsoid (11.38) is also called the *index ellipsoid.* To emphasize this interpretation, the equation for the ellipsoid (11.38) is rewritten

$$\frac{x^2}{n_{xx}^2} + \frac{y^2}{n_{yy}^2} + \frac{z^2}{n_{zz}^2} = 1 \tag{11.45}$$

where $n_{xx}^2 \equiv 1/\varepsilon_0 \kappa_{xx}$, and so on.

Normal (Index) Surface

We have found that two orthogonal polarizations of **D** may be assigned to each direction of propagation. The $\boldsymbol{k}$ value of these two directions is given by (11.37). By marking off two values of k, for every direction of $\boldsymbol{k}$, one may construct a surface of $\boldsymbol{k}$, or index $\boldsymbol{n} \equiv \boldsymbol{k}(c/\omega)$. This normal (index) surface is, in general, a fourth-order surface. In the case of a uniaxial crystal, the fourth-order surface is a sphere and an ellipsoid, respectively, as we now proceed to show.

The index ellipsoid of a uniaxial crystal (Fig. 11.4) has one axis of rotational symmetry, the z axis. Construct a plane perpendicular to $\boldsymbol{k}$. The minor and major axes of the intersection ellipse are in the plane of the paper and normal to the paper, respectively. A point of the normal (index) surface is obtained by putting onto the $\boldsymbol{k}$ direction the distance $1/\sqrt{\kappa''_{xx}\varepsilon_0}$, which is the minor axis of the intersection ellipse. Repeating the process for all orientations of $\boldsymbol{k}$ in the plane of the paper results in an ellipse with axes orthogonal to the index–ellipsoid intersection. The major axis $1/\sqrt{\kappa''_{yy}\varepsilon_0}$ is constant for all orientations $\boldsymbol{k}/\omega\sqrt{\mu\varepsilon_0}$ and results in a circle in the cross section shown.

The entire normal (index) surface is obtained by rotating the circle and ellipse around the axis of symmetry of the crystal, the z axis. The resulting sphere is the surface for the *ordinary* propagation; the rotationally symmetric ellipsoid is for *extraordinary* propagation. The equation of the index ellipsoid for the uniaxial crystal is a simplified version of (11.45):

$$\frac{x^2}{n_0^2} + \frac{y^2}{n_0^2} + \frac{z^2}{n_e^2} = 1 \tag{11.46}$$

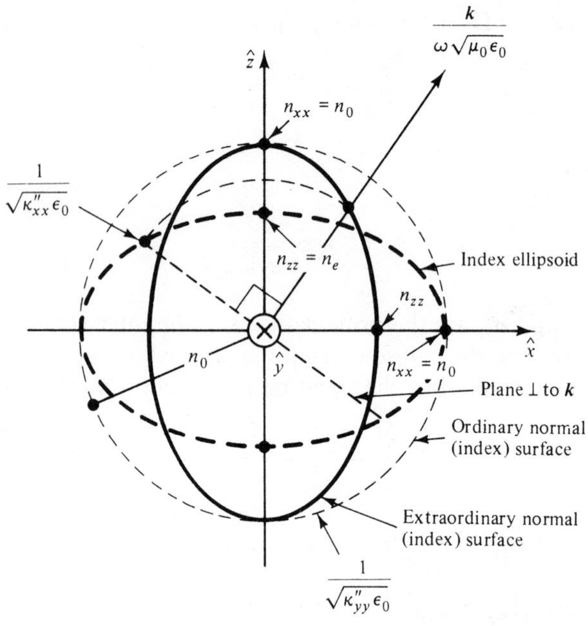

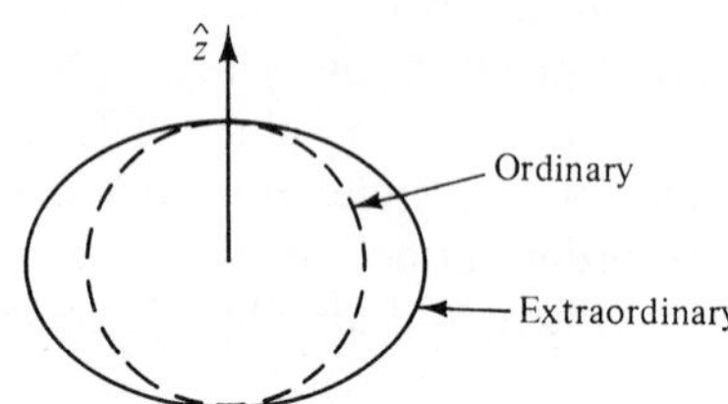

Figure 11.4 Construction of index surface. The surface is of rotational symmetry around the z axis. This is an example of a *positive* uniaxial crystal. A *negative* uniaxial crystal is shown in the inset.

where the "ordinary" index n_0 is defined by

$$n_0^2 = n_{xx}^2 = n_{yy}^2 = \frac{1}{\varepsilon_0 \kappa_{xx}} = \frac{1}{\varepsilon_0 \kappa_{yy}} = \frac{\varepsilon_{xx}}{\varepsilon_0} = \frac{\varepsilon_{yy}}{\varepsilon_0}$$

and the "extraordinary" index n_e by

$$n_e^2 = n_{zz}^2 = \frac{1}{\varepsilon_0 \kappa_{zz}} = \frac{\varepsilon_{zz}}{\varepsilon_0}$$

The normal (index) surface is composed of an ellipsoid

$$\frac{x^2 + y^2}{n_e^2} + \frac{z^2}{n_0^2} = 1 \tag{11.47}$$

and a sphere

$$\frac{x^2 + y^2 + z^2}{n_0^2} = 1 \tag{11.48}$$

The normal surface can be expressed as a surface in k space because it gives the magnitude of the vector $\boldsymbol{k}$ for every direction. For the extraordinary wave,

$$\frac{k_x^2 + k_y^2}{n_e^2} + \frac{k_z^2}{n_0^2} = \frac{\omega^2}{c^2} \tag{11.49}$$

and for the ordinary surface,

$$\frac{k_x^2 + k_y^2 + k_z^2}{n_0^2} = \frac{\omega^2}{c^2}$$

In a dispersive uniaxial medium, n_e and n_0 are functions of frequency.

The *normal* (*index*) *surface* used here is consistent with that used in Ref. 1. In the literature a *normal surface* is often used in which the inverse index is plotted in the direction of group velocity [2]. We distinguish the normal surface used here by adding the designation *index* in parentheses.

11.3 ENERGY DENSITY IN AN ANISOTROPIC DISPERSIVE DIELECTRIC

An anisotropic, lossless, reciprocal, dispersive dielectric possesses a dielectric tensor that is ω dependent, symmetric, and real for real ω. We shall now determine the time-averaged energy density in the dielectric. Because the polarization P_i,

$$\mathrm{P}_i = [\varepsilon_{ij}(\omega) - \varepsilon_0 I_{ij}]\mathrm{E}_j$$

is not an instantaneous function of E_j, it is not possible to claim that the time-averaged energy density in the dielectric is simply $\frac{1}{4}\mathbf{E}^* \cdot \bar{\bar{\varepsilon}} \cdot \mathbf{E}$, as it would be in a dispersion-free dielectric. In order to determine the time-averaged energy density, one needs to study carefully the power supplied to the dielectric as the field is built up. Because $\bar{\bar{\varepsilon}}$ is treated as a function of ω, the buildup of the field must maintain the quasi-monochromaticity (single frequency character) of the field.

We start with Maxwell's equation (11.13) and (11.14) with $\mathbf{J} = 0$:

$$\nabla \times \mathbf{E} = -j\omega\mu_0 \mathbf{H} \tag{11.50}$$

$$\nabla \times \mathbf{H} = j\omega\bar{\bar{\varepsilon}}(\omega) \cdot \mathbf{E} \tag{11.51}$$

In a thought experiment, we perturb the equations above with a small current drive $\delta\mathbf{J}$ at frequency $\omega + \delta\omega$, with $\mathrm{Im}\ \delta\omega < 0$ so that the time dependence is $\exp(|\mathrm{Im}\ \delta\omega|\, t) \exp[j(\omega_0 + \mathrm{Re}\ \delta\omega)t]$, corresponding to a rate of buildup of the field. This current drive is responsible for the exponential buildup of the field. The equations (11.13) and (11.14) perturbed to first order are ($\mathbf{E} \to \mathbf{E} + \delta\mathbf{E}$, $\mathbf{H} \to \mathbf{H} + \delta\mathbf{H}$, etc.)

$$\nabla \times \delta\mathbf{E} = -j\omega\mu_0\,\delta\mathbf{H} - j\delta\omega\mu_0\,\mathbf{H} \tag{11.52}$$

$$\nabla \times \delta\mathbf{H} = \delta\mathbf{J} + j\delta(\omega\bar{\bar{\varepsilon}}) \cdot \mathbf{E} + j\omega\bar{\bar{\varepsilon}} \cdot \delta\mathbf{E} \tag{11.53}$$

We dot-multiply the complex conjugate of (11.50) by $\delta\mathbf{H}$, the complex conjugate of (11.51) by $-\delta\mathbf{E}$, (11.52) by $\mathbf{H}^*$, and (11.53) by $-\mathbf{E}^*$ and add. With $\delta\omega$ treated as a differential, the result is a form of a perturbation Poynting's theorem:

$$\nabla \cdot [\mathbf{E}^* \times \delta\mathbf{H} + \delta\mathbf{E} \times \mathbf{H}^*] + j\,\delta\omega\left[\mu_0\mathbf{H} \cdot \mathbf{H}^* + \mathbf{E}^* \cdot \frac{\partial(\omega\bar{\bar{\varepsilon}})}{\partial\omega} \cdot \mathbf{E}\right] + \mathbf{E}^* \cdot \delta\mathbf{J} = 0 \tag{11.54}$$

$-(\frac{1}{2})$ Re $\mathbf{E}^* \cdot \delta\mathbf{J}$ is the power per unit volume supplied by the current perturbation. By virtue of its role in the equation above, $\frac{1}{2}[\mathbf{E}^* \times \delta\mathbf{H} + \delta\mathbf{E} \times \mathbf{H}^*]$ is identified as the perturbation of the Poynting vector caused by $\delta\mathbf{J}$. The real part of the remaining term,

$$-\text{Im}\,(2\delta\omega)\,\frac{1}{4}\left[\mu_0\mathbf{H} \cdot \mathbf{H}^* + \mathbf{E}^* \cdot \frac{\partial(\omega\bar{\bar{\varepsilon}})}{\partial\omega} \cdot \mathbf{E}\right]$$

must be the time rate of change of the time-averaged energy density. Indeed, the energy is quadratic in the field amplitudes. An exponential growth of the field $\exp\,[-\text{Im}\,(\delta\omega)t]$ is accompanied by a growth of the energy with the time dependence $\exp\,\{-2[\text{Im}\,(\delta\omega)t]\}$. The rate of growth, the derivative with respect to time, is equivalent to multiplication by $-\text{Im}\,(2\delta\omega)$. The time-averaged energy density in the dielectric is thus

$$\langle w_e \rangle = \frac{1}{4}\,\mathbf{E}^* \cdot \frac{\partial\omega\bar{\bar{\varepsilon}}}{\partial\omega} \cdot \mathbf{E} \tag{11.55}$$

In the case of an isotropic dispersion free dielectric, this reduces to $\langle w_e \rangle = \frac{1}{4}\varepsilon\mathbf{E} \cdot \mathbf{E}^*$ as previously derived.

The derivation of the expression for time-averaged energy density applies to any quasi-monochromatic process. Energy density in plane waves is given by (11.55) as a special case of this general relation. In Section 11.5 we shall rederive the above in the context of plane waves and shall obtain, in addition, an explicit expression relating group velocity, power flow density, and energy density.

11.4 GROUP VELOCITY IN AN ANISOTROPIC DISPERSIVE DIELECTRIC

It is possible to construct an optical beam of finite transverse dimensions by superposition of plane waves with different $\boldsymbol{k}$-vector orientations. This has been shown in the context of the Hermite–Gaussian beams propagating in free space. Here we shall extend the analysis to a superposition of waves with different $\boldsymbol{k}$-vector orientations and of different frequencies to construct a wave packet of finite dimensions in all three spatial directions. Instead of the superposition integral in terms of transverse components of $\boldsymbol{k}$ at one single frequency we use a superposition in terms of both transverse and longitudinal $\boldsymbol{k}$-vector components. The frequency is varied in accordance with the dispersion relation

$$|\boldsymbol{k}(\omega)| = \frac{\omega}{c}\, n(\boldsymbol{k}, \omega) \tag{11.56}$$

The dispersion relation defines a surface in $\boldsymbol{k}$ space for every value of ω. Indeed, the normal (index) surface is no more than the surface $c\boldsymbol{k}/\omega$ at constant ω.

We write a solution of the wave equation for the electric field $\boldsymbol{E}(\boldsymbol{r}, t)$ in an anisotropic medium as a superposition of plane waves clustered around $\boldsymbol{k}_0$, $\boldsymbol{k} = \boldsymbol{k}_0 + \Delta\boldsymbol{k}$, and in frequency around ω_0, $\omega = \omega_0 + \Delta\omega$:

$$\mathbf{E}(\boldsymbol{r}, t) = \int d^3\Delta\boldsymbol{k}\, \exp\{j[\omega t - (\boldsymbol{k}_0 + \Delta\boldsymbol{k})\cdot\boldsymbol{r}]\}\mathbf{E}(\Delta\boldsymbol{k}) \tag{11.57}$$

where $d^3\,\Delta\boldsymbol{k}$ indicates the three-dimensional integral over three components of $\Delta\boldsymbol{k}$. $\mathbf{E}(\boldsymbol{r}, t)$ is, in general, complex, because it is only the positive frequency part of the full time-dependent function [compare (1.85)]. One may factor out the space–time dependence $\exp[j(\omega_0 t - \boldsymbol{k}_0\cdot\boldsymbol{r})]$ and expand ω as a function of $\Delta\boldsymbol{k}$:

$$\omega = \omega_0 + \Delta\omega = \omega_0 + \frac{\partial\omega}{\partial\boldsymbol{k}}\cdot\Delta\boldsymbol{k} = \omega_0 + (\nabla_k\omega)\cdot\Delta\boldsymbol{k} \tag{11.58}$$

where $\partial\omega/\partial\boldsymbol{k}$ is the gradient of ω in $\boldsymbol{k}$ space; it is perpendicular to a surface of constant ω. With the expansion (11.58) one obtains from (11.57):

$$\mathbf{E}(\boldsymbol{r}, t) = \exp[j(\omega_0 t - \boldsymbol{k}_0\cdot\boldsymbol{r})] \cdot \int d^3\Delta\boldsymbol{k}\,\exp[-j\,\Delta\boldsymbol{k}\cdot(\boldsymbol{r} - \nabla_k\omega t)]\mathbf{E}(\Delta\boldsymbol{k}) \tag{11.59}$$

The wave packet progresses with the phase velocity $\omega_0/|\boldsymbol{k}_0|$ in the direction of $\boldsymbol{k}_0$, and with the group velocity $\boldsymbol{v}_g = \nabla_k\omega$ in the direction normal to the surface $\boldsymbol{k} = \boldsymbol{k}(\omega)$ at constant ω.

Consider first the dispersion-free case for which the index surface is frequency independent. The gradient $\nabla_k \omega$ is perpendicular to the scaled version of the index surface at $\boldsymbol{k} = \boldsymbol{k}_0$ and the magnitude of the group velocity is $v_g = c/n$.

When the index surface is a function of frequency, we proceed in several steps that show more clearly the procedure of wave-packet construction. The integral over $\Delta\boldsymbol{k}$ in (11.59) is split into two parts, $d\,\Delta k_\xi\, d\,\Delta k_\eta$ over the surface of constant ω in $\boldsymbol{k}$ space, and $d\,\Delta k_\zeta$ in the direction normal to the surface (see Fig. 11.5). The first integral over the surface of constant ω gives the beam shape that would be achieved in the steady state (constant ω). The integral over Δk_ζ is an integral over frequency because Δk_ζ is parallel to $\nabla_k \omega$; with a change of Δk_ζ is associated a corresponding change of ω. The integral over k_ζ constructs a wave packet of finite extent in time, and of finite extent along the k direction. This wave packet proceeds, according to (11.59), in the k_ζ direction with the velocity $|\nabla_k \omega|$.

The magnitude of the group velocity is obtained easily along the principal axes of the index ellipsoid. Thus, along x,

$$\omega = \frac{c}{n_{xx}} k_x$$

The derivative of ω with respect to k_x gives

$$\frac{\partial\omega}{\partial k_x} = -\frac{c}{n_{xx}^2}\frac{\partial n_{xx}}{\partial\omega}\frac{\partial\omega}{\partial k_x}k_x + \frac{c}{n_{xx}}$$

and thus, solving for the group velocity $v_g = \partial\omega/\partial k_x$

$$\frac{\partial\omega}{\partial k_x} = v_g = \frac{c}{n_{xx}}\frac{1}{1 + (\omega/n_{xx})(\partial n_{xx}/\partial\omega)} \tag{11.60}$$

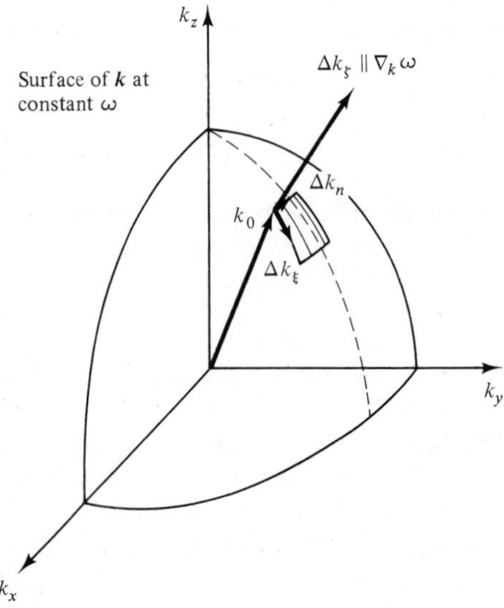

Figure 11.5 Coordinates $\Delta k_\xi\ \Delta k_\eta\ \Delta k_\zeta$.

We have defined an energy velocity in Section 6.9 relating the time-averaged power in a wave to the energy per unit length in the wave. This energy velocity was found to be equal to the group velocity by constructing wave packets from waves of slightly different frequencies. The same argument can be repeated with plane waves in an anisotropic medium with the conclusion that the time-averaged flow of power per unit area, $\frac{1}{2}\,\mathrm{Re}\,[\mathbf{E} \times \mathbf{H}^*]$ must be related to

$$\langle w \rangle = \frac{1}{4}\left[\mu_0 \mathbf{H} \cdot \mathbf{H}^* + \mathbf{E}^* \cdot \frac{\partial \omega \bar{\bar{\varepsilon}}}{\partial \omega} \cdot \boldsymbol{E}\right]$$

by

$$\tfrac{1}{2}\,\mathrm{Re}\,[\mathbf{E} \times \mathbf{H}^*] = \boldsymbol{v}_g \langle w \rangle \tag{11.61}$$

A detailed proof of this relation is given in the next section.

11.5 GROUP VELOCITY, POWER FLOW, AND ENERGY DENSITY IN AN ANISOTROPIC DISPERSIVE DIELECTRIC

In the preceding section we argued that the energy in a wavepacket is trapped and hence has to travel with the velocity of the wavepacket. This reasoning led to a relation between the energy density and the Poynting vector containing the group velocity. In this section we shall extend the analysis of Section 11.3 to the perturbation of a plane wave and rigorously establish relation (11.61).

Consider a plane wave with the space–time dependence $\exp\,[j(\omega t - \boldsymbol{k} \cdot \boldsymbol{r})]$ in a medium with the dielectric tensor $\bar{\bar{\varepsilon}}(\omega)$. Equations (11.50) and (11.51) applied to the wave give

$$-j\boldsymbol{k} \times \mathbf{E} = -j\omega\mu_0 \mathbf{H} \tag{11.62}$$

$$-j\boldsymbol{k} \times \mathbf{H} = j\omega\bar{\bar{\varepsilon}} \cdot \mathbf{E} \tag{11.63}$$

where ω and $\boldsymbol{k}$ are real for a plane wave in a lossless medium. Now let us perturb these equations in a thought experiment that introduces a driving current density $\delta\mathbf{J}$ at frequency $\omega + \delta\omega$ and with the propagation constant $\boldsymbol{k} + \delta\boldsymbol{k}$. The perturbations of (11.62) and (11.63), with $\mathbf{E} \to \mathbf{E} + \delta\mathbf{E}$, $\mathbf{H} \to \mathbf{H} + \delta\mathbf{H}$, are

$$-j\boldsymbol{k} \times \delta\mathbf{E} - j\,\delta\boldsymbol{k} \times \mathbf{E} = -j\,\delta\omega\mu_0 \mathbf{H} - j\omega\mu_0\,\delta\mathbf{H} \tag{11.64}$$

$$-j\boldsymbol{k} \times \delta\mathbf{H} - j\,\delta\boldsymbol{k} \times \mathbf{H} = j\delta(\omega\bar{\bar{\varepsilon}}) \cdot \mathbf{E} + j\omega\bar{\bar{\varepsilon}} \cdot \delta\mathbf{E} + \delta\mathbf{J} \tag{11.65}$$

We dot-multiply the complex conjugate of (11.62) with $\delta\mathbf{H}$, the complex conjugate of (11.63) with $-\delta\mathbf{E}$, (11.64) by $\mathbf{H}^*$, and (11.65) by $-\mathbf{E}^*$, and add. This manipulation leads to a Poynting perturbation theorem such as (11.54), but now involving both $\delta \boldsymbol{k}$ and $\delta\omega$:

$$-j\,\delta \boldsymbol{k} \cdot (\mathbf{E} \times \mathbf{H}^* + \mathbf{E}^* \times \mathbf{H}) + j\,\delta\omega \left[\mu_0 \mathbf{H} \cdot \mathbf{H}^* + \mathbf{E}^* \cdot \left(\frac{\partial \omega \bar{\bar{\varepsilon}}}{\partial \omega} \right) \cdot \mathbf{E} \right] + \mathbf{E}^* \cdot \delta \mathbf{J} = 0 \tag{11.66}$$

If we take the real part of the above, we get

$$(\operatorname{Im} \delta \boldsymbol{k}) \cdot (\mathbf{E} \times \mathbf{H}^* + \mathbf{E}^* \times \mathbf{H}) - \operatorname{Im} \delta\omega \left[\mu_0 \mathbf{H} \cdot \mathbf{H}^* + \mathbf{E}^* \cdot \frac{\partial \omega \bar{\bar{\varepsilon}}}{\partial \omega} \cdot \mathbf{E} \right] + \operatorname{Re} (\mathbf{E}^* \cdot \delta \mathbf{J}) = 0 \tag{11.67}$$

where we have taken into account that $\bar{\bar{\varepsilon}}$ is a Hermitean tensor for real ω. The above is a statement of energy conservation. The spatial rate of growth of the real part of the complex Poynting vector

$$\mathbf{S} = \tfrac{1}{2} \mathbf{E} \times \mathbf{H}^* \tag{11.68}$$

is $2(\operatorname{Im} \delta \boldsymbol{k}) \cdot \operatorname{Re} \mathbf{S}$. The temporal growth of the time-averaged energy density $\langle w \rangle$ is $-2(\operatorname{Im} \delta\omega)\langle w \rangle$, where [compare (11.55)]

$$\langle w \rangle = \frac{1}{4} \left(\mu_0 \mathbf{H} \cdot \mathbf{H}^* + \mathbf{E}^* \cdot \frac{\partial \omega \bar{\bar{\varepsilon}}}{\partial \omega} \cdot \mathbf{E} \right) \tag{11.69}$$

According to (11.67), the spatial rate of growth of Re **S** and the temporal growth of the energy density are driven by the power per unit volume supplied to the electric field by the drive current $\delta \mathbf{J}$.

Equation (11.66) can be used to relate the group velocity–energy density product to the Poynting vector. In the absence of a drive, $\delta \mathbf{J} = 0$, we obtain from (11.66) with (11.68) and (11.69), for a real $\delta \boldsymbol{k}$

$$\delta \boldsymbol{k} \cdot \operatorname{Re} \mathbf{S} = \delta\omega \langle w \rangle \tag{11.70}$$

In the absence of a drive, ω is related to $\boldsymbol{k}$. At a fixed frequency the normal (index) surface gives $|\boldsymbol{k}|$ for any orientation of $\boldsymbol{k}$, Fig. 11.6. When $\delta\omega = 0$, then $\delta \boldsymbol{k}$ must be parallel to the normal (index) surface. It follows from (11.70) that $\delta \boldsymbol{k} \cdot \operatorname{Re} \mathbf{S}$ must be zero; that is, the Re **S** must be perpendicular to the normal (index) surface.

Taking advantage of this information, we may rewrite (11.70)

$$\operatorname{Re} \mathbf{S} = \nabla_k \omega \langle w \rangle$$

where $\nabla_k \omega = \partial\omega / \partial \boldsymbol{k}$ is the gradient of ω in $\boldsymbol{k}$ space, in which ω is viewed as a function of $\boldsymbol{k}$. This confirms the wave-packet argument presented in the preceding section. It is also useful to point out that the Poynting vector, that is, the vector product

$$\mathbf{E} \times \mathbf{H}^* = \mathbf{E} \times \left(\frac{\boldsymbol{k}}{\omega \mu_0} \times \mathbf{E}^* \right)$$

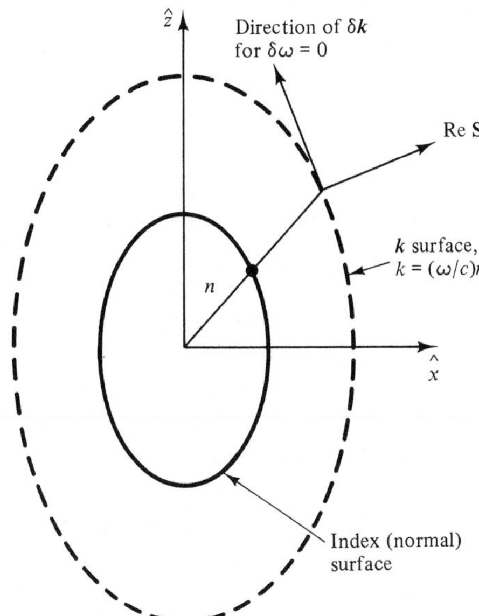

Figure 11.6 Normal (index) surface and surface of constant ω.

is perpendicular to the normal (index) surface, and thus **E** must be parallel to the surface.

11.6 HERMITE–GAUSSIAN BEAM PROPAGATION IN AN ANISOTROPIC MEDIUM

It is of interest to study the propagation of Hermite–Gaussian beams in anisotropic media, because Gaussian beams are used in most optical applications. We shall first concentrate on the extraordinary waves in a uniaxial crystal. The normal (index) surface for the wave (written in terms of propagation vector components) is (11.49). The scalar wave equation for a wave Ψ is obtained by identifying k_x^2 with $-\partial^2/\partial x^2$, k_y^2 with $-\partial^2/\partial y^2$, k_z^2 with $-\partial^2/\partial z^2$ operating on the inverse Fourier transform

$$\frac{\partial^2 \Psi}{n_e^2\,\partial x^2} + \frac{\partial^2 \Psi}{n_e^2\,\partial y^2} + \frac{\partial^2 \Psi}{n_0^2\,\partial z^2} + \omega^2 \mu_0\,\varepsilon_0\,\Psi = 0 \tag{11.71}$$

Ψ represents the amplitude of the vector potential of direction perpendicular to **D**. If we introduce new variables

$$\xi \equiv n_e x, \qquad \eta \equiv n_e y, \qquad \zeta = n_0 z \tag{11.72}$$

we obtain the wave equation

$$\left(\frac{\partial^2}{\partial \xi^2} + \frac{\partial^2}{\partial \eta^2} + \frac{\partial^2}{\partial \zeta^2}\right)\Psi + \omega^2 \mu_0\,\varepsilon_0\,\Psi = 0 \tag{11.73}$$

which is now spherically symmetric. Accordingly, all solutions applicable to free-space propagation may be taken over. In particular, the fundamental mode propagating in a general ζ' direction is

$$u_{00} \propto \frac{1}{(\zeta' + jb)} \exp\left[-jb \frac{\xi'^2 + \eta^2}{2(\zeta' + jb)}\right] \tag{11.74}$$

where $k = \omega/c$. The propagation in actual space is obtained by transforming the geometry of the beam axis and phase fronts into the x-y-z coordinates. The choice of ζ' determines the beam axis. In the coordinates with spherically symmetric propagation the beam axis and the normal to the phase front are coincident. It is of interest to study the relationship between the direction of the beam axis and the direction normal to the phase front after the transformation back into the anisotropic system. In particular, the analysis of Section 11.4 has identified the direction of the group velocity as the direction normal to the index surface. One would expect, therefore, that the Gaussian beam axis, which determines the direction of the energy propagation, would lie along the normal to the index surface. Pick a point x_0, z_0 along the axis of the Gaussian beam. This point transforms into a point ξ_0, ζ_0. We have for the angle θ of the beam axis (see Fig. 11.7)

$$\tan\theta = \frac{z_0}{x_0} = \frac{n_e}{n_0}\frac{\zeta_0}{\xi_0} = \frac{n_e}{n_0}\tan\psi \tag{11.75}$$

Pick intersection points of the phase plane which have coordinates ξ_1 and ζ_1 in transform space. They correspond to the points x_1 and z_1.

$$\frac{x_1}{z_1} = \frac{\xi_1}{\zeta_1}\frac{n_0}{n_e} \tag{11.76}$$

The direction perpendicular to the phase front (i.e., the direction of the propagation vector $\boldsymbol{k}$) is

$$\tan\beta = \frac{x_1}{z_1} = \frac{\xi_1}{\zeta_1}\frac{n_0}{n_e} \tag{11.77}$$

Because the line $\overline{\xi_1\zeta_1}$ is perpendicular to the vector (ξ_0, ζ_0), we have

$$\frac{\zeta_1}{\xi_1} = \frac{\xi_0}{\zeta_0}$$

Therefore, from (11.77) and (11.75),

$$\tan\beta = \frac{\zeta_0}{\xi_0}\frac{n_0}{n_e} = \frac{z_0}{x_0}\frac{n_0^2}{n_e^2} = \tan\theta \cdot \frac{n_0^2}{n_e^2} \tag{11.78}$$

Now consider the point on the $\boldsymbol{k}$ surface with the $\boldsymbol{k}$ vector inclined by β with respect to the x axis. The $\boldsymbol{k}$ ellipse obtained from the normal (index) surface by

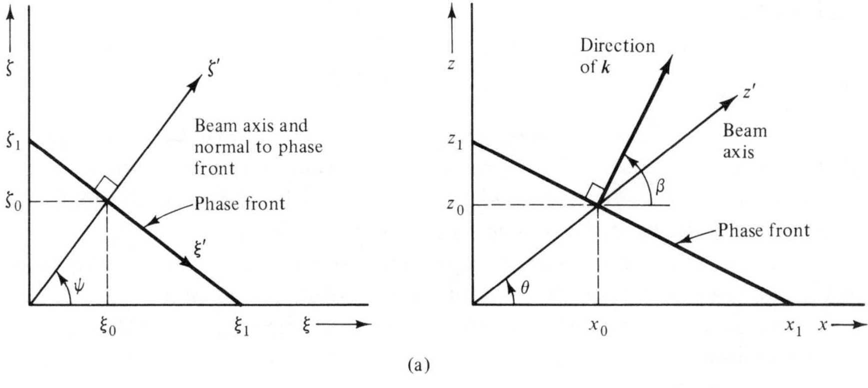

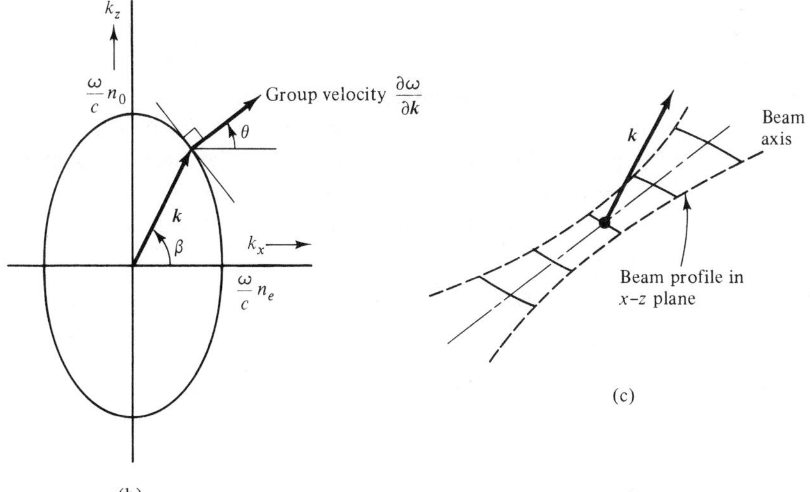

Figure 11.7 Relationship between beam axis and phase front normal. Figure is drawn for $n_0 = 1.5n_e$.

multiplication by ω/c, $k = (\omega/c)n(\boldsymbol{k}/|\boldsymbol{k}|)$, is described by the equation (see Fig. 11.7b)

$$\frac{k_x^2}{n_e^2} + \frac{k_z^2}{n_0^2} = \omega^2 \mu_0 \varepsilon_0 \tag{11.79}$$

The tangent to the ellipse at the intersection of the $\boldsymbol{k}$ vector with components k_x, k_z is

$$\frac{dk_z}{dk_x} = -\frac{k_x}{k_z}\frac{n_0^2}{n_e^2} \tag{11.80}$$

The normal to the tangent, which is parallel to the group velocity, is at the angle with the tangent

$$-\frac{dk_x}{dk_z} = \frac{k_z}{k_x}\frac{n_e^2}{n_0^2} = \tan\beta \cdot \frac{n_e^2}{n_0^2} = \tan\theta \tag{11.81}$$

where θ is the angle of the beam axis of (11.75). The vector $k_x\hat{x} + k_z\hat{z}$ points along the phase velocity. Therefore, we get the same relationship between the direction of *group velocity* and *phase velocity* as that between the Gaussian *beam axis* and the *normal to the phase surface on the beam axis*, (11.78).

Figure 11.7c shows in an exaggerated way the beam contour and phase fronts of a Gaussian beam propagating in an anisotropic crystal.

The Gaussian beam (11.74) transformed into the spatial coordinates x, y, z is not cylindrically symmetric. The minimum beam diameter ω_0 in the ξ, η, ζ coordinate system is, according to the solution (11.74),

$$\omega_0 = \sqrt{\frac{2b}{k}}$$

The distance ω_0 is shown along the phase front in Fig. 11.8. When transformed into the x-z coordinates it appears along the transformed phase front as w_0. Its projection into a direction perpendicular to the beam axis gives what one may call the radius of the Gaussian beam in the anisotropic medium, $w_{0x'}$. The transformations (11.72), eliminating the angles ψ and β with the aid of (11.75), and (11.78) give

$$w_{0x'} = \frac{\omega_0}{n_0}\sqrt{\frac{1 + (n_0^2/n_e^2)\tan^2\theta}{1 + \tan^2\theta}} \tag{11.82}$$

The radius in the y direction is found simply as

$$w_{0y} = \frac{\omega_0}{n_e} \tag{11.83}$$

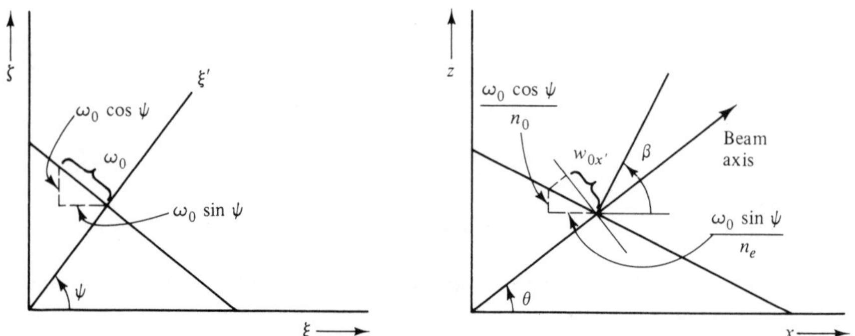

Figure 11.8 Beam radius in the two coordinate systems.

So far, we have treated only uniaxial crystals. More general crystal symmetries lead to more complicated (fourth order) index surfaces. However, paraxial beams involve $\boldsymbol{k}$ components that "bundle" around one particular spatial direction. Within a small solid angle, any continuous surface may be approximated by an ellipsoid, as long as both Gaussian radii of curvature are positive. Hence the present analysis applies to all such cases.

11.7 SUMMARY

An anisotropic medium is described by a tensor dielectric constant $\bar{\bar{\varepsilon}}$, or its inverse $\bar{\bar{\kappa}}$. The $\bar{\bar{\varepsilon}}$ tensor is Hermitean when the medium is lossless. A reciprocal medium has a symmetric $\bar{\bar{\varepsilon}}$ tensor which is real for a lossless medium. The displacement density vector of a plane wave propagating in a lossless reciprocal medium is perpendicular to the propagation vector $\boldsymbol{k}$. For every direction of propagation $\boldsymbol{k}/|\boldsymbol{k}|$, there are, in general, two orthogonal polarizations of $\mathbf{D}$ and two values of $|\boldsymbol{k}|$; anisotropic media are birefringent. The two values of $\boldsymbol{k}$ are found from the magnitudes of the major and minor axes of the ellipse of intersection with the index ellipsoid of a plane perpendicular to $\boldsymbol{k}$. The normal (index) surface is constructed from the index ellipsoid by marking off the two values of "index," $n = ck/\omega$ along every direction of propagation. For a uniaxial crystal, the normal surface consists of a sphere (ordinary) and an ellipsoid (extraordinary).

An anisotropic dispersive dielectric calls for a rederivation of the expression for energy density. We found the time-averaged electric energy density to be

$$\langle w_e \rangle = \frac{1}{4}\left[\mathbf{E}^* \cdot \frac{\partial}{\partial \omega}(\omega \bar{\bar{\varepsilon}}) \cdot \mathbf{E}\right]$$

In an isotropic dispersive medium the $\bar{\bar{\varepsilon}}$ tensor is to be replaced by a scalar, a function of frequency.

A wave packet propagating in an anisotropic medium has, in general, different directions for the phase propagation (phase velocity) and (packet) envelope propagation (group velocity). The direction of the group velocity coincides with the direction of the Poynting vector, which is perpendicular to the normal surface; the electric field is tangential to the normal surface. The time-averaged power flow density is given, in magnitude and direction, by the product of group velocity and time-averaged energy density. The beam axis of a Gaussian beam propagating in an anisotropic medium coincides with the group-velocity direction; the surfaces of constant phase are in general not perpendicular to the beam axis.

APPENDIX 11A

Transformation of Vectors and Tensors [3]

We use index notation to denote the three directions along a Cartesian coordinate system. We denote a unit vector by $\hat{\boldsymbol{x}}_i$, where $i = 1, 2, 3$, $x_1 = x$, $x_2 = y$, and $x_3 = z$. Any vector $\boldsymbol{E}$ may be written as

$$\boldsymbol{E} = \sum_i \hat{\boldsymbol{x}}_i E_i \tag{11A.1}$$

We shall omit the summation sign and imply the Einstein convention of summation over repeated indices.

$$\boldsymbol{E} = \hat{\boldsymbol{x}}_i E_i \tag{11A.2}$$

In a different coordinate system, $\hat{\boldsymbol{x}}'_j$:

$$\boldsymbol{E} = \hat{\boldsymbol{x}}'_j E'_j \tag{11A.3}$$

Dot-multiplying (11A.2) and (11A.3) by $\hat{\boldsymbol{x}}_i$, and equating the result, we get

$$\hat{\boldsymbol{x}}_i \cdot \boldsymbol{E} = E_i = \hat{\boldsymbol{x}}_i \cdot \hat{\boldsymbol{x}}'_j E'_j = A'_{ij} E'_j$$

or

$$E_i = A'_{ij} E'_j \tag{11A.4}$$

where

$$A'_{ij} \equiv \hat{\boldsymbol{x}}_i \cdot \hat{\boldsymbol{x}}'_j \tag{11A.5}$$

is a matrix of transformation between the two coordinate systems (see Fig. 11A.1). Conversely, dot-multiplying (11A.2) and (11A.3) by $\boldsymbol{x}'_j$:

$$E'_j = \hat{\boldsymbol{x}}'_j \cdot \hat{\boldsymbol{x}}_i E_i = A_{ji} E_i \tag{11A.6}$$

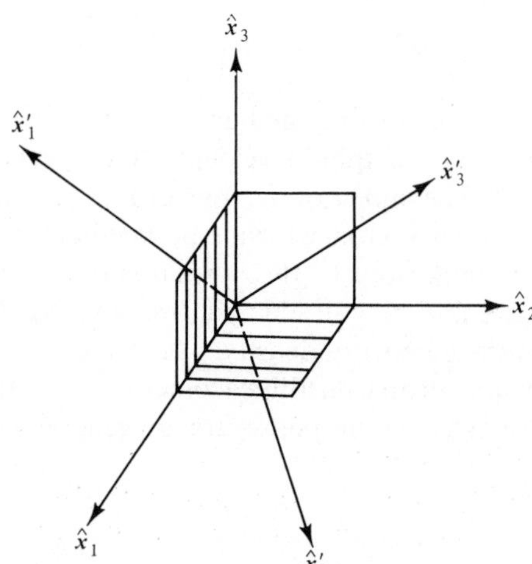

Figure 11A.1 Coordinate transformation.

where

$$A_{ji} = \hat{\boldsymbol{x}}'_j \cdot \hat{\boldsymbol{x}}_i \tag{11A.7}$$

The transformation matrices A'_{ij} and A_{ji} that transform E'_j into E_i, and vice versa, obey an important identity. Because the scalar product is commutative,

$$A_{ji} = A'_{ij} \tag{11A.8}$$

A transformation from the coordinate system $\hat{\boldsymbol{x}}_i$ into the coordinate system $\hat{\boldsymbol{x}}'_j$ and back again returns the vector components to their original value,

$$E_k = A'_{kj} A_{ji} E_i = E_i$$

and thus

$$A'_{kj} A_{ji} = A'_{kj} A'_{ij} = \delta_{ki} \tag{11A.9}$$

where δ_{ki} is the "Kronecker delta", $\delta_{ki} = 1$ for $k = i$, $\delta_{ki} = 0$ for $k \neq i$. The transformation of the $\bar{\bar{\kappa}}$ tensor follows from the transformation of the vector relation between $\boldsymbol{E}$ and $\boldsymbol{D}$, written in the primed coordinate system

$$E'_i = \kappa'_{ij} D'_j \tag{11A.10}$$

Using the transformation (11A.4) to express E'_i and D'_j in terms of E_l and D_k, we obtain with the aid of (11A.8)

$$E_l = A_{il} A_{jk} \kappa'_{ij} D_k = A'_{li} A'_{kj} \kappa'_{ij} D_k$$

But $E_l = \kappa_{lk} D_k$ and thus we have obtained the transformation law

$$\kappa_{lk} = A'_{li} A'_{kj} \kappa'_{ij} \tag{11A.11}$$

This law is easily generalizable to tensors of rank higher than second.

A vector dot-multiplied (scalarly multiplied) by a coordinate vector yields a scalar. Thus the phase factor of a plane wave, $\exp(-j\boldsymbol{k} \cdot \boldsymbol{r})$, contains the scalar phase

$$\phi = \boldsymbol{k} \cdot \boldsymbol{r} = k_i x_i \tag{11A.12}$$

This scalar set equal to a constant describes a plane of constant phase. In a new coordinate system k_i is replaced by k'_i, x_i by x'_i. The invariance of the equation of the plane with respect to a coordinate transformation is checked easily, using (11A.4) for the transformation of the vectors k_i and x_i:

$$\phi = k_i x_i = A'_{ij} k'_j A'_{il} x'_l = \delta_{jl} k'_j x'_l = k'_j x'_j \tag{11A.13}$$

where we have used (11A.9).

A tensor of second rank twice scalarly multiplied by a coordinate vector also yields a scalar. The scalar is a quadratic function of the coordinates. Set equal to a constant it yields the equation of a second-order surface. An example is the index ellipsoid

$$\kappa_{ij} x_i x_j = \frac{1}{\varepsilon_0} \tag{11A.14}$$

The surface is an ellipsoid because κ_{ij} is a positive-definite tensor. Indeed, the time-averaged energy of a nondispersive medium is

$$\langle w\rangle = \tfrac{1}{4}\mathrm{E}_i \varepsilon_{ij} \mathrm{E}_j^* = \tfrac{1}{4}\mathrm{D}_i \kappa_{ij} \mathrm{D}_j^*$$

a positive-definite quantity. Thus κ_{ij} is a positive-definite tensor. In a way analogous to (11A.13) one may prove the invariance of the ellipsoid equations with respect to a coordinate transformation

$$\frac{1}{\varepsilon_0} = \kappa_{ij} x_i x_j = A'_{ik} A'_{jl} \kappa'_{kl} A'_{im} x'_m A_{jn} x'_n$$

$$= \delta_{km} \delta_{ln} \kappa'_{kl} x'_m x'_n = \kappa'_{kl} x'_k x'_l$$

where we have applied (11A.4) to the coordinate vector x_i, and used (11A.9), and (11A.11).

We can always pick a coordinate system in which κ_{ij} is diagonal. To motivate this statement we have first recourse to a plausibility argument: A symmetric tensor, such as κ_{ij}, has six degrees of freedom; a diagonalized tensor has three degrees of freedom. Thus a quantity with six degrees of freedom may be specified by a diagonal tensor (three degrees of freedom) and the rotation of the coordinate system (rotation of, say, $\hat{\boldsymbol{x}}_3$ axis into $\hat{\boldsymbol{x}}'_3$ axis, two degrees of freedom, and rotation around new $\hat{\boldsymbol{x}}'_3$ axis, one degree of freedom). The formal proof proceeds as follows.

If we want to cast κ_{ij} into diagonal form, we ask for those directions of D_i which produce an electric field parallel to D_i:

$$\kappa_{ij} D_j = E_i = \lambda D_i \tag{11A.15}$$

where λ is a multiplier. From (11A.15) we conclude that

$$\det\,[\kappa_{ij} - \lambda I_{ij}] = 0 \tag{11A.16}$$

where I_{ij} is an identity tensor $I_{ij} = 1$, $i = j$; $I_{ij} = 0$, $i \neq j$. Solving this equation of third order we find three roots, eigenvalues, $\lambda^{(\alpha)}$, $\alpha = 1, 2, 3$. When we set

$$\kappa_{ij} D_j^{(\alpha)} = \lambda^{(\alpha)} D_i^{(\alpha)}, \qquad \alpha = 1, 2, 3 \tag{11A.17}$$

we find the directions of $D_i^{(\alpha)}$. They are the *eigenvectors* of the tensor relation $\kappa_{ij} D_j = \lambda D_i$. The three eigenvectors must be orthogonal with respect to each other. To prove this, we write down (11A.17) for two eigenvalues $\lambda^{(\alpha)}$ and $\lambda^{(\beta)}$:

$$\kappa_{ij} D_j^{(\alpha)} = \lambda^{(\alpha)} D_i^{(\alpha)} \tag{11A.18}$$

$$\kappa_{ij} D_j^{(\beta)} = \lambda^{(\beta)} D_i^{(\beta)} \tag{11A.19}$$

Multiply (11A.18) by $D_i^{(\beta)}$, (11A.19) by $D_i^{(\alpha)}$, and subtract:

$$D_i^{(\beta)} \kappa_{ij} D_j^{(\alpha)} - D_i^{(\alpha)} \kappa_{ij} D_j^{(\beta)} = (\lambda^{(\alpha)} - \lambda^{(\beta)}) D_i^{(\alpha)} D_i^{(\beta)}$$

The left-hand side is zero because $\kappa_{ij} = \kappa_{ji}$. Therefore,

$$D_i^{(\alpha)} D_i^{(\beta)} = 0$$

for

$$\lambda^{(\alpha)} \neq \lambda^{(\beta)}$$

The eigenvectors are orthogonal to each other. The κ_{ij} tensor in diagonal form has zeros for all off-diagonal elements. Because κ_{ij} is a positive-definite tensor, the remaining diagonal elements must be positive. The equation of the index surface is

$$\kappa_{11} x_1^2 + \kappa_{22} x_2^2 + \kappa_{33} x_3^2 = \frac{1}{\varepsilon_0}$$

which is the equation of an ellipsoid for positive coefficients κ_{ii}.

PROBLEMS

11.1. Consider a uniaxial crystal with the diagonal $\bar{\bar{\kappa}}$ tensor

$$\frac{1}{\varepsilon_0} \begin{bmatrix} \frac{1}{n_0^2} & 0 & 0 \\ 0 & \frac{1}{n_0^2} & 0 \\ 0 & 0 & \frac{1}{n_e^2} \end{bmatrix}$$

where two components are equal.

(a) Show how a circularly polarized wave with the field

$$\mathbf{E} = \hat{y} + j\hat{z}$$

at $x = 0$, propagating in the crystal along the x axis, becomes periodically linearly polarized. A plate of the appropriate thickness so as to transform circularly polarized light into linearly polarized light is called a *quarter-wave plate*.

(b) Find the thickness d of a quarter-wave plate in terms of n_0, n_e, and λ (free-space wavelength).

11.2. Show that the quarter-wave plate of Problem 11.1 can transform linear polarization at its "input" plane into circular polarization at its "output" plane. Determine the angle of the **E** polarization with respect to the y axis.

11.3. Using the result of Problem 11.2, show that the structure shown acts as an "isolator": Reflection at the output end does not make it back into input source.

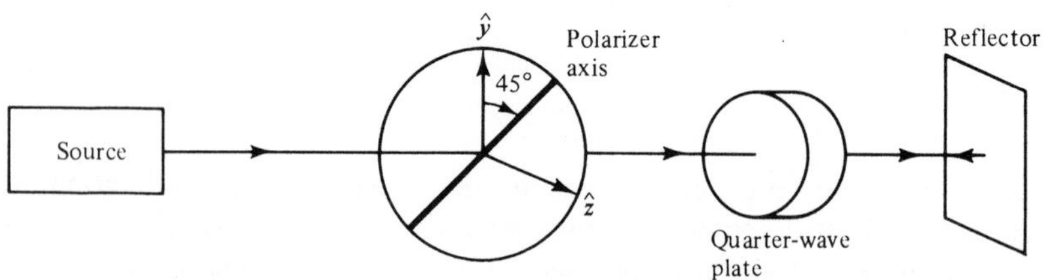

Figure P11.3

11.4. In the preceding problems we have ignored the reflection at the interfaces of the quarter-wave plates. In this problem consider the reflection and transmission of the circularly polarized wave of Problem 11.1a incident from air on the interface of a crystal with the $\bar{\bar{\kappa}}$ tensor of P11.1 and extending over $0 \leq x \leq \infty$.

11.5. The purpose of this problem is to show that the **E** field of a plane wave is parallel to the tangent to the normal surface at the point of intersection of the $\boldsymbol{k}$ vector. The equation of the index ellipsoid is

$$\frac{x^2}{n_0^2} + \frac{z^2}{n_e^2} = 1$$

or in terms of κ_{ij}

$$\kappa_{xx} x^2 + \kappa_{zz} z^2 = 1/\varepsilon_0$$

(a) Show that the slope of the tangent to the normal surface is

$$\tan \gamma = \cot \beta \frac{n_0^2}{n_e^2} = \cot \beta \frac{\kappa_{zz}}{\kappa_{xx}}$$

(b) Now show that the **E** field is parallel to the tangent.

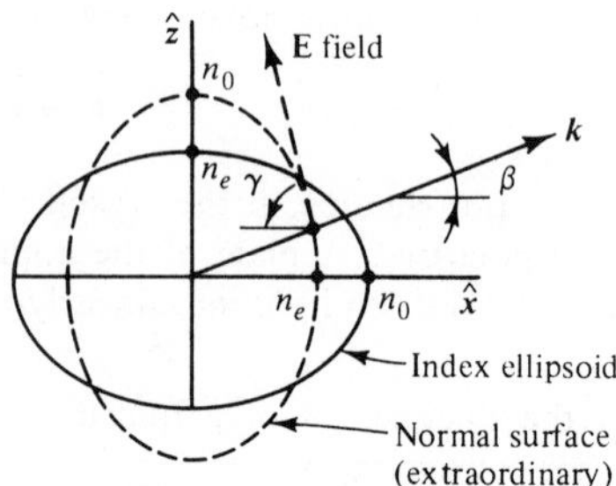

Figure P11.5

11.6. The purpose of this problem is to develop a method for constructing the equivalent of Snell's law for reflection and transmission at the interface of an isotropic to an anisotropic uniaxial medium. Suppose the normal (index) surface of the positive uniaxial crystal is as shown. The axis of the ellipsoid is in the x-z plane.

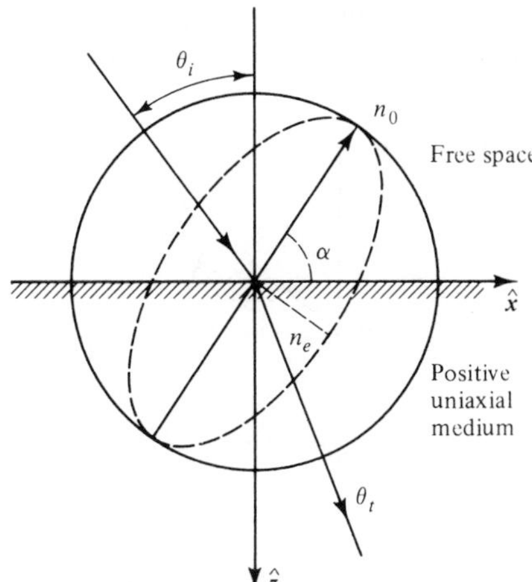

Figure P11.6

(a) The $\boldsymbol{k}$ vector of the incident wave is in the x-z plane and the wave is TE. Write an equation determining the angle of the transmitted wave.

(b) Repeat for TM.

(c) Suppose that the $\boldsymbol{k}$ vector is of general orientation. Describe a method for determining the angle of the transmitted wave. Is the plane of incidence still the same as the plane of transmission? Why?

11.7. The model of a dispersive medium introduced in Problem 1.4, can be used to check the meaning of time-averaged energy density. Because of the inertial force on the acceleration $\ddot{x}$ of the charged particle, kinetic energy $\frac{1}{2}m\dot{x}^2$ must be associated with a time-dependent polarization. Show that

$$\frac{1}{4}\frac{\partial \omega \varepsilon}{\partial \omega}|\mathbf{E}|^2$$

contains the time-averaged potential energy density stored in the elastic spring with the force on the particles $m\omega_0^2 x$ and the time-averaged potential energy density. Treat the loss-free case, $\sigma = 0$.

11.8. Two Gaussian beams with parallel phase fronts may separate if the group velocities are not collinear. Phase matching in harmonic generation discussed in Chapter 13 is achieved by alignment of the $\boldsymbol{k}$ vectors of two waves, one an ordinary, the other an extraordinary wave. In this problem evaluate the angle between the phase velocity and group velocity of the extraordinary wave of a uniaxial crystal of indices n_e and n_0, as a function of inclination β of the wave vector $\boldsymbol{k}$. Find the maximum of this angle in terms of n_0/n_e. With the values for $LiNbO_3$ of $n_0 = 2.29$ and $n_e = 2.20$, what is the value of the angle?

11.9. Show that $[(z + jb_x)(z + jb_y)]^{-1/2} \exp[-jkx^2/2(z + jb_x)] \exp[-jky^2/2(z + jb_y)]$ is a solution of the paraxial wave equation. This describes a Gaussian beam of elliptical cross section. Determine the beam "radii" in x and y directions. Compare Problem 5.3.

11.10. A Gaussian beam of circular cylindrical symmetry, polarized as shown in Fig. P11.10, with minimum radius at the interface is incident on a uniaxial crystal.

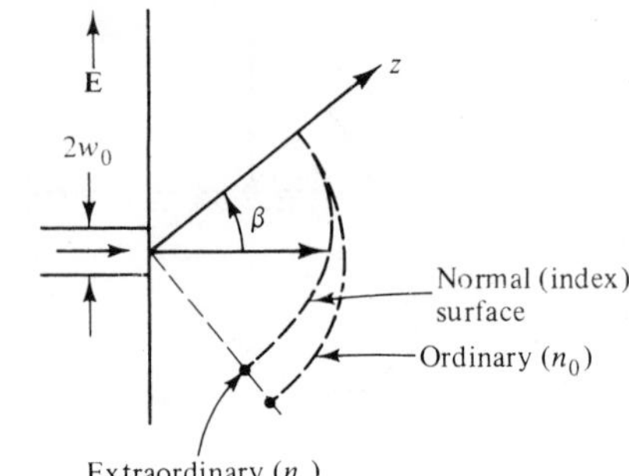

Figure P11.10

(a) Find the direction of the beam axis in the crystal.
(b) Find the major and minor axes of the Gaussian beam in the crystal at the interface.
(c) Find the reflection coefficient Γ (treating the beam as an infinite parallel plane wave).

11.11. In Problem 11.9 you have shown that the expression

$$\frac{1}{\sqrt{(z+jb_x)(z+ib_y)}}\exp\left[-\frac{jkx^2}{2(z+jb_x)}\right]\exp\left[-\frac{jky^2}{2(z+jb_y)}\right]$$

obeys the isotropic paraxial wave equation. Using this fact, find b_ξ and b_η in the transformed isotropic solutions, like (11.74), so that the Gaussian beam in the anisotropic medium is of circular cross section at $z' = 0$. Find the z' dependence of $w_{x'}$ and $w_{y'}$. Use the geometry of Fig. 11.8.

REFERENCES

[1] A. Yariv, *Quantum Electronics*, Wiley, New York, 1975.
[2] E. Born and E. Wolf, *Principles of Optics*, Macmillan, New York, 1964.
[3] J. F. Nye, *Physical Properties of Crystals, Their Representation by Tensors and Matrices*, Clarendon Press, Oxford, 1967.

12

ELECTRO-OPTIC MODULATORS

The electro-optic (or Pockels') effect is the simplest and one of the most important nonlinear optics effect. We take advantage of this fact and analyze several optical modulators based on the electro-optic effect before launching into a more detailed discussion of nonlinear optics in Chapter 13.

Electro-optic modulators use an electric field to modify the index of a dielectric medium. This principle may be employed both for amplitude as well as phase modulation of optical waves passing through the medium. In the design of an electro-optic modulator it is usually desirable to achieve a required depth of modulation with the smallest possible consumption of modulation power over the largest possible bandwidth. We shall address this issue in this chapter.

We start with the index ellipsoid describing propagation in an anisotropic medium. The electro-optic effect is expressed in terms of the modification of the index ellipsoid by an applied electric field. Next we describe an electro-optic modulator utilizing the electric field-induced birefringence. The same principle can be applied to phase modulation.

The modulating electric field may be applied either in the direction of the wave propagation or perpendicular to it. The latter configuration has the advantage that the voltage required to produce a given depth of modulation can be decreased by decreasing the transverse dimensions of the optical system. Finally, we consider an optical switch in a configuration suitable for integrated optics application.

12.1 THE LINEAR ELECTRO-OPTIC EFFECT

The dielectric tensor of some media, such as lithium niobate ($LiNbO_3$), can be modified by the application of an electric field. This is the linear electro-optic effect. The name seems inappropriate because the medium is, in fact, nonlinear. The name refers to the linear dependence of the $\bar{\bar{\kappa}}$ tensor, the inverse of the dielectric tensor, on the electric field.

One writes for the change of the tensor components κ_{ij}

$$\Delta(\varepsilon_0 \kappa_{ij}) = r_{ijk} E_k \tag{12.1}$$

where the r_{ijk} is a tensor of third rank, which, when multiplied by a vector, gives a tensor of second rank. The coefficients r_{ijk} are the coefficients of the linear electro-optic effect, characteristic of the dielectric medium. The index ellipsoid is a function of the applied *control field* $\mathbf{E}$. It can be written concisely in the form

$$(\varepsilon_0 \kappa_{ij} + r_{ijk} E_k) x_i x_j = 1 \tag{12.2}$$

where κ_{ij} now denotes the inverse dielectric tensor in the absence of an applied field. The tensor r_{ijk} is symmetric in the indices i and j. It is customary to combine these two subscripts into one subscript that assumes six values from 1 to 6. One makes the correspondence $r_{ijk} \leftrightarrow r_{Ik}$ with $I = 1, 2, \ldots 6$ and the following assignment:

$$\overset{ij}{\begin{bmatrix} 11 & 12 & 13 \\ 21 & 22 & 23 \\ 31 & 32 & 33 \end{bmatrix}} \leftrightarrow \overset{I}{\begin{bmatrix} 1 & 6 & 5 \\ & 2 & 4 \\ & & 3 \end{bmatrix}}$$

The behavior of the index ellipsoid as a function of electric field is best visualized by looking at a specific case. Consider, for example, potassium dihydrogen phosphate (KDP) with the composition KH_2PO_4. It is in the $\bar{4}2m$ crystallographic group [1]. In the absence of an applied field it behaves as a uniaxial optical medium with the dielectric tensor ellipsoid ($\varepsilon_0 \kappa_{ij} x_i x_j = 1$) written out in terms of squared indices

$$\frac{x^2}{n_0^2} + \frac{y^2}{n_0^2} + \frac{z^2}{n_e^2} = 1 \tag{12.3}$$

The index n_0 is that of ordinary wave propagation, n_e that of extraordinary wave propagation. In the presence of an applied field the index ellipsoid changes. The $\bar{4}2m$ crystallographic group has r_{Ik}'s that are zero except for r_{63} and $r_{41} = r_{52}$. The index ellipsoid equation (12.2) now reduces to

$$\frac{x^2}{n_0^2} + \frac{y^2}{n_0^2} + \frac{z^2}{n_e^2} + 2r_{41}E_x yz + 2r_{41}E_y xz + 2r_{63}E_z xy = 1 \tag{12.4}$$

Lithium niobate is in the $3m$ crystallographic group and has the nonzero r coefficients

$$r_{12} = -r_{22} = r_{61}, \quad r_{13} = r_{23}, \quad r_{33}, \quad r_{42} = r_{51}$$

Its index ellipsoid is written in the form

$$\left(\frac{1}{n_0^2} + r_{12}E_y + r_{13}E_z\right)x^2 + \left(\frac{1}{n_0^2} + r_{22}E_y + r_{23}E_z\right)y^2 + \left(\frac{1}{n_e^2} + r_{33}E_z\right)z^2 + 2(r_{42}E_y)yz + 2(r_{51}E_x)xz + 2(r_{61}E_x)xy = 1 \qquad (12.5)$$

12.2 APPLICATION TO AMPLITUDE MODULATOR

Consider a crystal like KDP of the $\bar{4}2m$ symmetry with its principal axes aligned with x, y, z as shown in Fig. 12.1. An electric field is applied along the z axis. This can be done by coating the end faces at $z = 0$ and $z = l$ with a thin conducting layer that is transparent to the optical radiation and by applying a voltage between the two layers. A polarizer precedes, and an analyzer follows, the crystal in the arrangement shown in Fig. 12.1. The polarizer (or analyzer) consists of a medium whose loss to a wave passing through it is polarization sensitive. It has low (ideally zero) loss for one polarization of the electric field shown schematically by the arrow, and high (ideally infinite) loss for the orthogonal polarization direction.

The z-directed field changes the index ellipsoid according to (12.4):

$$\frac{x^2}{n_0^2} + \frac{y^2}{n_0^2} + \frac{z^2}{n_e^2} + 2r_{63}E_z xy = 1 \qquad (12.6)$$

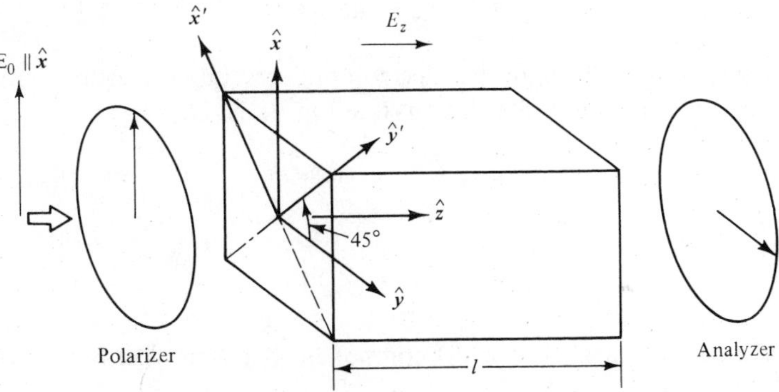

Figure 12.1 Electro-optic modulator.

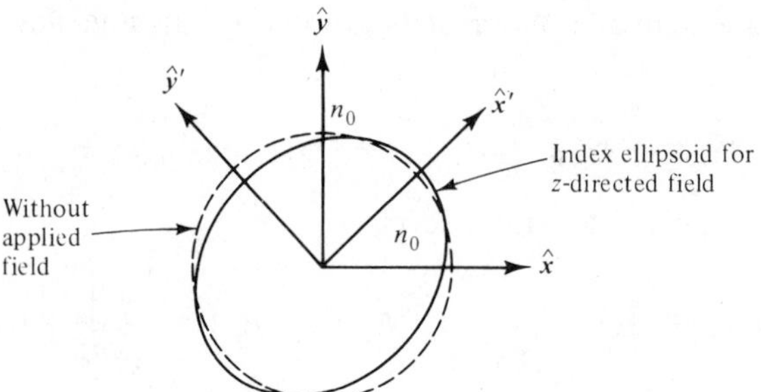

Figure 12.2 Intersection of index ellipsoid with x-y plane.

The ellipsoid has principal axes along the coordinates x', y', z (see Fig. 12.2), where

$$x = \frac{1}{\sqrt{2}}(x' + y') \tag{12.7}$$

$$y = \frac{1}{\sqrt{2}}(-x' + y') \tag{12.8}$$

Introducing the expressions above into (12.6), we obtain the equation of the index ellipsoid in the coordinate system aligned with the new principal axis:

$$\left(\frac{1}{n'_{xx}}\right)^2 x'^2 + \left(\frac{1}{n'_{yy}}\right)^2 y'^2 + \frac{z^2}{n_e^2} = 1 \tag{12.9}$$

where the indices n'_{xx} and n'_{yy} along the new principal axes are given approximately ($|r_{63} E_z| \ll 1$ in all practical cases) by

$$n'_{xx} \simeq n_0 + n_0^3 \frac{r_{63}}{2} E_z \tag{12.10}$$

$$n'_{yy} \simeq n_0 - n_0^3 \frac{r_{63}}{2} E_z \tag{12.11}$$

After passage through the electro-optic crystal, the field components along x' and y' experience retardations (see Fig. 12.1).

$$\mathrm{E}'_x = \frac{E_0}{\sqrt{2}} e^{-j(\omega/c)n'_{xx}l} \tag{12.12}$$

$$\mathrm{E}'_y = \frac{E_0}{\sqrt{2}} e^{-j(\omega/c)n'_{yy}l} \tag{12.13}$$

The **E** component parallel to $\hat{\mathbf{y}}$, the component that is passed by the analyzer, is

$$\mathrm{E}_y = \frac{1}{\sqrt{2}}(-\mathrm{E}'_x + \mathrm{E}'_y) = \frac{1}{2} E_0[-e^{-j(\omega/c)n'_{xx}l} + e^{-j(\omega/c)n'_{yy}l}] \tag{12.14}$$

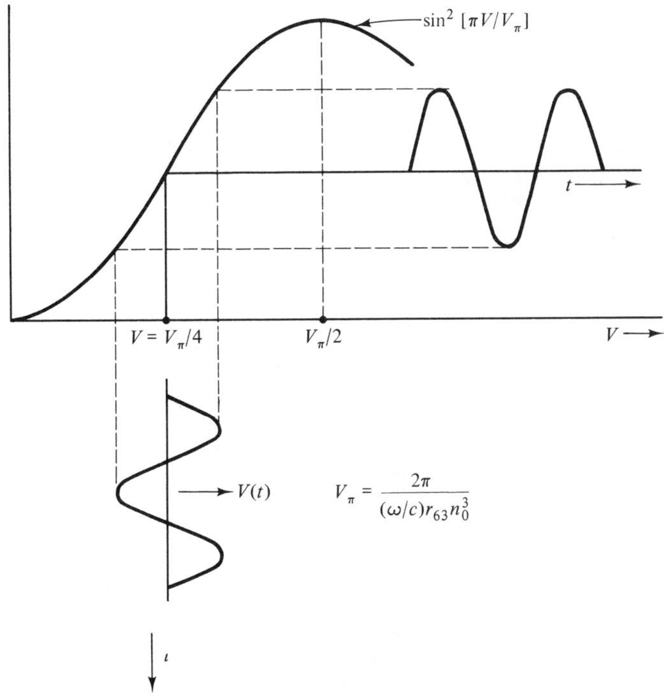

Figure 12.3 Graphical construction of modulator response.

and the ratio of transmitted (I_t) and incident (I_i) intensities is

$$\frac{I_t}{I_i} = \sin^2 \frac{\omega}{2c}(n'_{xx} - n'_{yy})l \equiv \sin^2\left(\frac{1}{2}\frac{\omega}{c} r_{63} n_0^3 E_z l\right) \tag{12.15}$$

The most linear response of the output intensity to an applied sinusoidal voltage is obtained for a bias voltage such that the argument of the sine function is $\pi/4$ (Fig. 12.3). The voltage $E_z l$ required to produce an argument of the sine in (12.15) of $\pi/2$ is a parameter typical of an electrooptic medium. For KDP at $\lambda = 0.55\ \mu$m it is 7.5 kV.

Transverse Modulator

The amplitude modulation by a field parallel to the direction of propagation requires a fixed voltage for a desired amplitude of modulation, regardless of the size of the crystal, and cannot be improved by miniaturization. The transverse modulator can be improved by miniaturization. Consider the configuration shown in Fig. 12.4. The crystal is of the $\bar{4}2m$ symmetry, its principal axes are oriented along x and y, the propagation direction is along y', at 45° with respect to x and y, in the x-y plane. The x', y' axes are along the principal axes of the index ellipsoid as modified by the applied electric field. The change

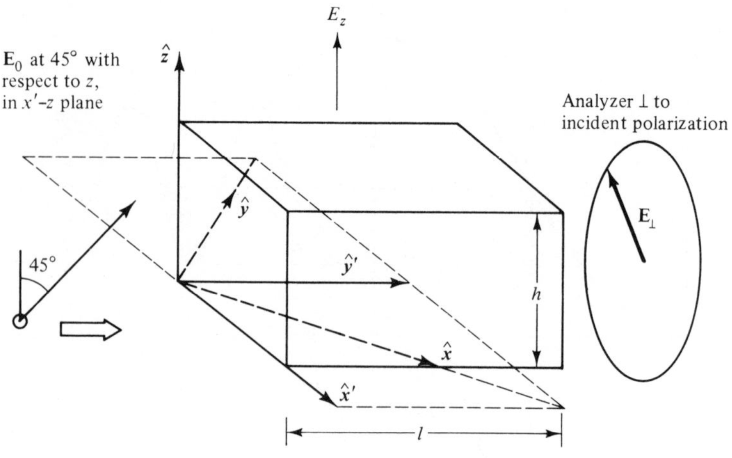

Figure 12.4 Transverse modulator with KDP crystal.

of the index ellipsoid produced by E_z is still given by (12.10-11). The z component of the electric field of the optical wave experiences no phase modulation, but the x' component does. The total electric field of the optical wave at the end of the crystal is

$$\frac{1}{\sqrt{2}}\,E_0[e^{-j(\omega/c)n'_{xx}l}\hat{x}' + e^{-j(\omega/c)n_e l}\hat{z}] = \frac{1}{\sqrt{2}}\,E_0 \exp -\left(j\,\frac{\omega}{c}\,\frac{n'_{xx}+n_e}{2}\,l\right)$$

$$\cdot\,[e^{-j(\omega/c)[(n'_{xx}-n_e)/2]l}\hat{x}' + e^{+j(\omega/c)[(n'_{xx}-n_e)/2]l}\hat{z}] \tag{12.16}$$

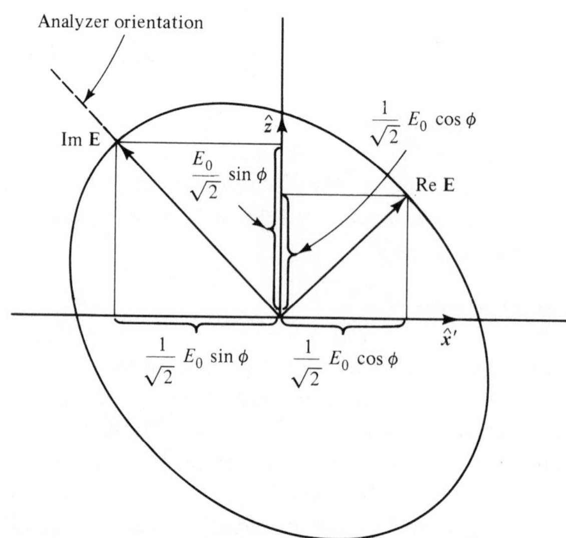

Figure 12.5 Polarization ellipse at output of transverse modulator.

The emerging field is elliptically polarized even in the absence of an applied field because of the natural birefringence. Defining

$$\phi \equiv \frac{\omega}{c}\frac{n'_{xx} - n_e}{2}\, l = \frac{\omega}{c}(n_0 - n_e)l + \frac{\omega}{c}\frac{r_{63}\, n_0^3}{4} E_z l \tag{12.17}$$

we can construct the diagram shown in Fig. 12.5 from the real and imaginary parts of the vectors in the brackets, and obtain the polarization ellipse. An analyzer with its transmission direction oriented at 45° with respect to the z axis as shown transmits the intensity

$$I_t = I_i \sin^2 \phi$$

The applied voltage $E_z h$ is now $\frac{1}{2}l/h$ times more effective than in the case of the electro-optical modulator with an applied field parallel to the direction of propagation.

12.3 PHASE MODULATOR

Systems similar to the amplitude modulators can be used as phase modulators. The phase delay of a wave traveling through the crystal of Fig. 12.6 with the crystal orientation and E polarization shown is (compare 12.10)

$$\phi = \frac{\omega}{c} n'_{xx} l = \frac{\omega}{c} l\left(n_0 + \frac{r_{63}}{2} n_0^3 E_z\right) \tag{12.18}$$

If the applied field E_z is modulated sinusoidally

$$E_z = E_{z0} \sin \omega_M t$$

the field at the output plane is

$$E'_x = E_0\, e^{-j(\omega/c)n_0 l} e^{j\omega t - j\delta \sin \omega_M t} \tag{12.19}$$

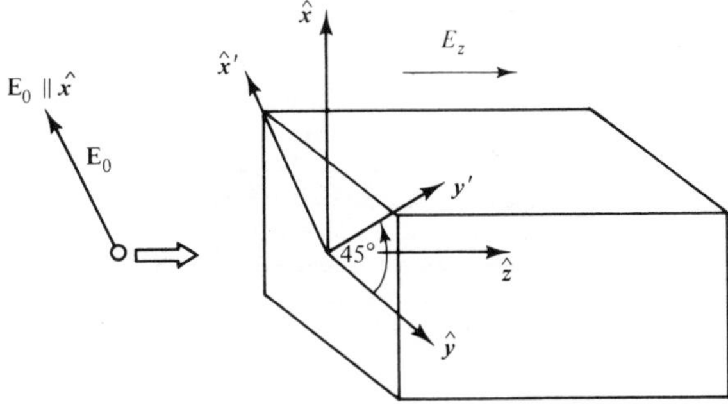

Figure 12.6 Schematic of phase modulator. The crystal orientation is the same as that in Fig. 12.1.

where

$$\delta = \frac{\omega}{c} \frac{r_{63}}{2} n_0^3 E_{z0} l \tag{12.20}$$

The output is phase modulated. The maximum phase deviation δ is equal in magnitude to the argument of the sine function in (12.15). We found there that the voltage required for the argument to become equal to $\pi/2$ is 7.5 kV. It takes a large voltage to produce a phase shift of $\pi/2$! For this reason practical realizations of phase modulators utilize modulation fields transverse to the direction of propagation in order to enhance the modulation for a given voltage by a length-to-width ratio.

The spectrum of the field at the end of the crystal is obtained from the Bessel function identity (9.20):

$$E'_x = E_0 e^{-j(\omega/c)n_0 l} e^{j\omega t} \sum_n (-1)^n J_n(\delta) e^{jn\omega_M t}$$

12.4 WAVEGUIDE MACH–ZEHNDER MODULATOR

A very effective way of modulating an optical signal has been proposed by Martin [2], and perfected by J. F. Leonberger [3]. The modulator consists of a waveguide interferometer of the Mach–Zehnder type (see Figs. 12.7 and 7.14). The incoming optical beam is separated equally in a "waveguide Y" and then recombined. If the "waveguide Y" is slowly tapered, this can be accomplished with low loss (less than 1 dB for both Y's in cascade has been measured). If the phase shift in both arms is identical, the optical waveguide modes interfere constructively and all the power reappears in the output (minus the unavoidable waveguide and Y losses, of course). When the net phase-shift difference between the two arms is π, the modes in the two waveguides are 180° out of phase and excite an antisymmetric mode in the output waveguide. If the output waveguide is a single-mode guide, no power emerges from it; the power radiates out into the substrate and is lost. Thus the output power can be

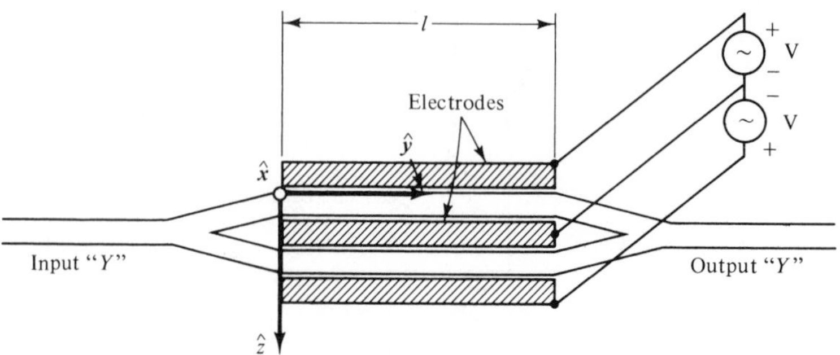

Figure 12.7 Mach–Zehnder waveguide interferometer in an x-cut crystal (x axis normal to plane of incidence).

controlled by a voltage applied to the waveguides. Because the electrode spacing w can be made very small ($\sim 3\ \mu$m), relatively small switching voltages are required [3].

Let us estimate the voltage required for full cancellation. If the surface of the crystal is normal to the x-directed principal axis of $LiNbO_3$, waveguides are along the y direction, and the E field in the waveguide is z-directed (TE wave), the r_{Ij} coefficient of $LiNbO_3$ used for the modulation is r_{33}. The applied field does not change the orientation of the principal axes of the $\bar{\bar{\kappa}}$ tensor ellipsoid (12.5). The phase difference $\Delta\phi$ between the two arms is

$$\Delta\phi = \frac{\omega}{c} r_{33} n_e^3 l E_z \simeq \frac{\omega}{c} r_{33} n_e^3 \frac{l}{w} V$$

where V is the applied voltage. In $LiNbO_3$, $r_{33} = 30.8 \times 10^{-12}$ m/V, and $n_e = 2.1694$ at $\lambda = 0.85\ \mu$m. Taking the length l of the electrode as 5 mm, the width of the guide $w = 3\ \mu$m and treating the mode pattern and E field as uniform in a very rough approximation, we obtain for the voltage required to produce a phase difference of π between the waves in the two waveguides

$$V_\pi = \frac{\lambda}{2r_{33} n_e^3 (l/w)} = 1.67 \text{ V}$$

The voltage measured in a physical device was 1.6 V [3].

The Mach–Zehnder interferometer offers the opportunity of broad-band modulation. The electrodes can be made into microwave strip lines, a form of transmission lines. If the characteristic impedance is Z_0, the power required in each of the waveguides to achieve a phase shift $\Delta\phi$ of π is

$$P_\pi = \frac{(V_\pi)^2}{2Z_0}$$

For $Z_0 = 50\ \Omega$, and using the value of the 1.6-V peak,

$$P_\pi = 26 \text{ mW}$$

The bandwidth is limited by the losses of the strip line, which increase with frequency and eventually prevent modulation at extremely high frequency, and by the transit time τ of the optical wave through the 5-mm section of the modulator $\tau = n_e l/c = 36$ ps. The transit-time limitation can be overcome by construction of the strip line so that its phase velocity is matched to the group velocity of the optical wave.

12.5 SUMMARY

Electro-optic modulators most commonly utilize the linear electro-optic effect. This effect is described in terms of a third-rank tensor, r_{ijk}, expressing the linear dependence of the coefficients of the index ellipsoid on the applied electric field. We presented examples of two media, KDP and $LiNbO_3$. Because the linear electro-optic effect is described by a third-rank tensor, it reveals more details of the crystallographic symmetry than the index ellipsoid.

Some amplitude modulators utilize the birefringence induced by the applied electric field. Arrangements of polarizers and analyzers transform a rotation of the polarization of the optical wave into an amplitude modulation. Phase modulators utilize the phase shift without polarization rotation. The transverse modulator has the advantage over the longitudinal modulator in that the induced phase changes can be increased, at fixed voltage, by an increase of the interaction length. All optical waveguide modulators utilize this principle. A useful amplitude modulator in waveguide geometry, which avoids the use of a polarizer and analyzer, is the Mach–Zehnder interferometer. The modulation is produced by interference of mode patterns in the waveguide Y at the output of the phase-shifting section. A net phase-shift difference of π produces an antisymmetric higher-order mode in the output waveguide. Because the single-mode waveguide does not propagate this mode, the output of the interferometer is zero for an applied voltage that produces such a phase shift.

PROBLEMS

12.1. Assuming wave propagation along one of the three principal axes in $LiNbO_3$, determine those combinations of modulation field directions and E-field polarizations that lead to phase modulation. Determine the one with the largest effect for a given modulation E field. The values for $LiNbO_3$ are at $\lambda = 1.06$ μm:

$$r_{33} = 30.8, \qquad r_{22} = 3.4$$

$$r_{13} = 8.6, \qquad r_{42} = 28$$

in units of 10^{-12} m/V.

12.2. In a medium (like CdTe) with a "threefold rotation axis" (a form of crystal symmetry) the electro-optic coefficients r_{Ij}, in a coordinate system with z parallel to the axis, are

$$r_{Ij} \leftrightarrow \begin{bmatrix} r_{11} & -r_{22} & r_{13} \\ -r_{11} & r_{22} & r_{13} \\ 0 & 0 & r_{33} \\ r_{41} & r_{42} & 0 \\ r_{42} & -r_{41} & 0 \\ -r_{22} & -r_{11} & 0 \end{bmatrix}$$

Suppose that we have an applied ("pumping") field

$$\boldsymbol{E}_p = E_p^0(\hat{x} \sin \omega_p t + \hat{y} \cos \omega_p t)$$

(a) Find the intersection of the index ellipsoid with the x-y plane.

(b) Find the principal axes x', y' of the ellipse by a coordinate transformation

$$x = x' \cos\theta - y' \sin\theta$$

$$y = x' \sin\theta + y' \cos\theta$$

(c) Show that θ rotates at a constant angular rate and find the rate [4].

12.3. The intent is to build a mirror of variable reflectivities. The arrangement is shown in Fig. P12.3. The incident wave is propagating through a crystal of KDP oriented as shown with the control field parallel to z.

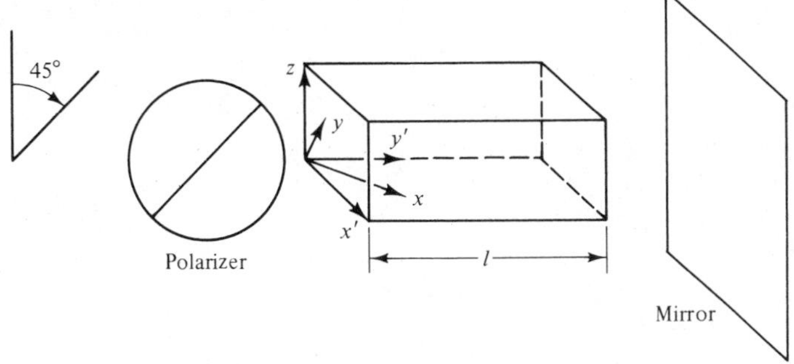

Figure P12.3

(a) Find the z and x' components of the field after first passage through the crystal.

(b) Find the amplitude of the reflected wave emerging from the polarizer as a function of E_z.

12.4. For a waveguide-coupled switch as described in Section 7.7, find an estimate for the electrode voltage required to prevent power transfer from one guide to the other. Assume x-cut $LiNbO_3$, with waveguides along y and applied field along z. The electrode spacing is 5μ, the wavelength is $\lambda = 1.06\ \mu$. Assume an interaction length of 5 mm. The electrode structure is shown in Fig. P12.4.

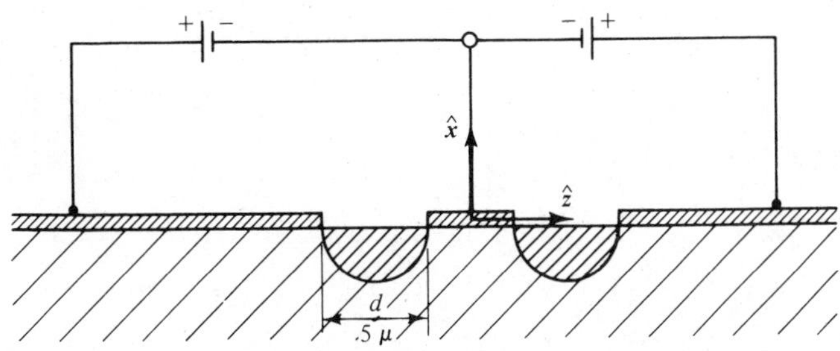

Figure P12.4

REFERENCES

[1] J. F. Nye, *Physical Properties of Crystals, Their Representation by Tensors and Matrices*, Clarendon Press, Oxford, 1967.

[2] W. E. Martin, "A new waveguide switch modulator for integrated optics," *Appl. Phys. Lett.*, *26*, 560–564, 1975.

[3] F. J. Leonberger, C. E. Woodward, and D. L. Spears, "Design and development of a high-speed electro-optic A/D converter," *IEEE Trans. Circuits Syst.*, *CAS-26*, 1125–1131, Dec. 1979.

[4] G. M. Carter and H. A. Haus, "Optical single sideband generation at 10.6 μm," *IEEE J. Quantum Electron.*, *QE-15*, no. 4, 217–224, Apr. 1979.

13

NONLINEAR OPTICS

In the preceding chapter, the electro-optic effect was treated as a special case of nonlinear optics. In this chapter we take up the general theory of nonlinear optics, a field that has been highly developed in recent years [1–5].

We define the basic susceptibility tensors of nonlinear optics and show how the electro-optic effect is a special case of nonlinear optics. We derive some interrelations among the second-order susceptibility tensors that are the consequence of the Manley–Rowe relations. We develop a model of a nonlinear material which illustrates the frequency dependence of the susceptibility tensors.

We analyze a parametric oscillator (frequency down-converter). Here the Manley–Rowe relations find a useful application. Then we take up the problem of optical frequency doubling, and derive the pertinent equations to judge the efficiency of the doubling process.

Finally, we look at third-order susceptibility tensors, the optical Kerr effect, and consider, as one of its applications, a picosecond time-resolved spectrometer. For more details on nonlinear optical phenomena, see Refs. 1–5.

13.1 CHARACTERIZATION OF A NONLINEAR MEDIUM

Optical nonlinearities that are used for modulation and frequency multiplication are most commonly associated with a reactive (lossless) parameter (e.g., the dielectric susceptibility). A linear isotropic medium has a polarization $\boldsymbol{P}$ proportional to the electric field

$$\boldsymbol{P} = \varepsilon_0 \chi \boldsymbol{E} \tag{13.1}$$

where χ is the dielectric susceptibility, frequency independent if the medium is nondispersive. Suppose that $\boldsymbol{P}$ is a nonlinear function of $\boldsymbol{E}$. To start, we shall treat $\boldsymbol{E}$ and $\boldsymbol{P}$ as time-independent (real) scalars. We may express P as a Taylor expansion in terms of powers of E:

$$P = \sum_n \varepsilon_0 \chi^{(n)} E^n \tag{13.2}$$

where $\chi^{(n)}$ is the coefficient associated with the nth power of E. $\chi^{(1)}$ is the linear susceptibility, and $\chi^{(2)}$ gives the first nonlinear term. If the medium has inversion symmetry (inversion about the origin of the coordinates of the atoms making up the medium, that is, the transformation $\boldsymbol{r} \rightarrow -\boldsymbol{r}$, does not alter the medium) there can be no second-order term in the expansion (13.2). To prove this, we imagine a polarization P produced by a field E, and then reverse E to $-E$. The reversal of E must reverse P, because the medium itself is invariant under such a reversal. All terms of even power in (13.2) do not reverse, when E is reversed. Hence all $\chi^{(2m)}$ coefficients with m an integer must vanish for media with inversion symmetry.

One may estimate the order of magnitude of the $\chi^{(n)}$'s of a solid in which the nonlinear polarizabilities are due to the nonlinear response of the bound electrons [4]. The electrons are under the influence of an internal electric field which we denote by E_{int}. When the applied optical field is imagined to reach a magnitude comparable to E_{int}, causing major distortion of the electronic wave function, it is reasonable to suppose that all terms in the expansion (13.2) would become comparable to each other. Thus

$$\chi^{(n)} E_{\text{int}}^n \simeq \chi^{(n+2)} E_{\text{int}}^{n+2}$$

or

$$\frac{\chi^{(n+2)}}{\chi^{(n)}} = \left(\frac{1}{E_{\text{int}}}\right)^2$$

assuming, of course that one of the χ^n's is not zero by virtue of the symmetry of the lattice. Because optical fields are usually much smaller than E_{int}, one may carry the expansion to the lowest-order nonlinear term of interest in the particular phenomenon under investigation, all higher-order terms will be negligible in comparison.

Consider a medium with no inversion symmetry. Suppose that $E(t)$ contains two frequency components:

$$E(t) = \tfrac{1}{2}[\mathrm{E}(\omega_\alpha)e^{j\omega_\alpha t} + \mathrm{E}(\omega_\beta)e^{j\omega_\beta t} + \text{c.c.}] \tag{13.3}$$

Here we treat explicitly the Fourier components at positive and negative frequencies, in distinction from (1.85), where we have introduced the definition appropriate for one-sided, positive-frequency, spectra. This is done in order to conform with usage in the literature of nonlinear optics. The electric fields produce polarizations at various frequencies. If only the first and second powers of E in (13.1) are taken into account, the polarization at the frequency

$\omega_\gamma = \omega_\alpha + \omega_\beta$ is produced by $E(\omega_\gamma)$ and the product $2E(\omega_\alpha)E(\omega_\beta)$ arising from E^2. With the definition for the polarization density at the frequency ω_γ,

$$P_\gamma(t) = \tfrac{1}{2}[P(\omega_\gamma)e^{j\omega_\gamma t} + P^*(\omega_\gamma)e^{-j\omega_\gamma t}] \tag{13.4}$$

one obtains for the complex amplitude $P(\omega_\gamma)$:

$$P(\omega_\gamma) = \varepsilon_0\,\chi^{(1)}E(\omega_\gamma) + \varepsilon_0\,\chi^{(2)}E(\omega_\alpha)E(\omega_\beta) \tag{13.5}$$

Polarization at the frequency $\omega_\delta = \omega_\alpha - \omega_\beta$ is produced by $E(\omega_\delta)$ and the product $2E(\omega_\alpha)E^*(\omega_\beta)$:

$$P(\omega_\delta) = \varepsilon_0\,\chi^{(1)}E(\omega_\delta) + \varepsilon_0\,\chi^{(2)}E(\omega_\alpha)E^*(\omega_\beta) \tag{13.6}$$

Thus far we have treated P and E as scalars. Since the polarization and the electric fields are vectors, one must use tensor notation to define **P** in terms of **E**. The polarization vector $P_i(\omega_\gamma)$ at frequency $\omega_\gamma = \omega_\alpha + \omega_\beta$ is written

$$P_i(\omega_\gamma) = \varepsilon_0\,\chi_{ij}^{(1)}E_j(\omega_\gamma) + \varepsilon_0\,\chi_{ijk}^{(2)}(\omega_\gamma;\,\omega_\alpha,\,\omega_\beta)E_j(\omega_\alpha)E_k(\omega_\beta) \tag{13.7}$$

Summation over j and k is implied by the Einstein summation convention. This is an example of frequency up-conversion. Frequency down-conversion with $\omega_\delta = \omega_\alpha - \omega_\beta$ has the polarization density

$$P_i(\omega_\delta) = \varepsilon_0\,\chi_{ij}^{(1)}E_j(\omega_\delta) + \varepsilon_0\,\chi_{ijk}^{(2)}(\omega_\delta;\,\omega_\alpha,\,-\omega_\beta)E_j(\omega_\alpha)E_k^*(\omega_\beta) \tag{13.8}$$

$\chi_{ijk}^{(2)}(\omega_\gamma;\,\omega_\alpha,\,\omega_\beta)$ is the second-order susceptibility tensor, a tensor of third rank. For both fields directed along x, and considering only the x component of **P**, we can relate the newly defined tensor to the expansion coefficients (13.2):

$$\chi_{xxx}^{(2)}(\omega_\gamma;\,\omega_\alpha,\,\omega_\beta) = \chi^{(2)}$$

In general, the tensors $\chi_{ijk}^{(2)}(\omega_\gamma;\,\omega_\alpha,\,\omega_\beta)$ are functions of ω_α and ω_β. A special case of interest is the case of second harmonic generation (SHG) when $\omega_\alpha = \omega_\beta = \omega$. Then $\omega_\alpha = \omega_\alpha + \omega_\beta = 2\omega$, and the polarization is at 2ω. Because the square of the field does not bring in cross products as in (13.4), a factor of 2 is lost and

$$P_i(2\omega;\,\omega,\,\omega) = \varepsilon_0\,\chi_{ij}^{(1)}\,E_j(2\omega) + \tfrac{1}{2}\varepsilon_0\,\chi_{ijk}^{(2)}(2\omega;\,\omega,\,\omega)E_j(\omega)E_k(\omega) \tag{13.9}$$

Processes involving products of two electric fields are called *second-order processes*.

13.2 CONNECTION WITH ELECTRO-OPTIC COEFFICIENTS

We have introduced the $\bar{\bar{\chi}}$ tensors in a general way for any nonlinear material. This is the modern method of analysis of nonlinear materials. The electro-optic effect has been described in terms of the r_{Ij} coefficients, the traditional description of this effect. In this section we establish the correspondence between the susceptibility tensor of second order and the r_{Ij} coefficients. The term $r_{Ij}E_j$ gives the change of κ_{ij} proportional to the applied field E_j, (12.2). The first subscript of r_{Ij} assumes six values; the second subscript assumes three values. For the following derivation, let us use two subscripts for I in the spirit of tensor notation. By definition, the application of an electric field results in

$$\varepsilon_0 \kappa_{ij} \rightarrow \varepsilon_0 \kappa_{ij} + \varepsilon_0 \Delta\kappa_{ij} = \varepsilon_0 \kappa_{ij} + r_{ijk} E_k \tag{13.10}$$

The linear electro-optic effect describes the change of optical index as caused by a dc electric field or a field of frequency much smaller than the optical frequency. The $\chi^{(2)}_{ijk}$ tensor to be compared with r_{ijk} is $\chi^{(2)}_{ijk}(\omega; \omega, 0)$ which expresses the change of the dielectric tensor by an applied dc electric field E_k.

$$\varepsilon_{ij} \rightarrow \varepsilon_{ij} + \varepsilon_0 \chi^{(2)}_{ijk} E_k \tag{13.11}$$

The inverse of the $\bar{\bar{\kappa}}$ tensor gives the dielectric tensor. Since $\Delta\kappa_{ij}$ is usually very small compared with κ_{ij}, the inverse is

$$[\bar{\bar{\kappa}} + \Delta\bar{\bar{\kappa}}]^{-1}_{ij} \simeq [\bar{\bar{\kappa}}^{-1}]_{ij} - [\bar{\bar{\kappa}}^{-1}]_{il}\, \Delta\kappa_{lm}[\bar{\bar{\kappa}}^{-1}]_{mj} = \varepsilon_{ij} + \varepsilon_0 \chi^{(2)}_{ijk} E_k \tag{13.12}$$

The correctness of the second expression in (13.12) can be checked by dot-multiplying both sides by $\bar{\bar{\kappa}} + \Delta\bar{\bar{\kappa}}$ and showing that the identity tensor is obtained to first order in $\Delta\bar{\bar{\kappa}}$.

Let us consider a uniaxial crystal as an example. In the coordinate system aligned with the principal axes the $\bar{\bar{\kappa}}$ tensor for a uniaxial medium is

$$\varepsilon_0 \bar{\bar{\kappa}} = \begin{bmatrix} \dfrac{1}{n_0^2} & 0 & 0 \\ 0 & \dfrac{1}{n_0^2} & 0 \\ 0 & 0 & \dfrac{1}{n_e^2} \end{bmatrix} \tag{13.13}$$

The change of the $\bar{\bar{\kappa}}^{-1}$ tensor is

$$[\bar{\bar{\kappa}}^{-1}]_{il}\, \Delta\kappa_{lm}[\bar{\bar{\kappa}}^{-1}]_{mj} = \varepsilon_0 \begin{bmatrix} n_0^4\, \Delta\kappa_{11} & n_0^4\, \Delta\kappa_{12} & n_0^2 n_e^2\, \Delta\kappa_{13} \\ n_0^4\, \Delta\kappa_{21} & n_0^4\, \Delta\kappa_{22} & n_0^2 n_e^2\, \Delta\kappa_{23} \\ n_0^2 n_e^2\, \Delta\kappa_{31} & n_0^2 n_e^2\, \Delta\kappa_{32} & n_e^4\, \Delta\kappa_{33} \end{bmatrix}$$

$$= -\varepsilon_0 \chi^{(2)}_{ijk} E_k \tag{13.14}$$

For instance,

$$\varepsilon_0 n_0^2 n_e^2\, \Delta\kappa_{13} = n_0^2 n_e^2 r_{13k} E_k = -\chi^{(2)}_{13k} E_k \tag{13.15}$$

13.3 DISPERSIVE MEDIUM—SECOND-ORDER PROCESSES

If the medium is dispersive, the $\bar{\bar{\chi}}$ tensors are frequency dependent. We develop here a simple scalar model of a nonlinear medium to illustrate the physics underlying dispersion. Consider a medium consisting of polarizable particles of zero net charge of density N. Suppose that the displacement x of the + charge q within each particle, from its neutral position (the position coincident with the center of opposite charge), follows the law

$$m\left(\frac{d^2x}{dt^2} + \sigma\frac{dx}{dt} + \omega_0^2 x\right) + m\,Dx^2 = qE \tag{13.16}$$

Here m is the mass of the charge, σ a phenomenological loss term, and $\omega_0^2 x + Dx^2$ expresses the restoring force.

Let us ignore the nonlinear term Dx^2 at first, because it is usually very small compared with $\omega_0^2 x$. Then, an applied field

$$E = \tfrac{1}{2}[\mathrm{E}(\omega_\alpha)e^{j\omega_\alpha t} + \mathrm{E}(\omega_\beta)e^{j\omega_\beta t} + \text{c.c.}] \tag{13.17}$$

produces the complex amplitude of the displacement at frequency ω_α:

$$\mathrm{x}(\omega_\alpha) = \frac{(q/m)\mathrm{E}(\omega_\alpha)}{(\omega_0^2 - \omega_\alpha^2) + j\omega_\alpha\sigma} \tag{13.18}$$

A similar expression holds for the displacement $\mathrm{x}(\omega_\beta)$.

We may correct x to include the effect of D to first order in D by noting that the terms of frequencies other than ω_α and ω_β must cancel on the left-hand side of (13.16). If we denote ω_γ by (frequency up-conversion)

$$\omega_\gamma = \omega_\alpha + \omega_\beta \tag{13.19}$$

we find for $\mathrm{x}(\omega_\gamma)$

$$\mathrm{x}(\omega_\gamma) = -\frac{D\mathrm{x}(\omega_\alpha)\mathrm{x}(\omega_\beta)}{\omega_0^2 - \omega_\gamma^2 + j\omega_\gamma\sigma} \tag{13.20}$$

Introducing $\mathrm{x}(\omega_\alpha)$ and $\mathrm{x}(\omega_\beta)$ and using the fact that $\mathrm{P}(\omega_\gamma) = qN\mathrm{x}(\omega_\gamma)$, where N is the particle density, we find

$$\chi^{(2)}(\omega_\gamma;\,\omega_\alpha,\,\omega_\beta) = -\frac{D(q^2N/\varepsilon_0 m)(q/m)}{(\omega_0^2 - \omega_\gamma^2 + j\omega_\gamma\sigma)(\omega_0^2 - \omega_\alpha^2 + j\omega_\alpha\sigma)(\omega_0^2 - \omega_\beta^2 + j\omega_\beta\sigma)} \tag{13.21}$$

with $\omega_\gamma = \omega_\alpha + \omega_\beta$. This shows one possible frequency dependence of $\chi^{(2)}$. It also shows the possibility for the enhancement of the nonlinearity if the frequency of one of the applied fields is at, or near, the resonance frequency ω_0.

We may consider a frequency down-conversion for which

$$\omega_\gamma = \omega_\beta - \omega_\alpha \tag{13.22}$$

In this case, the displacement $x(\omega_\gamma)$ is

$$x(\omega_\gamma) = -\frac{D x^*(\omega_\alpha) x(\omega_\beta)}{\omega_0^2 - \omega_\gamma^2 + j\omega_\gamma \sigma} \tag{13.23}$$

and

$$\chi^{(2)}(\omega_\gamma;\ -\omega_\alpha, \omega_\beta)$$
$$= -\frac{D(q^2 N/\varepsilon_0 m)(q/m)}{(\omega_0^2 - \omega_\gamma^2 + j\omega_\gamma \sigma)(\omega_0^2 - \omega_\alpha^2 - j\omega_\alpha \sigma)(\omega_0^2 - \omega_\beta^2 + j\omega_\beta \sigma)} \tag{13.24}$$

13.4 MANLEY–ROWE RELATIONS—SECOND-ORDER PROCESSES

The Manley–Rowe relations were first derived classically for lossless nonlinear circuit elements [6]. They were generalized to include nonlinear continuous media [7]. They were shown to hold for any system possessing a Hamiltonian [8]. We shall not derive them here but simply state them in the appealing form of photon number generation and annihilation [9].

Suppose that we illuminate a volume element of a lossless nonlinear material with radiation at frequency ω_γ and ω_β, generating radiation at the frequency ω_α in the process

$$\omega_\alpha = \omega_\gamma - \omega_\beta \tag{13.25}$$

(frequency down-conversion). One views the process as a conversion of a flux of photons at frequency ω_γ, carrying energy $\hbar\omega_\gamma$ each, with

$$\hbar\omega_\gamma = \hbar\omega_\alpha + \hbar\omega_\beta \tag{13.26}$$

into photons at frequencies ω_α and ω_β. The photon of energy $\hbar\omega_\gamma$ is annihilated to produce photons of energies $\hbar\omega_\alpha$ and $\hbar\omega_\beta$, the number of photons produced at ω_α is equal to the number of photons produced at frequency ω_β. The photon at frequency ω_γ "splits" to give the two photons at lower frequencies (and energies) ω_α and ω_β ($\hbar\omega_\alpha$ and $\hbar\omega_\beta$). This is the basic content of the Manley–Rowe relations applied to a second-order process.

Consider now the process in greater detail. Polarization densities $\mathbf{P}(\omega_\alpha)$, $\mathbf{P}(\omega_\beta)$, and $\mathbf{P}(\omega_\gamma)$ are generated in the nonlinear material at frequencies ω_α, ω_β, and ω_γ. The time-averaged power per unit volume produced by the polarization current density $j\omega_\alpha \mathrm{P}_\alpha$ at frequency ω_α divided by $\hbar\omega_\alpha$,

$$-\frac{\frac{1}{2}\mathrm{Re}\,[j\omega_\alpha \mathbf{E}^*(\omega_\alpha)\cdot\mathbf{P}(\omega_\alpha)]}{\hbar\omega_\alpha}$$

is proportional to the number of photons produced at frequency ω_α per unit volume. This rate of photon production must be equal to the rate of photon production at frequency ω_β.

$$-\tfrac{1}{2}\text{Re}\,[j\mathbf{E}^*(\omega_\alpha)\cdot\mathbf{P}(\omega_\alpha)] = -\tfrac{1}{2}\,\text{Re}\,[j\mathbf{E}^*(\omega_\beta)\cdot\mathbf{P}(\omega_\beta)] \tag{13.27}$$

This is one of the Manley–Rowe relations that can be derived for the particular process (13.25). It is fundamental to parametric oscillators (see Section 13.5). It implies that absorption at frequency ω_β (i.e., annihilation of photons at ω_β) makes available photons for absorption at frequency ω_α.

The Manley–Rowe relation (13.27) implies a restriction upon the $\chi^{(2)}_{ijk}$ susceptibility tensors of a lossless system. Introduce the expressions for $P_i(\omega_\alpha)$ and $P_i(\omega_\beta)$, (13.8), into (13.27) and note that the contribution of $\chi^{(1)}_{ij}$ to the time-averaged power is zero.

$$\text{Re}\,\{j\chi^{(2)}_{ijk}(\omega_\alpha;\,\omega_\gamma,\,-\omega_\beta)E^*_i(\omega_\alpha)E_j(\omega_\gamma)E^*_k(\omega_\beta)\}$$
$$= \text{Re}\,[j\chi^{(2)}_{ijk}(\omega_\beta;\,\omega_\gamma,\,-\omega_\alpha)E^*_i(\omega_\beta)E_j(\omega_\gamma)E^*_k(\omega_\alpha)]$$

Because $E_i(\omega_\alpha)$, $E_j(\omega_\gamma)$, and $E_k(\omega_\beta)$ are arbitrarily adjustable in phase and amplitude, we must have

$$\chi^{(2)}_{ijk}(\omega_\alpha;\,\omega_\gamma,\,-\omega_\beta) = \chi^{(2)}_{kji}(\omega_\beta;\,\omega_\gamma,\,-\omega_\alpha) \tag{13.28}$$

In the case of collinear fields and no loss ($\sigma = 0$), one may verify (13.28) from the scalar model of a dispersive medium of Section 13.3 (disregarding tensor subscripts).

Another parametric process is also of interest, frequency up-conversion. In this process a photon of frequency ω_α and one of frequency ω_γ are annihilated to give a photon of larger frequency ω_β:

$$\hbar\omega_\beta = \hbar\omega_\gamma + \hbar\omega_\alpha \tag{13.29}$$

The production rate of photons per unit volume by the polarization current density at frequency ω_β must lead to a consumption rate of photons per unit volume at frequency ω_α; two photons combine to give a photon at larger frequency (and energy).

$$\text{Re}\,[j\chi_{ijk}(\omega_\alpha;\,-\omega_\gamma,\,\omega_\beta)E^*_i(\omega_\alpha)E^*_j(\omega_\gamma)E_k(\omega_\beta)]$$
$$= -\text{Re}\,[j\chi^{(2)}_{ijk}(\omega_\beta;\,\omega_\gamma,\,\omega_\alpha)E^*_i(\omega_\beta)E_j(\omega_\gamma)E_k(\omega_\alpha)]$$

Because $E_i(\omega_\alpha)$, $E_j(\omega_\gamma)$, and $E_k(\omega_\beta)$ are arbitrary

$$\chi^{(2)}_{ijk}(\omega_\alpha;\,-\omega_\gamma,\,\omega_\beta) = \chi^{*(2)}_{kji}(\omega_\beta;\,\omega_\gamma,\,\omega_\alpha) \tag{13.30}$$

Again we may verify this relation for the scalar dispersive medium of Section 13.3 for the specific case of collinear fields and no loss ($\sigma = 0$).

13.5 PARAMETRIC OSCILLATORS

The adjective "parametric" derives from electronics, where parametric amplifiers were developed as low-noise amplifiers in the 1950s. The interaction between two frequencies, the "signal" and "idler" frequencies, are achieved by variation of a circuit parameter (e.g., capacitance) [10]. The process of parametric interaction is a very familiar one. A child operating a swing changes the distance of the center of gravity from the suspension point(s) of the "pendulum" (consisting of swing and child) twice within one period. By this action, power is supplied to the pendulum, twice within one cycle.

The electric analog is described by a LC circuit whose capacitance is changed with time. This change can be produced electrically. A simpler picture is provided by mechanical changes of the spacing between the plates of a

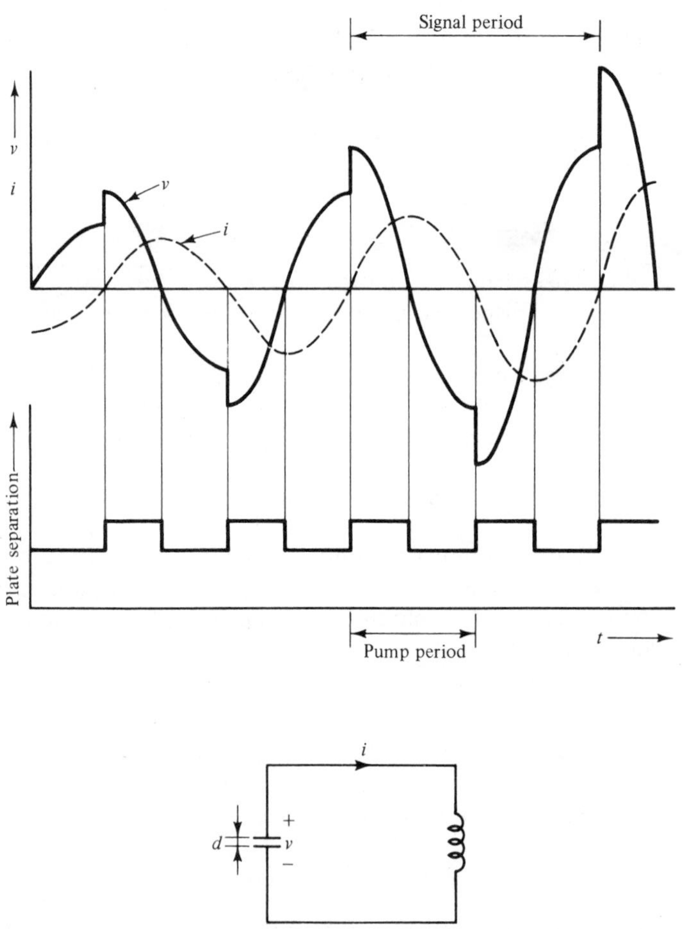

Figure 13.1 Degenerate parametric oscillator.

parallel-plate capacitor. The capacitor plate-separation is increased each time the voltage across the capacitor is in a maximum, decreased when it is zero (see Fig. 13.1). The voltage is increased in the first instance. The ratio of peak voltage to peak current is $\sqrt{L/C}$ within each quarter-cycle. The envelopes of the voltage and current waveforms grow exponentially with time; the energy is supplied by the work done in separating the capacitor plates. Conversely, if the separation is decreased at the instant of peak voltage and increased when the voltage is zero, the envelopes decay exponentially with time. An initial excitation in the circuit excites, in general, both the exponentially growing and decaying solutions; the growing one eventually predominates. This is an example of a *degenerate parametric oscillator*; the word "degenerate" refers to the fact that the signal and idler frequencies are equal to each other and both are equal to half the fundamental pump frequency.

In this section we shall analyze a nondegenerate parametric oscillator with signal frequency (ω_α) different from the idler frequency (ω_β), the sum of the two equal to the pump frequency ω_γ:

$$\omega_\gamma = \omega_\alpha + \omega_\beta \tag{13.31}$$

The Coupling-of-Modes Equation of Parametric Oscillator

Consider a resonator possessing modes $\mathbf{e}(\omega_\alpha)$ and $\mathbf{e}(\omega_\beta)$ at frequencies ω_α and ω_β, respectively (Fig. 13.2). We use the coupling-of-modes formalism developed in Chapter 7. The mode α of amplitude a_α is coupled to mode β of amplitude a_β by a source s_α and similarly for mode β.

$$\frac{da_\alpha}{dt} = j\omega_\alpha a_\alpha - \frac{a_\alpha}{\tau_\alpha} + s_\alpha \tag{13.32}$$

$$\frac{da_\beta}{dt} = j\omega_\beta a_\beta - \frac{a_\beta}{\tau_\beta} + s_\beta \tag{13.33}$$

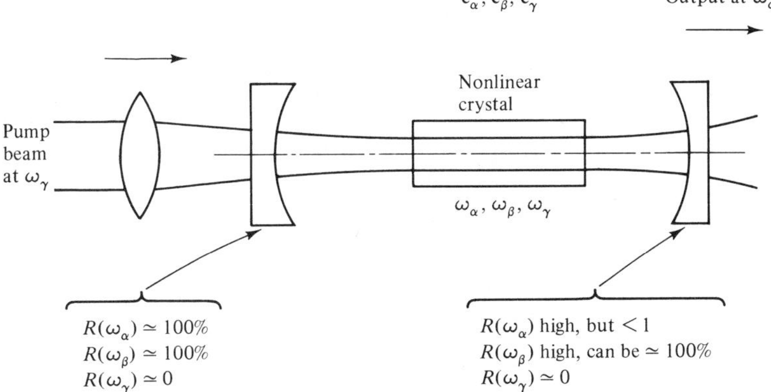

Figure 13.2 Parametric oscillator.

The sources are due to the interactions of the fields $\mathbf{E}(\omega_\alpha)$ and $\mathbf{E}(\omega_\beta)$ with the pump field, and can be evaluated from power considerations. The rate of change of energy in mode α is equal to the power $p_{\alpha\beta}$ fed into the mode by mode β:

$$\frac{d}{dt}|\mathrm{a}_\alpha|^2 + 2\frac{|\mathrm{a}_\alpha|^2}{\tau_\alpha} = \mathrm{a}_\alpha^* \mathrm{s}_\alpha + \mathrm{a}_\alpha \mathrm{s}_\alpha^* = p_{\alpha\beta} \tag{13.34}$$

But this power is produced by the flow of the polarization current density against the electric field $\mathbf{E}(\omega_\alpha)$

$$p_{\alpha\beta} = -\tfrac{1}{4}j\omega_\alpha \int \mathbf{E}^*(\omega_\alpha)\cdot\mathbf{P}(\omega_\alpha)\,dv + \text{c.c.} \tag{13.35}$$

Denote the field pattern of the pump field by $\mathbf{e}(\omega_\gamma)$. Then the power $p_{\alpha\beta}$ fed into the mode α is (compare (13.8))

$$p_{\alpha\beta} = -\frac{j\omega_\alpha}{4}\int \varepsilon_0\,\chi_{ijk}^{(2)}(\omega_\alpha;\,\omega_\gamma,\,-\omega_\beta)\mathrm{e}_i^*(\omega_\alpha)\mathrm{e}_j(\omega_\gamma)\mathrm{e}_k^*(\omega_\beta)\,dv\;\mathrm{a}_\alpha^*\,\mathrm{a}_\gamma\,\mathrm{a}_\beta^* + \text{c.c.} \tag{13.36}$$

because the "self-term" due to $\chi_{ij}^{(1)}(\omega_\alpha)$ is zero, and only the polarization density produced at the frequency ω_α by the pumping field and the field of mode β contribute to the power. The source term s_α is thus, using (13.34),

$$\mathrm{s}_\alpha = -\mathrm{a}_\gamma\,\mathrm{a}_\beta^*\frac{j\omega_\alpha}{4}\varepsilon_0\int \chi_{ijk}^{(2)}(\omega_\alpha;\,\omega_\gamma,\,-\omega_\beta)\mathrm{e}_i^*(\omega_\alpha)\mathrm{e}_j(\omega_\gamma)\mathrm{e}_k^*(\omega_\beta)\,dv \tag{13.37}$$

The source s_β is found by interchanging the subscript α and β.

Next we introduce slowly time varying "envelope" quantities A_α by

$$\mathrm{a}_\alpha(t) \equiv \mathrm{A}_\alpha(t)\exp(j\omega_\alpha t)$$

and so on. The source s_α can be seen to have the time dependence of $\mathrm{A}_\beta^*\,\mathrm{A}_\gamma \exp[j(\omega_\gamma - \omega_\beta)t]$, which is equal to $\mathrm{A}_\beta^*\,\mathrm{A}_\gamma\exp(j\omega_\alpha t)$.

We shall limit ourselves to the case of a strong "pump" at frequency ω_γ so that depletion of the pump by the "signals" at frequencies ω_α and ω_β can be neglected. In this "small-signal" limit, the pump amplitude may be considered prescribed (i.e., a constant). In this limit the equations relating a_α and a_β are linearized. Defining

$$\kappa_{\alpha\beta} \equiv -\frac{j\omega_\alpha}{4}\mathrm{A}_\gamma\,\varepsilon_0\int \chi_{ijk}^{(2)}(\omega_\alpha;\,\omega_\gamma,\,-\omega_\beta)\mathrm{e}_i^*(\omega_\alpha)\mathrm{e}_j(\omega_\gamma)\mathrm{e}_k^*(\omega_\beta)\,dv \tag{13.38}$$

and a corresponding expression $\kappa_{\beta\alpha}$ with β and α interchanged, we may write (13.32) and (13.33) in the form

$$\frac{d\mathrm{A}_\alpha}{dt} = -\frac{\mathrm{A}_\alpha}{\tau_\alpha} + \kappa_{\alpha\beta}\,\mathrm{A}_\beta^* \tag{13.39}$$

$$\frac{d\mathrm{A}_\beta^*}{dt} = -\frac{\mathrm{A}_\beta^*}{\tau_\beta} + \kappa_{\beta\alpha}^*\,\mathrm{A}_\alpha \tag{13.40}$$

These are the coupling-of-modes equations in time in their familiar form [compare (7.59) and (7.60)]. The coupling coefficients obey the relation

$$\frac{\kappa_{\alpha\beta}}{\omega_\alpha} = \frac{\kappa_{\beta\alpha}}{\omega_\beta} \tag{13.41}$$

which can be proven using (13.28) and the defining equation (13.38). The coupling coefficients obey a different relation than in the case of a linear lossless system. In the absence of loss, $1/\tau_\alpha = 1/\tau_\beta = 0$, (13.39) and (13.40) satisfy the conservation law

$$\frac{d}{dt}\frac{|\mathrm{A}_\alpha|^2}{\hbar\omega_\alpha} = \frac{d}{dt}\frac{|\mathrm{A}_\beta|^2}{\hbar\omega_\beta} \tag{13.42}$$

Since $|\mathrm{A}_\alpha|^2$ and $|\mathrm{A}_\beta|^2$ are the energies in modes α and β, $|\mathrm{A}_\alpha|^2/\hbar\omega_\alpha$ and $|\mathrm{A}_\beta|^2/\hbar\omega_\beta$ are the photon numbers. Relation (13.42) is a consequence of the generation in pairs of photons at frequencies ω_α and ω_β.

We have encountered a similar weighting factor in the acousto-optic coupler case in Section 9.1. There, we found that the coefficients of coupling between the incident and diffracted waves were not equal, they differed by a factor equal to the ratio of the two frequencies. The Manley–Rowe relations are "at work" there as well. The incident wave at frequency ω_i creates a photon at energy $\hbar\omega_d$ and creates or annihilates a phonon of energy $\hbar\omega_s$. The photon number is conserved, because one diffracted photon results for each incident photon.

If one assumes the time dependence $\exp(-t/\tau)$ for A_α and A_β, one finds for $1/\tau$, from (13.39) and (13.40),

$$\frac{1}{\tau} = \frac{1}{2}\left(\frac{1}{\tau_\alpha} + \frac{1}{\tau_\beta}\right) \pm \sqrt{\frac{1}{4}\left(\frac{1}{\tau_\alpha} - \frac{1}{\tau_\beta}\right)^2 + \kappa_{\alpha\beta}\kappa^*_{\beta\alpha}} \tag{13.43}$$

Because of (13.41), $\kappa_{\alpha\beta}\kappa^*_{\beta\alpha}$ is real and positive. Thus $1/\tau$ may become negative, the system has a solution that grows exponentially, if

$$|\kappa_{\alpha\beta}|^2 \frac{\omega_\beta}{\omega_\alpha} > \frac{1}{\tau_\alpha \tau_\beta}$$

With the $>$ sign replaced by an equality sign, this is the threshold condition for "startup" of the parametric oscillator. It indicates the magnitude of $|\kappa_{\alpha\beta}|^2$ [i.e., the magnitude of the pump field $|\mathrm{a}_\gamma e_j(\omega_\gamma)|^2$ required to provide sufficient "parametric" gain in order to exceed the resonator losses].

When threshold is exceeded, the determinantal equation predicts an exponentially growing solution. The system can (and does) start from the unavoidable noise background (even at zero temperature there are zero point field fluctuations [11]) and the fields build up at both frequencies ω_α and ω_β. When the amplitudes of the fields grow to levels such that their depleting effect on the pump radiation cannot be ignored, the coupling coefficients $\kappa_{\alpha\beta}$ decrease in magnitude so that a steady state of zero growth or decay is reached.

Momentum Conservation and Tuning of Parametric Oscillator

In optical resonators, the interaction media are usually many wavelengths long, the mode patterns $e_i(\omega_\alpha)$, and so on, are wave-like. In order to achieve coherent superposition in the integrals (13.37) and (13.38), one requires that the wave vectors associated with the fields obey the relation:

$$\pm \mathbf{k}_\alpha \pm \mathbf{k}_\beta + \mathbf{k}_\gamma = 0 \tag{13.44}$$

The double sign appears because a resonator has (usually) standing waves, rather than traveling waves. Equation (13.44) is the *momentum conservation relation* for the parametric process. This relation provides an opportunity for frequency tuning of the parametric oscillator. The normal (index) surface is, generally, a function of frequency. Under normal dispersion, n increases with frequency. Suppose that the crystal is positive uniaxial. If we choose $\mathbf{k}_\gamma$—the pump wave—to be an extraordinary wave and $\mathbf{k}_\alpha$ and $\mathbf{k}_\beta$ to be ordinary waves the phase matching as a function of frequency can be carried out as shown in Fig. 13.3a. The isotropic ordinary-wave propagation constants k_α and k_β are plotted against frequency, the propagation constant curve k_β is mirrored around $\omega_\gamma/2$, and added to k_α. In this way $k_\alpha + k_\beta$ is constructed as a function of ω_α and $\omega_\beta = \omega_\gamma - \omega_\alpha$. For phase matching, $\mathbf{k}_\gamma = \mathbf{k}_\alpha + \mathbf{k}_\beta$. The extraordinary normal (index) surface is drawn in terms of $k_\gamma = (\omega_\gamma/c)n_e(\omega_\gamma)$ (Fig. 13.3b) and the values of $k_\alpha + k_\beta$ are entered from the graph of Fig. 13.3a on the horizontal axis as shown. The intersection of the sphere with the ellipsoid gives the cone of phase-matching directions.

The frequency at which phase matching occurs has highest gain. If the mode spacing of the resonator is sufficiently small, a resonance will exist at (or very near) the frequencies ω_α and ω_β corresponding to the phase-matching direction and this pair of modes will grow faster than the modes that are not phase matched. In the steady state the pump photon flux is depleted so as to make the gains of these modes equal to their losses, whereas the gains of all other modes are lower so that they are suppressed. The oscillator is frequency tuned by a change of the orientation of the crystal which changes the phase matching direction, thereby changing the frequencies ω_α and ω_β.

Parametric Up-Conversion

The Manley–Rowe relations imposed the constraint (13.41) between the coupling coefficients of the fields at frequencies ω_α and ω_β for parametric "down-conversion," both radiations involved in the coupling had frequencies lower than the pump frequency. Another process of interest is parametric up-conversion, in which ω_α and ω_β obey the following relation with respect to the pump frequency ω_γ

$$\omega_\beta = \omega_\gamma + \omega_\alpha \tag{13.45}$$

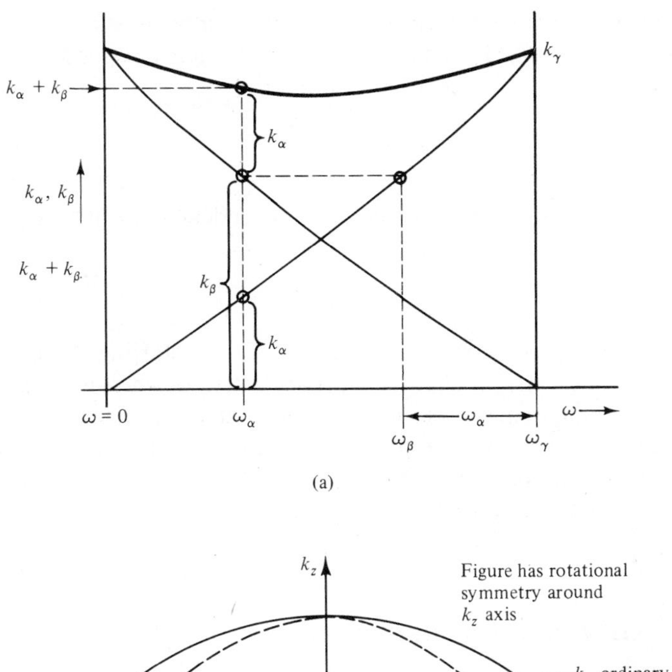

(a)

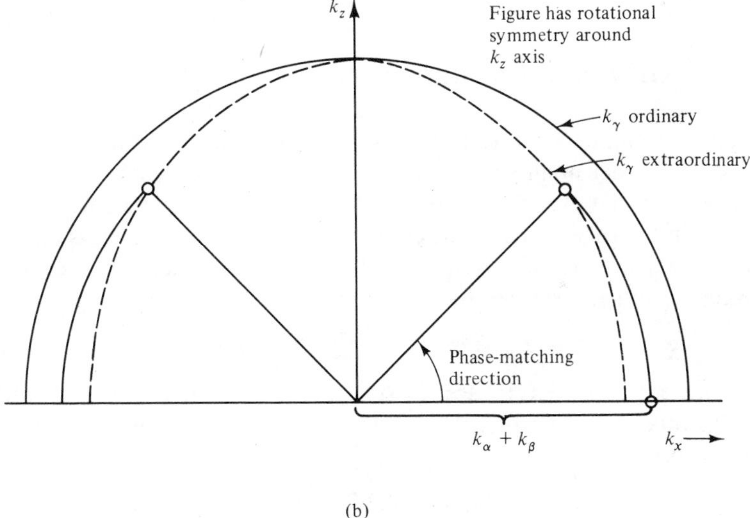

(b)

Figure 13.3 Construction of phase-matching direction for parametric oscillation.

The analysis is identical to the one presented above, except that (13.39) and (13.40) have to be replaced by

$$\frac{d\mathrm{A}_\alpha}{dt} = -\frac{\mathrm{A}_\alpha}{\tau_\alpha} + \kappa_{\alpha\beta}\,\mathrm{A}_\beta \tag{13.46}$$

$$\frac{d\mathrm{A}_\beta}{dt} = -\frac{\mathrm{A}_\beta}{\tau_\beta} + \kappa_{\beta\alpha}\,\mathrm{A}_\alpha \tag{13.47}$$

The Manley–Rowe constraint on the $\chi^{(2)}$-tensor components (13.30) combined with an adaptation of (13.38) to the up-conversion case impose the relation between the coupling coefficients

$$\frac{\kappa_{\alpha\beta}}{\omega_\alpha} = -\frac{\kappa^*_{\beta\alpha}}{\omega_\beta} \tag{13.48}$$

The determinantal equation for an assumed dependence exp $(j\Omega t)$ is

$$j\Omega = -\frac{1}{2}\left(\frac{1}{\tau_\alpha} + \frac{1}{\tau_\beta}\right) \pm \sqrt{\frac{1}{4}\left(\frac{1}{\tau_\alpha} - \frac{1}{\tau_\beta}\right)^2 - |\kappa_{\alpha\beta}|^2 \frac{\omega_\beta}{\omega_\alpha}} \tag{13.49}$$

There is no growing solution—the system is not self-starting. Parametric up-conversion is inherently different from parametric down-conversion. In the latter process both frequencies can be generated spontaneously from noise with no external excitation. The photons at ω_α and ω_β are generated in pairs, and the power in the two modes' exponentiates. In parametric up-conversion a photon of frequency ω_α has to be *absorbed* to *generate* a photon at $\omega_\beta = \omega_\gamma + \omega_\alpha$.

13.6 SECOND HARMONIC GENERATION [12]

An important nonlinear optical process is frequency multiplication. A medium with a second-order nonlinearity tensor $\chi^{(2)}_{ijk}(2\omega; \omega, \omega)$ excited by a field $\mathbf{E}(\omega)$ at frequency ω can be used for frequency doubling, second harmonic generation (SHG). As in the case of the linear electro-optic effect, in which it is customary to tabulate r_{ijk} instead of $\chi^{(2)}_{ijk}(\omega; \omega, 0)$, the tensor coefficients for second harmonic generation are denoted by

$$d_{ijk} = d_{iI} \equiv \tfrac{1}{2}\varepsilon_0\, \chi^{(2)}_{ijk}(2\omega; \omega, \omega) \tag{13.50}$$

The coefficients for KDP are [13] $d_{36} = d_{14} = 0.45 \pm 0.03$ in units of $\frac{1}{9} \times 10^{-22}$ (mks), and for $LiNbO_3$, in the same units, the following two coefficients are listed:

$$d_{31} = 4.76 \pm 0.5$$

$$d_{22} = 2.3 \pm 1.0$$

The symmetry of the d_{ijk} tensor is the same as that of the r_{jki} tensor, with one-to-one correspondence between letter indices as written here.

Consider the system shown in Fig. 13.4. A uniaxial crystal with its z axis oriented as shown has propagating within it a plane wave at frequency ω with an electric field in the x-z plane with a $\boldsymbol{k}$ vector parallel to y, of magnitude $k(\omega) = (\omega/c)n(\omega)$. This plane wave interacts with the nonlinearity of the crystal to produce a polarization at 2ω [compare (13.9)]:

$$\mathrm{P}_i(2\omega) = \tfrac{1}{2}\varepsilon_0\, \chi^{(2)}_{ijk}(2\omega; \omega, \omega)\mathrm{E}_j(\omega)\mathrm{E}_k(\omega) \tag{13.51}$$

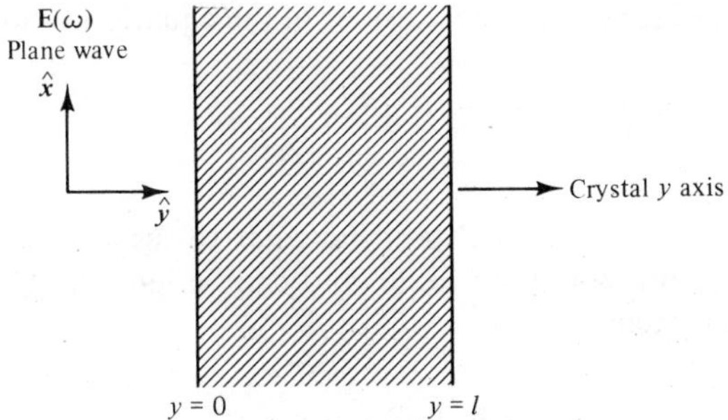

Figure 13.4 Frequency doubling arrangement.

The polarization current density is a plane-wave-like source distribution at the frequency 2ω with the spatial dependence $\exp[-j2\boldsymbol{k}(\omega)] \cdot \boldsymbol{r} = \exp[-j2k(\omega)y]$. Such a source distribution radiates forward along the principal axis $\hat{y}$ of the crystal, producing a field $\mathbf{E}(2\omega)$ polarized in the x-y plane. The radiated field at 2ω interferes constructively only when $2\boldsymbol{k}(\omega) = \boldsymbol{k}(2\omega)$, that is the radiation is phase matched to the second harmonic. To analyze the general case, we write down the equation for the propagation of $\mathbf{E}(2\omega)$ in the presence of a source $\mathbf{s}(2\omega)$

$$\frac{d}{dy}\mathbf{E}(2\omega) = -jk(2\omega)\mathbf{E}(2\omega) + \mathbf{s}(2\omega) \tag{13.52}$$

The source can be evaluated from power considerations, using the expression for the polarization generated at the second harmonic (13.51). The power per unit area at the second harmonic is

$$S_y(2\omega) = \frac{1}{2}\sqrt{\frac{\varepsilon_0}{\mu_0}}\, n(2\omega)\,|\,\mathbf{E}(2\omega)\,|^2 \tag{13.53}$$

The spatial rate of change of the Poynting flux $S_y(2\omega)$ is obtained from (13.52) by dot multiplication with $\mathbf{E}^*(2\omega)$ and addition of the complex conjugate

$$\frac{d}{dy}[S_y] = \frac{1}{2}\sqrt{\frac{\varepsilon_0}{\mu_0}}\, n(2\omega)[\mathbf{s}(2\omega)\cdot\mathbf{E}^*(2\omega) + \text{c.c.}] \tag{13.54}$$

On the right-hand side is the power supplied per unit volume by the source $\mathbf{s}(2\omega)$. This is provided by the induced polarization current density $j2\omega\mathbf{P}(2\omega)$ working against $\mathbf{E}(2\omega)$

$$\frac{1}{2}\sqrt{\frac{\varepsilon_0}{\mu_0}}\, n(2\omega)[\mathbf{s}(2\omega)\cdot\mathbf{E}^*(2\omega) + \text{c.c.}] = -\tfrac{1}{4}[j2\omega\mathbf{P}(2\omega)\cdot\mathbf{E}^*(2\omega) + \text{c.c.}] \tag{13.55}$$

Comparison of terms on the two sides of the equation yields

$$\mathbf{s}(2\omega) = -j\sqrt{\frac{\mu_0}{\varepsilon_0}}\frac{\omega}{n(2\omega)}\mathbf{P}_\perp(2\omega) \tag{13.56}$$

where $\mathbf{P}_\perp$ is the component of $\mathbf{P}$ perpendicular to the y-axis. The component of $\mathbf{P}$ along y does not couple to $\mathbf{E}(2\omega)$ and can be ignored. Using the second-order susceptibility tensor,

$$\mathbf{s} = -\frac{1}{2}j\frac{\omega}{cn(2\omega)}\chi_{ijk}^{(2)}(2\omega;\, \omega,\, \omega)\tilde{\mathrm{E}}_j(\omega)\tilde{\mathrm{E}}_k(\omega)\exp[-j2k(\omega)y],\ i = 1, 3 \tag{13.57}$$

where we have factored out explicitly the spatial dependence of the electric field at the fundamental frequency, $\mathrm{E}_j(\omega) = \tilde{\mathrm{E}}_j(\omega)\exp[-jk(\omega)y]$. Equation (13.57) introduced in (13.52) gives the equation for the second harmonic field:

$$\frac{d\mathrm{E}_i(2\omega)}{dy} = -jk(2\omega)\mathrm{E}_i(2\omega) - j\frac{\omega}{2cn(2\omega)}\chi_{ijk}^{(2)}(2\omega;\, \omega,\, \omega)\tilde{\mathrm{E}}_j(\omega)\tilde{\mathrm{E}}_k(\omega) \cdot \exp[-j2k(\omega)y, \qquad i = 1, 3 \tag{13.58}$$

The vector $\chi_{ijk}^{(2)}\tilde{\mathrm{E}}_j(\omega)\tilde{\mathrm{E}}_k(\omega)(i = 1, 3)$ may have both an x and a z component. The second harmonic field points in the direction of this vector. Equation (13.58) could have been obtained directly from the wave equation in a manner analogous to the derivation of wave diffraction by an acoustic wave (Section 9.1). We used the perturbation approach, because the excitation of a mode in space, by a spatially distributed source, is analogous to the excitation of a mode in time, by a time-dependent source. The second harmonic generation example complements the perturbation approach of Section 7.2 and shows its flexibility. It is an approach that can be extended to processes other than electromagnetic (e.g., frequency doubling in acoustic wave interaction).

Integration of (13.58) gives the excitation of the second harmonic field, with the boundary condition $\mathrm{E}_i(2\omega) = 0$ at $y = 0$:

$$\mathrm{E}_i(2\omega, y) = -\frac{1}{2}\frac{\omega}{cn(2\omega)}\chi_{ijk}^{(2)}(2\omega;\, \omega,\, \omega)\tilde{\mathrm{E}}_j(\omega)\tilde{\mathrm{E}}_k(\omega)e^{-jk(2\omega)y} \cdot \frac{\exp\{j[k(2\omega) - 2k(\omega)]y - 1\}}{k(2\omega) - 2k(\omega)}, \qquad i = 1, 3 \tag{13.59}$$

The power per unit area in the second harmonic at the output of the crystal, $y = l$, is

$$\mathrm{S}_y(2\omega) = \frac{1}{8}\sqrt{\frac{\varepsilon_0}{\mu_0}}\frac{\omega^2 l^2}{c^2 n(2\omega)}\sum_{i=1,3}|\chi_{ijk}^{(2)}|^2|\mathrm{E}_j(\omega)\mathrm{E}_k(\omega)|^2\left[\frac{\sin(\Delta kl/2)}{\Delta kl/2}\right]^2$$

where $\Delta k \equiv k(2\omega) - 2k(\omega)$ is the mismatch of the wave propagation constants.

This can be expressed in terms of the squares of the powers per unit area in the fundamental if E_j is linearly polarized along x.

$$S_y(2\omega) = \frac{1}{2}\sqrt{\frac{\mu_0}{\varepsilon_0}}\frac{\omega^2 l^2}{c^2 n(2\omega)n^2(\omega)}\sum_{i=1,3}|\chi_{ixx}^{(2)}|^2[S_y(\omega)]^2\left[\frac{\sin(\Delta kl/2)}{\Delta kl/2}\right]^3 \tag{13.60}$$

The power is maximized when the second harmonic is phase-matched to the dependence exp $[-2jk(\omega)y]$ of the polarization density driving source (i.e., when $\Delta k = 0$). Phase matching along one of the principal axes of the crystal can be accomplished in some materials by utilizing the different temperature dependences of n_0 and n_e. Usually, n_e is a stronger function of temperature than n_0. Thus, if $E_x(\omega)$ propagates as an ordinary wave and $E_i(2\omega)$ as an extraordinary wave, heating or cooling can bring $(2\omega/c)n_e(2\omega)$ into coincidence with $(2\omega/c)n_0(\omega)$.

Finite Beam Size in SHG [12]

The power in the second harmonic is proportional to the square of the power in the fundamental and increases with the square of the length of the crystal according to (13.60). This is correct for a beam of sufficiently wide cross section so that diffraction does not occur within the distance l. If one focuses the beam in an effort to increase the fundamental intensity, one may decrease the minimum beam radius w_0 only down to a value such that the beam spreading due to diffraction within the crystal is not excessive. A beam of radius w_0 can be kept confined at w_0 over a distance of the order

$$l \simeq \left(\frac{\pi w_0^2}{\lambda}\right)n \tag{13.61}$$

where λ is the free-space wavelength.

Consider a beam of radius w_0 and compute the powers that correspond to this beam radius. At the fundamental

$$P(\omega) = \int_0^\infty S_y(\omega, \rho)2\pi\rho\, d\rho = \frac{1}{2}\sqrt{\frac{\varepsilon}{\mu_0}}\int_0^\infty |\mathbf{E}|^2 e^{-2\rho^2/w_0{}^2}2\pi\rho\, d\rho$$

$$= \frac{\pi w_0^2}{2} S_y(\omega, \rho = 0) \tag{13.62}$$

The second harmonic power has the profile exp $[-(4\rho^2/w_0^2)]$ and thus its power is

$$P(2\omega) = \frac{\pi w_0^2}{4} S_y(2\omega, \rho = 0) \tag{13.63}$$

When we use (13.61), (13.62), and (13.63) in (13.59), we obtain in the phase-matched case,

$$P(2\omega) \simeq \frac{1}{4\pi}\sqrt{\frac{\mu_0}{\varepsilon_0}}(\omega/c)^3 \frac{l}{n(2\omega)n(\omega)} \sum_{i=1,3} |\chi_{ixx}|^2 P^2(\omega) \tag{13.64}$$

In the focused configuration, the second harmonic intensity is proportional to l, not l^2. Of course, the power cannot increase indefinitely with increasing l. Depletion of the fundamental "pumping" power limits the second harmonic output.

Pulse-Width Detection by SHG

Second harmonic generation is used for the determination of the width of pulses that are too short for direct detection by a fast photodetector. An experimental arrangement is shown in Fig. 13.5. The train of pulses from a modelocked laser is separated by the half-silvered mirror of a Michelson interferometer into two trains that are recombined with a variable delay τ of one of the arms of the interferometers. The interferometer uses corner reflectors for ease of alignment. The recombined pulses are focused into an SHG crystal and the output is detected by a photomultiplier. If extra sensitivity is required, one may chop the signal and synchronously detect as shown.

Denote the fields after superposition in the SHG crystal by $E_{x1} + E_{x2}$, where

$$E_{x1} = \tilde{E}(t) \exp\left[-jk(\omega)y\right] \tag{13.65}$$

$$E_{x2} = \tilde{E}(t-\tau) \exp\left[-jk(\omega)y\right] \exp(-j\omega\tau) \tag{13.66}$$

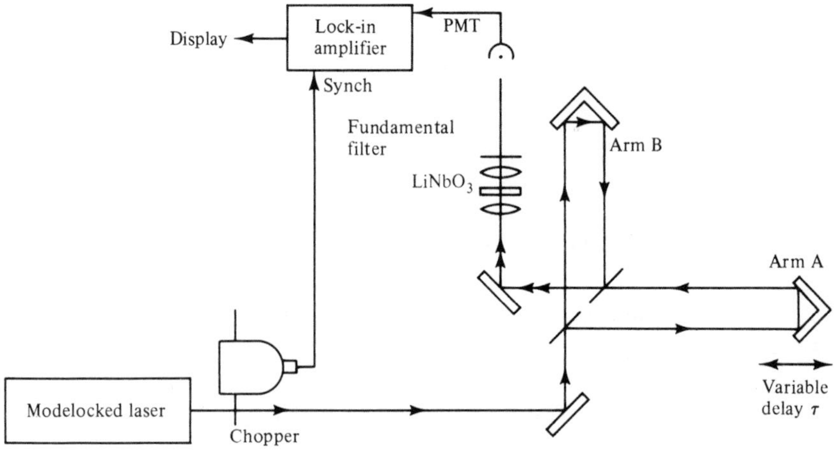

Figure 13.5 Experimental arrangement for pulse-width detection by SHG.

Here $\tilde{E}(t)$ gives the envelope of the electric field and the factor $\exp(-j\omega\tau)$ gives the phase delay of the mirror. The electric field at the output of the SHG crystal under phase-matched conditions is proportional to $[E_{x1} + E_{x2}]^2$ according to (13.59). The second harmonic power, detected by the photomultiplier, is proportional to the time average of the second harmonic field squared

$$P(2\omega) \propto \langle | E_{x1} + E_{x2} |^4 \rangle \tag{13.67}$$

Now, this is a function of the delay τ because of the phase factor $\exp(-j\omega\tau)$. The exponent changes by 2π for motion of the mirrors by one half-wavelength. If the time constant of the detection circuit is long enough, this rapid variation with translation of the mirror at normal speed can be averaged out and all terms that contain the factor $\exp[-(j\omega\tau)]$ raised to any power average out. We denote by brackets with a subscript τ this short-term average. Carrying out the products, we find for $P(2\omega)$,

$$\langle P(2\omega) \rangle_\tau \propto \langle\langle | E_{x1} + E_{x2} |^4 \rangle\rangle_\tau = \langle | \tilde{E}(t) |^4 \rangle + \langle | \tilde{E}(t-\tau) |^4 \rangle + 4\langle | \tilde{E}(t) |^2 | \tilde{E}(t-\tau) |^2 \rangle \tag{13.68}$$

The time average of $| \tilde{E}(t+\tau) |^4$ is independent of τ and thus

$$P(2\omega) \propto 2\langle | \tilde{E}(t) |^4 \rangle + 4\langle | \tilde{E}(t)\tilde{E}(t-\tau) |^2 \rangle \tag{13.69}$$

We can now explain how the pulse shape is traced out by translation of the mirror in the Michelson interferometer. When $\tau = 0$ we get for the output

$$P(2\omega) \propto 6 | \tilde{E}(\tau) |^4$$

When the pulse trains are shifted so that the individual pulses do not overlap,

$$| \tilde{E}(t)\tilde{E}(t-\tau) |^2 = 0$$

and the power is

$$P(2\omega) \propto 2\langle | \tilde{E}(t) |^4 \rangle$$

The peak-to-background ratio is 3 to 1. Because $| \tilde{E}(t) |^2$ is proportional to the fundamental intensity $I(t)$, we find that the second harmonic power detects the autocorrelation function of the intensity:

$$P(2\omega) \propto \langle I(t)^2 \rangle + 2\langle I(t)I(t-\tau) \rangle \tag{13.70}$$

More recently, SHG traces have been published that remove the background by the use of crossed beams E_1 and E_2 in the SHG crystal [14]. The angles can be so adjusted that phase matching is achieved with $\boldsymbol{k}_1(\omega) + \boldsymbol{k}_2(\omega) = \boldsymbol{k}(2\omega)$ and SHG power is produced only in the presence of both fields E_{x1} and E_{x2}, in the directions $\boldsymbol{k}_1(\omega) + \boldsymbol{k}_2(\omega)$. Figure 13.6 shows an SHG trace obtained in this way.

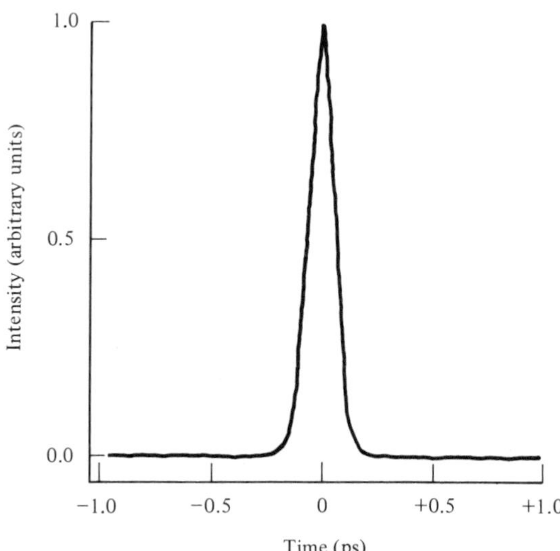

Figure 13.6 Second harmonic generation trace with no background. (From R. L. Fork, B. I. Greene, and C. V. Shank, "Generation of optical pulses shorter than 0.1 psec by colliding pulse modelocking," *Appl. Phys. Lett.*, *38*, 671–672, 1981.)

13.7 THE KERR EFFECT—THIRD-ORDER PROCESS

The so-called linear electro-optic effect is associated with an index change proportional to the applied electric field. It is described by the tensor $\chi^{(2)}_{ijk}(\omega_\alpha; \omega_\alpha, 0)$. The Kerr effect is associated with an index change proportional to the square of the applied electric field. Media with inversion symmetry can exhibit this effect, whereas they do not possess the linear electro-optic effect that depends on the lack of inversion symmetry. Liquids such as nitrobenzene and carbon disulfide, which consist of anisotropic molecules, possess particularly large Kerr coefficients, because the molecules tend to align under the influence of the electric field. The tensor responsible for the dc Kerr effect is $\chi^{(3)}_{ijkl}(\omega_\alpha; \omega_\alpha, 0, 0)$, where ω_α is the optical frequency.

The optical Kerr effect utilizes a change of index produced by an optical "drive" at frequency ω_β. The tensor describing this effect is $\chi^{(3)}_{ijkl}(\omega_\alpha; \omega_\alpha, \omega_\beta, -\omega_\beta)$. The nonlinear optical phenomena described in Chapter 10 were based on the optical Kerr effect. Indeed, the index change caused by I in (10.6) is proportional to the square of the optical field.

The optical Kerr effect in liquids can be due, in part, to alignment of molecules under the influence of $E_i(\omega_\beta)E_j^*(\omega_\beta)$, in part due to rearrangement of the electronic configuration of the molecules. The former effect is usually much larger than the latter in liquids of anisotropic molecules, but is also much slower. Its recovery time after excitation is limited by the time it takes for the molecules to reorient themselves, a time generally in the range of a few picoseconds.

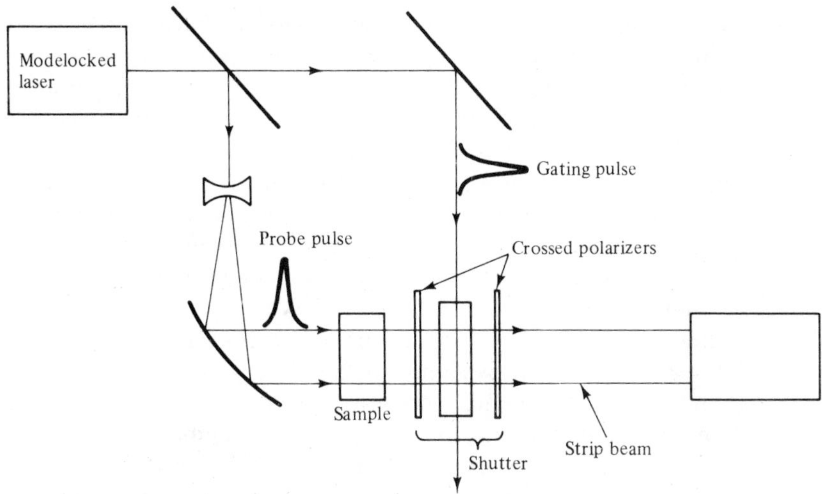

Figure 13.7 Kerr cell used for picosecond time-resolved spectroscopy.

A picosecond time-resolved spectroscopy experiment is illustrated in Fig. 13.7. A high-power, modelocked laser emits a picosecond (or subpicosecond) pulse which is separated by a beam splitter. One part of the pulse excites a sample whose fluorescence emission properties are to be studied. The other part of the pulse (with an adjustable time delay with respect to the first one) is used to "drive" the shutter. In the absence of the "gating" pulse, light from the sample cell is prevented from entering the spectrograph because of the pair of crossed polarizers. The picosecond "gating" pulse which passes through the Kerr cell creates a localized region of birefringence which acts as a slit moving along with the gating pulse through which light can be transmitted. The moving gating slit selects, at different delays with respect to the exciting pulse, the radiation entering the input slit of the spectrograph. The slit is in the plane of the gating pulse. The output of the spectrograph is two dimensional—one dimension displays the delay, the other dimension the frequency of the radiation. This arrangement gives a "one-shot" display of spectrum versus time.

13.8 SUMMARY

Optical nonlinearities were characterized by an expansion of the polarization vector in terms of products of the applied field. The expansion is the generalization of a Taylor series and "works" because optical nonlinearities are generally weak. Fields of magnitudes that are practical in signal-processing applications are small compared with the internal (atomic) fields and a nonlinearity of order n is of the order of the ratio of the applied field to the internal field raised to the nth power. A nonlinearity of nth order is described by a tensor of $(n + 1)$st rank. The linear electro-optic effect is a special case of a nonlinearity of second order.

A mass–spring model with a nonlinear restoring force was used as a model of a dispersive second-order nonlinearity. The Manley–Rowe relations put constraints on the $\bar{\bar{\chi}}$ tensors that underlie the operation of parametric oscillators and parametric up-converters. Parametric oscillators, as the name implies, are self-starting. They operate only as down-converters, the frequency pair ω_α and ω_β obeys the relation $\omega_\alpha + \omega_\beta = \omega_\gamma$, the frequency of the pump. Frequency up-conversion is not self-starting. It requires a source at frequency ω_α to produce an output at $\omega_\gamma + \omega_\alpha$, where ω_γ is the frequency of the pump. Parametric oscillators can be tuned by phase matching of the waves in the nonlinear crystal.

The different behaviors of the parametric down-converter (oscillator) and the parametric upconverter are explained by the Manley–Rowe relations. They imply two forms of photon number conservation in second-order processes: (1) generation or annihilation in pairs at frequencies ω_α and ω_β, and (2) generation of one photon at ω_α, absorption of one at ω_β, or vice versa. The latter process is akin to conventional energy conservation in the coupling of two modes of a single-frequency process where the sums of energies, $W_1 + W_2$, are conserved (time independent). The former process is akin to energy conservation of a single frequency process when one of the energies (say W_2) is negative; [16] then the sum of energies $W_1 + W_2$ reduces to $W_1 - |W_2|$. Such processes are unstable, they can grow exponentially with time and are described by the same formalism as the parametric down-converter.

Second harmonic generation is a special case of the second-order nonlinearity. For efficient second harmonic generation phase matching is required. Second harmonic generation is the most important method for the measurement of pulsewidths of modelocked pulses that are too short for direct detection. Higher-order nonlinearities are described by tensors of higher rank. The optical Kerr effect is a third-order process. We described a picosecond time-resolved spectroscopy setup as an example of the use of an optical nonlinearity for high-speed instrumentation.

PROBLEMS

13.1. In a parametric down-conversion one-pump photon is annihilated when one photon each are generated at ω_α and ω_β. Derive relations among

$$\chi^{(2)}_{ijk}(\omega_\gamma;\, \omega_\alpha,\, \omega_\beta)$$

$$\chi^{(2)}_{ijk}(\omega_\alpha;\, \omega_\gamma,\, -\omega_\beta)$$

and

$$\chi^{(2)}_{ijk}(\omega_\beta;\, \omega_\gamma,\, -\omega_\alpha)$$

analogous to (13.28).

13.2. Derive the coupling coefficient expression analogous to (13.38) for the parametric frequency up-conversion and prove (13.48).

13.3. Find the phase-matching direction for a parametric oscillator crystal of normal surface as shown in Fig. P13.3 for

$$\omega_\alpha = \tfrac{1}{3}\omega_\gamma, \qquad \omega_\beta = \tfrac{2}{3}\omega_\gamma$$

Show a graphical construction that will find the direction for $\boldsymbol{k}$ for which

$$\boldsymbol{k}_\alpha + \boldsymbol{k}_\beta = \boldsymbol{k}_\gamma$$

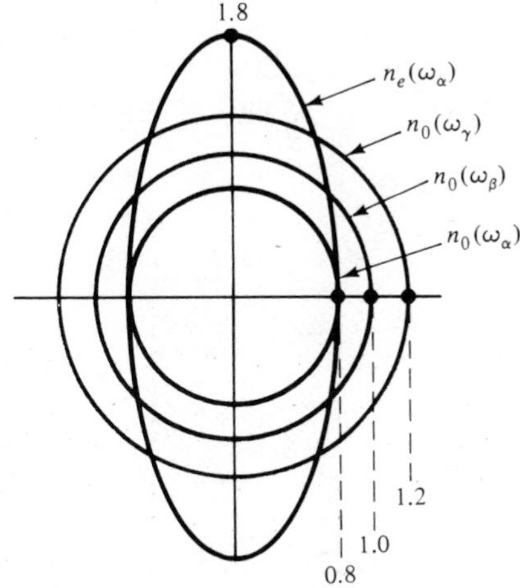

Figure P13.3

13.4. Consider the parametric amplifier equations for detuning of the resonator from the condition $\omega_\gamma = \omega_\alpha + \omega_\beta$. Prove that they are

$$\frac{dA_\alpha}{dt} = -\frac{A_\alpha}{\tau_\alpha} + \kappa_{\alpha\beta} A_\gamma A_\beta^* \exp\left[+j(\omega_\gamma - \omega_\beta - \omega_\alpha)t\right]$$

$$\frac{dA_\beta^*}{dt} = -\frac{A_\beta^*}{\tau_\beta} + \kappa_{\beta\alpha}^* A_\gamma^* A_\alpha \exp\left[-j(\omega_\gamma - \omega_\beta - \omega_\alpha)t\right]$$

Introduce the new variables

$$A_\alpha \equiv \tilde{A}_\alpha \exp\left[\frac{+j(\omega_\gamma - \omega_\beta - \omega_\alpha)t}{2}\right]$$

and

$$A_\beta^* = \tilde{A}_\beta^* \exp\left[\frac{-j(\omega_\gamma - \omega_\beta - \omega_\alpha)t}{2}\right]$$

Write equations for $\tilde{A}_\alpha$ and $\tilde{A}_\beta$. Now solve for the growth constant Re $[-s]$ as a function of detuning for an assumed exp $(-st)$ dependence. Assume that $1/\tau_\alpha = 1/\tau_\beta$. Plot Re $[-s\tau_\alpha]$ versus $(\omega_\gamma - \omega_\alpha - \omega_\beta)\tau_\alpha$.

13.5. If a crystal were dispersion-free, the SHG coefficient d_{ijk} and the electro-optic coefficient r_{ijk} would be related uniquely [(13.14) and (13.50)]. For example:

$$\varepsilon_0 n_0^2 n_e^2 \Delta\kappa_{13} = n_0^2 n_e^2 r_{13k} E_k = -\chi_{13k}^{(2)} E_k = -2\left(\frac{d_{13k}}{\varepsilon_0}\right) E_k$$

For $LiNbO_3$, which has the parameters

$$r_{22} = 3.4 \times 10^{-12} \text{ m/V}$$

$$d_{22} = 2.3 \pm 1 \times \tfrac{1}{9} \times 10^{-22} \text{ in mks units}$$

$$n_0 = 2.29, \qquad n_e = 2.20$$

find how well the equality holds. The numbers are taken from Tables 8.1 and 9.2 of Yariv [13]. He does not mention the frequency for Table 8.1. Table 9.2 is for 5500 Å.

13.6. A slab of material n is reflection coated to give reflectivities r_1 and r_2 at the interfaces (Fig. P13.6). Assume beams of uniform intensities over a cross-sectional area $\mathscr{A}$.

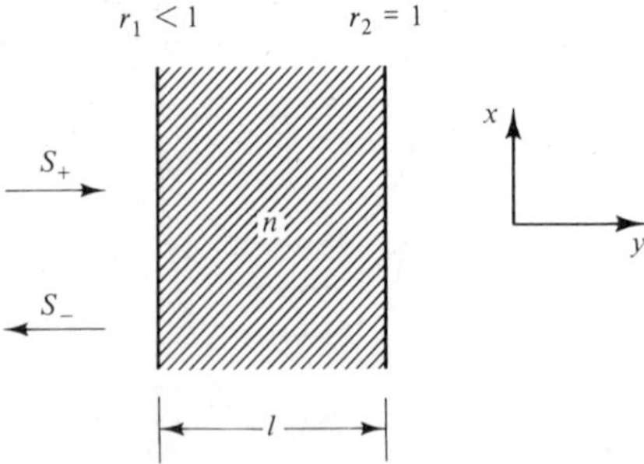

Figure P13.6

(a) Find an expression for the parameter $1/\tau_e$ of this resonator at the resonance frequency $\omega_m = m\pi c/nl$.

(b) Now suppose a traveling-wave pump field is produced inside the resonator of magnitude $\mathbf{E}_\gamma = \hat{y}E_\gamma \cos(\omega_\gamma t - k_\gamma z)$ and the medium has a nonlinear susceptibility $\chi_{131}(\omega_\alpha;\ \omega_\gamma,\ -\omega_\beta)$. Assume that the medium is phase matched. Write the coupling of modes equations for ω_α and ω_β, both assumed resonant in the Fabry-Perot. If $s_+(\omega_\alpha)$ is incident upon the resonator

$$s_+(\omega_\alpha) = S_+ e^{j\omega_\alpha t}$$

what is $s_-(\omega_\alpha)$ reflected from the resonator in the steady state (no growth or decay)? Normalize all quantities to unit area (in the x-y plane) ($|s_+|^2$ = power per unit area, $|a|^2$ is energy per unit area). Use the phenomenological coupling parameters $\kappa_{\alpha\beta}$ and $\kappa_{\beta\alpha}$.

(c) Evaluate $\kappa_{\alpha\beta}$ in terms of χ_{131} assuming the field at ω_α and ω_β to be polarized along the x direction.

(d) Estimate the threshold intensity for ω_α near ω_β, $\omega_\alpha \simeq \omega_\beta$. Phase-matched second harmonic generation along the principal axis is possible in $LiNbO_3$ for a pump wavelength of 0.5 μm via temperature "tuning." The value of $\chi^{(2)}_{131}(\omega_\alpha;\ \omega_\gamma,\ -\omega_\beta)$, which is equal to $\chi_{311}(\omega_\gamma;\ \omega_\alpha,\ \omega_\beta)$ according to Problem 13.1, can be estimated from the relation $\chi^{(2)}_{311}(2\omega;\ \omega,\ \omega) = 2d_{31} = 2d_{311} = 2 \times 6.25 \times 10^{-12}$ m/V according to Ref. 17. Note that the d_{31} given in this reference is the d_{31}/ε_0 of Yariv. Use $r_1^2 = 0.99$, $l = 2$ cm.

13.7. Derive the acousto-optic diffraction equation by the method used for the derivation of (13.58) via power conservation.

13.8. In Section 13.6 we ignored depletion of the power in the fundamental. In this problem we want to take it into account.

(a) Set up a pair of coupled equations for the second harmonic and fundamental when they are phase matched along the principal axes of $LiNbO_3$ propagating along y and polarized along z and x, respectively. Show that they must be of the form

$$\frac{d\tilde{E}_3(2\omega)}{dy} = -jK_{21}\tilde{E}_1^2(\omega)$$

$$\frac{d\tilde{E}_1(\omega)}{dy} = -jK_{12}\tilde{E}_3(2\omega)\tilde{E}_1^*(\omega)$$

From power conservation establish a relation between K_{21} and K_{12}.

(b) Show that phases are maintained so that the equations above can be rewritten

$$\frac{dA_1}{dy} = -\kappa A_1^2, \qquad \frac{dA_2}{dy} = \kappa A_2 A_1$$

where κ, A_1, and A_2 are real:

$$A_2 \propto \sqrt{n_0}\,\tilde{E}_3(2\omega), \qquad A_1 \propto \sqrt{n_e}\,\tilde{E}_1(\omega)$$

except for phase factors of magnitude unity.

(c) Solve the equations above and find the required intensity in $\tilde{E}_1(\omega)$ to achieve 90% power transfer within a distance of 2 cm. Use the value for d_{31} quoted in Problem 13.6, $\lambda = 1$ μm for the fundamental. Ignore beam spreading and nonuniform intensity profiles.

13.9. The rotating index ellipsoid of Problem 12.2 in a circularly polarized electric field at frequency ω_0 can be used to generate a single sideband at frequency $\omega_p \pm \omega_0$ depending on the sense of the circular polarization. To prove this statement, evaluate the polarization perturbation

$$\Delta P_i(t) = \varepsilon_0 \Delta\chi_{ij}(t) \cdot E_j(t)$$

where $\Delta\chi_{ij}(t)$ is the time-dependent polarizability tensor obtained from Problem 12.2.

REFERENCES

[1] Y. R. Shen, "Recent advances in nonlinear optics," *Rev. Mod. Phys.*, *48*, no. 1, 1–32, Jan. 1976.

[2] D. C. Hanna, M. A. Yuratich, and D. Cotter, *Nonlinear Optics of Free Atoms and Molecules*, Springer-Verlag, New York, 1979.

[3] P. G. Harper and B. S. Wherrett, eds., *Nonlinear Optics*, Proc. of the Sixteenth Scottish Universities Summer School in Physics, 1975, Academic Press, New York, 1977.

[4] N. Bloembergen, *Nonlinear Optics*, W. A. Benjamin, Menlo Park, Calif., 1965.

[5] P. N. Butcher, *Nonlinear Optical Phenomena*, Bull. 200, Ohio State University, Engineering Experimental Station, 1965.

[6] J. M. Manley and H. E. Rowe, "Some general properties of nonlinear elements—Part I. General energy relations," *Proc. IRE*, *44*, 904–914, July 1956.

[7] H. A. Haus, "Power flow relations in lossless nonlinear media," *IRE Trans. MTT*, *6*, 317–324, July 1958.

[8] P. Penfield, Jr., *Frequency-Power Formulas*, Technology Press, MIT, Cambridge, Mass./Wiley, New York, 1960.

[9] M. T. Weiss, "Quantum derivation of energy relations analogous to those for nonlinear reactances," *Proc. IRE*, *45*, 1012–1013, July 1957.

[10] W. H. Louisell, *Coupled Mode and Parametric Electronics*, Wiley, New York, 1960.

[11] D. Marcuse, *Engineering Quantum Electrodynamics*, Harcourt Brace and World, New York, 1970.

[12] G. D. Boyd and D. A. Kleinman, "Parametric interaction of focused Gaussian light beams," *J. Appl. Phys.*, *39*, no. 8, 3597–3639, July 1968.

[13] A. Yariv, *Introduction to Optical Electronics*, Holt, Rinehart and Winston, New York, 1976.

[14] E. P. Ippen and C. V. Shank, in *Ultrashort Light Pulses, Topics in Applied Physics*, Vol. 18, S. L. Shapiro, ed., Springer-Verlag, New York, 1977.

[15] R. L. Fork, B. I. Greene, and C. V. Shank, "Generation of optical pulses shorter than 0.1 psec by colliding pulse mode locking," *Appl. Phys. Lett.*, *38*, 671–672, 1981.

[16] P. A. Sturrock, "Kinematics of growing waves," *Phys. Rev. 112*, no. 5, 1488–1503, Dec. 1958.

[17] R. L. Byer, in *Nonlinear Optics*, P. G. Harper and B. S. Wherret, eds., Academic Press, New York, 1972, p. 148.

14

OPTICAL DETECTION

The observation and measurement of optical signals is often reduced to the problem of detection of a signal in a noisy environment. We have studied some functions with statistical properties when we introduced the concept of temporal and spatial coherence. In the brief discussion of detection of optical signals we shall concentrate on the photon detector, a detector that generates electric currents directly by photoexcitation, and study the statistical properties of the output of the photon detector.

The carrier generation process is statistical, and hence "noisy." We study the noise characteristics and derive the *shot-noise formula.* Detectors are characterized in terms of their noise equivalent power (NEP), which describes the "noisiness" of the detector. We look at two limiting cases, one in which the NEP is caused by the noise generated by the signal itself via the shot noise of the current produced by the signal, the other in which the NEP is determined by the thermal background. In the former case we study a specific detection system—heterodyne detection and determine its sensitivity. In the latter case we determine the ultimate sensitivity achievable in a thermal environment.

We do not study in detail the physics of any particular detector and refer the reader to the literature for further details [1,2].

14.1 QUANTUM EFFICIENCY

Photodetectors produce an electric current proportional to the incident photon flux via generation of carriers in a semiconductor, or electron emission from the cathode in a vacuum diode. These detectors are characterized by a quantum efficiency η which expresses the ratio of the charged-particle current collected at the terminals to the rate of incident photons. The current due to an optical power P is

$$i = \eta e \frac{P}{\hbar\omega} \tag{14.1}$$

where $P\hbar/\omega$ is equal to the rate of photons incident on the detector, ω is the radian frequency of the radiation and e is the electron charge. If $\mathbf{E}$ is the complex amplitude of the electric field of a plane wave and its propagation constant is $\boldsymbol{k} = k\boldsymbol{n}$, so that

$$\boldsymbol{E} = \tfrac{1}{2}(\mathbf{E}e^{j\omega t - j\boldsymbol{k}\cdot\boldsymbol{r}} + \text{c.c.})$$

then

$$P = \int_{\substack{\text{receiving area}\\ \text{of detector}}} \boldsymbol{n}\cdot d\boldsymbol{a}\,\frac{1}{2}\sqrt{\frac{\varepsilon_0}{\mu_0}}\,\mathbf{E}\cdot\mathbf{E}^* \tag{14.2}$$

Note that the definition of P, (14.2), gives the optical power averaged over one cycle of the optical radiation; the double-frequency component contained in the instantaneous Poynting vector is omitted. Of course, the output current of the photodetector could not contain frequency components of the second harmonic of the incident optical radiation, even if they were excited within the photodetector, because of the limited bandwidth of the electrical circuit. However, more fundamentally, such second harmonic components are not produced within the photodetector as shown by a careful quantum analysis of photodetection [3]. Thus the use of the time-averaged power in (14.2) is an appropriate one.

The quantum efficiency is less than unity. If each "primary" carrier produces "secondary" carriers, as by secondary emission in a photomultiplier, or by avalanche multiplication in a semiconductor diode, the process is taken into account by a gain factor G:

$$i = G\eta e \frac{P}{\hbar\omega} \tag{14.3}$$

The noise associated with the gain mechanism and that associated with the primary carrier generation call for separate analyses. We shall consider only the latter in detail.

14.2 SHOT NOISE

When an electron of charge $-e$ is emitted from the photocathode of a vacuum photodiode at the time instant t_i, and travels to the anode, a current $i(t)$ flows in the external leads connecting the cathode and anode (Fig. 14.1). The con-

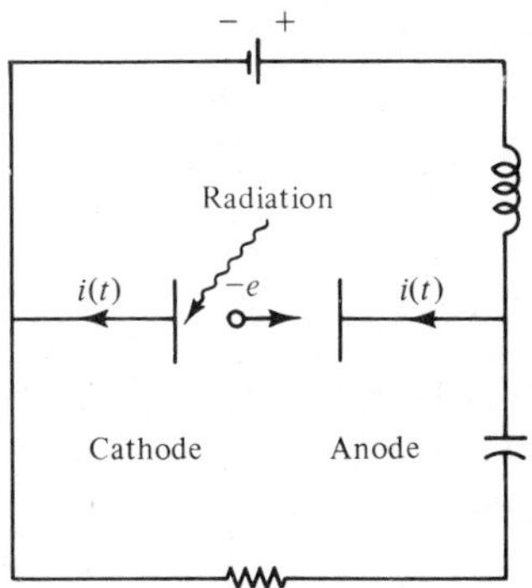

Figure 14.1 Vacuum photodiode with biasing battery and load resistor.

tinuous current during travel of the photo-emitted electron may be understood as the continuous rate of change of the image charge on the two electrodes that has to be supplied by $i(t)$. The electron emerging from the cathode leaves a net charge imbalance $+e$ behind. This image charge imbalance in the cathode decreases as the electron travels away from the cathode, and is accompanied by a corresponding increase of the image charge in the anode. When the electron arrives at the anode it encounters an equal and opposite image charge with which it is neutralized upon impact. During the time of traversal, a current flows in the leads to the photodiode from cathode to anode that integrates over time to $+e$, so that

$$i(t) = eh(t - t_i) \tag{14.4}$$

with

$$\int_{-\infty}^{\infty} h(t - t_i)\, dt = 1 \tag{14.5}$$

The process is not essentially different in the generation by photons of hole electron pairs in a photoconductor placed between a pair of electrodes. The hole travels to the negative electrode, the electron to the positive electrode. The net effect is the same as if only an electron had traveled from cathode to anode; the shape of the $h(t)$ function must be corrected of course for the, in general, different mobilities and thus velocities of holes and electrons.

The net current $i(t)$ produced by the flow of charges is

$$i(t) = e \sum_i h(t - t_i) \tag{14.6}$$

where t_i is a random variable.

The spectrum of the current can be obtained by the approach outlined in Section 1.5. One treats the process as if it were periodic of period T, and derives the associated discrete spectrum. The spectrum is squared, the limit $T \to \infty$ is taken, and the spectral density is defined by (1.81).

The discrete Fourier transform $I(n)$ of (14.6) is

$$I(n) = e\,\frac{1}{T}\int_{-T/2}^{T/2} e^{-j\omega t}h(t - t_i)\,dt$$
$$= e\sum_i e^{-j\omega t_i}\,\frac{1}{T}\int_{-T/2-t_i}^{T/2-t_i} e^{-j\omega t'}h(t')\,dt' \qquad (14.7)$$

If T is very large, we do not have to worry about the few current pulses near the end of the interval. The integrals can be replaced by the integral Fourier transform of $h(t)$, $H(\omega)$, where

$$H(\omega) \equiv \frac{1}{2\pi}\int_{-\infty}^{\infty} h(t)e^{-j\omega t}\,dt \qquad (14.8)$$

Thus the Fourier component $I(n)$ of the current can be written

$$I(n) = e\,\frac{2\pi}{T}\sum_i e^{-j\omega t_i}H(\omega) \qquad (14.9)$$

The spectrum Φ is given by (1.81).

$$\Phi = \lim_{T\to\infty}\frac{T}{2\pi}\,\overline{|\mathrm{I(n)}|^2} = e^2\lim_{T\to\infty}\left\{\frac{2\pi}{T}\,|H(\omega)|^2\,\overline{\sum_{i,k} e^{-j\omega(t_i-t_k)}}\right\} \qquad (14.10)$$

If the photodector is illuminated by light of constant intensity, the only case we shall be concerned with in this chapter, the t_i's are random variables that are uncorrelated with each other. The statistical average gives zero for $t_i \neq t_k$ and $\omega \neq 0$ so that the double sum is replaced by the single sum for $i = k$. For $\omega = 0$, the dc component of the spectrum, the double sum must be carried out.

Consider first the spectrum for $\omega \neq 0$. If the rate of flow of the photoelectrons is I_0/e, where I_0 is the dc current produced by the illumination, then the sum becomes

$$\overline{\sum_{i,k} e^{-j\omega(t_i-t_k)}} = \overline{\sum_{i=k} e^{-j\omega(t_i-t_k)}} = T\,\frac{I_0}{e} \qquad (14.11)$$

and the spectrum at $\omega \neq 0$ is, combining (14.10) and (14.11)

$$\Phi(\omega) = e^2\lim_{T\to\infty}\,2\pi\,|H(\omega)|^2\,\frac{I_0}{e} = 2\pi\,e\,I_0\,|H(\omega)|^2 \qquad (14.12)$$

For frequencies low enough, so that $h(t)$ can be treated as a unit impulse function, the Fourier transform $H(\omega)$ is

$$H(\omega) = \frac{1}{2\pi}\int e^{-j\omega t}h(t)\,dt \simeq \frac{1}{2\pi} \qquad (14.13)$$

and the spectrum Φ is essentially flat and has the value

$$\Phi(\omega) = \frac{e}{2\pi} I_0 \tag{14.14}$$

In the literature on noise it is customary to introduce the concept of mean square current fluctuations in the frequency increment Δf. One imagines an observation of the mean square current fluctuations through a narrow band filter centered at ω_0 of width $\Delta\omega = 2\pi\Delta f$. The mean square noise current fluctuations passing the filter, denoted by $\overline{|i_n|^2}$, are, according to (14.14)

$$\overline{|i_n|^2} = 2eI_0\,\Delta f \tag{14.15}$$

where we have combined the contributions of the spectral density of both positive and negative frequencies, because experimentally, one deals with positive frequencies only.

Before proceeding it is worth-while to check for the spectrum at the origin of ω, $\omega = 0$. There the double sum in (14.10) is equal to the square of the number of photo-electrons, $(TI_0/e)^2$. Using the value of $H(\omega = 0) = 1/2\pi$ one has

$$\Phi(\omega = 0) = \lim_{T\to\infty} I_0^2 \frac{T}{2\pi} \tag{14.16}$$

an infinite contribution. However, one must remember that (14.16) represents a discrete spectrum, with the frequency spacing of the components equal to $2\pi/T$. Thus (14.16) can be considered to be an impulse function of area I_0^2, the square of the current.

If $H(\omega)$ cannot be approximated by an impulse, the spectrum is not flat but falls off beyond a characteristic frequency. Consider as an example the idealized current waveform for a hole-electron pair in a photoconductor for equal molobilities of holes and electrons. Then $h(t)$ is rectangular of width τ_0, where τ_0 is the transit time through the photoconductor. The spectrum $H(\omega)$ is

$$H(\omega) = \frac{1}{2\pi}\int_0^{\tau_0} e^{-j\omega t}\frac{1}{\tau_0}\,dt = \frac{1}{2\pi}\frac{\sin(\omega\tau_0/2)}{\omega\tau_0/2}e^{-j(\omega\tau_0/2)} \tag{14.17}$$

When (14.17) is introduced in (14.12) one obtains the shot noise spectrum for this idealized case. The shot noise spectrum is shown in Fig. 14.2. It is of interest to define an effective bandwidth of the noise spectrum that is defined by the width $4\pi B$ of the rectangle of the same peak height and of equal area. The area of the function $\sin^2(x/2)/(x/2)^2$ is 2π so that

$$\int |H(\omega)|^2\,d\omega = \frac{1}{\tau_0}\frac{1}{2\pi} = 4\pi B\left(\frac{1}{2\pi}\right)^2 \tag{14.18}$$

or

$$B = \frac{1}{2\tau_0} \tag{14.19}$$

We use this expression below and in Section 14.6.

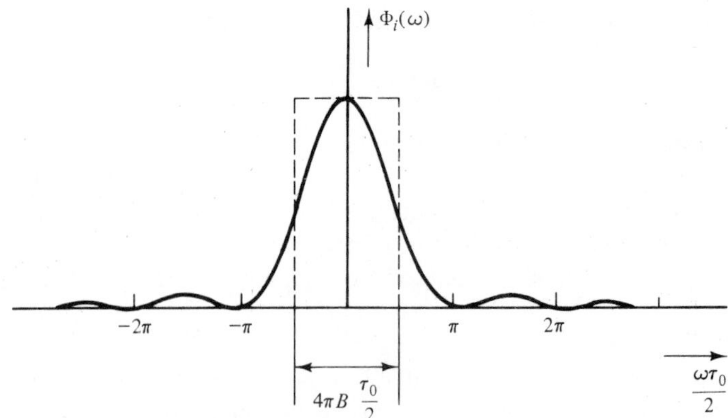

Figure 14.2 Spectrum of shot noise for rectangular $h(t)$.

To develop some insight into the physical meaning of the shot-noise formula, consider an experiment in which the current fluctuations are monitored via the voltage developed across a resistor R which is filtered by a filter centered at the frequency ω_0 of bandwidth $B \ll \omega_0/2\pi$, and then displayed on a scope. If the scope is in the single-sweep mode, each time a sample is displayed it looks like a sinusoid of frequency ω_0, provided that the time interval of the sweep $T_s \ll 1/B$. Within this time interval the phase ϕ of the sinusoid, and hence its frequency $\omega_0 + d\phi/dt$ cannot change. Successive sweeps display sinusoids of different amplitude and different phase. The mean-square voltage amplitude of the display is

$$\overline{|v|^2} = 2eI_0 R^2 B \tag{14.20}$$

One may derive the shot-noise formula (14.15) via a faster, more intuitive route, that may serve as a mnemonic to the reader. Suppose that we observe the mean-square current fluctuations $\overline{i_n^2}$ in the leads to the detector through a filter of bandwidth B. According to (14.19) we require an observation time $\tau_0 = 1/(2B)$ for each of our observational samples. Within this time τ_0, $\bar{N}$ charges pass on the average through the detector, where

$$\bar{N} = \frac{\tau_0 I_0}{e} = \frac{I_0}{2Be} \tag{14.21}$$

For a random arrival of particles, N is a fluctuating quantity and the mean-square deviation is given by [4] (see Appendix 14A)

$$\overline{N^2} - \bar{N}^2 = \bar{N} \tag{14.22}$$

The mean-square current fluctuations are

$$\overline{i^2} - \bar{i}^2 = \overline{\left(\frac{eN}{\tau_0}\right)^2} - \left(\frac{e\bar{N}}{\tau_0}\right)^2 = \frac{e^2\bar{N}}{\tau_0^2} = 2eI_0 B$$

as derived above in (14.15) for $B = \Delta f$. This derivation arrives at the shot-noise formula in a succinct manner. It gives the spectrum around $\omega = 0$. It does not give directly the spectrum at $\omega = \omega_0 \neq 0$, even though it happens to be correct for this case as well as in the ideal case of an impulse-like current response, because the spectrum is flat.

14.3 NOISE EQUIVALENT POWER

The signal-to-noise ratio S/N of detection is defined as the ratio of the mean-square signal current, $\overline{|i_s|^2}$ to the mean square noise current $\overline{|i_n|^2}$:

$$\frac{\mathrm{S}}{\mathrm{N}} = \frac{\overline{|i_s|^2}}{\overline{|i_n|^2}} \tag{14.23}$$

If S/N < 1, the signal gets "buried" in the noise: with S/N $\gg$ 1 the signal is easily detectable. Denote the power of the incident optical radiation by P_s. The mean-square signal current delivered by the detector is from (14.1)

$$\overline{|i_s|^2} = \left(\frac{e\eta}{\hbar\omega}\right)^2 \overline{P_s^2} \tag{14.24}$$

where ω is the frequency of the incident photons. Even if the optical signal is modulated and thus possesses a finite bandwidth, the bandwidth is very narrow compared with the optical frequency, so that one value of ω can be used to describe the incident photons.

The mean-square noise current for a bandwidth B of the detector is given by (14.15) with $\Delta f = B$. The signal-to-noise ratio is thus

$$\frac{\mathrm{S}}{\mathrm{N}} = \left(\frac{e\eta}{\hbar\omega}\right)^2 \frac{\overline{P_s^2}}{2eI_0 B} \tag{14.25}$$

The dc current I_0 is produced by the signal power and by the background radiation. Denote the average rate of photons of the background radiation impingent on the detector by $\bar{r}_b$. Then

$$I_0 = \frac{e\eta}{\hbar\omega} \bar{P}_s + e\eta\bar{r}_b \tag{14.26}$$

The signal-to-noise ratio is thus

$$\frac{\mathrm{S}}{\mathrm{N}} = \frac{(\eta/\hbar\omega)^2 \overline{P_s^2}}{2[(\eta/\hbar\omega)\bar{P}_s + \eta\bar{r}_b]B} \tag{14.27}$$

If the electron current due to the background photon flux is large compared to that due to the signal photon flux, the important case for detection of small signals in a thermal-radiation background, the signal-to-noise ratio is

$$(S/N)_{\mathrm{BL}} = \frac{\eta}{(\hbar\omega)^2} \frac{\overline{P_s^2}}{2\bar{r}_b B} \tag{14.28}$$

where the subscript BL stands for background limited. The noise equivalent power (NEP) is defined as that steady state (time-independent) signal power for which the signal-to-noise ratio is unity

$$(\text{NEP})_{\text{BL}} = \hbar\omega \sqrt{\frac{2\bar{r}_b}{\eta} B} \tag{14.29}$$

If the area A of the detector is increased, $\bar{r}_b$ increases proportionately. Also, if B increases, the NEP increases as $\sqrt{B}$. It is customary to define a quantity such that it is independent of the detector area and bandwidth. This quantity is thus characteristic of the detector type and less dependent on its geometry. This quantity is D^*:

$$D^* \equiv \frac{\sqrt{AB}}{(\text{NEP})_{\text{BL}}} = \frac{1}{\hbar\omega}\sqrt{\frac{\eta A}{2\bar{r}_b}} \frac{\text{cm}\cdot\text{Hz}^{1/2}}{\text{watt}} \tag{14.30}$$

In Section 14.5 we shall obtain an estimate for the magnitude of the obtainable D^*. For this purpose, we must evaluate the rate of carrier production due to the thermal background radiation.

14.4 PLANCK'S FORMULA FOR BLACKBODY RADIATION AND THE NYQUIST FORMULA

In order to determine the background radiation, we start by evaluating the energy density within the frequency interval ω, $\omega + d\omega$ inside a thermal enclosure at temperature T. The energy of each mode is quantized so that the energy of a mode of frequency ω is [5] $n\hbar\omega$, where n is the quantum number. The probability of excitation of one electromagnetic mode in the enclosure to an energy $n\hbar\omega$ is proportional to the Boltzmann factor [6] $\exp[-(n\hbar\omega/kT)]$. The probability $p(n)$ of exciting a mode to level n must be normalized so that $\sum_n p(n) = 1$. Therefore, the probability $p(n)$, properly normalized, is

$$p(n) = \frac{e^{-n\hbar\omega/kT}}{\sum_{n=0}^{\infty} e^{-n\hbar\omega/kT}} = (1 - e^{-\hbar\omega/kT})e^{-n\hbar\omega/kT} \tag{14.31}$$

The average energy is

$$\begin{aligned}\hbar\omega \sum_{n=0}^{\infty} p(n)n &= \hbar\omega(1 - e^{-\hbar\omega/kT}) \sum_{n=0} ne^{-n\hbar\omega/kT} = \frac{e^{-\hbar\omega/kT}}{1 - e^{-\hbar\omega/kT}}\hbar\omega \\ &= \frac{\hbar\omega}{e^{\hbar\omega/kT} - 1}\end{aligned} \tag{14.32}$$

In evaluating the sum over all n we have taken advantage of the following identities:

$$\frac{d}{dx}\sum_{n=0}^{\infty} e^{-nx} = -\sum_{n=0}^{\infty} ne^{-nx} = \frac{d}{dx}\left(\frac{1}{1-e^{-x}}\right) = -\frac{e^{-x}}{(1-e^{-x})^2} \tag{14.33}$$

The enclosure supports electromagnetic modes whose number is found as follows. Under the assumption that the perfectly conducting enclosure is a cube of side length L, the frequency ω_{mnp} of the mode m, n, p is given by

$$\left(\frac{\pi}{L}\right)^2 (m^2 + n^2 + p^2) = \frac{\omega_{mnp}^2}{c^2} \tag{14.34}$$

The quantity $(\pi/L)\sqrt{m^2 + n^2 + p^2}$ is equal to the length of the radius vector from the origin in a space in which each mode is represented by a point of coordinates π/L times m, n, p. The points are corners of cubic cells of volume $(\pi/L)^3$ (see Fig. 14.3). The number of modes contained in a volume V in k-space, in units of m^{-3}, is equal to VL^3/π^3. In the limit $L \to \infty$, the cells become differential and the summation over cell volumes can be replaced by an integral. The number of modes within the frequency range 0, ω is the volume of one-eighth of the sphere of radius ω/c divided by $(\pi/L)^3$. The number of modes in the frequency range ω, $\omega + \Delta\omega$, is given by one-eighth of the volume of a spherical shell of radius ω/c and thickness $\Delta\omega/c$, divided by $(\pi/L)^3$, which is $\omega^2\, \Delta\omega\, L^3/2c^3\pi^2$. The energy assigned to this number of modes per unit volume is the product of the number of modes per unit volume,

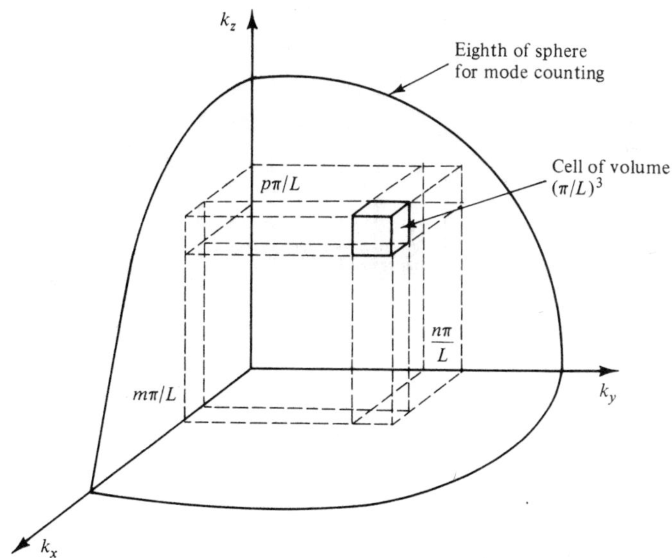

Figure 14.3 Construction for mode counting.

$\omega^2 \, \Delta\omega / 2c^3\pi^2$, and the average energy (14.32) per mode:

$$\frac{\omega^2 \, \Delta\omega}{2c^3\pi^2} \frac{\hbar\omega}{e^{\hbar\omega/kT} - 1}$$

Because each mode exists with two polarizations, the energy above has to be multiplied by a factor of 2. We denote the energy density by $\rho(\omega) \, d\omega$:

$$\rho(\omega) \, d\omega = \frac{\omega^2}{c^3\pi^2} \, d\omega \, \frac{\hbar\omega}{e^{\hbar\omega/kT} - 1} \tag{14.35}$$

This is the formula derived by Planck for the radiation density. The size of the box has disappeared, a gratifying result. This energy density is due to waves traveling in all different directions. The thermal electromagnetic power crossing an area A within a solid angle $d\Omega$ is

$$\frac{d\Omega}{4\pi} \, Ac\rho(\omega) \, d\omega = d\Omega \, A \, \frac{\omega^2 \, d\omega}{4\pi^3 c^2} \frac{\hbar\omega}{e^{\hbar\omega/kT} - 1} \tag{14.36}$$

We apply this formula in the next section to derive the expression for D^* of an ideal detector. Before we do this we show that an argument parallel to the one just presented arrives at the formula for thermal noise emitted by a resistive termination of transmission line, the quantum generalization of the Nyquist formula for thermal noise [7]. Instead of the three-dimensional box

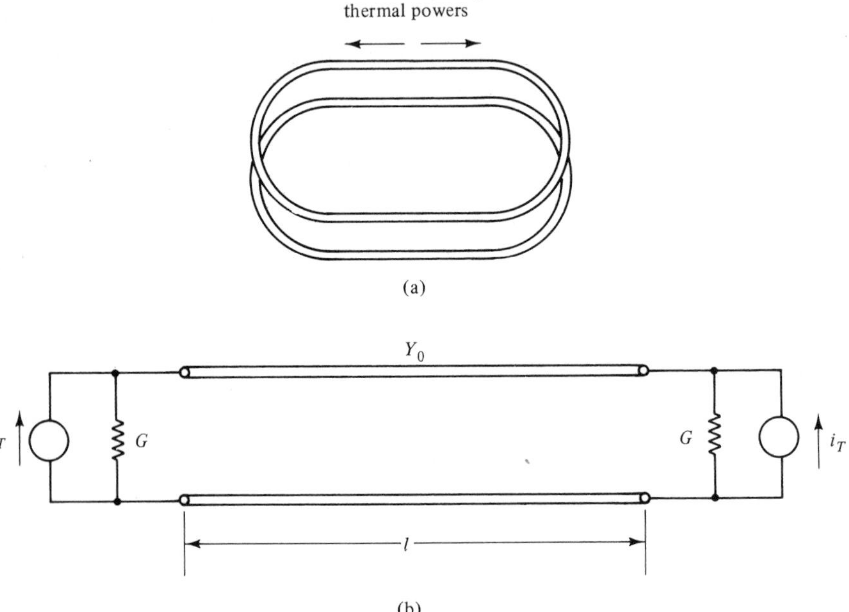

Figure 14.4 Schematic of transmission line of length l, (a) as ring resonator (b) opened and terminated.

with the natural frequencies ω_{mnp}, consider a one-dimensional transmission line of length l closed in a ring (Fig. 14.4). The natural frequency of the mth mode of frequency ω_m traveling in one direction is given by

$$\frac{\omega_m l}{c} = 2m\pi \tag{14.37}$$

The number of modes in frequency increment $\Delta\omega$ is

$$\Delta m = \frac{l}{2\pi c} \Delta\omega \tag{14.38}$$

The energy within this frequency interval is

$$\frac{l}{c} \frac{\hbar\omega}{e^{\hbar\omega/kT} - 1} \frac{\Delta\omega}{2\pi}$$

and the power ΔP in this frequency interval traveling in one direction is c times the energy per unit length:

$$\Delta P = \frac{\hbar\omega}{e^{\hbar\omega/kT} - 1} \frac{\Delta\omega}{2\pi} \tag{14.39}$$

This is the quantum generalization of the Nyquist formula, which is obtained from the equation above in the limit when $kT \gg \hbar\omega$. Then

$$\Delta P = kTB \tag{14.40}$$

with $B \equiv \Delta\omega/2\pi$, the bandwidth in hertz. In the classical limit, $kT \gg \hbar\omega$, the Nyquist formula can be adapted to evaluate the mean-square current fluctuations from a short-circuited conductor of conductance G and temperature T. Open up the ring resonator and terminate it at the two ends in conductances $G = Y_0$, where Y_0 is the characteristic admittance of the transmission line (Fig. 14.4). The waves traveling in opposite directions are absorbed by the two matched loads. If the loads are at the same temperature T as the transmission line, and thermal equilibrium is to be maintained, the loads have to deliver the same amount of power as they are receiving (i.e., kTB within a bandwidth B). A conductor alone is a passive element and hence incapable of delivering power. The equivalent circuit of a linear conductor at thermal equilibrium with a transmission line at temperature T must contain an internal noise source; its v–i characteristic must be generalized

$$i = Gv + i_T \tag{14.41}$$

where i_T is the internal current source (see Fig. 14.4). The power delivered by the conductance to the transmission line is

$$\overline{\left|\frac{i_T}{G + Y_0}\right|^2} Y_0 = \frac{\overline{|i_T|^2}}{4G} = kTB \tag{14.42}$$

for $G = Y_0$. Thus

$$\overline{|i_T|^2} = 4GkTB \tag{14.43}$$

14.5 EVALUATION OF D^* FOR IDEAL DETECTOR

The analysis in the preceding section can be used to derive the D^* parameter of an ideal detector. Consider a cooled photodetector of area A. The detector collects thermal radiation from the cone-like solid angle $2\pi \sin\theta\, d\theta$ passing the area $A \cos\theta$ (see Fig. 14.5). The net thermal power dP_b collected by the detector within the frequency interval $d\omega$ follows from (14.36):

$$dP_b = \frac{A}{4\pi^3 c^2} \frac{\hbar\omega^3\, d\omega}{e^{\hbar\omega/kT} - 1} \int_0^{\theta_0} 2\pi \sin\theta \cos\theta\, d\theta \tag{14.44}$$

The detector detects radiation from some lower cutoff frequency ω_c set by the work function of the photosurface, in the case of a vacuum photodiode, or the energy gap, in the case of a photoconductive semiconductor. We idealize the detection characteristic of the detector as a function of frequency by equating the quantum efficiency to unity times a step function extending from $\omega = \omega_c$ to infinity. The rate of collection of photons, $\bar{r}_b$, is

$$\begin{aligned}\bar{r}_b &= \int_{\omega_c}^{\infty} \frac{dP_b}{\hbar\omega} = \frac{A}{(2\pi c)^2} \sin^2\theta_0 \int_{\omega_c}^{\infty} \frac{\omega^2\, d\omega}{e^{\hbar\omega/kT} - 1} \\ &\simeq \frac{A}{c^2} \sin^2\theta_0\, 2\pi\left(\frac{kT}{\hbar}\right)^3 \int_{x_c}^{\infty} x^2 e^{-x}\, dx \\ &= \frac{A}{c^2} \sin^2\theta_0\, 2\pi\left(\frac{kT}{\hbar}\right)^3 (x_c^2 + 2x_c + 2)e^{-x_c}\end{aligned} \tag{14.45}$$

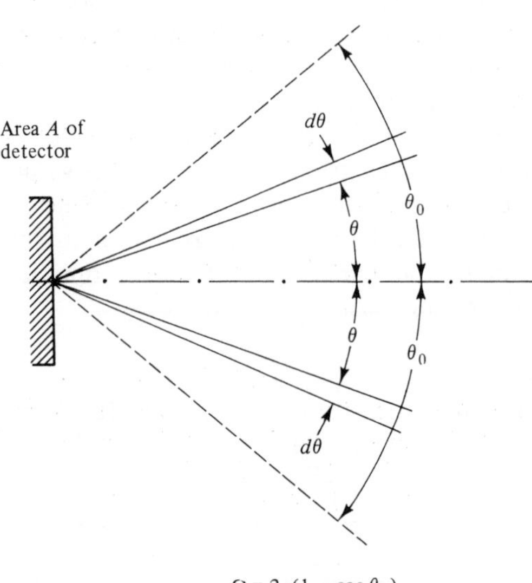

Figure 14.5 Detector surface and the solid angle of radiation collection.

where we assumed that $\hbar\omega_c/kT \gg 1$. We have for D^*, combining (14.30) with (14.45), in the ideal case of unity quantum efficiency,

$$D^* = \left[\sin^2\theta_0 \frac{4\pi(kT)^5}{c^2\hbar^3} x_c^2(x_c^2 + 2x_c + 2)e^{-x_c}\right]^{-1/2}$$

where we pick ω by definition as the cutoff frequency (which picks the largest possible D^*). A numerical evaluation yields

$$D^* = 1.3 \times 10^{11}(300/T)^{5/2} \frac{1}{\sin\theta_0} \frac{e^{x_c/2}}{x_c(x_c^2 + 2x_c + 2)^{1/2}} \tag{14.46}$$

Figure 14.6 shows D^* for various detectors. The D^* for an ideal detector,

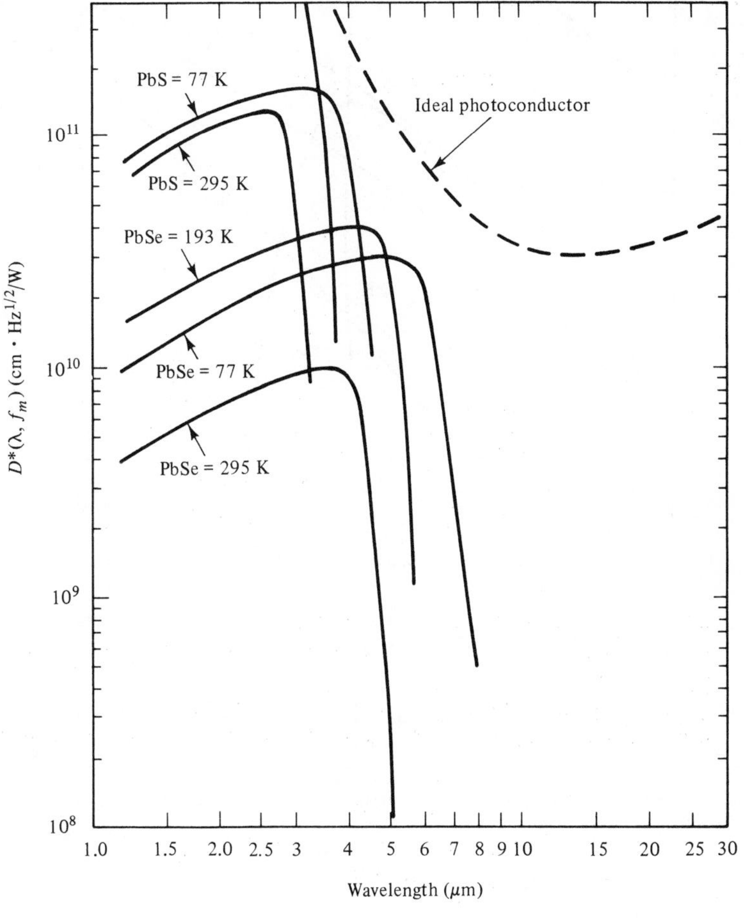

(a)

Figure 14.6 D^* of some infrared detectors. (Courtesy of Santa Barbara Research Center.)

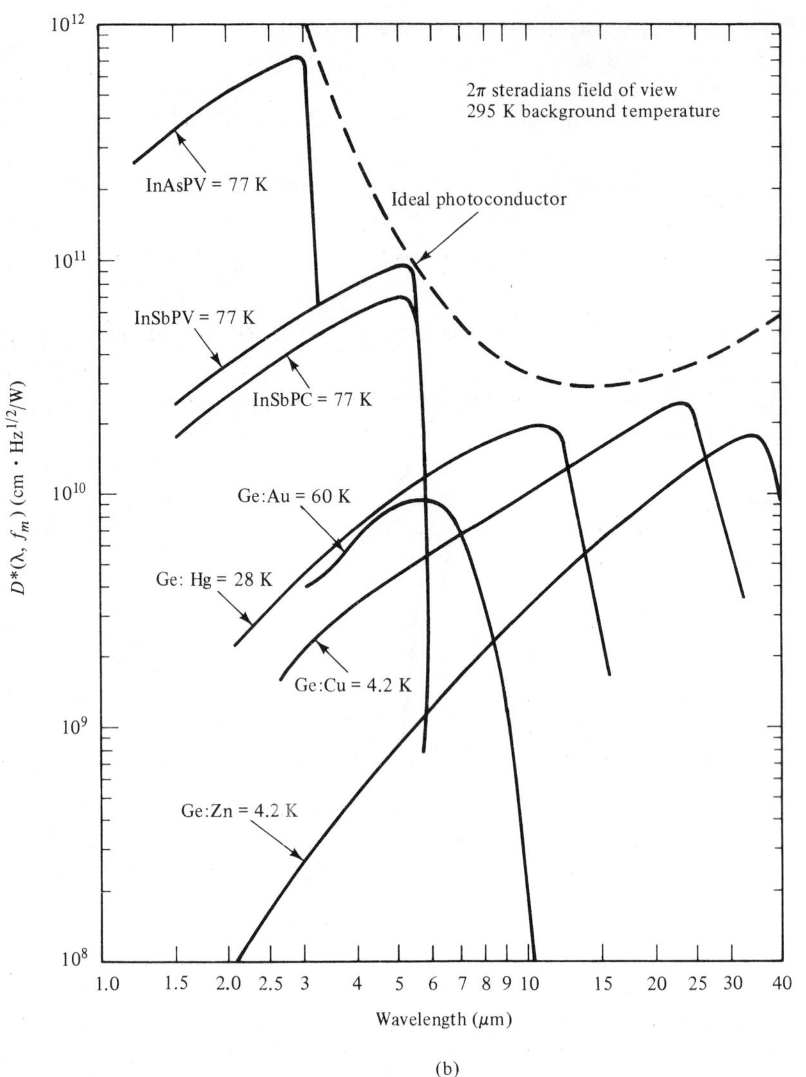

(b)

Figure 14.6 *(continued)*

(14.46), is shown for comparison; the following values are taken:

$$\theta_0 = \frac{\pi}{2}, \qquad T = 295 \text{ K}$$

The D^* drops abruptly at the wavelength corresponding to the band-gap energy of the semiconductor. Note the relative proximity of the D^* of physical detectors to the D^* of the ideal detector of unity quantum efficiency.

14.6 HETERODYNE DETECTION

A photodetector illuminated by optical radiation produces a current proportional to the incident power; phase modulation of the incident radiation is not detected. If the radiation incident on the photodetector is combined with the radiation from a stable laser oscillator, phase modulation of the incident radiation can be detected; this method of detection is called *heterodyne detection*, a name adopted from conventional communication technology. Figure 14.7 shows a schematic of the arrangement. The incident optical signal is passed through a transmitting mirror which reflects very little, so that the power in the incident radiation is only slightly reduced. The laser "local oscillator" illuminates the same mirror from the other side, is weakly reflected, and is combined with the incoming signal radiation. The combined radiation is detected and the electric current from the detector is filtered and applied to the input of the amplifier. Here we analyze the signal-to-noise ratio of heterodyne detection, ignoring thermal background radiation.

Suppose that the signal radiation incident upon the detector has the linearly polarized field

$$\tfrac{1}{2}(\mathbf{E}_s\, e^{j\omega_s t - j\boldsymbol{k}_s \cdot \boldsymbol{r}} + \text{c.c.})$$

The detector is also irradiated by a local oscillator with the electric field

$$\tfrac{1}{2}(\mathbf{E}_0\, e^{j\omega_0 t - j\boldsymbol{k}_0 \cdot \boldsymbol{r}} + \text{c.c.})$$

We assume the waves to be plane. The current produced in the detector is,

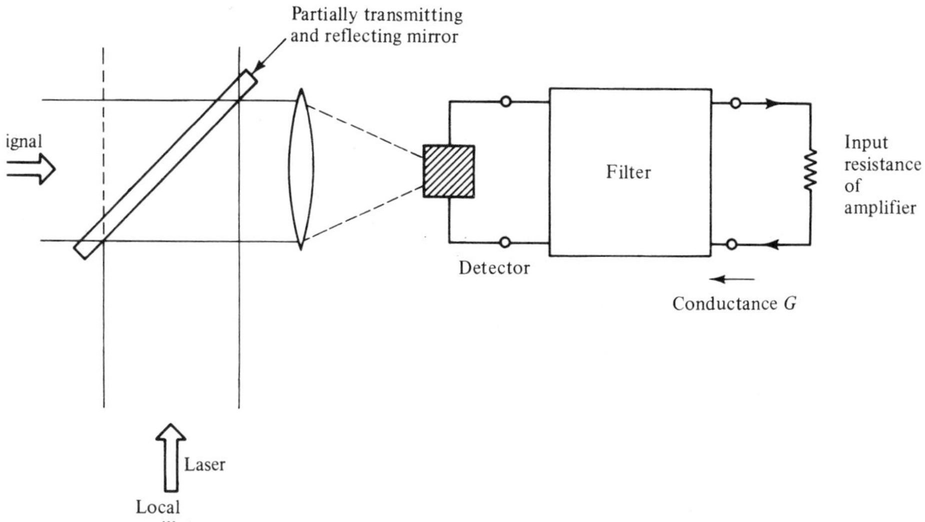

Figure 14.7 Heterodyne detector.

according to (14.1) and (14.2), for $\boldsymbol{da} \parallel \boldsymbol{n}$:

$$i_0 + i_s = e\frac{\eta}{\hbar\omega}\int da\,\frac{1}{2}\sqrt{\frac{\varepsilon_0}{\mu_0}}\,|\mathbf{E}_0 e^{j\omega_0 t - j\boldsymbol{k}_0\cdot\boldsymbol{r}} + \mathbf{E}_s e^{j\omega_s t - j\boldsymbol{k}_s\cdot\boldsymbol{r}}|^2$$

$$\Rightarrow \frac{e\eta}{\hbar\omega}\frac{1}{2}\sqrt{\frac{\varepsilon_0}{\mu_0}}\int da\,[|\mathbf{E}_0|^2 + |\mathbf{E}_s|^2$$

$$+ \mathbf{E}_0\cdot\mathbf{E}_s{}^* e^{j(\omega_0-\omega_s)t - j(\boldsymbol{k}_0-\boldsymbol{k}_s)\cdot\boldsymbol{r}} + \text{c.c.}] \tag{14.47}$$

The term $|\mathbf{E}_0|^2 + |\mathbf{E}_s|^2$ produces a dc current. We see that a severe cancellation of the signal current at frequency $\omega_0 - \omega_s$ occurs if the signal and local oscillator waves are not aligned (i.e., the factor exp $[-j(\boldsymbol{k}_0 - \boldsymbol{k}_s)\cdot\boldsymbol{r}]$ cannot be treated as a constant). We shall assume henceforth that $(\boldsymbol{k}_0 - \boldsymbol{k}_s)\cdot\boldsymbol{r} =$ constant. Then the mean-square signal current is

$$\overline{|i_s|^2} = e^2\left(\frac{\eta}{\hbar\omega}\right)^2\left(\frac{1}{2}\sqrt{\frac{\varepsilon_0}{\mu_0}}\right)^2\frac{1}{2}\left|\int da\,2\mathbf{E}_0\cdot\mathbf{E}_s^*\right|^2$$

$$= 2e^2\left(\frac{\eta}{\hbar\omega}\right)^2 P_0 P_s \cos^2\theta \tag{14.48}$$

P_0 and P_s are the powers of local oscillator and signal respectively intercepted by the detector, and θ is the angle between the local oscillator and signal polarizations. Note that the mean square signal current contains P_0 as a multiplier, a high local oscillator power can produce a high gain. The gain may be defined as the current that flows in the heterodyne detector compared to the current that would flow if the signal power by itself were incident on an ideal detector of unity efficiency. For $\cos\theta = 1$, the gain $\mathscr{G}$ is

$$\mathscr{G} = \frac{2e^2(\eta/\hbar\omega)^2\,P_0 P_s}{(e/\hbar\omega)^2\,P_s^2} = 2\eta^2\frac{P_0}{P_s}.$$

If the signal power is small compared with the local oscillator power, the noise is due to the shot noise of the dc current produced by the local oscillator

$$\overline{|i_n|^2} = 2eI_0 B \simeq 2e^2\left(\frac{\eta}{\hbar\omega}\right)P_0 B \tag{14.49}$$

where we neglect the contribution of the signal power compared with the local oscillator power. The signal-to-noise ratio is, therefore,

$$\frac{\text{S}}{\text{N}} = \frac{\overline{|i_s|^2}}{\overline{|i_n|^2}} = \frac{\eta P_s\cos^2\theta}{\hbar\omega B} \tag{14.50}$$

For $\cos\theta = 1$, the signal-to-noise ratio is η times the number of signal photons collected within the time $1/B$.

If the conductance of the system is G, the thermal noise of the conductance must be added to $\overline{|i_n|^2}$. The thermal noise will be more important than the shot noise if according to (14.49) and (14.43):

$$4GkT > 2e^2 \frac{\eta}{\hbar\omega} P_0$$

or

$$G > \frac{e^2(\eta P_0/\hbar\omega)}{2kT} \tag{14.51}$$

For small optical powers, the thermal noise may dominate. The optical power P_0, for which the thermal noise equals the shot noise, is

$$P_0 = \frac{1}{\eta} 2\left(\frac{kT}{e}\right)\left(\frac{\hbar\omega}{e}\right)G \tag{14.52}$$

kT/e and $\hbar\omega/e$ have units of voltage.

One electron volt corresponds to an optical wavelength of $\lambda = 1.2$ μm and kT/e at room temperature is 1/25 eV. Normally, $G = \frac{1}{50}$ mho, the value used for a high-speed detector feeding a 50-Ω line. Thus for $G = \frac{1}{50}$ mho and $\lambda = 1$ μm,

$$P_0 = \frac{1}{\eta} 1.2 \times \frac{1}{25} \times \frac{2}{50} = \frac{1}{\eta} 1.9 \text{ mW}$$

For a local oscillator power less than this value, the thermal noise is dominant. We have not included the noise contributed by the amplifier stage(s) if any. For a discussion of this issue the reader is referred to the literature [8].

14.7 SUMMARY

Vacuum photodiodes and photoconductive detectors are characterized, in part, by their quantum efficiency. The dc current produced by a steady-state illumination is accompanied by shot noise which puts a lower limit on the level of a detectable optical signal. We have presented two derivations of the shot noise, one from the spectral density, the other from the mean-square fluctuations of the number of current pulses in a time interval T. The latter gave very simply the formula for the mean-square fluctuations near zero frequency.

The noise equivalent power of a detector gives the optical power required to produce a signal-to-noise ratio of unity. For background-limited detection, one requires Planck's formula to evaluate the rate of thermal photons impingent on the detector. A useful special case of the Planck formula is the Nyquist formula for the mean-square fluctuations of the current in a conductance. The D^* of a detector was defined in a way so as to make it independent of detector area and detection bandwidth and the D^* of an ideal detector was evaluated and compared with those of some available detectors. Finally, we considered heterodyne detection and determined the signal-to-noise ratio achieved by it.

APPENDIX 14A

Poisson Statistics

We derive the probability of occurrence of N photons, or photoelectrons in a time interval $0 \leq t \leq T$, when the rate of occurrence is r, and the probability of occurrence in any time interval is independent of the probability of occurrence in any other interval. Denote by $P_0(t)$ the probability of no occurrence in the time interval 0, t. The probability $P_0(t + \Delta t)$, where Δt is taken as very small, is given by the product of the probability $P_0(t)$, and the probability of no occurrence in Δt, $(1 - r\,\Delta t)$.

$$P_0(t + \Delta t) = P_0(t)(1 - r\,\Delta t) \tag{14A.1}$$

Expanding to first order in Δt and dividing by Δt, we obtain a differential equation for $P_0(t)$, in the limit $\Delta t \rightarrow 0$.

$$\frac{dP_0(t)}{dt} = -rP_0(t) \tag{14A.2}$$

with the solution

$$P_0(t) = e^{-rt} \tag{14A.3}$$

Consider next the probability of one occurrence, within the interval $t + \Delta t$, $P_1(t)$. One occurrence is possible in two ways; the probability is the sum of two probabilities: the probability of no occurrence within 0, t and one occurrence in Δt; and of one occurrence in 0, t and no occurrence in Δt.

$$P_1(t + \Delta t) = P_0(t)\, r\,\Delta t + P_1(t)(1 - r\,\Delta t) \tag{14A.4}$$

The differential equation for $P_1(t)$ is

$$\frac{dP_1(t)}{dt} = -rP_1(t) + rP_0(t) \tag{14A.5}$$

Using (14A.3), we find the solution

$$P_1(t) = (rt)e^{-rt} \tag{14A.6}$$

The argument can be generalized to determine the probability of N occurrences.

$$P_N(t + \Delta t) = P_{N-1}(t)\, r\,\Delta t + P_N(t)(1 - r\,\Delta t) \tag{14A.7}$$

with the differential equation

$$\frac{dP_N(t)}{dt} = -rP_N + rP_{N-1} \tag{14A.8}$$

Successive solutions of (14A.8), starting with $P_1(t)$, give

$$P_N = \frac{(rt)^N}{N!}\, e^{-rt} \tag{14A.9}$$

Within the time interval $0 \le t \le T$, the probability of N occurrences is

$$P_N = \frac{(\bar{N})^N}{N!} e^{-\bar{N}} \tag{14A.10}$$

where $\bar{N} = \sum_N NP_N = rT$ is the average number of occurrences. The probability distribution (14A.10) is known as the Poisson distribution. The mean-square deviation can be found to be

$$\overline{N^2} - \bar{N}^2 = \bar{N} \tag{14A.11}$$

PROBLEMS

14.1. The purpose of this problem is to evaluate $h(t)$ describing the current induced by an electron that moves at a velocity $v(t)$, $0 \le t \le \tau_0$ between two parallel conducting plates, separated by the distance d.

$$\int_0^{\tau_0} v(t)\, dt = d$$

(a) Treat the electron as if it were a sheet of charge, parallel to the plates of area A ($\gg d^2$). With the plates short-circuited, find the induced charge as a function of distance x ($0 \le x \le d$) from one of the two plates.

(b) Find the induced short-circuit current.

(c) Present an argument showing that the derivation is not limited to charge sheets, but applies to point charges as long as the transverse dimensions of the plates are much larger than d.

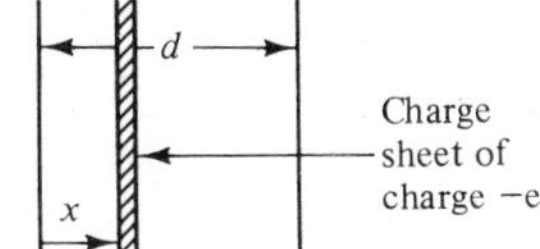

Figure P14.1

14.2. Consider the Michelson interferometer of Fig. P14.2. An incident pulse train of well-separated pulses, each having an amplitude $E_0(t)$ and phase factor $e^{j\omega_0 t}$, is split into two pulse trains, each with an amplitude $E_0(t)/\sqrt{2}$. The two pulse trains are made to traverse separate orthogonal arms of the interferometer. After traversing their respective paths, $2l_a$ and $2l_b$, the pulses are recombined on a photodetector.

(a) Determine the intensity incident on the detector as a function of τ.

(b) Show that if the response of the detector is slow compared to the duration of each pulse in the train, then the output current $i(\tau)$ of the detector is given by

$$i(\tau) = \frac{1}{2}\frac{\eta e}{\hbar\omega} f_r W(1 + \gamma(\tau))$$

where W is the pulse energy, $\gamma(\tau)$ is the normalized autocorrelation function of the pulse amplitude, that is,

$$W \propto \int_{-\infty}^{\infty} E_0^2(t)\, dt$$

and

$$\gamma(\tau) = \frac{\left| \int_{-\infty}^{\infty} E_0(t) E_0^*(t - \tau)\, dt \right|}{\int_{-\infty}^{\infty} E_0^2(t)\, dt} \cos\left[\omega_0 \tau + \phi(\tau)\right]$$

where $\phi(\tau)$ is the phase of $E_0(t)E^*(t - \tau)$ and f_r is the repetition frequency of the pulses. How could this interferometer be utilized to measure the duration of the pulse?

(c) What information on a statistical waveform is not revealed by this measurement, but does show up in SHG, Section 13.6?

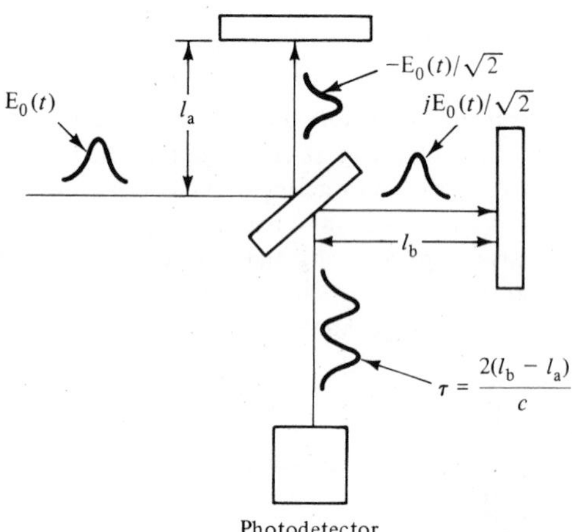

Figure P14.2 Michelson interferometer.

14.3. What power P_s is required to obtain a signal-to-noise ratio of 40 dB using a detector with $D^* = 10^{12}$, a cross-sectional area $A = 1\ \text{cm}^2$, $\eta = 0.1$, and a bandwidth $B = 100$ MHz. Assume that $\overline{P_s^2} = \bar{P}_s^2$. Treat the background limited case.

14.4. Prove (14.22) using the Poisson distribution of Appendix 14A.

14.5. A modulated laser source emits 0.1 W in the fundamental Gaussian mode with $w_0 = 0.5$ mm at $\lambda = 1.06\ \mu$m. It is at a far-field distance l from a detector with $D^* = 10^{12}$ and area $A = 5\ \text{mm}^2$. Find the maximum distance of the laser source for the achievement of S/N = 40 dB with a bandwidth of 100 MHz. If a parabolic dish antenna is used to produce a Gaussian beam with $w_0 = 5$ cm, how does l change?

14.6. Repeat Problem 14.5 for a heterodyne system with local oscillator power of 0.1 mW. Ignore the Nyquist noise. Use $\eta = 10\%$ and $\theta = 0$.

14.7. The reciprocity theorem of electromagnetic theory leads to the relation between the gain of an antenna, $G(\theta, \phi)$, and its receiving cross section, $A(\theta, \phi)$, for radiation in the direction θ, ϕ of a spherical coordinate system [9].

$$A(\theta, \phi) = \frac{\lambda^2}{4\pi} G(\theta, \phi)$$

This relation can be proven solely from thermodynamics, using Planck's radiation formula. To show this, require that the thermal power $G(\theta, \phi)\,(kTB/4\pi)$, transmitted in the direction θ, ϕ, by the antenna terminated in a matched load at temperature T be equal to the power received due to thermal background ($\hbar\omega \ll kT$). Note that $G(\theta, \phi)$ and $A(\theta, \phi)$ refer to one polarization only.

REFERENCES

[1] A. Yariv, *Introduction to Optical Electronics*, Holt, Rinehart and Winston, New York, 1976.

[2] R. M. Kingston, *Detection of Optical and Infrared Radiation*, Springer-Verlag, Heidelberg, 1978.

[3] P. L. Kelley and W. H. Kleiner, "Theory of electromagnetic field measurement and photoelectron counting," *Phys. Rev.*, *136*, no. 2A, A316–A334, Oct. 1969.

[4] W. Feller, *An Introduction to Probability Theory and Its Application*, Wiley, New York, 1950.

[5] L. I. Schiff, *Quantum Mechanics*, McGraw-Hill, New York, 1949.

[6] R. C. Tolman, *The Principles of Statistical Mechanics*, Oxford University Press, New York, 1955.

[7] W. B. Davenport, Jr., and W. L. Root, *An Introduction to the Theory of Random Signals and Noise*, McGraw-Hill, New York, 1958.

[8] A. van der Ziel, *Noise in Measurements*, Wiley, New York, 1976.

[9] S. Silver, "Microwave theory and design," *Rad. Lab. Series*, McGraw-Hill, New York, 1949.

EPILOGUE

Simple concepts are most easily remembered and applied. Fortunately, important concepts have engaged the talents of creative people with the result that, in the course of time, many of them have been simplified and made easier to grasp and use. There is a limited number of truly important concepts in optical electronics, many of which have been developed in other research areas and applied to this newly emerging field. The concept of wave impedance developed in transmission line and waveguide theory is useful for the analysis of reflection from multiple layers and the derivation of the modes of slab waveguides. The Fabry–Perot interferometer emerged in optics, but is conveniently treated in terms of the scattering matrix formalism of waveguide theory.

Optical phenomena discussed in this text are all of narrow bandwidth. The analysis of wave packets containing many wavelengths is a minor perturbation of the steady-state analysis. The concept of group velocity—envelope propagation velocity—is always appropriate. Also, all optical beams of interest are wide—of diameters much greater than the optical wavelength. For this reason, an infinite parallel plane wave analysis provides much of the necessary information, certainly in all those cases in which the wave is changed within distances much smaller, than or comparable to, the beam diameter, as in refraction, reflection, and multiple-layer coatings.

For the propagation over distances long compared with the wavelength, the paraxial wave equation is adequate. This equation treats diffraction, optical guidance, fiber optics, and self-focusing. The paraxial approximation of the wave equation solution $(1/r) \exp(-jkr)$ provides the impulse response function for the diffraction problem when r is assigned a real origin on the axis; it serves for the Gaussian beam solution when the origin is picked imaginary.

The Gaussian beam solution and the higher-order Hermite Gaussians contain all the information of Fresnel diffraction theory. The complicated diffraction patterns produced in Fresnel diffraction are due, solely, to the relative phase shifts of the propagating Hermite–Gaussians in terms of which the field at the input aperture is expanded. Spatial Fourier transformation by lenses can also be fully understood from this simple model. Parabolic index fiber modes are closely related to the free-space solutions except that the spreading of the beam is prevented by total internal reflection in the layered medium. The parabolic index profile preserves the Hermite–Gaussian profiles through propagation along the fiber. Other index profiles produce different mode patterns and, in particular, lead to a finite number of eigenmodes, if bounded at infinity as physically required. The most important method of analysis of optical phenomena treated in this book is the perturbation approach. Dispersion of fibers was analyzed by modifying the propagation equation of a wave packet by the contribution $d\beta^2/d\omega^2$, due to the curvature of the β–ω dispersion diagram. Coupling of modes, originally developed for microwave beam tubes, is also a perturbation approach which treats the excitation of composite structures as coupling of a mode to the "outside," or as coupling between two, or more, modes of the component structures. Transient excitation of a Fabry–Perot resonator can be handled easily by this approach. Coupling between two optical waveguides in mutual proximity is understood in this way and simple models of optical switches can be developed from it. The analysis of nonlinear optical phenomena is simplified by the fact that optical nonlinearities are weak. They cause very small changes per optical wavelength and hence perturbation approaches are eminently applicable. We were able to discuss self-focusing, soliton formation, and saturable absorber modelocking, all this by use of the perturbation approach.

Because most nonlinear optical phenomena occur in anisotropic crystals, we had to discuss propagation in anisotropic media. One virtue of anisotropy is that it enables one to match the propagation constants of the interacting waves, to "phase match," without which frequency multiplication and parametric oscillations could either not be realized, or not be made efficient. Here again coupling of modes was used to understand parametric interactions. With the aid of the Manley–Rowe relations it was easily demonstrated that self-excitation is possible if, and only if, the pump frequency is above the two parametrically excited frequencies.

Optical detection requires an understanding of noise. Noise is a statistical process, not unlike the statistical character of temporally incoherent radiation. Such processes are quantified in terms of correlation functions or power spectral densities. The autocorrelation function and power spectral density of shot noise was derived, and the Planck blackbody radiation formula was obtained. The Nyquist formula for thermal noise was shown to be the one-dimensional, low-frequency form of the Planck formula.

The topics covered in this text are fundamental to the researcher and inventor of optical signal-processing devices. The author believes that high-speed optical signal processing is presently at the threshold of major developments fostered by the perfection of microstructure fabrication techniques and the discovery of materials with high nonlinearities.

GENERAL INDEX

C

D

E

I

K

L

M

Q

R

S

T

U

V

W

Z

INDEX OF AUTHORS